From Alchemy to Quarks

The Study of Physics as a Liberal Art

From Alchemy to Quarks

The Study of Physics as a Liberal Art

Sheldon L. Glashow

Harvard University

Brooks/Cole Publishing Company
Pacific Grove, California

Brooks/Cole Publishing Company
A Division of Wadsworth, Inc.

Copyright © 1994 Sheldon L. Glashow.
All rights reserved. No part of this book may be reproduced,
stored in a retrieval system, or transcribed, in any form or
by any means—electronic, mechanical, photocopying, recording,
or otherwise—without the prior written permission of the
publisher, Brooks/Cole Publishing Company, Pacific Grove,
California 93950, a division of Wadsworth, Inc.

Printed in the United States of America

10 9 8 7 6 5 4 3 2 1

Library of Congress Cataloging-in-Publication Data

Glashow, Sheldon L.
    From alchemy to quarks : The study of physics as a liberal art /
Sheldon L. Glashow
      p. cm.
    ISBN 0-534-16656-3
    1. Physics.   I. Title.
    QC21.2.G58   1993
    530–dc20

Executive Editor: Harvey Pantzis
Sponsoring Editor: Lisa Moller
Project Development Editors: Margaret Kuntz and Suzanne Ewing
Editorial Associate: Beth Wilbur
Production Editor: Penelope Sky
Production Assistant: Katharyn Graham
Manuscript Editor: Carol Dondrea
Permissions Editor: May Clark
Interior and Cover Design: Sharon L. Kinghan
Cover Illustration: John K. Heyl
Art Coordinator: Lisa Torri
Photo Editor: Diana Mara Henry
Photo Researcher: Joan Meyers-Murie
Indexers: George and Andrea Olshevsky
Typesetting and Interior Illustration: Electronic Technical Publishing Services
Printing and Binding: R.R. Donnelley & Sons Company/Crawfordsville
Cover Printing: Lehigh Press Lithographers

Credits
Page vii: Photo copyright 1993, *The Daily Cougar*, University of Houston
Page 431: Excerpt from "A Wish to Comply," by Robert Frost, in *The Poetry of
Robert Frost*, edited by Edward Connery Latham. Copyright ©1969 by
Henry Holt and Company, Inc. Reprinted by permission of Henry Holt and
Company, Inc.

About the Cover
Orobouros is the ancient symbol of a snake swallowing its tail. Today it
symbolizes the convergence of particle physics and cosmology.

For Joan

About the Author

Sheldon Lee Glashow was born in Manhattan in 1932, the year in which positrons and neutrons were discovered. He was educated in New York public schools, attended the Bronx High School of Science, studied physics at Cornell University, and received his graduate degrees from Harvard.

Dr. Glashow's teaching career began after three years of postdoctoral research at the Niels Bohr Institute in Denmark, CERN in Switzerland, and the California Institute of Technology. He has taught at Boston University, the University of Aix-Marseille, Texas A & M University, the University of Houston, Stanford University, and the University of California at Berkeley. Since 1966, he has been a professor of physics at Harvard, where he is Mellon Professor of the Sciences and Higgins Professor of Physics.

The author of more than 250 research papers, Dr. Glashow has also written two popular books, *The Charm of Physics* and, with Ben Bova, *Interactions*. The basis for this book is the Harvard core curriculum course, "From Alchemy to Quarks," which Dr. Glashow has taught since 1979.

Among the many honors Dr. Glashow has received are the J. Robert Oppenheimer Prize, the Nobel Prize in Physics, and membership in the National Academy of Sciences. He played key roles in developing the unified theory of weak and electromagnetic forces and in predicting the charmed quark.

With his wife Joan, Dr. Glashow roams about his house in Brookline, a 40-minute walk along the Charles River from his office at Harvard. He is immensely proud of his four grown children, even though none of them is likely to choose a career in physics.

Preface

Some people regard scientists as cultural illiterates, unable to write, unwilling to read, captives of their narrow expertise and deserving candidates of humanists' contempt. They are mistaken. Most of us are reasonably well read and can hold our own with historians, literary critics, and philosophers. Some humanists, on the other hand, are scientifically and mathematically inept and proudly so. Our conversations must turn on matters of their concern, not ours: we must match wits on their turf. As my wife demands, "No physics at the dinner table!"

There is but one culture of which science is an essential part. Membership in the community of educated men and women demands competence in science and awareness of its history. How far we have strayed far from the path set by Franklin and Jefferson, who admired and appreciated Lavoisier as much as they did Shakespeare! The plagues of innumeracy and scientific illiteracy that beset our society must be eradicated.

I have taught "From Alchemy to Quarks" at Harvard College since the inception of our core curriculum program. According to Harvard's Official Register:

> The philosophy of the Core Curriculum rests on the conviction that every Harvard graduate should be broadly educated, as well as trained in a particular academic specialty or concentration. It assumes that students need some guidance in achieving this goal, and that the faculty has an obligation to direct them toward the knowledge, intellectual skills, and habits of thought that are the hallmarks of educated men and women . . .

All Harvard is divided into three parts: social sciences, humanities, and natural sciences. Within these branches of knowledge are scores of possible fields of concentration. The Harvard student achieves a broad general education and focuses in depth on a chosen specialty. For those with concentrations elsewhere, a one-semester course in the physical sciences (chosen from a dozen offerings) is mandatory. This book, like the core course from which it evolved, is the result of a decade's effort to formulate a terminal introduction to the study of physics.

A one-semester course aimed at bright students of law, language, or letters—some without love for physics or aptitude in mathematics—cannot make them scientifically literate. Fifteen weeks are not enough! Instead, I tip-toe about the world of physics, chemistry, and their history, discussing how each scientist addresses a small part of the puzzle of nature and, if lucky, finds a few pieces that fit together. Eureka! "We live under an ocean of air!" or "Combustion is oxidation!" or "We shall call it Radium!" or "A particle is a wave is a particle!" or "We have found a new flavor of quark!"

Effective pedagogy and sound scholarship are not always compatible. We focus, perhaps unduly, on the accomplishments of certain scientists. The American physicist Robert A. Millikan—himself a Nobel laureate—warned against such a view:

> It seems to be necessary to link every new theory, every great discovery, every important principle, with the name of a single individual. But it is an almost universal rule that developments in physics actually come about in a very different way. A science, like a plant, grows in the main by a process of infinitesimal accretion. Each research is usually a modification of a preceding one; each new theory is built like a cathedral through the addition by many builders of many different elements.

From Alchemy to Quarks is addressed to the student who has not studied physical science or mathematics in depth. He or she must be conversant with algebra and have studied high-school chemistry or physics. Many quantitative exercises are offered. The book—in whole or in part—can be the text for a course lasting a quarter, a semester, or a full year. It may be used in various modalities:

1. As a quantitative and historical survey of the evolution of physics and chemistry from ancient times to today. I have striven for balance between the concepts of physics and their mathematical representation: words and formulæ—this is physics for poets who can count.

2. As a qualitative study of the history of the subject, either for liberal arts students or for science majors. In the former case, the more technical sections, as well as many of the exercises, could be replaced by required readings and essays. Students with AP physics could begin at Chapter 6.

3. As a follow–up course for those students who sampled the physical sciences and seek a deeper insight into modern physics. Such a course may be based on the later chapters, beginning at or about Chapter 10. The more technical exercises may be useful.

4. As a source book for a seminar focusing on a defined topic, such as the evolution of our understanding of matter, the growth of big science, or the relation between pure and applied science.

Because of the several ways this book can be used, cross–references between chapters are kept to a minimum. Almost any ordered selection of chapters (or fragments of chapters) may serve as a text.

From Alchemy to Quarks is not an appropriate introduction to physics for scientists, doctors, or engineers. Who fixes my plumbing had best follow a more comprehensive course of study. I include many irrelevant asides, loose strings, and unanswered questions. My approach is quantitative, semi-historical, and anecdotal. I find it helpful to recall that Newton was born in the year Galileo died, that phosphorus was discovered in a city to be destroyed by incendiary bombs centuries later, that Lavoisier met the guillotine in the aftermath of the French Revolution in the same year that the rare earth elements were found, and that Rumford wedded Lavoisier's widow.

The reader may not acquire a through understanding of physics, but should emerge with curiosity and imagination intact. He or she may learn to think quantitatively, become fascinated by physics, and acquire a compulsion to learn more. We succeed if we ignite a spirit that continues to burn well beyond the college years.

From Alchemy to Quarks is an unfinished historical novel with exercises, the saga of the search for the ultimate constituents of the universe and the rules by which they combine. The characters are physicists and chemists, astronomers and mathematicians. Their lives interweave with the wars and revolutions some call history. The plot tells of their attempts to understand the world they were born to. Only incidentally do I describe how their discoveries, for better or worse, change the course of human affairs. Generations of puzzle solvers assembled a remarkably correct, consistent, and constantly improving picture of how the world works. In our treasure hunt for the ultimate building blocks of matter, we find clues within clues: from atoms to nuclei to elementary particles to quarks and, perhaps, to the superstring. We have come a long way, but the story is not done: it has probably barely begun. The quest was described with earthy elegance by Richard P. Feynman in the 1963

Lectures on Physics (Volume 2). His thoughts form the gist of my book.

> Imagine that the world is something like a great chess game being played by the gods, and we are observers of the game. We do not know what the rules of the game are; all we are allowed to do is to watch the playing. Of course, if we watch long enough, we may eventually catch on to a few of the rules. The rules of the game are what we call fundamental physics. Even if we knew every rule, however, we might not be able to understand why a particular move is made in the game, merely because it is too complicated and our minds are limited. If you play chess you must know that it is easy to learn all the rules, and yet it is often very hard to select the best move or to understand why a player moves as he does. So it is in nature, only much more so; but we may be able at least to find all the rules. Actually, we do not have all the rules now. Aside from not knowing all of the rules, what we really can explain in terms of those rules is very limited, because almost all situations are so enormously complicated that we cannot follow the plays of the game using the rules, much less tell what is going to happen next. We must, therefore, limit ourselves to the more basic question of the rules of the game. If we know the rules, we consider that we "understand" the world . . .

Feynman is among the immortals whose genius led us to today's world view. I trace their footsteps with trepidation, for the leaps of their boundless imaginations lie far beyond my reach.

Acknowledgments

I am deeply indebted to my core curriculum students, who played unwitting roles in the development of this book.

I appreciate the contributions of James B. Whitenton, who wrote the *Solutions Manual* and also checked the problems in the text, and of the following reviewers: Royal Albridge, Vanderbilt University; Robert Boughton, Bowling Green State University; Anthony Buffa, California Polytechnic State University at San Luis Obispo; Paul Frampton, University of North Carolina; Roger Freedman, University of California at Santa Barbara; James Kettler, Ohio University, Eastern Campus; Roger Ludlin, California Polytechnic State University at San Luis Obispo; Allen Miller, Syracuse University; Marvin Morris, San Jose State University; Leo Takahashi, Pennsylvania State University; and Joseph Priest, Miami University.

Finally, I thank my friends at Brooks/Cole for their patience and fortitude.

Sheldon L. Glashow

Readings

Here is an eclectic list of books I have enjoyed. They are related (often tangentially) to the material of each chapter.

1 Victor F. Weisskopf, *The Privilege of Being a Scientist* (W.H. Freeman, 1989).

Davis S. Landes, *Clocks and the Making of the Modern World* (Belknap Press, 1983).

2 Robert K. Adair, *The Physics of Baseball* (Harper & Row, 1989).

Russell McCormmach, *Night Thoughts of a Classical Physicist* (Harvard University Press, 1989).

3 Arthur Koestler, *The Sleepwalkers* (Hutchinson, 1959).

4 Aaron J. Ihde, *The Development of Modern Chemistry* (Harper & Row, 1964).

5 Sanford Connor Brown, *Benjamin Thompson, Count Rumford* (MIT Press, 1979).

6 Mary Elvira Weeks, *The Discovery of the Elements, Seventh Edition* (Journal of Chemical Education, 1968).

7 Robert Darnton, *Mesmerism* (Harvard University Press, 1968).

Maurice Crossland, *The Society of Arcueil: A View of French Science at the Time of Napoleon I* (Harvard University Press, 1967).

8 Bern Dibner, *Oersted and the Discovery of Electromagnetism* (Blaisdell Publishing Company, 1962).

9 Howard Georgi, *The Physics of Waves* (Prentice-Hall, 1993).

10 Thaddeus J. Trenn, *The Self-Splitting Atom: A History of the Rutherford-Soddy Collaboration* (Taylor & Francis Ltd., 1977).

Robert A. Millikan, *Electrons . . .* (University of Chicago Press, 1947).

11 Walter Moore, *Schroedinger: Life and Thought* (Cambridge University Press, 1989).

Jeremy Bernstein, *Quantum Profiles* (Princeton University Press, 1990).

Heinz R. Pagels, *The Cosmic Code* (Simon and Schuster, 1981).

12 Albert Einstein, *The Meaning of Relativity, Fifth Edition* (Princeton University Press, 1955).

Abraham Pais, *Subtle Is the Lord . . .* (Clarendon Press, 1982).

13 Richard Rhodes, *The Making of the Atomic Bomb* (Simon and Schuster, 1986).

Leona Marshall Libby, *The Uranium People* (Crane, Russak, Inc., 1970).

A. Zee, *Fearful Symmetry: The Search for Beauty in Modern Physics* (Macmillan, 1986).

14 Sheldon L. Glashow (with Ben Bova), *Interactions* (Warner Books, 1988).

Abraham Pais, *Inward Bound* (Clarendon Press, 1986).

Richard P. Feynman, *QED: The Strange Theory of Light and Matter* (Princeton University Press, 1985).

Frank Wilczek and Betsy Devine, *Longing for the Harmonies: Themes and Variations from Modern Physics* (W.W. Norton & Company, 1987).

15 Robert N. Cahn and Gerson Goldhaber, *The Experimental Foundations of Particle Physics* (Cambridge University Press, 1989).

Laurie M. Brown and Lillian Hoddeson, editors, *The Birth of Particle Physics* (Cambridge University Press, 1983).

Steven Weinberg, *The First Three Minutes: A Modern View of the Origin of the Universe* (Basic Books, 1977).

Lawrence M. Krauss, *The Fifth Essence* (Basic Books, 1989).

Michael Reardon and David Schramm, *The Shadows of Creation: Dark Matter and the Structure of the Universe* (W.H. Freeman and Co., 1991).

Brief Contents

Contents

1 Introduction

Our considerations range from the smallest particles of matter to the universe as a whole. In this chapter we explain our motivations and set the tone of the book.

2 The Science of Motion

Mechanics, the study of the motion of bodies in space and time, provided the first demonstration of the unity of science. Planets and moons follow exactly the same laws of physics as balls and bullets.

3   Energy and Momentum   103

Much of what we know of the microworld comes from studying the collisions of subatomic particles. In this chapter, we extend the study of mechanics to describe these phenomena.

4   The Behavior of Gases   151

Earthly matter appears in three states: liquid, gas, and solid. Gases were found to obey simple and universal laws of physics. The realization that air is a tangible substance and consists of a mixture of elements was the first essential step in the development of chemistry.

5   Heat Is a Form of Motion   193

The structure of matter could not begin to be understood until scientists realized that heat is not a substance but a reflection of the random motions of the molecules making up matter. A quantitative theory of heat emerged from the study of gases and their thermal properties.

6   Atoms and Elements   239

The histories of physics and chemistry are inextricably linked. Decisive evidence of the reality of atoms came from the quantitative study of chemical reactions, and suggestive evidence for structure within atoms came from the periodic law.

7   **Tell Me What Electricity Is, and I'll Tell You Everything Else**   283

Electrical phenomena are more than a technological marvel. Electrical forces hold together atoms, molecules, rocks, trees, and us. The law governing electrical forces was found to be tantalizingly like the law governing gravity. The first hint of the existence of particles of electricity (ions and atoms) came from the study of electrolysis.

8   The Marriage of Electricity and Magnetism   331

Electricity and magnetism are two facets of the electromagnetic force. Light itself is an electromagnetic phenomenon. The pretty colors that salts impart to a flame inspired spectroscopy, a powerful tool for the study of the structure of atoms and of the universe.

9   **Waves**   369

Any disturbance that propagates from place to place can be thought of as a wave. Sound waves travel through matter, but light waves can travel through empty space. Matter itself is described by quantum-mechanical waves.

10   **Inside the Atom**   413

The classical theories of mechanics and electrodynamics did not anticipate and could not deal with the dramatic discoveries of the parts of atoms: electrons and atomic nuclei. The stage was set for a revolution in science.

11   Quantum Mechanics   449

A new theoretical framework was established. Quantum uncertainty replaced the determinism of classical mechanics. The new theory works, the structure of the atom is revealed, and many of nature's wonders are explained.

14   Elementary Particles   571

The wealth of phenomena associated with elementary particles led to a marriage between the theory of relativity and quantum mechanics. Its issue, quantum electrodynamics, described the weird behavior of electrons and the antiparticles. Particle physics is a paradigm for the creation of a theory of all subatomic phenomena.

15   The Standard Model—Where We Are Today   607

We describe the triumph and tragedy of the so-called theory of the microworld.
Most of the puzzles of the past have been solved. Nature presents us with many
newer and deeper puzzles to challenge future generations of scientists.

1 Introduction

The scientist does not study nature because it is useful; he studies it because he delights in it, and he delights in it because it is beautiful. If nature were not beautiful, it would not be worth knowing, and if nature would not be worth knowing, life would not be worth living.

H. Poincaré, French mathematician

"Would you like to learn physics?" asked the teacher.

"What does physics do?"

"Physics explains the properties of natural bodies and the properties of matter; it discourses on the nature of the elements, minerals, plants, rocks and animals, and teaches us the causes of all the meteors, the rainbow, the aurora, comets, lightning, thunder, thunderbolts, rain, snow, hail, winds and whirlwinds.

J. Molière, French playwright

Laboratory of the Alchemist. Engraving by H. Cock after a painting by Peter Brueghel. *Source:* Metropolitan Museum of Art. The Bettmann Archive.

I am amazed at how much scientists have learned about the nature of matter and the universe. At the same time, I am dismayed that so many nonscientists know so little about what has been accomplished. This chapter examines some of the many ways in which science enlightens us and enriches our lives. Section 1.1 extols the virtues of scientific literacy and tells why everyone should learn about physics and its history. This introductory chapter is otherwise a largely historical account of the charms of our language, the measures of time, the evolution of the universe, and the tribulations of a nonmetric nation whose citizens are often afraid of mathematics and untutored in science. Section 1.2 examines some of the precursors of modern science: magic, myth, astrology, and alchemy. Many traces of these pseudosciences remain in our language. The many measures of time and their origins are discussed in Section 1.3, which also describes the origin of our calendar. In our wanderings through the subatomic and astronomical sciences, we encounter numbers that are very large and very tiny. A method for dealing with them, the logarithmic display, is developed in Section 1.4. The chapter concludes with Section 1.5, a brief introduction to the use of numbers in science and to the dimensional units with which things are measured.

1.1 In Praise of Physics

What encouraged you to take this course or one like it? Your specialty may be Celtic languages or semigroups, social psychology or classical archaeology. You may be pre-law, pre-medical, pre-dental, or pre-business. You probably couldn't care less about Coulomb's Law, the conservation of energy, or quark confinement. Here are several reasons for you to invest a semester or more of your college experience studying the physical sciences:

- *To become a responsible citizen of your country and the world.* Familiarity with the methods and substance of physics is essential if democracy is to thrive. How often have you been "informed" about the following things by those who may know little more about them than you do?

Strategic Defense Initiative	Radioactive waste disposal
The space station	Solar, wind, and tidal power
The Mars mission	Supersonic transport
The energy crisis	Superconducting supercollider
High-definition television	Magnetically levitated trains
High-temperature superconductors	Homeopathic medicine
The ozone hole	The greenhouse effect
Radon in homes	Acid rain

You are asked to support, oppose, deplore, or enthuse about these items. In addition, you are warned of the dangers (real or imagined) of cellular telephones, soft-boiled eggs, nuclear power, television screens, and electric blankets. On election day, your choice may depend on candidates' views on these issues. Your opinions should be sane, sound, and sensible, and they require an appreciation of physical science and an understanding of what questions it can and cannot answer. Furthermore, the pursuit of pure science has become very expensive. Ambitious endeavors such as the Space Station, the Superconducting Supercollider, the Genome Project, and orbiting observatories must be weighed against more evident and immediate social needs. It is you who must do the weighing.

- *To understand modern technology and its impact on society.* History and politics are inextricably tied to the progress of science. Cures for diseases, new weapons, the plow, the cotton gin, the electric battery, and the wireless telegraph were more significant to human society than all the world's kings, admirals, and politicians. Here are some of the recent fruits of science:

Artificial skin	Beepers	Camcorders
Disposable diapers	Electronic ignition	Freeze-dried food
Graphics terminals	High-definition TV	ICBMs
Jet aircraft	Kryptonite locks	Lasers and masers
Astronauts	Nuclear medicine	Open-heart
on the moon	Quartz watches	surgery
Pacemakers	Telefax	AZT and RU 486
Scanners	Walkman	Ultrasonic dental
Valium and Zantac	Copying machines	drills

These and their predecessors—refrigerators, cars, phones, and electric lights among them—are so much part of everyday experience that we rarely recall their origins. The things of science structure our lives as individuals, nations, and as a species. Some may be harmful, most are beneficial—all are parts of the human experience. The harvest of our imaginations has been with us since first we built fires and gazed at the sky in wonder. It will remain so for as long as civilization endures. No historian of the past, politician of the present, or social seer of the future can afford to neglect the impact of science. Science is an integral part of our heritage. Without understanding it and its child, technology, we cannot know why we are what we are or do what we do.

- *To bridge the gap between science and the humanities.* C.P. Snow, the British novelist, claimed that there are two contrary and conflicting cultures: the humanist and the scientific. However, Snow's two cultures are different aspects of a coherent whole, of which science forms an essential part. "Between the two cultures," wrote Primo Levi, "there is no incompatibility; on the contrary, there is, at times, when there is good will, mutual attraction." Science is one form of the search for knowledge; the humanities are another. The two are not mutually exclusive. No study of philosophy or quest for understanding of the meaning of life or of our role in the universe can ignore the triumphs of the human intellect.

- *To participate in an exciting adventure.* To live is to watch a magic show, surely the greatest show on Earth. Some people are content to sit back and watch the tricks: as day follows night, spring follows winter, tides ebb and flow, and rainbows trace the rain. Others are too concerned about spouses, mortgages, or careers to appreciate the extravaganza. But there are always a few scientists who must try to understand nature's wonders as best they can. Just as there must be both authors of books and readers, so also there must be both scientists and fans of science. And, just as everyone should be a reader of books, I believe that every educated person should be an aficionado of science. Not only does science build sound minds in strong bodies, but it can and should be fun! I hope that you will find it so.

"The eternal mystery of the universe is its comprehensibility," said Albert Einstein. Yet, the great fascination of science is that it is, was, and always will be incomplete. As questions are answered, still deeper questions are exposed for future thinkers to tackle. Most scientists have an unquenchable and unjustifiable faith in the underlying simplicity and inevitability of natural law. The more we learn, the more our faith is proven. Or, as my more pragmatic Harvard colleague, Percy Bridgman, put it:

> Whatever may be one's opinion as to the simplicity of either the laws or the material structure of matter, there can be no question that the possessors of some such convictions have a real advantage in the race for physical discovery. Doubtless there are many simple connections still to be discovered,

and he who has a strong conviction of the existence of these connections is much more likely to find them than he who is not at all sure they are there.

Science contributes immeasurably to human health, welfare, and fulfillment. We have yet to realize its full potential. Although enormous strides have been taken to decipher the mysteries of nature, life, and mind, even grander challenges remain. Perhaps the wise application of science can help us to live in peace, comfort, health, and harmony as the crew of spaceship Earth. Perhaps, by understanding science, we can more easily face our own mortality and that of our species and our planet.

1.2 Linguistic Relics of Ancient Sciences

To most of us, science, superstition, and religion are clearly distinguished, but long ago they were jumbled together. Why could some things (like seasons and sunrises) be predicted with certainty, and others (like tomorrow's weather or the outcome of a battle) not? Could one's fate be read from the moles on the face, lines on the hand, or the moment of one's birth? Did there exist a cure for all diseases, a panacea? Could the gods be appeased by sacrifice? Gradually, people learned that some tricks work but most do not. Cleanliness, not godliness, led to the eradication

Ancient seers foretold one's fate from lines on the palm (chiromancy) or on the forehead (metoposcopy). In each case, the lines were compared with the canonical positions of the seven heavenly bodies (Saturn, Jupiter, Mars, Venus, Mercury, the Sun, and the Moon), as shown in the figures. Today's doctors ascertain your health with far more success by looking at your tongue and listening to your heart. *Source: The History of Magic,* Kurt Seligmann, Pantheon, 1948, New York. Plates 175 and 217. The Bettmann Archive.

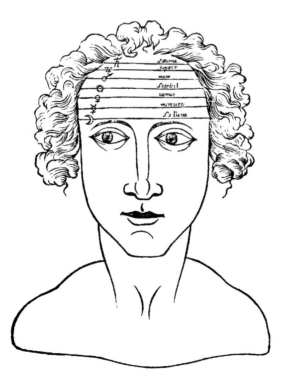

of childbirth fever. Tuberculosis is not spread by a mystical miasma but by lowly germs. An eclipse is not a sign of heavenly displeasure, but a calculable consequence of the motions of Earth and the Moon. Over many centuries, and as a result of the collective experience of many generations, science emerged from its humble beginnings.

The origins of science are lost in the sands of time, but many relics of old protosciences lie hidden in our language. I spent many amusing hours collecting 100 English words whose origins trace back to alchemy, astrology, magic, and myth. They are presented in Table 1.1.

Many words come to us from the practice of magic. The Roman *auspex*, for example, was a "scientist" who observed the flight of birds to foretell the future. *Au-* or *av-* means bird, as in "avian"; *spex* refers to an observer, as in "spectator." Hence, *auspex* = birdwatcher. From this ancient trade come *auspices* (propitious influences) and *auspicious* (betokening success). *Augurs* predicted future events from the flight, song, and feeding of birds, or from the appearance of their entrails. *Gur-* means talk, as in "garrulous"; thus, *augur* = bird talk. Today we have *augur* (to forebode), *augury* (a divination), and *inauguration* (the formal ceremony of induction for soothsayers and their ilk). From more recent times, we have *mesmerize*, which honors an eighteenth-century practitioner of alternative medicine, and *charm*, a word I coopted to describe a type of quark.

From astrology we have *influence*—originally an occult fluid flowing

TABLE 1.1   Words from Alchemy, Astrology, Magic, and Myth

Alchemy	Astrology	Magic	Myth
Alcohol	Almanac	Amulet	Avatar
Alkali	Ascendant	Augur	Cobalt
Amalgamate	Astral	Auspicious	Cyclopean
Antimony	Consider	Bewitch	Easter
Arsenic	Constellation	Charm	Enthusiasm
Bain-Marie	Desire	Contemplate	Gargantuan
Bilious	Disaster	Divination	Genius
Chemistry	Dismal	Effigy	Halcyon
Choleric	Firmament	Enchant	Harpy
Elixir	Influence	Exorcism	Hermetic
Gas	Influenza	Fetish	Ingenious
Humor	Jovial	Hex	Lethargy
Melancholy	Lunacy	Inaugurate	Museum
Miasma	Lunatic	Incantation	Narcissism
Panacea	Martial	Ingenious	Nickel
Phlegmatic	Mercurial	Mesmerize	Nymph
Phosphorus	Moonstruck	Mystery	Oedipal
Quicksilver	Nadir	Ominous	Panic
Quintessence	Occident	Seer	Psychology
Sanguine	Orient	Spellbound	Salamander
Soda	Projection	Totem	Siren
Sublime	Saturnine	Vamp	Sprightly
Touchstone	Trepidation	Voodoo	Stygian
Transmute	Venereal	Weird	Tantalize
Vitriolic	Zenith	Zombie	Volcano

from the stars and affecting one's fate. The word *dismal* comes from *dies mal*, or bad days—unlucky days, according to Egyptian astrology. *Disaster* comes from *dis* for bad and *aster* for star. The horoscope determined which of the seven gods of the heavenly bodies lent us their dispositions: the Sun (sunny), the Moon (lunatic), Mercury (mercurial), Venus (venereal), Mars (martial), Jupiter (jovial), or Saturn (saturnine). Medieval medicine was founded on a belief in four vital fluids, or humors, in the body, an excess of which was believed to make people bilious, choleric, sanguine, or phlegmatic. Illness was the result of an imbalance of these humors, and the treatment—in order to restore the balance—was bleeding (with cups or leeches).

Some astrological words survive in modern astronomy: A *constellation* is a group of stars, the *zenith* is a point directly above the observer, and an *ephemeris* is a listing of the locations of a heavenly body at various times. Other words, like *consider* (to take the stars into account) and *desire* (to wish upon a star), became everyday words.

Perhaps not surprisingly, chemistry is chock full of old words from its predecessor, alchemy. *Alcohol* and *alkali*, as well as the science itself, were named by Arab scholars. The *bain-marie* (a fancy word for a double boiler) recalls the invention of an ancient alchemist known only as Mary the Jewess, and *phosphorus* was first isolated by a seventeenth-century German alchemist.

Hermetic (now meaning airtight) comes from mythology, originally alluding to the followers of the legendary Egyptian god Hermes Trismegistrus.

Numbers, as well as words, played a part in the earliest strivings toward science. Pythagoras and his followers held the view that all things may be represented by numbers. The first number, 1, signified reason; the number 2 implied discord. Justice was represented by the number 4: the first perfect square, the product of equals. The sum of the first male number and the first female number, $2 + 3$, gives 5, the number of marriage; 7 indicated virginity.

Although these early beliefs may seem silly to us, the Pythagoreans nevertheless made decisive contributions to mathematics and to pedagogy. They proved the well-known relation between the hypotenuse of a right triangle and its legs, thereby establishing the first link between algebra and geometry. They also showed that there exist precisely five regular polyhedra: the tetrahedron, cube, octahedron, dodecahedron, and icosahedron (illustrated in Figure 1.1). These geometrical figures were sometimes regarded as the shapes of atoms. A thousand years later, they formed the basis of Kepler's model of the solar system.

In addition, the followers of Pythagoras pinpointed the difference between the discrete "How many?" and the continuous "How much?" These questions exhibit the distinction between discrete quantities (like the number of peas in a pod) and continuous quantities (like the weight of a pea). Discrete phenomena were subdivided into the absolute (arith-

FIGURE 1.1
The regular polyhedra. There are only five solids with regular and identical faces and vertices. Ancient scientists believed that this mathematical fact had to have profound consequences for astronomy and physics.

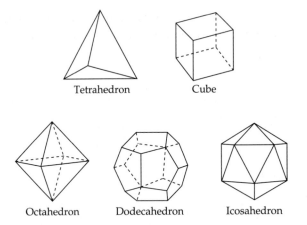

Tetrahedron   Cube

Octahedron   Dodecahedron   Icosahedron

metic) and the relative (music). This distinction makes sense because 2+2 is always 4, but the discrete notes of a song may be played in any key. Continuous phenomena were subdivided into the stable (geometry) and the variable (astronomy). Arithmetic, music, geometry, and astronomy formed the fourfold way to knowledge: the *quadrivium.* In the Middle Ages, the quadrivium was enriched by three more disciplines: rhetoric, grammar, and logic. These three formed what was regarded as the less important *trivium* (from which we get the word *trivial*). Along with the more recent additions of sociology and economics, these subjects comprise the liberal arts curriculum.

The absolute certainty of mathematical reasoning made it a paradigm for an all-encompassing world view. Many societies found a mystical significance in numerology. The number 42, for example, was of paramount importance to ancient Egyptians, while the number 60 was especially important to the Babylonians and Chinese. The Jewish Kabbalah identified each letter of the Hebrew alphabet with a number. The numbers, associated with words, had mystical significance. For example, we read in the Apocalypse of the New Testament: "Let the reader, if he has the skill, cast up the sum of the figures in the beast's name, after our human fashion, and the number will be 666." Perhaps the most important number in human history, however, is 7:

Wonders of the ancient world	Hills of Rome	Seas
Last words of Christ	Days of creation	Samurai
Orifices in the head	Heavenly bodies	Deadly sins
Tones in the diatonic scale	Primary metals	Continents
Colors of the rainbow	Pillars of wisdom	Dwarfs
Worthies of China	Odes of Mu'allaqat	League boots
Years of feast and famine until my next sabbatical		

Ancients saw that seven heavenly bodies in the night sky—the five planets visible to the naked eye (excluding Earth), the Sun, and the Moon—move relative to the background of stars. Table 1.2 shows the associations that ancient Babylonians made among these bodies, the seven known metals, and the parts of the head.

TABLE 1.2 Babylonian Connections

Sun	Gold	Forehead
Moon	Silver	Brain
Mercury	Quicksilver	Tongue
Venus	Copper	Left nostril
Mars	Iron	Right nostril
Jupiter	Tin	Right eye
Saturn	Lead	Left eye

Two thousand years later, a poem by Chaucer recalled the ancient relationships between primary metals and heavenly bodies:

The bodies seven, eek lo! Hem heer anoon:
Sol gold is, and Luna silver we threpe,
Mars yren, Mercurie quik-silver we clepe,
Saturnus leed and Jupiter is tin,
And Venus coper, by my fader kin

The sense of the last phrase is, "as my old man knew." From the earliest times, philosophers were convinced a fundamental link existed between phenomena on Earth and in the heavens. Another hypothetical connection is shown in Figure 1.2, which attempts to find a geometrical relation

FIGURE 1.2
The unity of science in the medieval period. This figure shows the relationship between the four "elements" of which matter was assumed to be made (fire, water, earth, and air) and the 12 astrological signs of the zodiac.

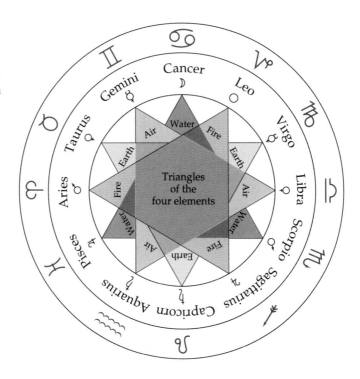

between the 12 signs of the zodiac and the 4 Greek elements. Numerological superstitions such as these both contributed to and obstructed the development of rational science, and are a part of its history.

1.3 A Brief History of Time

Our measures of time have historically derived from the periodic happenings of nature: The alternation between light and dark defines a day; the cyclic phases of the Moon a month; and the annual flow of the seasons, a year. The lengths of the day and the month vary with the seasons, and, as we shall see, there are several possible astronomical definitions of the year. For these reasons, the second was chosen to be the fundamental unit of time. It is not defined in terms of astronomical obervations, but in terms of a more precisely measurable microphysical process.

Every atom vibrates at certain discrete frequencies. This fact is the basis for the atomic clock. Since 1967, the second has been defined as the duration of 9,192,631,770 cycles of a particular oscillation of the cesium-133 atom. Definitions of the minute, hour, calendar day, and week are obtained from that of the second by sequential multiplications by 60, 60, 24, and 7. A calendar year (not a leap year) contains precisely 365 days and is exactly 31,536,000 atomic seconds long.

The word *second* has an older meaning as the counting number between first and third. How did it come to designate a time interval as well? The answer dates back to the ancient Babylonians. We have not always counted by tens as we do today. The British word *score* and the French word *quatre-vingt* recall a time when Europeans counted by twenties. Ancient Babylonians counted by sixties. They called the sixtieth part of something a *minute* (meaning a minute part) of that thing. A sixtieth of a 1-degree angle is a minute of arc. A sixtieth of an hour is a minute of time. Smaller yet is the minute of the minute of a thing, or its *second minute*. The second minute of a degree (a sixtieth of its sixtieth part) is usually referred to as an arc second. "Second minute of an hour" was shortened to "second minute" and, eventually, to just "second."

The Day of Time   Although both minute and hour are now unambiguously specified in terms of the atomically defined second, the same is not true for longer measures of time. Calendar years and days are not quite the same as astronomically defined years and days, which are linked to Earth's motions. A year is the period of time in which the Earth completes an orbit about the Sun. A day is the period of time in which Earth rotates once about its north-south axis. The word *day* comes from the Sanskrit root *dah*, meaning to burn. Day originally meant the period of sunlight, the interval between dawn and dusk. Later, the word came to mean the time interval between two successive sunsets—one day and one night to-

The precise measurement of time was one of the paramount challenges faced by early scientists. Two ancient water clocks are shown here. In the simpler model at the left, water dribbles into a cylinder, causing the piston to rise and activate a gear turning the hour hand. The clock had to be reset every day. The improved version at the right resets itself each day. During the course of a year, it adjusts the lengths of the hours in such a way that there are 12 hours of daylight and 12 hours of night. Clever, those Greeks! *Source: Revolution in Time,* David Landes, Belknap Press of Harvard University, 1983. *Left:* The Bettmann Archive. *Right:* Historical Pictures/Stock Montage, Inc.

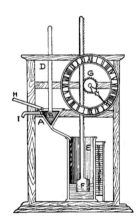
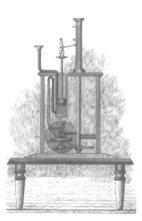

gether. We have adopted this latter sense of the word. However, when we ask for a mathematically precise definition, we are forced to consider three distinct and slightly different alternatives:

- The *mean solar day* is the time between consecutive sunsets averaged over the year.
- The *calendar day* is exactly equal to 24 hours or 86,400 atomic seconds.
- The *sidereal day* is the time it takes for Earth to rotate once about its axis relative to the stars.

Let's look at the origins of these differing definitions. The mean solar day came into being because Earth's axis of rotation is tilted relative to the plane of its orbit. For this reason, the Sun in the Northern Hemisphere sets a little bit later each day in the first half of the year and a little bit earlier in the last half. The time between successive sunsets varies cyclically over the year, being a few minutes shorter in the fall than in springtime. A day so defined is changeable and is not a reliable unit of time. Scientists, however, demand a more stable standard. They average the time between sunsets for a whole year to obtain the *mean solar day*.

Although the mean solar day is a fairly precise standard, its length does vary by a few milliseconds from year to year as a result of the changing size of the polar ice caps, earthquakes, hurricanes, and unknown events happening deep underground. In addition to these unpredictable variations, Earth's rate of rotation is very gradually but systematically slowing down. This effect is caused by the tides, which are due to the Moon. Energy is lost (i.e., converted into heat and other forms of energy) as the oceans slosh about and Earth itself is deformed. The energy is extracted from Earth's store of rotational energy. Thus, Earth is spinning down and the mean solar day is systematically lengthening by about two milliseconds per century—not much for us, to be sure. However, if we look backward far enough, the effect adds up. We find that the day was only about 21 of our hours long at the beginning of the Paleozoic Era, before fish had crawled out of the sea.

"My goodness, it's 12:15:0936420175! Time for lunch." © 1993 by Sidney Harris.

Because of these variations in the mean solar day, scientists often use another definition of the day: the *calendar day.* Unlike the mean solar day, the calendar day is invariable. It is defined in terms of the atomically defined second. Now, as it was in Chaucer's time, it is "foure and twenty houres," or precisely 86.4 kiloseconds. The size of the second was chosen with the greatest of care to make the calendar day coincide as closely as possible with the mean solar day. However, the two do not coincide exactly. For example, the mean solar day in 1970 was 86,400.0026 seconds long. Multiplying this number by 365, we get 31,536,000.95 seconds for the length of that year, which was about one second longer than 365 calendar days. Thus, a perfect calendar watch set at the beginning of the year was a second behind the Sun at year's end. If a correction were not made, our standard of time would diverge from solar time. For this reason, international timekeepers add one or two extra leap seconds at the end of each calendar year, at midnight on December 31. Thanks to their wisdom, our clocks will never go awry and the Sun will always rise in the morning.

Now that I've explained the need for and the difference between the mean solar day and the calendar day, I must complicate things with yet a third definition of the day—the *sidereal day*. Whereas the mean solar day is the mean interval between sunsets, the sidereal day is the mean interval between starsets. It is the time it takes for the celestial sphere of the fixed stars to appear to turn once about the Earth. More precisely, the sidereal day is the time it takes the Earth to complete one rotation about its axis. It's not difficult to see why the sidereal day is shorter than the mean solar day. Suppose that Earth rotated about its axis just once a year. The Sun would remain fixed in the sky. Half the world would always be sunlit and half would stay dark. If Earth rotated twice a year, there would be one sunrise each year. If it rotated three times a year, there would be two sunrises, and so on. Because there are 365 sunsets in a year, Earth spins about its axis 366 times as it completes a revolution about the Sun. Therefore, the sidereal day is 365/366 of a mean solar day, or about 236 seconds (about 4 minutes) shy of it.

Sidereal time is important to astronomers. Suppose that a star appears at a certain position in the sky at midnight. It appears at the same position the next night exactly one sidereal day later, at about 11:56 P.M. because the sidereal day is about four minutes shorter than the mean solar day.

The Year of Time

Just as there are three definitions of the day—with respect to atomic time, to the Sun, and to the stars—so also there are three definitions of the year: the calendar year, the sidereal year, and the tropical year. These three standards are very nearly equal to one another and are ordinarily interchangeable.

The *calendar year* is a unit completely divorced from astronomical reckoning. It is defined to be exactly 365 calendar days. Consequently, there are 31,536,000 atomically defined seconds in a calendar year. A leap year, of course, is one calendar day (or 86,400 seconds) longer.

The *sidereal year* is the time it takes Earth to complete one revolution about the Sun. It is equivalently and more conveniently defined in terms of the apparent motion of the Sun relative to the stars. A brief astrological digression should clarify this definition.

The starry sky seems to rotate rigidly about us. Each star, in the course of an evening, traces an arc of a circle. The ancients imagined that all the stars were attached to a celestial sphere. By connecting the dots, they drew dragons, whales, heroes, and lovers on it—the constellations. The celestial sphere has two poles and an equator, which are simply extensions of the corresponding aspects of Earth. The north star (Polaris) is located close to the north celestial pole. The Sun and the stars cannot both be seen at the same time, but we can easily deduce the position of the Sun on the celestial sphere. In the course of a year, the apparent path of the Sun relative to the stars is a great circle called the *ecliptic*. Because the axis of Earth is tilted relative to the plane of its orbit, the ecliptic is not the same as the celestial equator (Figure 1.3).

There are about 12 lunar months in a year, so 12 of the constellations were assigned a special significance. They were designated as the signs of the zodiac, one for each month.* The sidereal year is the time it takes the Sun to pass through all 12 signs to complete its circuit about the celestial sphere. Although the sidereal year is used by astronomers and

FIGURE 1.3
The seasons and the signs of the zodiac. The Earth spins about its axis, which now points toward Polaris, the north star. In addition to this daily motion, the Sun appears to complete one turn about the stars in the zodiac belt every year. The Sun's apparent motion defines the colored band, the plane of the ecliptic. The darker oval is the plane defined by Earth's equator, which separates northern and southern celestial hemispheres. The two points of intersection between the apparent path of the Sun and the equatorial plane define the spring and fall equinoxes.

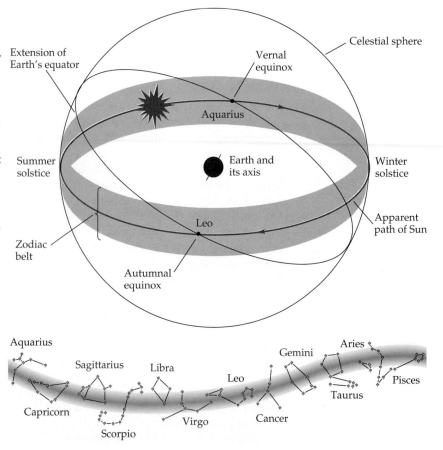

*Because there are twelve months in the year, it seemed reasonable to the ancients to divide the periods of both daylight and dark into twelve equal parts. That's why there are 24 hours in the day.

DO YOU KNOW
The Reasons for Seasons?

Why is it cold and dark in the wintertime? What makes the seasons change? A survey of graduating Harvard seniors showed that hardly any of them knew! Some said that Earth's orbit is an ellipse, with the Sun at one focus.* So far so good, but they concluded that winter is colder than summer because the Sun is then furthest away—a plausible hypothesis, but a false one. It does not explain why winter in the Northern Hemisphere coincides with summer in the Southern Hemisphere, or why there is more sunshine in a summer day than a winter day. In fact, Earth is closest to the Sun in early January and furthest away in July, but the difference is too small to affect the weather noticeably.

The seasons change because of the inclination of Earth's axis of rotation, the imaginary line passing through its two poles. The axis points at an angle of about 23.5° relative to the direction perpendicular to the plane of Earth's orbit. As Earth revolves about the Sun, the angle affects the amount of sunlight hitting an area (Figure 1.4). In summer, the Northern Hemisphere leans toward the Sun so that the period of daylight is longer. The Sun is high in the sky and the weather is warm. At the summer solstice, around 21 June, the Sun is highest in the sky. In wintertime, the Northern Hemisphere leans away from the Sun. On or about 21 December, at the winter solstice, the Sun is lowest in the sky. The days are shortest and our winters begin. At two intermediate times, the spring and fall equinoxes, Earth's axis is momentarily perpendicular to the sunward direction. These are the moments marking the beginnings of spring and fall. At either equinox, the Sun is directly overhead at midday on the equator.

FIGURE 1.4
The motion of Earth around the Sun. At the summer solstice, the North Pole leans toward the Sun; at the winter solstice, it leans away from the Sun. At the two equinoxes, the Sun is directly overhead at midday on the equator.

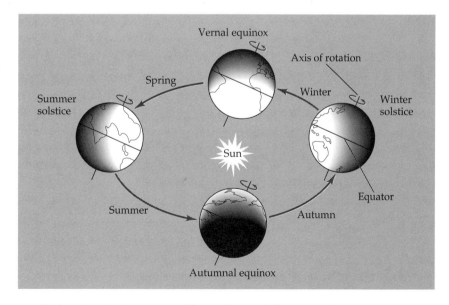

*A *focus* is a fixed point. An *ellipse* is the set of all points such that the sum of the distances to two fixed points or foci is the same.

astrologers, farmers, for sound reasons, prefer a definition of the year in terms of the seasons—the tropical year.

During summer days in the Northern Hemisphere, the Sun is north of the celestial equator and rides high in the sky. The summer solstice marks the longest period of daylight. The winter solstice marks the longest night, when the Sun is furthest south of the celestial equator. Twice a year—at the equinoxes—the Sun crosses the celestial equator. The *tropical year* is defined to be the time between successive spring equinoxes.

The sidereal and tropical years are not exactly equal to one another although they are close enough for most purposes. The sidereal year—the time it takes Earth to complete a turn relative to the stars—is about 20 minutes longer than the tropical year. Because their ratio is 1.00004, we may ordinarily ignore the difference. Nonetheless, it is important to keep in mind that these two measures of the year do not coincide.

The tropical and sidereal years differ because the direction in which Earth's axis points is changing. Today it points at Polaris (the north star), but it will not always do so. Earth's orientation in space is slowly

---

DO YOU KNOW

Why the Moon Always Shows Us the Same Face?

One side of the Moon is always hidden from us. Does this mean that the Moon does not rotate about its own axis? Certainly not! The Moon turns precisely once about its axis on each of its trips around Earth. Consequently, to us on Earth, it does not seem to turn at all (see Figure 1.5).

Is it just an incredible coincidence that the Moon's rotation period equals its orbital period? Why does Earth not always show the same face to the Moon? Once upon a time and long ago, the Moon spun about its axis much faster than it was revolving about Earth. That is, its rotational period was much shorter than its orbital period. As it traveled around Earth, however, the intense tides produced by Earth on the Moon acted as a brake to slow down the Moon's rate of rotation.

Once its rotation period equaled its orbital period, Earth became fixed in the lunar sky. Tides no longer traveled around the Moon to consume its rotational energy. At that time, billions of years ago, the stable and familiar configuration we know today came about. And that is why the Man in the Moon looks right at you.

But the Moon still produces moving tides on Earth, which act to slow down Earth's rotation. For this reason, and as happened with the Moon, in a few billion years, Earth will always present the same face to the Moon! In that incredibly distant era, our day and our lunar month will coincide in duration. Both will be about 50 of our present days long. The Moon will be forever fixed in Earth's sky—perhaps over Miami if the North American continent still exists.

changing. Its axis, regarded as an infinitely long line, is gradually tracing out a cone among the background of fixed stars. In the year 14,980, Earth's axis will point toward Vega, which will then be our north star. Not until 26 millenia from now will the direction of Earth's axis complete its loop through the sky. In that distant era, Polaris will once again become our guiding star.

Because of the motion of Earth's axis, the points along Earth's orbit at which the equinoxes take place are slowly changing. This effect is known as the *precession of the equinoxes* (see Figure 1.6, on page 18). When the rules of astrology were written, the Sun was in the constellation Aries at the spring equinox, but today it lies in Pisces. Your astrological sign corresponds to the constellation the Sun was in at the time of your birth, had you been born two millenia sooner! In another half century, the Sun will enter the next sign of the zodiac, and the Age of Aquarius will begin.

The Irish satirist Jonathan Swift, centuries ago, shared my distaste for the false science of astrology. He published a parody entitled *Predictions for the year 1708, by Isaac Bickerstaff, Esq.*, in which he pretended to believe

FIGURE 1.5
The Moon always presents the same face to Earth because it rotates once about its axis as it revolves about Earth.

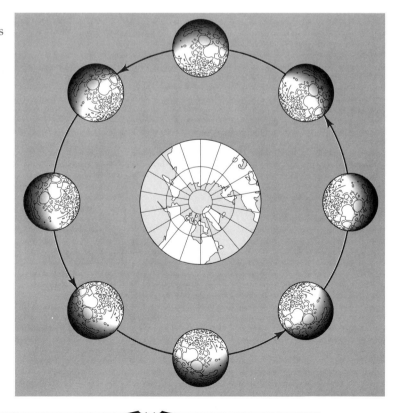

FIGURE 1.6
The precession of the equinoxes. The direction in which Earth's axis points is very slowly changing. It remains at an angle of 23.5° relative to the direction perpendicular to Earth's orbit, but it rotates or precesses about this direction once every 26,000 years. As the axis precesses, the Sun's position in the sky at the spring equinox passes through all of the signs of the zodiac.

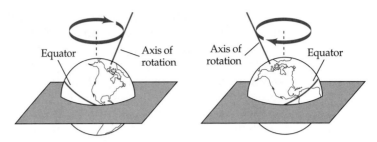

in this occult art. In the parody, the fictional Bickerstaff dismisses his own first astrological prediction as just a trifle, and continues:

> It refers to Partridge the almanac maker. I have consulted the star of his nativity by my own rules, and find he will infallibly die upon the 29th of March next about eleven at night of a raging fever. Therefore I advise him to consider of it and settle his affairs in time.

This is followed by a letter from Bickerstaff, addressed to a famous man, giving a full but fictional account of Partridge's death on the day and nearly the hour mentioned. Partridge the astrologer tried to prove that he was still alive by publishing a new edition of his almanac. Swift replied, disputing this evidence:

> Gadbury, Poor Robin, Dove and Way do yearly publish their almanacs, though several of them have been dead since before the Revolution.

The Twelve Months of the Year

In addition to the Sun, the ancients knew of six other heavenly bodies that move relative to the celestial sphere. They are the Moon and the five planets that are visible to the naked eye: Mercury, Venus, Mars, Jupiter, and Saturn. Of these, the Moon is the most conspicuous, and its changing appearance led to the unit of time we call the month. (The Mayans, who were obsessed with the measurement of time, developed an exceedingly precise calendar based on the motions of the Sun, the Moon, and Venus.)

The words *moon, menstruate,* and *month* share a common Germanic root. The month is linked to the phases of the Moon, just as the day is tied to the daily voyage of the Sun, and the year to the passage of the seasons. The lunar month, defined as the period between consecutive full moons, is about 29.5 days long. The earliest Roman calendar was based on this lunatic unit, despite the fact that the year does not consist of a whole number of lunar months.

EXAMPLE 1.1

The time it takes for the Moon to revolve about Earth is known as its sidereal period, T_s. What is T_s if the time between full moons (the lunar month) is $T_f = 29.5$ days?

Solution

Suppose we set a clock to zero at new moon, when the Sun, Earth, and the Moon are approximately in a line, with Earth in the middle, as shown by A in Figure 1.7. At time T_s, the Moon has completed one turn about Earth and is oriented as shown in B. The Sun, Earth, and the Moon are not yet aligned with one another and the moon is not yet new. We see that T_f is somewhat larger than T_s. At time T_f of the next new moon, the Moon has completed T_f/T_s orbits, but meanwhile Earth has turned about the Sun by a fraction, $T_f/365$, of a full circle. Since the new moon occurs when the three bodies align, the Moon must complete more than one revolution to catch up with Earth:

$$\frac{T_f}{T_s} = 1 + \frac{T_f}{365}$$

from which we get

$$T_s = \frac{T_f}{1 + T_f/365}$$

From this equation, with $T_f = 29.5$, we find T_s (the sidereal period of the Moon) $\simeq 27.3$ days. ∎

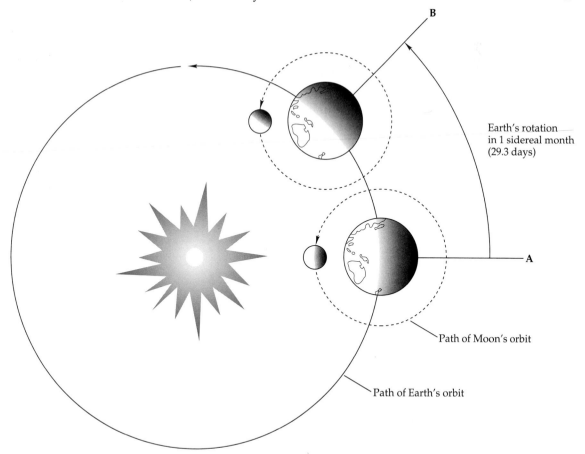

FIGURE 1.7 The Sun, Earth, and the Moon are aligned in A and the Moon is new. The Moon completes one revolution about Earth about 27.3 days later (its sidereal period). During this time, Earth completes about a twelfth of its path about the Sun. The three bodies are not yet in line and the moon is not yet new again. When about 29.5 days have elapsed, the next new moon occurs.

EXAMPLE 1.2   What is the time T_m between successive moonrises on Earth, as seen by an observer on the equator? This is not a purely academic question because the ocean tides are caused mostly by the Moon, and the time between high tides is roughly half the time between moonrises.

Solution   Again, we set our clock to zero at moonrise, as shown in a of Figure 1.8. When one sidereal day has passed, Earth has rotated once, as shown in b, but it is not yet moonrise on Earth because the Moon has moved ahead. Thus, the time between moonrises is a bit longer than a day. When the next moonrise does occur, at time T_m, Earth has rotated about its axis by one full turn plus a bit more to make up for the Moon's motion meanwhile. The extra fraction of its orbit that the Moon has traversed in time T_m is T_m/T_s. If times are expressed in sidereal days, we find:

$$T_m = 1 + \frac{T_m}{T_s}$$

from which we get

$$T_m = \frac{1}{1 - 1/T_s}$$

FIGURE 1.8
To the observer in a, the crescent moon is just rising in the eastern sky. Exactly one day later, Earth completes one rotation about its axis, but the Moon has moved clockwise in the sky and has not yet risen. The time between moonrises is about an hour longer than a day.

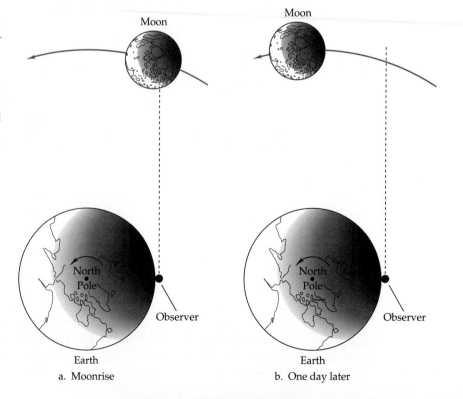

Moon

Moon

North
Pole

North
Pole

Observer

Observer

Earth

Earth

a. Moonrise

b. One day later

We know from Example 1.1 that $T_s = 27.3$ sidereal days. Using this result in our expression for T_m, we find

$$T_m \simeq 1.04 \text{ sidereal days}$$

That is, moonrises and high tides occur about one hour later every day.

        ■

    It's hard to remember how many days there are in each month. To aid us in this matter, we mumble the mnemonic rhyme:

Thirty days hath September,
    April, June and November.
All the rest have thirty-one,
    Except for February . . .

which normally has 28 days but gets one more in leap years. Let's trace the curious origin of our cockeyed calendar.

    The first calendar of the Western world is alleged to have been a gift of Romulus, one of the two mythical wolf-bred founders of Rome. It had 10 lunar months and didn't work well at all. The last four months of our current calendar, once numbered seven (September) through ten (December), recall the wretched Romulian calendar. Never trust a son of a wolf.

    In 452 B.C., a 12-month calendar consisting of 355 days was introduced, with months alternating between 29 and 30 days. An extra day was thrown in somewhere because odd numbers were thought to be luckier than even ones. Because the solar year is significantly longer than 12 lunar months, a thirteenth month had to be added every three years or so. The occasional interloper was called *Mercedinus,* as in "mercenary." It was the time when domestic workers would be paid. How and when to insert the extra month became a political football and a source of social unrest.

    Much later, Julius Caesar called upon the Egyptian astronomer, Sosigenes, to devise a better system. Sosigenes came up with a 365-day year, with an extra day inserted into the calendar each fourth year. In those days, one might have said:

Thirty days hath Sextember,
    April, June, October and December.
All the rest have thirty-one,
    Except for February . . .

which (as usual) got shortchanged. February was to have 29 days, with an extra one added in leap years. The result was a better calendar than the one we are stuck with today. In Caesar's calendar, long and short months simply alternate throughout the year. What could be simpler? Furthermore, February was not quite the dwarf it has since become. For his contribution to science and society, Caesar was awarded a timeless honor: The month Quintember was renamed July for Julius, his first name.

According to legend, Romulus and Remus were abandoned as infants, but found and cared for by a wolf. As young men, in 716 B.C., they founded the city of Rome on the spot where they had been saved. Romulus later killed Remus, became king, and presented the Romans with their first calendar. *Source:* Historical Pictures/Stock Montage, Inc.

So how did we get from Caesar's calendar to our current one? The priests (who were instructed to follow Sosigenes's directions) made a disastrous blunder! They leaped every third year rather than every fourth, as Caesar had ordained. By the time Caesar's nephew Augustus became the first Roman Emperor, the calendar was a mess. One of Augustus's first tasks, therefore, was to get the calendar back in step with the Sun. He abolished a few days from the then-current year and reprogrammed the priests to give February an extra day every four years. (Emperors can do such things with abandon.) That was the good news. The bad news was that Augustus demanded as his due his very own eponymous month, the one just after July. Since his award could be no less than that of his uncle, August had to be given its full share of 31 days. I don't know why, but the extra day was stolen from February. That left three consecutive months of 31 days. To fix this, one September day was given to October and one November day to December. That's how we got today's calendar. Blame it on imperialism!

Since the fall of Rome, there have been a few minor changes in the calendar. In the sixth century, the Church changed the starting point (i.e., the year 1) to coincide with the date of birth of Jesus Christ. The

46th Julian Year was declared to be the first Anno Domini or year of the Lord. Once again, however, someone erred. According to Christian scripture, Christ was born in the same year that Herod died, a date now believed to be 4 B.C.

The last change took place a millennium after this. The average length of a year, according to Sosigenes's algorithm, was 365.25 calendar days. But the tropical year is 365.2422 days. Thus, a considerable error accumulated during the Dark Ages. To remedy this, in 1582, Pope Gregory XIII extinguished the days between October 5 and October 15—and risked being lynched as a thief of time. He furthermore declared that in the future, centennial years would be leap years only if they were divisible by 400. His prescription leads to an average year of 365.2425 days, which is in excellent agreement with the length of the tropical year. It will be many centuries before today's calendar gets out of synch by one day.

The Seven Days of the Week

Of the many measures of time, the week is the most arbitrary one, having no obvious relation to motions of the Sun or the Moon. The word *week* originates from the old Germanic *wikon*, meaning change. Many societies have used a week of anywhere from three to ten days to regulate their markets. Our seven-day week may have begun in ancient Egypt, where the calendar had hours of the day tied to the seven heavenly bodies. The Egyptians listed the heavenly bodies in descending order of distance. Saturn, farthest from us, headed their list. Then came Jupiter (2), Mars (3), Sun (4), Venus (5), Mercury (6), and, finally, the nearby Moon (7). They had it almost right. Today we recognize some ambiguities and discrepancies in this sequence. Mars, for example, is sometimes much closer to us than the Sun. At other times, it is more than twice as far away.

The consecutive hours were linked to these seven bodies and to their position on the list. Saturn, for example, controlled the 1st hour (and the 8th, 15th, and so on); Jupiter the 2nd, 9th, 16th, and so on; Mars the 3rd, 10th, 17th, and so on. Each day was linked to the heavenly body (or deity) that governed its first hour. Because the Sun, for instance, controlled the 25th hour (the first hour of the second day), it became linked to that day. By this curious procedure, exhibited in Table 1.3, Egyptian priests established the sequence of the days of the week.

When 7 days (or 168 hours) pass, lord Saturn begins the cycle anew. Although the Egyptians began their week with Saturday, the sequence

TABLE 1.3   Establishing the Order of the Days

Saturn,	lord of the first hour, controls	Saturday.
Sun,	lord of the 25th (and 4th) hour, controls	Sunday.
Moon,	lord of the 49th (and 7th) hour, controls	Monday.
Mars,	lord of the 73rd (and 3rd) hour, controls	Tuesday.
Mercury,	lord of the 97th (and 6th) hour, controls	Wednesday.
Jupiter,	lord of the 121st (and 2nd) hour, controls	Thursday.
Venus,	lord of the 145th (and 5th) hour, controls	Friday.

of days they established has remained unchanged for millenia. On their exodus from Egypt, the Jews put Saturday last to spite their oppressors. Christians kept this tradition, but made Sunday their day of rest. Later yet, Islam returned to the original system.

The relationship of the days of the week to the heavenly bodies and their assigned deities persists today in many languages. The English, French, Spanish, and Latin versions of the days of the week are compared in Table 1.4.

TABLE 1.4   Four Ways to Say the Day

English	French	Spanish	Latin
Sunday	Dimanche	Domingo	Dies Solis
Monday	Lundi	Lunes	Dies Lunae
Tuesday	Mardi	Martes	Dies Martis
Wednesday	Mercredi	Miércoles	Dies Mercurii
Thursday	Jeudi	Jueves	Dies Jovis
Friday	Vendredi	Viernes	Dies Veneris
Saturday	Samedi	Sábato	Dies Saturni

The association between the old gods and the days of the week is clearest in Latin. In French and Spanish, the five business days of the week retain their connection to the Roman deities. English is further from the original. Monday comes from the Teutonic word *mond* meaning moon, which is a cognate of *month*. From Tuesday to Friday, the Saxon gods have usurped their Latin predecessors: Tiw, Wodin (or Odin), Thor, and Odin's wife Frigg replace Mars, Mercury, Jove, and Venus.

Exercises

1. By how many days does the 365-day year exceed twelve 29.5-day lunar months?

2. What is the mean distance between the 24 time zones on the equator? (Use 6400 km for Earth's radius.)

3. Suppose that the tilt of Earth's axis were 90°. Describe briefly what you think the seasons would be like in such an imaginary world. Give reasons for your answers.

4. Assume that the day has been growing at the uniform rate of 2 milliseconds per century. How long ago was the day 21 hours long?

5. We discussed two rules for determining which years are leap years. The Julian calendar uses Caesar's rule of one leap year every four years. Most countries use the Gregorian calendar in which centennial years must be divisible by 400 to be leap years. How many more leap years have there been since the year A.D. 1 according to the Julian calendar?

6. How long will it take before our calendar slips by a single calendar day?

7. Is the Moon closest to the Sun when it is new or full? Explain your answer.

8. When does Earth move faster around the Sun, at midnight or midday? Explain your answer.

9. From the fact that the sidereal year is about 20 minutes longer than the tropical year, find out how long it takes the Sun to traverse one sign of the zodiac.

10. The Mercurian year (the time for planet Mercury to circle the Sun) is about 90 days. Its sidereal day (the time in which it makes a full rotation about its axis) is about 60 days. What is its solar day (the time between sunsets on Mercury)?

1.4 Life on Log Time

Consider the growth and development of a human being from the moment of conception onward—the miracle of life. At least nine stages of development may be distinguished. The fertilized egg, after about a day, gets its genes together and begins to divide furiously. It is a free blastocyst, a spherical body of cells traveling through its mother's reproductive tract. Late in its sixth day, its wanderings done, it cleaves to the uterine wall, becoming an attached blastocyst. After about a month, the placenta develops and the hungry embryo feeds on its mother's bloodstream. Beyond the first trimester, the person-to-be fills the uterus as a fetus. The earliest postnatal stage of human life is infancy, followed by babbling childhood, rebellious adolescence, and finally, ID-free maturity. The stages of life are presented in Table 1.5.

Can we show this information in the form of a picture or a graph? Suppose we portray the story of life as a linear plot. By this, I mean

TABLE 1.5 Stages of Human Development

Stage of Life	Beginning at
1. Fertilized egg	Conception
2. Free blastocyst	1 day
3. Attached blastocyst	6 days
4. Embryo	4 weeks
5. Fetus	3 months
6. Infant	9 months
7. Child	2.5 years
8. Teenager	10 years
9. Adult	21 years

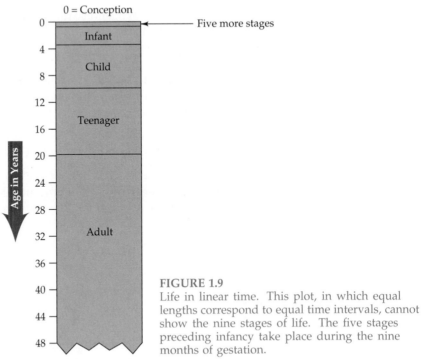

FIGURE 1.9

Life in linear time. This plot, in which equal lengths correspond to equal time intervals, cannot show the nine stages of life. The five stages preceding infancy take place during the nine months of gestation.

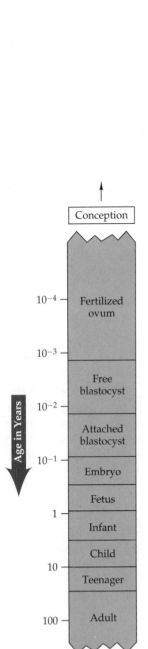

FIGURE 1.10

Life in log time. Equal intervals in this plot correspond to equal multiples of the time since conception. The nine stages of life are clearly distinguished.

that equal lengths correspond to equal periods of time. In Figure 1.9, we choose four-year increments as a convenient time unit. What a useless picture results! The five first stages of life are squeezed together at the very top and cannot be distinguished.

This kind of linear or arithmetic display is fine for a discussion of American politics over the past two centuries. Congressional representatives come and go every two years, presidents every four years, and senators every six years, like clockwork. But it is inappropriate for a discussion of human development, where things happen much more quickly at the beginning. Children learn to ride bicycles and teenagers erupt in pimples, but they never sprout arms, legs, or livers as they did when they were younger. From a psychological point of view as well, the linear progression of time is flawed. Time passes more quickly as we grow older. A month is an eternity to a six-year-old child, but it passes in a twinkling to a pensioner. Perhaps subjective time is felt in terms of time already spent on Earth.

Because of this limitation, clever mathematicians invented another way to present such information—a method more appropriate to problems like this: a *logarithmic* display. Each unit of length in such a display corresponds to a fixed multiplicative factor in elapsed time (or whatever the relevant variable). The data in Table 1.5 are shown as a logarithmic display in Figure 1.10. Each rung of the ladder signifies a factor of 10 in time since conception. The display is called logarithmic because the data are presented versus the logarithm (to base 10) of the elapsed time—that is, $\log 10^4 = 4$, $\log 10^5 = 5$, and so on. By displaying the information logarithmically, we create a panoramic view of human development from its earliest stages to maturity.

Although the moment of conception no longer appears in this display (it has been pushed off to infinity), see how effectively the other nine stages of life are arranged. The last eight stages are of comparable size, just as they are of comparable embryological or psychological significance.

Embryology has little to do with physics, but our discussion illustrates the virtue of a logarithmic display. It is a device often used in the sciences. After a brief digression, I show the development and evolution of the entire universe in a logarithmic display—surely as grand an anthropomorphic analogy as may be imagined.

A View of What's to Come

The logarithmic display is an important device for discussing physical quantities that span many powers of 10 in magnitude, such as the stages of life, the sizes of physical bodies from subatomic particles to clusters of galaxies, and so on. In my view, the most remarkable accomplishment of physical science is the extent to which we are beginning to understand the birth, development, and nature of the universe. Many of the silly seeming questions of children, like "How did the Earth begin?" and "What makes the stars shine?" are being answered. The questions that remain are even more profound: "How did matter evolve and why does it have the properties that it does?" and "How did the universe begin and why did it beget stars, planets, and galaxies?"

That we may, in all seriousness, ask such questions is amazing. That we realistically expect answers is even more astonishing. Now is a time of great excitement and discovery. Sciences as disparate as particle physics (the study of the smallest things) and cosmology (the study of the largest things) have recently converged on one another. We cannot understand the universe without understanding its least particles, and the properties of particles themselves may have been established in the course of the birth of the universe. A new breed of scientist has emerged to investigate the earliest and least observable events in universal history. Its practitioners have deduced how some of the chemical elements found on Earth were synthesized just a few minutes after the Big Bang, how others are the ashes of long-dead stars that were once not so different from the Sun, and how still others were made in tremendous cosmic catastrophes. Scientists are on the verge of finding out whether the universe will expand forever, or will, in the very distant future, collapse upon itself.

Our discussion of embryology, and the display of life on log time, illustrates how human life is an almost miraculous culmination of many earlier and enormously complex biological processes. In a similar way, the sketch to come of the development of the universe illustrates how most of the processes that made life on Earth possible took place in the infancy of the universe. Long ago and near its birth, the visible universe was tiny and incredibly hot. At that time, all of the stars and galaxies we see would have fit on a pinhead. Today, the universe is very old, very large, and very cold. Its temperature was first taken in 1964. It is –270° C, only three Celsius degrees above absolute zero, the temperature at which all molecular motions cease. It is not so cold on Earth, not even in a New England winter. Luckily, we live near a modest star. Should our planet be severed from the solar system and put halfway to nowhere, we would soon find out how cold it is out there.

The Universe on Log Time

According to cosmological theory, the universe arose in a titanic explosion approximately 15 billion years ago. The universe's moment of conception is called the *Big Bang*. Subsequent developments may be segregated into seven stages. In the first stage, the universe was exceedingly hot and full of every kind of particle. This was the "Age of Quarks and Gluons." (Neonatal cosmologists define several earlier stages, but we begin our story at this point.)

As time passed, the universe expanded and cooled. A point was reached when quarks could combine with one another to form neutrons, protons, and their antiparticles. This "Age of Matter and Antimatter" lasted from a microsecond to a hundredth of a second after the Big Bang, when the universe became cool enough to permit the mutual annihilation of nucleons and antinucleons. Only a tiny excess of nucleons survived. Much later they would form matter as we know it, but at that time they were a negligible contaminant in a universe full of electrons, positrons, photons, and neutrinos. The "Age of Leptons"* lasted until the universe was about 100 seconds old.

The happenings of the next few minutes were of great importance to the development of the universe. During this time, most of the smallest atomic nuclei (such as helium and lithium) were formed. This "Age of Nucleosynthesis" lasted until the universe was about 1000 seconds old.

With the end of this fourth stage of universal history, most of the complex reorganization of the universe's matter and energy was complete and the scene was set for the production of stars, galaxies, planets, and the more exotic bodies that make up the heavens. However, it was still too hot for these bodies to form. Indeed, it was even too hot for atoms to exist! During the long "Age of Ions," the hot, homogeneous, and opaque ionized plasma that formed the universe expanded and cooled. Over a period of 500,000 years, the temperature dropped to below 3000° C, hotter than your oven but cooler than the Sun. Ions and electrons could then *recombine* with one another (a curious usage, since

---

Lepton, from the Greek for small, was coined in 1948 to denote either an electron or a neutrino. Today, it refers to any of six apparently elementary particles that are immune to the nuclear force.

they had never been married before) and neutral atoms form. The "Age of Atoms" began.

Suddenly, amazingly, throughout the universe and all at once, the sky became transparent. Light could pass freely from one point to another although nothing was to be seen and nobody was watching. The universe was a tenuous gas of neutral hydrogen and helium, still cooling and expanding. Eventually, a kind of condensation occurred. Small irregularities in the giant gas cloud grew larger and became more pronounced. When the universe reached maturity at an age of a few hundred million years, the primordial fluctuations became enormous. They evolved into the luminous heavenly bodies that still inspire wonder in a curious species living on a planet called Earth. The seventh and still current stage started. This "Age of Stars and Galaxies" has lasted some fifteen billion years. The stages of the history and development of the universe are listed in Table 1.6.

TABLE 1.6   Stages in the Development of the Universe

Stage of Development	Beginning at
1. Quarks and gluons	The Big Bang
2. Matter and antimatter	1 microsecond
3. Leptons	$\frac{1}{100}$ second
4. Nucleosynthesis	100 seconds
5. Ions	1000 seconds
6. Atoms	500,000 years
7. Stars and galaxies	500,000,000 years

As you might guess, a linear display would be useless in showing the stages in the development of the universe. Instead, let's look at these stages in log time, as shown in Figure 1.11 (on page 30). In this display, the Age of Ions is by far the longest of the seven stages. It lasted for a full ten powers of 10: from the age of about 1000 seconds to about half a million years. Although the Age of Atoms lasted a thousand times longer than the Age of Ions in real time, it is represented by a segment only about a third as long in log time.

Important events in the history of the universe (many of which took place in its infancy) are sensibly separated. The long Age of Ions divides universal history into two regimes: the very early universe, whose description involves the exotic terminology of particle physics, and the astronomical universe, whose signs we see at night.

Later in the universe's history, all sorts of cosmic beasts evolved—not only stars and galaxies but quasars, pulsars, bursters, supernovæ, and black holes. The study of these often baffling entities is in the midst of a renaissance due to the development of many new techniques for astronomical observation. The telescope is no longer the astronomer's only tool. Optical astronomy has been joined by radio astronomy, X-ray astronomy, γ (gamma)-ray astronomy, infrared astronomy, microwave as-

FIGURE 1.11
The universe in log time. In this logarithmic display, of the seven stages of the development of the universe are clearly indicated. A linear display could not accomplish this.

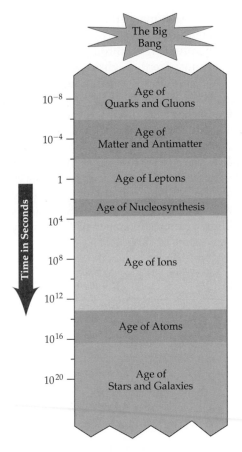

tronomy, and ultraviolet astronomy.* These are all forms of electromagnetic radiation, as we shall learn in Chapter 8. These seven astronomies differ only in the frequency (or equivalently, wavelength) of the radiation that is studied. Red light has nearly twice the wavelength of violet light. Radio waves can be many meters in length, while some γ-rays have wavelengths of 10^{-18} meters. The entire range of types of electromagnetic radiation, organized according to their wavelengths, forms the *electromagnetic spectrum*. Figure 1.12 conveniently shows this spectrum as a logarithmic display.

Visible light occupies a narrow window of wavelengths from violet (approximately 3.5×10^{-7} m) to red (approximately 6.5×10^{-7} m). Adjacent to this band are the infrared rays that warm us and the ultraviolet rays that tan us. X rays are shorter than ultraviolet, while γ rays are

---

*And that's not all! Neutrinos and cosmic rays provide even more information about the heavens.

shorter yet. On the other side, microwaves (used for cooking or to relay information) are longer than infrared rays of radiant heat, and radio waves are even longer.

Sound is another area in which logarithmic displays are useful. Sound consists of waves that propagate through the air at a speed of about 1000 feet per second. Sound waves are characterized by their frequencies: sopranos and piccolos produce sounds of high frequency, basses and bassoons produce predominantly low frequencies. A youngster hears sounds with frequencies between 30 and 20,000 Hz. (The symbol Hz means cycles per second. It is pronounced "Hertz" and honors the discoverer of radio waves.) The ear responds to sound logarithmically. Sounds with frequencies differing by a factor of 2 are pleasingly consonant: We identify them as the same tone an octave apart. The piano keyboard, spans more than seven octaves of frequency. Its highest note has well over 100 times the frequency of its lowest. A song played on the treble side of the piano is recognized as the same song when played again at the bass end because equal distances on the keyboard correspond to equal ratios of frequency: The piano is a logarithmic instrument.

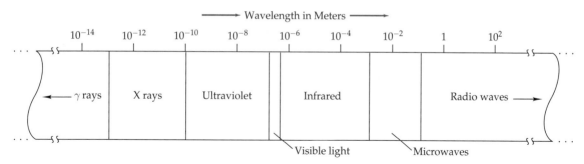

FIGURE 1.12 The electromagnetic spectrum. Visible light consists of electromagnetic waves with wavelengths lying in the octave from 3.8×10^{-7} to 7.6×10^{-7} meters. Electromagnetic waves with much longer wavelengths are used for radio and television transmission. X rays, with much shorter wavelengths, are used by physicians. γ (gamma) rays, with exceedingly short wavelengths, allow physicists to study the properties of subatomic particles.

All Features of the Universe, Great and Small

To understand the nature of matter, we must study the smallest particles of matter. Atoms are inconceivably tiny to us, but they have a complex structure. We shall examine the parts of atoms and even the parts of those parts. To understand the nature of the universe, we must study the largest systems of matter. Stars are inconceivably large to us, but billions of them are grouped together as galaxies, which, in turn, are grouped into clusters and clusters of clusters. The laws of physics are the same for all systems, be they large or small. Figure 1.13, another logarithmic display, encompasses the gamut of sizes of the things of nature.* Human affairs usually concern things a bit larger or smaller than a meter, but our display—and the subject material of this course—ranges over 45 powers of 10, from the smallest particles studied at giant accelerators to the furthest realms of the universe.

Because we will be dealing with numbers that may be very large or very small, we must get used to the scientists' use of exponential notation. Rather than saying that the distance to Earth is about 150 million kilometers, we say that is about 1.5×10^{11} m. Rather than saying that the radius of the proton is about 0.0000000000000012 meter, we say that it is about 1.2×10^{-15} m.

Each science deals with a range of sizes: biology extends from the tiny DNA molecule to the giant whale. Geologists contemplate minerals, mountains, and continents; astrophysicists study stars; and cosmologists concern themselves with galaxies, clusters of galaxies, and the universe itself. Chemists ask about the atoms within molecules, while nuclear physicists pry into the substructure of the atomic nucleus. Particle physics is today's term for the search for the ultimate components of all matter, the smallest things of all.

FIGURE 1.13
Science considers all features of the universe, however large or small. Each of its many disciplines focuses on a relatively narrow range of sizes.

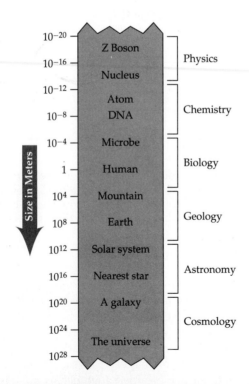

Gamut means an entire range or series and arises from *gamma + ut*, the Greek and Latin names for the lowest note of the musical scale.

For reasons I hope to make clear, the study of the immeasurably small and the study of the inconceivably large are closing in on one another. The more we learn about the microworld, the more we understand about the universe. Nuclear physics tells us how stars shine and particle physics tells us what the universe was like when it was young and burning bright. Conversely, what we see in the sky constrains and helps determine our theory of elementary particles. *Orobouros*, the ancient image of a snake swallowing its tail, once symbolized a universe without beginning or end and the unity of all of nature. Somewhat deformed, it became ∞, the mathematical symbol for infinity. In Figure 1.14, we show the ladder of sizes turned into a circle to symbolize the astonishing new unity of particle physics and cosmology.

Logarithmic displays, which are unbounded in both directions, are appropriate for distances, frequencies, times, or masses, but not for speeds. Einstein taught us that no material body or form of radiation can exceed the speed of light in a vacuum, $c \simeq 3 \times 10^8$ m/s. We may imagine arbitrarily slow speeds, but there is an inviolable maximum speed limit.

FIGURE 1.14
Orobouros is the ancient symbol of a snake swallowing its tail. Today it symbolizes the convergence of particle physics and cosmology.

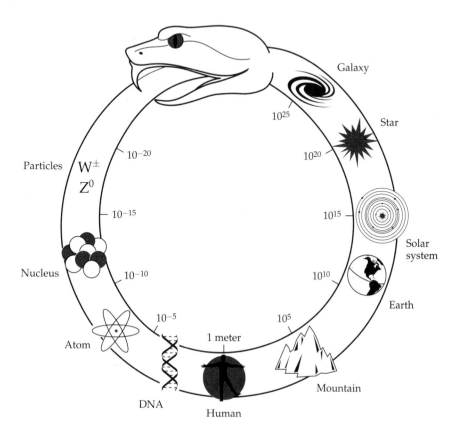

Table 1.7 shows the speeds of a number of things expressed as fractions of the speed of light. Light travels through empty space at the limit of an unbreakable law. In Chapter 11 we shall learn why there never will be a faster-than-light drive.

For more than a century, scientists measured the speed of light in a vacuum with greater and greater precision, but the game has finally come to an end. Our basic unit of time, the second, is defined in terms of a specific atomic process. The constant c, the speed of light in a vacuum according to Einstein and everyone else, is a universal constant. In the past, the meter was defined as the length of a platinum standard set on a pedestal in Sèvres. The velocity of light used to be the number of such meters that light traverses in one second. Not any longer. The

---

A BIT MORE
About Musical Scales

Music has played a larger role than you might expect in the history of physics. We have seen how music was included among the four disciplines of the quadrivium, which was the "core curriculum" a thousand years ago. In Chapter 2, we discuss the three laws of planetary motion deduced by Johannes Kepler. His third law relates the motions of different planets to one another. He found that law while looking for an analogy between planetary motion and music, and published it in a book called *Harmonies of the World*. Centuries later, as we discuss in Chapter 9, a tentative version of the periodic table of the elements was created by another musical analogy. It is not surprising to me that most theoretical physicists, who continue to search for the harmonies of the universe, are music lovers as well.

Those of you who play musical instruments are familiar with the notion of a musical scale. Its history and development are subjects for comparative anthropological study and a demonstration of the unity of science and art. Any C and subsequent G, no matter where they are on the keyboard, have frequencies in the same ratio. They form the musical interval known as a fifth (because it spans five notes of the scale). After the octave (frequency ratio of 2), most cultures agree that the fifth is the most consonant interval. It corresponds to the frequency ratio 3/2. Can we understand why it sounds so pleasing? The answer has to do with harmonics. An oscillating string or air column usually vibrates at several different frequencies at the same time, all of which are integer multiples of its fundamental frequency f. Suppose f is the fundamental of the musical note C. It is accompanied by the frequencies $2f$ or C' (an octave above C), $3f = $ G', $4f = $ C'', and so on. The sound of C often involves a significant admixture of G, and that may be the reason for the consonance of the fifth.

The music of some societies is based on a *pentatonic scale* obtained by the method of tuning by fifths. A fifth above C is G, and a fifth above G is D. Continuing in this vein, we obtain the tones indicated in Table 1.8.

TABLE 1.8   The Pentatonic Scale

Note of Scale	Frequency	Reduced Frequency
C	f	$1.000 \times f$
G	$(3/2)f$	$1.500 \times f$
D	$(3/2)^2 f$	$1.125 \times f$
A	$(3/2)^3 f$	$1.688 \times f$
E	$(3/2)^4 f$	$1.266 \times f$

TABLE 1.7   The Ladder of Speed

Form of Motion	Speed Relative to Light
Growth in height of a teenager	$\sim 10^{-17}$
A snail's pace	$\sim 10^{-12}$
An aerobic jog	$\sim 10^{-8}$
Automobile speed limit	$\sim 10^{-7}$
Speed of a jet aircraft	$\sim 10^{-6}$
Escape velocity from Earth	$\sim 10^{-5}$
Earth's speed about the Sun	$\sim 10^{-4}$
The Sun's speed through the galaxy	$\sim 10^{-3}$
An electron in an atom	$\sim 10^{-2}$
The speed of light in glass	~ 0.5
An electron in a TV tube	~ 0.9999
Highest energy electron beam	~ 0.999999999

Stopping at this point, we obtain a five-tone scale with the notes C, D, G, D, and E. Pentatonic music can be played on the black keys of the piano.   Further divisions of the octave led to today's diatonic, or 12-tone, scale.

By "reduced frequency," we mean the same note transposed downward by an octave or two so as to lie in the octave above C between the frequencies f and $2f$. The interval C–E spans three tones of the scale and is called a third. By the method just described, it corresponds to the frequency ratio 81/64. At the end of the Middle Ages, European musicians replaced this interval by the just third, which corresponds to the simpler ratio 5/4, or 80/64, and has a more pleasing sound. Similarly, A could be changed to the simpler fraction $5/3 = 1.667$. Simpler ratios sound better.

Note that danger lurks in these ideas. A musical scale may be erected on any desired note. Thus, there are 12 distinct major scales, and the different notes of the various scales do not coincide. This presents insuperable difficulties for a keyboard instrument, which would have to be specially tuned to play a piece in each key. The notion of equal temperament arose around the time of the great composer Johann Sebastian Bach (1685–1750). The 12 tones of the new scale were equally spaced logarithmically. Adjoining tones were assigned frequencies in the universal and fixed ratio of the twelfth root of 2, or about 1.0595. In Table 1.9, the 12 notes of the musical scale—as determined by various schemes—are expressed in terms of multiples of the frequency of a given C. The equal-tempering system is a compromise. The interval C–G is not an ideal fifth, nor does the interval C–E coincide with either traditional definition of the third. However, the disparities are the same in any musical scale. Bach demonstrated the virtues of equal temperament by composing 24 marvellous preludes and fugues for the well-tempered clavier, one in each of the major and minor modes of the 12-tone diatonic scale.

TABLE 1.9   The Diatonic Scale

Note of Scale	Equal-Tempered	Simple Fractions	Successive Harmonics	Successive Fifths
C	1.000	1.000	1.000	1.000
C♯	1.059	...	...	...
D	1.122	...	1.125	1.125
E♭	1.189	...	...	...
E	1.260	1.250	1.250	1.266
F	1.335	1.333	...	...
F♯	1.414	...	...	1.424
G	1.498	1.500	1.500	1.500
A♭	1.587	...	...	...
A	1.682	1.667	1.625	1.688
B♭	1.782	...	1.75	...
B	1.888	...	...	1.898
C	2.000	2.000	2.000	2.000

Conference Générale des Poids et Mesures, in October 1983, redefined the meter to be the distance traveled by light in a vacuum in the time interval 1/299,792,458 second. As a consequence of this definition, the speed of light in a vacuum is now exactly $c \equiv 299\,792\,458$ m/s.

Exercises

11. How old are you in years and days a hundred million seconds after your birth? How old are you at your billionth second postpartum? How long would you have to live to celebrate your 10^{10} second "birthday"?

12. How long does it take light to travel from the Sun to Earth? (The distance from the Sun to Earth is about 1.5×10^{11} m.)

13. Choose a dozen historical events whose dates you can remember. Include at least one event that took place today, last week, last year, a decade ago, a century ago, and over a millenium ago. Make a logarithmic display indicating how far in the past each happening happened—that is, 1–10 days, 10–100 days, 10^2–10^3, and so on.

14. The world population is now about six billion, and it doubles approximately every 80 years. Assume (falsely) that it has always done so, and graph the world population versus time. According to the assumption we made, roughly how long ago were there only two people on Earth?

15. According to Table 1.6, for what percentage of the present age of the universe did stars and galaxies exist?

1.5 On Numbers and Their Units

The numerical quantities that scientists deal with are necessarily imprecise. They are the results of observation, measurement, or deduction. The population of the United States is 247.1 million according to my almanac, an obviously "rounded-off" number. Its exact population is unknown and unknowable. Some people are born, others die. Some people immigrate, others emigrate. Some people are homeless and hard to find, others live abroad. Some are illegal aliens, others dual citizens. Some people are kidnapped or missing in action, others are on holiday. No census of the American population can be, or ever will be, reliable to a precision of better than one part in a thousand.

The same holds true for any measured quantity, such as the distance between New York and Boston. What are the exact points between which we measure? Spires shake in the wind, the ground trembles as trucks or subways rumble past. Continents move so that relative distances change. How accurate are our instruments, and how well known is the standard mile? The precise distance between any two points on Earth is an utterly meaningless concept. Learn to live with imprecision, for it is integral to all of the sciences and to everyday life. Learn to say with confidence that our population is about 250 million, that the distance to the Sun is approximately 150 million kilometers, that you are nearly 2 meters tall and weigh about 100 kilograms. We often do not know, do not care, or cannot ascertain a number with precision. In the physical sciences, as in the social sciences, ballpark estimates often suffice.

We may express ourselves more precisely by specifying the degree of precision to which a quantity is known. A typical bathroom scale is a not very accurate instrument. If it reads 200 pounds, I can be reasonably confident that my true weight lies between 198 and 202 pounds. That is, it determines my weight to be 200 ± 2 lb. For a more precise determination, I may use the scale in my doctor's office to find that my weight is 198.75 ± 0.25 lb. However, no matter what method I use to weigh myself, there will be a small (and knowable) measurement error.

Consider a measurement more relevant to physics. The mass of a proton, m_p, is now known with 90 percent confidence to lie within the interval

$$1.6726231 \pm .0000010 \ \times 10^{-27}\,\mathrm{kg}$$

That is, there is not more than a 10 percent chance that m_p is greater than 1.6726241 or less than $1.6726221 \times 10^{-27}$ kg. Although the proton mass is known very well, notice that it is not known exactly. In fact, no measurement can be absolutely precise, and, moreover, our unit of mass itself is slightly uncertain. The standard kilogram is a real physical object. It is defined as the mass of a cylinder of platinum-iridium, which is kept under very controlled conditions. Each time the cylinder is used to calibrate our scales, a few of its atoms are scraped off or a few extraneous atoms cling to it. And the calibration itself is necessarily imperfect.

In the future, a new system of units may be adopted in which the kilogram is defined to be such that the mass of the proton is exactly $1.672\,623 \times 10^{-27}$ kg. If this comes to pass, all other masses may be compared to the proton mass and we shall have no further need of a standard kilogram. All masses except the proton mass will remain intrinsically uncertain and imprecise, if only very slightly so.

Scientific Notation and Significant Figures

Most of the calculations carried out in this book and in science are approximations or crude estimates. This is because the numbers we deal with are not known exactly or because we are not always interested in extreme precision. For this reason, we often use the following convention, which I explain through an example. We may indicate the mass of a proton by $\sim 1.67 \times 10^{-27}$ kg, or write it as an equation:

$$m_p \simeq 1.67 \times 10^{-27} \text{ kg} \tag{1}$$

There are four essential parts to this way of writing approximate numbers.

1. In text, the appearance of the symbol $\sim$ is to be read "about." In a formula, the symbol $\simeq$ is to be read "is approximately equal to." The appearance of the squiggle signifies that an approximation is being made whose precision is determined by the number of digits that are presented afterward.

2. A number between 1 and 10, expressed in ordinary decimal notation, follows the symbol. The number of digits that are specified is called the number of significant figures. In equation 1, there are three significant figures, and we are told that the quantity m_p lies somewhere between 1.665 and 1.675×10^{-27} kg. We never specify more digits than are needed for the problem at hand, and we never specify more digits than are known or knowable.

3. The exponential factor (a power of 10) corrects for the fact that the number to be specified may be larger than 10 (where the exponent is a positive integer) or smaller than 1 (where the exponent is a negative integer). If the number lies between 1 and 10, the exponential factor is omitted. Thus, to four significant figures, we write $\pi \simeq 3.142$.

4. Some numbers, such as π or the ratio of the mass of Earth to the mass of the Sun, are dimensionless. However, most of the numbers we deal with are expressed in dimensional units of length, velocity, and so on. For these quantities, the number itself is meaningless unless it is appended to an appropriate unit.

Combining Numbers Given in Scientific Notation

Two numbers, A and B, may be combined together by the arithmetical operations of multiplication, division, addition, or subtraction. The precision of the result is determined by the precision to which A and B are specified. The following rules of thumb should be used when manipulating approximate numbers. (To be explicit, we choose $A \simeq 2.2 \times 10^{-1}$ and $B \simeq 5.6789 \times 10^2$.)

- *Multiplication and division:* If A is specified to a significant figures and B is specified to b significant figures, then their product or quotient is known to a number of significant figures equal to the lesser of a and b. Following this rule, we find:

$$A\,B \simeq 1.2 \times 10^2, \qquad A^2 \simeq 4.8 \times 10^{-2}, \qquad B^2 \simeq 3.2250 \times 10^5, \qquad \text{and}$$

$$A/B \simeq 3.9 \times 10^{-4}, \qquad B/A \simeq 2.6 \times 10^3.$$

- *Addition and subtraction:* For these operations, the precision of the result depends on the absolute precision of A and B and not directly on a and b. For the case at hand, we revert to normal notation so that

$A \simeq 2.2$ and $B \simeq 567.89$. B is specified to two decimal places, but A is specified to just one. Consequently, their sum and difference are meaningful to one decimal place:

$$A + B \simeq 570.1, \qquad B - A \simeq 565.7$$

EXAMPLE 1.3   The volume of a tank is ~ 23 liters. It is completely filled by ~ 27.172 kg of a certain liquid. What is the density of the liquid in g/cm^3?

Solution   We must divide the mass of the fluid in grams by the volume in cubic centimeters to obtain its density. If we blindly use a calculator, we get 1.18139. This is not the right answer because the volume of the liquid is specified to just two significant figures. It doesn't matter that its mass is known to five figures: The ratio cannot be determined to more significant figures than either numerator or denominator. The proper answer is ~ 1.2 g/cm^3. ∎

Most of the numbers we encounter in physics or mathematics are approximate, but a few really are exact. The number of possible five-card poker hands is precisely 2,598,960. The number of dimensions in space is 3 not 3.001. There are exactly 5 Platonic solids, 9 observed planets in the solar system, 11 types of crystal lattices, and 92 electrons in a uranium atom. Numbers such as π, $\log 2$, and $\sqrt{3}$ are also exact numbers, even though we cannot write them as finite decimals. We must be aware of this dichotomy of numbers: the *exact* and the *approximate*.

Dimensions and Units

Numbers are also divided into the *pure*, or dimensionless, and the *impure*, or dimensional. The population of a nation, the number of possible dice throws, and the value of π are pure numbers because their specification does not depend on a choice of units. However, in the phrases, "55 miles per hour," "3-minute egg," or "100-watt light bulb," the numbers carry dimensions of velocity, time, and power. Expressed in meters per month, centuries, or horsepower, the numbers change but the physical quantities being expressed do not.

Table 1.10 (on page 40) lists some large pure numbers.* The numbers are indicated approximately: π to seven significant figures (although over a million are known), the Z^0 mass to three significant figures (which are all that are known), the possible orders of a deck of cards to two significant figures (although the order is known exactly), and the rest to zero significant figures. Some of these numbers are exact in principle but subject to unavoidable measurement error (e.g., the ratio of particle masses). Some are intrinsically imprecise (e.g., the mass of the Sun depends on the value of the standard kilogram, and, moreover, changes with time.)

---

*Why do I present only large numbers? Because small numbers can be represented by their large reciprocals.

TABLE 1.10   Some Large Numbers

Quantity	Its Value
Protons in the universe	$\sim 10^{79}$
Shuffles of a 52-card deck	$\sim 8.1 \times 10^{67}$
Protons in the Sun	$\sim 10^{57}$
Neutron star to universe density	$\sim 10^{45}$
Electric to gravitational force	$\sim 10^{39}$
Contract bridge setups	$\sim 10^{29}$
Stars in the heavens	$\sim 10^{22}$
Neurons in the brain	$\sim 10^{10}$
Z^0-boson to electron mass ratio	$\sim 1.78 \times 10^{5}$
Circumference of circle in diameters	~ 3.141593

The possible arrangements of a 52-card deck of cards is easy to determine precisely. There are 52 possibilities for the first card, 51 for the second, and so on. Thus, the answer is $52 \times 51 \times 50 \times \ldots \times 3 \times 2 \times 1$, or 52! The ratio of the circumference to the diameter of a circle is π. Although it cannot be expressed as a finite decimal expansion, it may be computed to any desired degree of precision. The number of protons (or stars) in the universe is only roughly known, and will never be known with any precision. The number of neurons in the brain varies considerably from person to person and with age. Ratios of particle masses are, in principle, exact numbers. However, until we have a better theory, they will be determined by experiments that are necessarily of limited precision.

Impure numbers require a dimensional unit. You may measure your height in feet, meters, or any of hundreds of historical and contemporary units of length, some of which are listed in Table 1.11.

TABLE 1.11   Lengths Around the World and Through the Alphabet

Unit of Length	Country of Origin	Value in Meters
Arshin	Estonia	0.7112
Braza	Argentina	1.73
Chang	China	3.5814
Diraa	Saudi Arabia	0.58
Elle	Switzerland	0.6
Foot	USA	0.3048
Guz	India	0.6858
Hat'h	India	0.452
Inch	USA	0.0254
Ken	Japan	1.820
Link	USA	0.2012
Mkono	East Africa	0.4572
Nin	Thailand	0.0212
Pecheus	Greece	0.648
Rod	USA	0.9144
Sagene	Russia	2.134
Toise	France	1.949
Vara	Mexico	0.838
Wah	Thailand	2
Yard	USA	0.9144
Zoll	Switzerland	0.03

Americans insist on retaining all sorts of crazy units: the apothe-caries' system to weigh drugs, the troy system for precious metals, and the avoirdupois system for most other things. In this last system, there are $27\frac{11}{32}$ grains in a dram, 16 drams in an ounce, 16 ounces in a pound, 112 pounds in a long hundredweight, and 20 long hundredweights in a ton. All told, there are 1,568,000 grains in an avoirdupois ton, but please don't ask me how many troy grains make a troy ton of gold, or how many five (apothecaries')-grain aspirins can be made from a ton of acetylsalicylic acid.

When it comes to volume, we buy oil in 42-gallon barrels, wine in 31.5-gallon barrels, and dry materials in 30.5-gallon barrels. (Curiously, most of the barrels I've seen contain 55 gallons.) We also use a wide variety of differing bushels: wet and dry, heaped and stricken. For lengths, many of our historical units are mercifully obscure, but Table 1.12 shows the whole proud English system with the inch chosen to be the basic unit.

TABLE 1.12   English System of Lengths

Unit	Equivalent to	
Foot	= 12 inches	= 12 in.
Yard	= 3 feet	= 36 in.
Rod	= $5\frac{1}{2}$ yards	= 198 in.
Chain	= 4 rods	= 792 in.
Furlong	= 10 chains	= 7,920 in.
Mile	= 8 furlongs	= 63,360 in.
League	= 3 miles	= 190,080 in.

The International System of Units

Things can easily get out of hand for merchants, manufacturers, repair-men, tailors, or travelers. It would save immense time and trouble if all nations could agree to abandon the tribal and traditional units that are all too often based on the physiognomy of long-dead tyrants. And, in fact, most nations have done just that! The metric system was de-vised in France in the aftermath of their Revolution, and was nationally mandated in 1801. Its current version, called SI (standing for *Système International*), is used in this book.

The United States is the only major nation that has not gone metric. Elsewhere (possibly excepting Burma and Yemen) people live, work, and sell in grams, meters, and seconds, and take their temperatures in Celsius degrees. One orders *sto gram*, or a hundred grams, of booze in Russia; counts kilojoules when dieting in Europe; and usually drives under 100 kilometers per hour in Japan. (This last is a compromise: Meters per second—the physicists' unit—is more faithful to the SI system.)

Metric units (usually those of the SI system) are always used by scien-tists, who must be able to communicate with their colleagues throughout the world. Lengths are specified in meters, denoted by m. The meter, the equivalent of 3.281 feet, is a bit more than a yard. The centimeter (a hundredth of a meter) is about the width of your pinkie. Exactly 2.54 centimeters make an inch.

There are seven basic units in the SI system, including seconds (s), meters (m), and kilograms (kg). The second is defined in terms of atomic vibrations (see Section 1.3) and the meter in terms of the atomic second and the speed of light (see Section 1.4). The standard kilogram was described earlier in this section. The other four units correspond to the quantity of electric current (the ampere), the measure of temperature (the kelvin), the chemical equivalent of matter (the mole), and the luminous intensity (the candela). All measured quantities may be expressed in terms of these units, or combinations of them. For example, the SI unit of velocity is m/s, while the SI unit of force is the $kg\,m/s^2$ (also known as the newton). Other common derived units are shown below.

Acceleration is the rate of change of velocity.
It is measured in m/s^2 (meters per second per second).

Pressure is force per unit area.
It is measured in $kg/m\,s^2$ (kilograms per meter per second per second) or N/m^2 (newtons per square meter).

Density is mass per unit volume.
It is measured in kg/m^3 (kilograms per cubic meter).

Energy is the product of force and distance.
It is measured in $kg\,m^2/s^2$ (joules or J).

Power is the rate of change of energy.
It is measured in $kg\,m^2/s^3$ (watts or W).

The meter is no better a measure of length than the foot, the gram no better a measure of mass than the grain—but the SI system has two great virtues. The first is simply that it is there: Most people (and all scientists) have already adopted it. Why should they switch?

Second, the SI system has a real edge over its rivals. By placing a prefix before the unit name, we create larger or smaller units. These different size units are convenient for different circumstances. In the case of lengths, we use *kilo*meters to drive, *centi*meters to dilate, and *milli*meters to adjust. For masses, we use metric tons (*mega*grams) of coal, *kilo*grams of cocoa, *milli*grams of medicines. All of these diverse units involve conversion factors that are powers of 10. Other systems of weights and measures also have a variety of units. However, conversion among inches, feet, yards, and miles is painfully difficult. So is the conversion between ounces, pounds, and English tons, or fluid ounces, pints, and gallons. There are all those nasty conversion factors to remember, such as 32 or 5280. Table 1.13 shows some of the more common SI prefixes.

Sometimes the compound units of the metric system have their own names. Area is measured in *ares* or hectares, but an are is simply 100 m^2.

TABLE 1.13   Metric Prefixes

Prefix	Multiplies by	Comes from	Means
atto–	10^{-18}	Danish	Eighteen
femto–	10^{-15}	Danish	Fifteen
pico–	10^{-12}	Italian	Small
nano–	10^{-9}	Greek	Dwarf
micro–	10^{-6}	Greek	Small
milli–	10^{-3}	Latin	Thousand
kilo–	10^{+3}	Greek	Thousand
mega-	10^{+6}	Greek	Great
giga–	10^{+9}	Greek	Giant
tera–	10^{+12}	Greek	Monster
peta–	10^{+15}	Latin	?

Our acre, in case you didn't remember, is a somewhat hard to remember 43,560 square feet.

The liter, a metric unit of volume, is a thousandth of a cubic meter. The U.S. gallon is about 0.13367 cubic foot. A French schoolgirl, asked how many liters of gasoline can be put into a cubic tank whose side measures 30 cm, says "Twenty-seven!" in a trice. Few American college students could figure out how many gallons of gasoline fit in a tank one foot on a side.

In the SI system, there is no distinction between dry measure and liquid measure and no more heaping teaspoons, fluid ounces, "fifths" of liquor, and stricken bushels. Volumes are expressed in liters, or in simple decimal multiples or fractions of liters. Energy is measured in joules; not kilowatt-hours, horsepower-minutes, calories, British thermal units, or foot-pounds. Let's lose our calculators and get metric, where conversion is trivial. Let's forget, if we ever knew, how many inches make a mile!

Using Dimensions

In all sciences, especially physics, we must keep careful track of the dimensions of quantities. A numerical result is usually meaningless unless it is linked to a dimensional unit. Every time you examine an equation, say $AB = CD$, you must be certain that the dimensions of the right and left sides agree with one another. Otherwise, the equation is senseless. When you determine a velocity or a force or a mass, be sure that your result bears the proper units.

Dimensions are a help more than a hindrance. A dimensional check of your result is an easy way to identify careless errors. Dimensional considerations can guide you to the solution of a problem, at least to within a dimensionless multiplicative constant, which is rarely very much larger or smaller than 1. Thus, the area of a sphere of radius R, which has dimensions of length squared, must be a pure number multiplied by R^2. (The pure number is 4π.) The volume of a sphere, which has dimensions of length cubed, is a pure number $(4\pi/3)$ times R^3. Here are two examples, taken from material that we discuss in Chapters 3 and 9, that illustrate how dimensions may be used to guess the right answer.

1. You will learn that the formula for the kinetic energy of a moving body is $\frac{1}{2}mv^2$, where m is its mass and v is its speed. How do we remember this formula? There is one and only one way to construct a quantity with dimensions of energy ($kg\,m^2/s^2$) from a mass m and a velocity v, namely mv^2. (You may forget mv, m^2v, or mv^3 because they have the wrong dimensions.) What you do have to remember is that the correct expression for kinetic energy involves a curious factor of one-half.

2. Newton was the first scientist to express the speed of sound in terms of the density and pressure of air. He could have arrived at his result by thinking dimensions. Pressure P is measured in units of $kg\,m/s^2$, while density ρ is measured in kg/m^3. The only way to make a velocity of these quantities is to take the square root of their ratio. The expression $\sqrt{P/\rho}$, which has the dimensions of velocity, is Newton's result. It is the speed of sound, except for a dimensionless numerical factor of ~1.2, which not even Newton knew about. This pure number factor was not calculated until 1816, but the point is that an analysis of the question in terms of the dimensions of the relevant quantities suffices to make an estimate of the speed of sound.

Exercises

16. Show that one newton per square meter has the same meaning as one joule per cubic meter.

17. A lump of metal is measured to have a volume of $\sim1.1111 \times 10^3$ cm^3 and a mass of 8.5 kg. What is its density in grams per cubic centimeter? Be sure to give your answer to the appropriate number of significant figures.

18. The mass of the Sun is $\sim 1.9891 \times 10^{30}$ kg. The mass of all other bodies in the solar system is $\sim5.67 \times 10^{27}$ kg. What is the total mass of the solar system? Be sure to give your answer to the appropriate number of significant figures.

19. In the world's smallest oil spill, ~1 cubic millimeter of oil is released upon a tranquil lake. The slick expands to fill an area of ~2 square meters. It is then one molecular diameter thick. Estimate from this information the mean diameter of an oil molecule.

20. If a car consumes 9 liters of gasoline every 100 kilometers, how many miles can it travel on a single gallon of gasoline?

21. A light-year is the distance light travels in one year. How far is this in meters? (The speed of light is $\sim 3 \times 10^8$ m/s.)

22. The density of mercury is ~ 13.6 grams per cubic centimeter. What is it in kilograms per liter?

23. The Fahrenheit temperature equals $\frac{9}{5}$ the Celsius temperature plus 32. At what unique temperature do the two scales coincide?

24. If the speed of sound in air is 320 m/s, what is it in helium at atmospheric pressure, whose density is about one seventh that of air?

25. The cgs system of units is based on grams, centimeters, and seconds. Its unit of energy is the erg or g cm^2/s^2. Its unit of force is the dyne or g cm/s^2. How many ergs make a joule? How many dynes make a newton?

26. A poker hand is a selection of five cards from a 52-card deck. Show that the number of different hands is 2,598,960. (*Hint:* There are 52 possibilities for the first card you are dealt, 51 for the second, down to 48 for the last. Since order is immaterial, the number of possible dealings must be divided by 5! to get the number of distinct hands.)

Where We Are and Where We Are Going

In this chapter, we examined several of the ideas that led to the development of physics and several of the tools used by scientists around the world. In ancient societies, science was a medley of myth, religion, and pragmatism that emerged from necessity and curiosity. The explosive growth of modern science started five centuries ago, at a time when explorers were making their voyages of discovery, the invention of printing was having its first impact, and the Renaissance was changing the Western world.

The theme of this book is the search for the ultimate constituents of matter and for the rules by which they interact with one another to produce the phenomena of nature. Part One deals with the origins and substance of classical physics. The science of motion, for bodies on Earth and in the heavens, is introduced in Chapters 2 and 3. The next three chapters explain how scientists learned that matter is composed of atoms of different elements that react with one another and combine to form molecules. The chaotic motions of these molecules give rise to the form of energy that we know as heat. Chapters 7 through 9 complete the picture of physical science at the close of the nineteenth century.

The second half of the book begins with a description of three remarkable experimental discoveries that foreshadowed the revolutionary developments to come. Because classical physics failed to describe things that are too small or move too fast, quantum mechanics and the theory of relativity became the pillars on which modern science was built. With these tools, the structure of atoms and their nuclei was explained and the mechanisms underlying chemical, physical, and biological processes could be explored. The foundations of today's technology were set. Furthermore, scientists today may be on the verge of creating a unified theory to describe all the phenomena of the microworld. The most recent developments enable them to ask not only *how* things happen as they do, but *why* nature is as it is. The final chapter touches on the ultimate unification of physical science: the coalescence of the studies of the incredibly small and the inconceivably large. Modern cosmology dares to explain the origin and evolution of the universe and all its wonders in terms of what has been learned from an examination of its smallest parts.

2 The Science of Motion

The basic constructs of classical mechanics—mass, velocity, acceleration, and force—are essential tools in the search for the constituents of matter and the rules they obey. Classical mechanics is an imperfect description of motion: Einstein's special theory of relativity shows that time and space are not as absolute and distinct as they seem to us. His general theory supplanted Newton's theory of gravity. Moreover, the notion of bodies moving in precisely defined trajectories is denied by quantum mechanics. In other words, relativity and quantum theory turned mechanics topsy-turvy. Why, then, should you study a discredited discipline? Because Newton's laws are still valid. Within

Baseball is as much fun for the physicist as it is for the fan. The bat–ball collision may not be as "fundamental" as an electron–proton collision (even at the hands of Toronto's Joe Carter), but it can be analyzed by the same physical laws. *Source:* Reuters/Bettmann.

their domain, they were true and will remain true forever. Science is a vertical discipline: Alterations or additions sit firmly on what is already in place. The new theory incorporates the old: In the workaday world, quantum theory and relativity reduce to classical physics. We must understand the things about us before we dream of the innermost secrets of matter and the majesty of the universe.

The discipline of physics is like a building, and mechanics is its foundation. *Mechanics* deals with the motions of bodies on Earth and in the heavens. Its laws are deceptively simple, but their truth is far from evident. Newton's first law of motion states that the uniform motion of an isolated body persists indefinitely. It describes neither bullets, which lose energy as they plow through the air, nor billiard balls, which slow down and stop because of friction. Isolated systems are either very tiny, such as molecules of air; or very large, such as heavenly bodies. And even these objects are not truly isolated: Molecules collide from time to time and gravity influences the motions of planets. Centuries ago, scientists knew little about molecules; the science of motion began when they pondered the wonders of the starry skies.

As astronomers developed precision instruments, they learned to describe the motions of heavenly bodies in the same fashion as those of birds, balls, and falling apples on Earth: in terms of positions, velocities, and accelerations. These quantities are associated with directions in space: they are *vectors*.

The first section of this chapter introduces vectors and applies them to the description of positions. In Section 2.2, vectors are applied to velocity and acceleration. These notions are prerequisite to an understanding of mechanics and are put to good use later. The contributions of Copernicus and Kepler to astronomy are described in Section 2.3; those of Galileo to the study of motion on Earth, in Section 2.4. The discoveries of these scientists laid the foundations for Newton's science of motion, which we discuss in Section 2.5.

2.1 Finding Your Way in Space with Vectors

Physical quantities are often linked to directions in space. Things fall *down*, compasses point *north*, and rivers flow to the sea. A *vector* is a mathematical construct describing a quantity linked to a direction in space: It is a magnitude that points. The distance between New York and Boston is simply a number with the dimension of length. The vector distance, however, tells us both the distance and the direction of the direct route from New York to Boston. A vector can describe motion as well as position. The speedometer of a car registers what we call the *speed* of the car—a number with dimensions of m/s. If the car is equipped with a compass, we may determine both the speed and its direction. Together, they determine a vector called the *velocity* of the car relative to the road.

A vector is an arrow whose length and direction indicate its magnitude and direction. (I use the word *magnitude* because vectors describe all sorts of things with different dimensions.) In equations, a vector is put in boldface, as in **A**. Its magnitude is indicated by A, which can be any positive number or zero. Every vector points in a definite direction except for one—the *null vector* **0**, whose magnitude is zero.

We distinguish between vectors (such as velocity, which has a direction in space) and other quantities (such as mass, which does not).

Nonpointing numbers are often called *scalars* in contradistinction to vectors. Vectors may be multiplied by scalars to give other vectors. If **A** is a vector and p is a scalar, their product is a vector with the following properties: If p is positive, p**A** has magnitude pA and points in the direction of **A**. If p is negative, p**A** has magnitude (recall that magnitude is positive or zero) $-pA$ and points in the direction opposite to **A**. Here are examples of both cases:

- Later, you will learn that the force **F** exerted on an object equals its mass m (a positive scalar) multiplied by its acceleration **a** (a vector). That is, **F**= m**a**. The magnitude of **F** is the product of m with a (the magnitude of **a**). Its direction is the same as the direction of the acceleration.

- The vector distance **D** from New York to Boston points from New York to Boston. Its magnitude is the distance between the two cities. If **D** is multiplied by -1, the vector $-$**D** is obtained. It has the same length as **D**, but it points in the opposite direction. It is the vector distance from Boston to New York.

Two vectors, **A** and **B**, describing dimensionally identical quantities such as two velocities, can be combined by an operation called *vector addition* to make a third vector. That is, **C** $=$ **A**$+$**B**. Because vectors have directions, the operation of vector addition is different from ordinary addition. In vector addition, the sum is obtained by the geometrical prescription shown in Figure 2.1a. First draw **A** as an arrow with a foot and a tip. Then place **B** with its foot at the tip of **A**. The sum **A**$+$**B** corresponds to an arrow whose foot is at the foot of **A** and whose tip is at the tip of **B**.

FIGURE 2.1
a. The sum of two vectors, **A** + **B**, is obtained by placing **B** at the tip of **A** and drawing the vector from the foot of **A** to the tip of **B**. *b.* The sum **B** + **A** is obtained by placing **A** at the tip of **B** and drawing the vector from the foot of **B** to the tip of **A**. The illustration shows that **A**+**B** = **B**+**A**. *c.* The difference **A** − **B** is obtained by placing the feet of **A** and **B** together and drawing a vector from the tip of **B** to the tip of **A**.

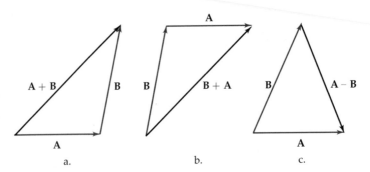

a.   b.   c.

In ordinary addition, the order of the terms being added together doesn't matter. The same is true for vector addition. To obtain **B**$+$**A**, we place **A** with its foot at the tip of **B**. The vector extending from the foot of **B** to the tip of **A** is the required sum. The two orders of addition are shown in Figure 2.1a and 2.1b. Both constructions yield the same vector. That is, the sum of two vectors is the same in either order:

$$\mathbf{A} + \mathbf{B} = \mathbf{B} + \mathbf{A}$$

Vector addition reduces to ordinary addition when the vectors being added point in the same or opposite directions. In particular, if we add the vector **A** to the vector −**A**, we obtain the null vector:

$$\mathbf{A} + (-\mathbf{A}) = 0 \quad \text{or more simply} \quad \mathbf{A} - \mathbf{A} = 0$$

This straightforward manipulation of signs suggests a natural definition for vector subtraction. The vector **A** − **B** is the vector obtained by adding −**B** to **A**:

$$\mathbf{A} - \mathbf{B} \quad \text{means} \quad \mathbf{A} + (-\mathbf{B}) \quad \text{or} \quad -\mathbf{B} + \mathbf{A}$$

The result is indicated in Figure 2.1c. To find **A** − **B**, we place the feet of the two vectors together and draw the vector from the tip of **B** to the tip of **A**.

EXAMPLE 2.1
Adding Vectors

An expedition sets out from an oasis in a trackless desert. After trekking 40 km due east, the explorers find a frayed letter in a cave. It reveals that a stolen treasure is buried exactly 40 km northeast of the oasis. How far and in what direction is the treasure from the travelers?

Solution

The direct route from the oasis to the cave is the vector **E**, and the direct route from the oasis to the treasure is **T** (Figure 2.2). The shortest way from the cave to the treasure is the difference between these two vectors: **D** = **T** − **E**. The three vectors form an isosceles triangle with equal legs of 40 km and an apex angle of 45°. The rest is just geometry, so I have omitted the details. The length of the unequal side of the isosceles triangle (the distance between the cave and the treasure) is $40\sqrt{2 - \sqrt{2}}$ or ∼ 30.6 km. The direction the explorers must walk is 22.5° west of due north.   ∎

FIGURE 2.2
E is the vector position of the cave, and **T** is the vector position of the treasure, both relative to the oasis. **D**, which equals the vector difference **T** − **E**, is the vector position of the treasure relative to the cave.

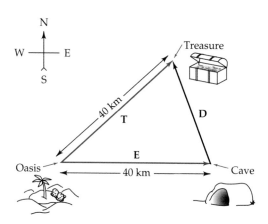

Cartesian Coordinate Systems

Vectors may be added, subtracted, and multiplied by scalars. Although we could describe physical processes using only geometry, it is simpler to introduce a device that lets us use the techniques of ordinary algebra. The device is known as a *coordinate system*. A *Cartesian coordinate system* is one whose coordinate axes are fixed in direction and perpendicular to one another—it is the kind of coordinate system we use in this book.

Every *where* question requires the specification of four quantities. For a pilot, three of these are the latitude, longitude, and altitude of the aircraft. For an astronomer, they are two angles specifying the direction of a star and another telling its distance from Earth. For someone with a toothache, they are the street, the house number, and the floor of the dentist's office. A fourth number, the time of the event, must be specified as well, since the aircraft is moving, Earth is turning, and the dental appointment is at 3 P.M. We live in a four-dimensional universe.

In this section, however, we ignore time. That is, we consider the attributes of bodies at a particular moment of time. (The subject of motion is dealt with in section 2.2). When considering such attributes, we need a standard way to describe positions and other physical quantities. Because happenings in three-dimensional space are hard to visualize, we often deal with simpler situations in which one or two of the dimensions of space play no essential role. Billiards is an example of motion in two dimensions. At any time, the position of a ball on the table is specified by two numbers. A falling brick is an example of motion in one dimension. At any time, its position is specified by one number: the distance it has fallen. Let's begin our study of coordinate systems in two dimensions.

A fixed point O in a plane is singled out as the origin of the coordinate system. A directed line passing through O is called the x-axis. Another line through O perpendicular to the x-axis is called the y-axis. Each point on each axis is labeled with a number whose magnitude is proportional to its distance from O. The number is positive on one side of O and negative on the other. These numbers are referred to as x- and y-coordinates.

Let P be a point on the plane. In Figure 2.3, it is the tip of the vector **A**. Its coordinates are determined in the following fashion. A

FIGURE 2.3
The coordinates of point P are (x_1, y_1). Those of Q are (x_2, y_2). The vector **A** extends from Q to P. Its components are $A_x = x_1 - x_2$ and $A_y = y_1 - y_2$.

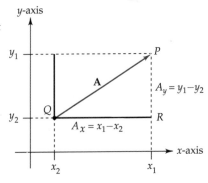

line through P and perpendicular to the x-axis intersects the x-axis at a certain coordinate value x_1. This is the x-coordinate of P. Another line through P and perpendicular to the y-axis intersects the axis at the coordinate value y_1. The location of the point P is specified by its two coordinates (x_1, y_1). Similarly, the point Q at the foot of $\mathbf{A}$ is specified by the coordinates (x_2, y_2).

The vector $\mathbf{A}$ is indicated by an arrow in the plane extending from Q to P. It has the same meaning no matter where it is drawn on the plane. Its length and its direction are determined by the right triangle $\triangle QRP$, which in turn is determined by its legs, the quantities $A_x = x_1 - x_2$ and $A_y = y_1 - y_2$. These quantities are called the *components* of $\mathbf{A}$. If the vector $\mathbf{A}$ is translated to another location in the plane, its components are unchanged. Because a vector is characterized by its components, we often specify a vector $\mathbf{A}$ by specifying its components (A_x, A_y).

We can use a coordinate system in examining the position of one object relative to another. Figure 2.4 shows a coordinate system with its origin at the Sun. On a particular day, $\mathbf{E}$ is the position vector of Earth and $\mathbf{M}$ is the position vector of Mars, both of them relative to the Sun. The position of Mars relative to Earth is the vector $\mathbf{A}$ extending from Earth to Mars. Our earlier discussion of vectors tells us that

$$\mathbf{A} = \mathbf{M} - \mathbf{E}$$

The tip of $\mathbf{A}$ is at the tip of the vector whose components are (M_x, M_y). The foot of $\mathbf{A}$ is at the tip of the vector whose components are (E_x, E_y). Thus, the components of $\mathbf{A}$ are

$$A_x = M_x - E_x \qquad \text{and} \qquad A_y = M_y - E_y$$

A Cartesian coordinate system may be used to describe physical quantities other than position—for example, velocity. Velocity has a magnitude (speed) and a direction in space. The origin corresponds

FIGURE 2.4
$\mathbf{M}$ is the vector position of Mars relative to the Sun. Its components are M_x and M_y. $\mathbf{E}$ is the vector position of Earth relative to the Sun. Its components are E_x and E_y. Their difference, $\mathbf{A} = \mathbf{M} - \mathbf{E}$, is the vector position of Mars relative to the Earth. Its components are $M_x - E_x$ and $M_y - E_y$.

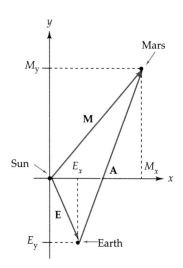

to zero velocity. Let **V** be the velocity of a river relative to land and **U** be that of a boat relative to the water. What is the velocity **W** of the boat relative to land? If the boat heads downstream, its speed is the sum of the boat speed and the river speed: $W = U + V$. If it heads upstream, its speed is $W = U - V$. Except in the very special case in which vectors lie on a line, vector addition must be used to add vectors: $\mathbf{W} = \mathbf{U} + \mathbf{V}$. The components w_x and w_y are found by summing the components of **U** and **V**:

$$W_x = U_x + V_x \qquad \text{and} \qquad W_y = U_y + V_y$$

In general, the sum of two vectors is obtained by adding their corresponding components. The difference of two vectors is obtained by subtracting their corresponding components. Moreover, the product of a scalar and a vector is the vector whose components are the products of the scalar and the components of the vector.

Notice that everything we have said about vectors and Cartesian coordinate systems can be readily extended to deal with three-dimensional space. With three dimensions, a third axis, the z-axis, passes through the origin and is perpendicular to both the x- and y-axes. In this case, a general vector is characterized by three components, as indicated in Figure 2.5. Now let us summarize the algebraic rules of vector algebra:

Adding vectors by components:

$$\text{If} \quad \mathbf{C} = \mathbf{A} + \mathbf{B}, \qquad \text{then} \quad \left\{ \begin{array}{l} C_x = A_x + B_x \\ C_y = A_y + B_y \\ C_z = A_z + B_z \end{array} \right. \qquad (1)$$

FIGURE 2.5
The vector **A** extends from the origin to a point P in space. Its three components, A_x, A_y, and A_z, are indicated on the coordinate axes.

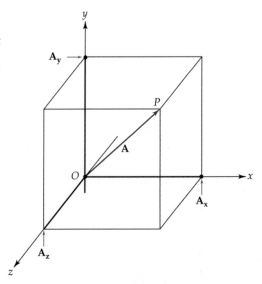

Subtracting vectors by components:

$$\text{If }\quad \mathbf{C} = \mathbf{A} - \mathbf{B}, \qquad \text{then} \quad \begin{cases} C_x = A_x - B_x \\ C_y = A_y - B_y \\ C_z = A_z - B_z \end{cases} \qquad (2)$$

Multiplying a vector by a scalar :

$$\text{If }\quad \mathbf{F} = m\mathbf{a}, \qquad \text{then} \quad \begin{cases} F_x = m\,a_x \\ F_y = m\,a_y \\ F_z = m\,a_z \end{cases} \qquad (3)$$

One last property of vectors you need to be aware of at this point concerns magnitude. From the Pythagorean theorem dealing with right triangles, it follows that the square of the magnitude of a vector is the sum of the squares of its components. Thus, the magnitude of a vector **A** is given by the formula:

$$A = \sqrt{A_x^2 + A_y^2 + A_z^2} \qquad (4)$$

Vectors have many other important properties, some of which we discuss in Chapter 3.

EXAMPLE 2.2
On an Ocean Liner

Figure 2.6 shows a ship sailing due east at 12 mph relative to land. A girl on the ship's deck runs north at a speed of 8 mph relative to the deck. As she runs, she fires a BB gun upward. Relative to her, the speed of the BB is 24 mph. What is the speed V of the BB relative to land?

FIGURE 2.6
The ship is sailing east at 12 mph relative to the water; the girl is running north at 8 mph relative to the deck. She fires a BB up at 24 mph relative to her. The speed of the BB relative to the water is the vector sum of these velocities. Its magnitude is 28 mph.

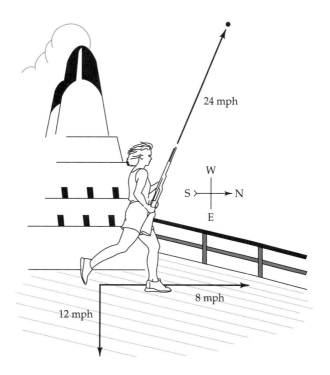

Solution   The answer is the vector sum of three velocities. Let's use a coordinate system with the x-axis pointing east, the y-axis north, and the z-axis vertically upward. Componentwise, the ship's velocity relative to land is (12, 0, 0); the girl's velocity relative to the ship is (0, 8, 0); and BB's velocity relative to the girl is (0, 0, 24), where all numbers are in miles per hour. Adding corresponding components, we find that the components of the velocity $\mathbf{V}$ of the BB relative to land are (12, 8, 24). The sum of the squares of the components of $\mathbf{V}$ is 784. The land speed of the bullet, according to equation 4, is $\sqrt{784}$, or 28 mph.   ■

EXAMPLE 2.3   A jet cruises at an airspeed of $v = 850$ km/hr. A favorable jet stream
On a Jet Aircraft   blows from 30° south of due west at a speed of $w = 260$ km/hr. The pilot heads in a direction such that the plane's velocity relative to the ground, $\mathbf{s} = \mathbf{v} + \mathbf{w}$, points due east (Figure 2.7a). What is the pilot's ground speed s?

FIGURE 2.7
a. The jet stream blows at $w = 260$ km/hr from a direction 30° south of due west. The airspeed of the plane is $v = 850$ km/hr. The velocity of the plane with respect to the ground is $\mathbf{s} = \mathbf{v} + \mathbf{w}$.
b. The components of the wind velocity, $\mathbf{w}$.

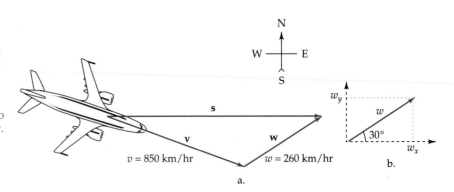

Solution   Let's use a coordinate system with the x-axis pointing east and the y-axis north. The components of the wind velocity, shown in Figure 2.7b, are

$$(w_x, w_y) = (260 \cos 30°, 260 \sin 30°) = (\sim 225, 130)$$

The pilot heads somewhat south of his intended eastward direction to ensure that the y component of his ground speed vanishes: $s_y = v_y + w_y = 0$, or $v_y = -130$. Because $v^2 = 850^2 = v_x^2 + v_y^2$, it follows that $v_x = \sqrt{850^2 - 130^2} = 840$. The velocity of the aircraft relative to the ground is then $\mathbf{s} = (v_x + w_x, 0) = (1065, 0)$. The plane heads east toward its destination at a ground speed of 1065 km/hr.   ■

Exercises

1. A gourmet restaurant is 24 mi south and 10 mi east of Duluth. How long is the direct route from the city to the restaurant?

2. A bat flies toward the northeast at 4 mph while a bug flies to the southeast at 3 mph. Both creatures have horizontal flight paths. What is the speed of the bug relative to the bat?

3. An airplane is 2 mi east, 6 mi north, and 3 mi above the airport tower. What is the distance between the plane and the tower?

4. The wind blows from the northwest at 3 mph. A bird is flying at 4 mph in a northeasterly direction relative to the wind. What is the ground speed of the bird? (Use the geometric method to add vectors.)

5. Object A has a momentum (a vector) of magnitude 10 in some unit. Object B has a momentum of the same magnitude but points in a direction 60° away from that of A. What is the magnitude of the sum of the two momenta. (Choose a Cartesian coordinate system and use the component method to add vectors.)

6. A straight, mile–wide river flows at a speed of 3 mph. A motor launch has a cruising speed of 5 mph. The launch starts at the shore and travels in a straight line across the river to the point on the shore directly opposite. Its heading is such that the vector sum of the river's velocity (relative to land) and the ship's velocity (relative to the water) is perpendicular to the shore. How long does the trip take?

2.2 Velocity and Acceleration

The science of mechanics deals with moving objects, such as a housefly as it zooms around a room in acrobatic arcs. It may flit about erratically, but at any given time t it is surely somewhere. Choose a fixed point O in the room as an origin, relative to which we may define position vectors. The location of any other point X is specified as a vector extending from O to X. Let $\mathbf{R}(t)$ be the vector that extends from O to the fly at time t. As the fly moves, this vector changes. Its tip traces the fly's trajectory.

In Figure 2.8, the fly's trajectory is shown as it sweeps by your nose N and performs a graceful loop before settling on a slice of jellied toast T. The fly passes N at time t_1. $\mathbf{R}(t_1)$ is the vector position of the fly at t_1; it extends from O to N. After swerving about, the fly reaches T at the later time t_2. $\mathbf{R}(t_2)$, extending from O to T, is its vector position at t_2. In the time interval $\Delta t = t_2 - t_1$, the fly travels from N to T. The vector distance $\mathbf{D}$ traversed is the difference between the fly's final and initial position vectors:

$$\mathbf{D} = \mathbf{R}(t_2) - \mathbf{R}(t_1)$$

FIGURE 2.8
a. The vector position of the fly relative to the origin *Ó* is $\mathbf{R}(t_1)$ at the time t_1, and $\mathbf{R}(t_2)$ at the later time t_2. The vector distance flown between t_1 and t_2 is $\mathbf{D}$. The average velocity of the fly in this time interval is $\mathbf{D}/(t_2 - t_1)$.
b. The average velocity of the fly in the time interval from t_1 (when it is at *N*) to t_A (when it is at *A*) is the vector distance from *N* to *A* divided by $t_A - t_1$. Its average velocity in the shorter interval from t_1 to t_B is the vector distance from *N* to *B* divided by $t_B - t_1$. As we examine shorter time intervals, the ratio approaches the limiting value $\mathbf{v}(t_1)$. It is the instantaneous velocity of the fly at the time t_1.

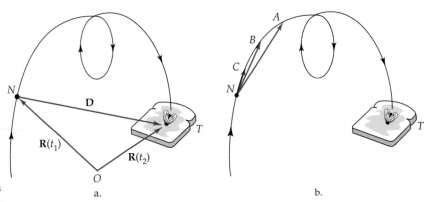

a.

b.

The *average velocity* of a moving object in a given time interval is defined as the vector distance traveled divided by the time it took. The average velocity of the fly in getting from nose (*N*) to toast is

$$\mathbf{V}_{av} = \frac{\mathbf{D}}{\Delta t} = \frac{\mathbf{R}(t_2) - \mathbf{R}(t_1)}{t_2 - t_1}$$

We define the *average speed* of the fly during its short voyage as the magnitude of its average velocity: V_{av}, or more simply, $D/\Delta t$.

Average speed and average velocity are useful concepts, but they have two drawbacks. First, average velocity depends on three different things: the starting point, the end point, and the duration of the trip. Second, average speed is rarely equal to the actual speed at any one time. If the fly had made ten loops before settling on the toast, it would have traveled further through the air on its way and taken a longer time to get there. Its average speed would have been smaller. If the fly had changed its mind and flown back to your nose, its average speed would have been zero. However, the idea of average velocity leads us to the more essential concept of *instantaneous velocity*.

Drivers know the difference between average speed and instantaneous speed. It may take four hours to drive from Boston to New York, a distance of 180 miles. The average speed is 45 mph, but the instantaneous speed (indicated on the speedometer) is sometimes larger, sometimes smaller. The average velocity points from Boston to New York, but the instantaneous velocity follows the curves in the road.

Figure 2.8b focuses on a series of points *A*, *B*, and *C* lying closer and closer to point *N* on the fly's trajectory. The average velocity of the fly as it goes from *N* to *A* is the vector distance from *N* to *A* divided by the time it takes for the fly to get from *N* to *A*. In the same way, we may determine the average velocity of the fly as it travels from *N* to *B* and from *N* to *C*. As we proceed to shorter time intervals beginning at t_1, the distance traversed becomes smaller as well. However, the ratio of these two quantities gets closer and closer to a vector that we call the instantaneous velocity of the fly at the time t_1, or $\mathbf{V}(t_1)$. It is the rate of change of the position vector at time t_1 and it points in a direction tangent to the trajectory at t_1. Its magnitude is the ratio of the distance traveled by the fly in a small amount of time divided by that time.

At each point along its trajectory, the fly moves at a certain velocity **V**. (From this point on, we use the terms *speed* and *velocity* instead of instantaneous speed and instantaneous velocity.) Each of its three components (V_x, V_y, and V_z) changes with time. Sometimes **V** points up, sometimes down, sometimes toward the toast, and sometimes away. The instantaneous speed of the fly is $V = (V_x^2 + V_y^2 + V_z^2)^{1/2}$, which changes with time as well. The motion of an object in space (such as a fly) can be complicated. For this reason, we limit ourselves to a discussion of three special but important types of motion: uniform motion, constant acceleration in one dimension, and circular motion.

Uniform Motion   With uniform motion, the velocity of an object does not change. It moves at a constant speed along a straight line, which we may choose to be the x-axis (Figure 2.9). The action takes place in one dimension. The position of the body at time t is specified by its location on the axis, which we denote simply as the function $x(t)$. For uniform motion, vectors are not needed. The position coordinate satisfies a simple algebraic equation:

$$x(t) = x(0) + Vt \qquad \text{for uniform motion} \qquad (5)$$

where V is constant velocity and $x(0)$ is the position of the object at the start. The distance $d(t)$ traveled from the starting point is $x(t) - x(0)$. For this quantity, we obtain the simpler equation:

$$d(t) = Vt \qquad \text{for uniform motion} \qquad (6)$$

FIGURE 2.9
Position versus time for an object in uniform motion. The graph intersects the x-axis at the initial position $x(0)$. The slope of the graph is the constant velocity V of the object.

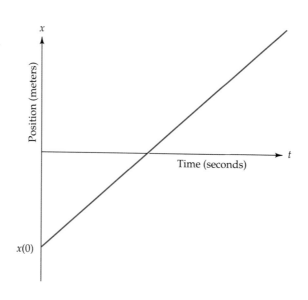

EXAMPLE 2.4
Uniform Motion

A car starts somewhere in New Mexico and drives east on Route 40 at 50 mph. The motion is uniform because both speed and heading are constant. The car passes Albuquerque at noon. We take Albuquerque as the origin of our one-dimensional coordinate system and noon as the starting point of our clock. The car's position is determined by a single coordinate, $x(t) = 50t$ miles, which tells how far east of Albuquerque the car is. The trajectory of the car is depicted as a graph of position versus time in Figure 2.10. Determine the distance from Amarillo to Oklahoma City from the data shown.

FIGURE 2.10
Position versus time for a car traveling from Albuquerque to Oklahoma City at 50 mph. The trip begins at 12:00 P.M. and ends at 11:00 P.M. The car passes Amarillo at 5:42 P.M.

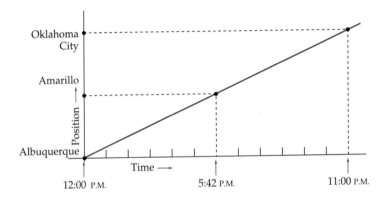

Solution

The car passes Amarillo at 5:42 P.M. and reaches Oklahoma City an hour before midnight. Thus, the drive between these cities took 5 hours and 18 minutes. Because distance traveled is the car's speed (50 mph) with the elapsed time (5.3 hours), the distance between Amarillo and Oklahoma City is 265 miles. ∎

Accelerated Motion

Uniform motion is rare in everyday life. In the real world, things often speed up, slow down, or turn. This kind of nonuniform motion involves acceleration. Acceleration results in a change of a body's speed, direction, or both.

Acceleration is the rate at which velocity changes, just as velocity is the rate at which position changes. The average acceleration in a time interval Δt is defined as the change in velocity divided by the time interval, $\mathbf{a} = \Delta\mathbf{v}/\Delta t$. It is a vector with a magnitude (measured in m/s²) and a direction. It is characterized by its three components, each with dimension m/s². If its acceleration of a body points in the same direction as its velocity, the body speeds up or accelerates. If its acceleration points in the opposite direction, it slows down or decelerates. If its acceleration points in any other direction relative to its velocity, the body describes a curved trajectory.

We begin our study of nonuniform motion with the case of constant acceleration in one dimension. The vectors corresponding to position, velocity, and acceleration point in the same direction and the motion takes place on the x-axis. A graph showing velocity versus time is a straight line (Figure 2.11), whose slope is the acceleration a. In a time interval t, the velocity changes by at. If the velocity is $V(0)$ when $t = 0$, then its velocity at another time t is:

$$V(t) = V(0) + at \qquad \text{for constant acceleration} \qquad (7)$$

FIGURE 2.11
The velocity of an object experiencing constant positive acceleration. In the case shown, the initial velocity is zero, $v(0) = 0$, and the acceleration is 0.75 m/s². At $t = 40$s, the object attains an instantaneous velocity of 30 m/s. However, its average velocity in this period of time is 15 m/s.

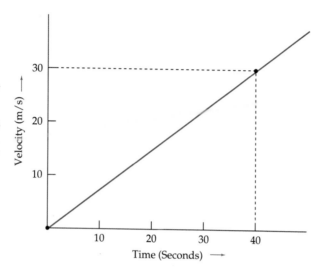

How far has the the body traveled up to time t? The average velocity V_{av} of a body experiencing constant acceleration equals half the sum of its initial and final velocities, $V_{av} = \frac{1}{2}(V_i + V_f)$. In the case at hand, $V_i = V(0)$, and from equation 7, $V_f = V(0) + at$. Average velocity, therefore, is

$$V_{av}(t) = V(0) + \tfrac{1}{2}at \qquad \text{for constant acceleration} \qquad (8)$$

The distance $d(t)$ that a body travels is the product of average velocity and elapsed time. Multiplying both sides of equation 8 by t, we find:

$$d(t) = V(0)t + \tfrac{1}{2}at^2 \qquad \text{for constant acceleration} \qquad (9)$$

If $a = 0$, velocity is constant and $d(t)$ is a linear function of t, as in equation 6. Otherwise, equation 9 is a quadratic function of t. Figure 2.12 is a graph of distance traveled versus time elapsed for an object moving (on page 62) with constant positive acceleration. The graph is not a straight line but a parabola whose slope continually increases as the object moves faster.

EXAMPLE 2.5
Constant Acceleration   A car at rest begins to move at $t = 0$ with a constant acceleration of 0.75 meters per second per second, or m/s². Its velocity increases by 0.75 m/s per second, so that at any time t (in seconds), $V(t) = 0.75t$ m/s (see Figure 2.11). How far does the car travel in 40 s?

FIGURE 2.12
The distance traveled by the object whose velocity is shown in Figure 2.11. Starting at rest, and accelerating at 0.75 m/s², the object travels 600 m in the first 40 s.

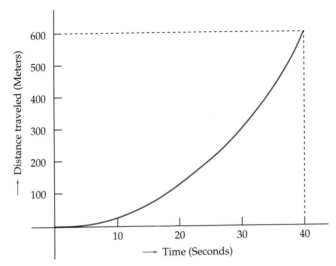

Solution   To answer this question, we need to know the average velocity of the car in the interval $0 < t < 40$ s. The velocity starts at zero and reaches 30 m/s at $t = 40$ s. In the case of constant acceleration, the average velocity in any time interval equals the average of the initial and final velocities. In this case, it is $\frac{1}{2}(0 + 30)$ or 15 m/s. The distance d traveled is the product of average velocity and elapsed time, $d = 600$ m.   ■

EXAMPLE 2.6
The Peril of Paul   The conductor sees Paul strapped to the tracks 1 km ahead (Figure 2.13). The train is hurtling toward disaster at velocity $v = 60$ m/s. What is the least deceleration of the train in m/s² that averts an accident?

Solution   A decelerating train experiences negative acceleration, $-b$. According to equation 7, with $a = -b$, the train's velocity, once it begins to brake, is $v - bt$. The velocity continues to decrease until the train stops at $t = v/b$. The average velocity of the train during this time is half the

FIGURE 2.13
The Peril of Paul. A train moving at a velocity of 60 m/s approaches a person strapped to the tracks 1 km ahead.

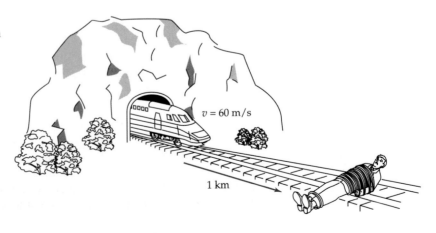

$v = 60$ m/s

1 km

sum of its initial and final velocities, or $\frac{1}{2}v$. The distance it travels before stopping is the product of its average velocity and the time interval; therefore, $d = v^2/2b$. To save Paul, d must be less than 1 km:

$$v^2/2b < 1000 \text{ m} \qquad \text{or} \qquad b > (60)^2/2000 = 1.8 \text{ m/s}^2$$

∎

Uniform Circular Motion

Many things move in circles: the valve of a spinning bicycle wheel, your seat on a Ferris wheel, the wooden horses at a merry-go-round, and so on. Circular motion is important in astronomy and space science as well. Planets move around the Sun in almost circular orbits, just as the Moon and artificial satellites move in circles about Earth. Indeed, any point on Earth's surface describes a circle as Earth turns on its axis. Circular motion is just as important in the microworld of atoms and elementary particles. Charged particles move in circles when they find themselves in a constant magnetic field. So do atomic electrons (according to classical physics) under the influence of the electrical attraction of an atomic nucleus.

Suppose that a body moves at a constant speed V, in a direction that changes continually, as in Figure 2.14. In particular, suppose it moves along the periphery of a circle of radius R. The time the body takes to complete one circuit around the central point is called its *period P*. The period equals the circumference of the circle divided by the body's speed, or $P = 2\pi R/V$.

FIGURE 2.14
An object moves at a constant speed in a circular orbit. The direction of its instantaneous velocity is shown at several points by arrows.

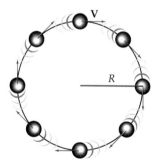

In uniform circular motion, the magnitude of the acceleration stays constant. The quantity V^2/R not only has the dimension m/s^2 of acceleration, but it is the acceleration experienced by a body moving in a circle of radius R and with speed V.

$$a = \frac{V^2}{R} \qquad \text{for circular motion} \tag{10}$$

More precisely, a is the magnitude of the acceleration vector. Its direction changes such that it is perpendicular to the velocity at every moment. If a body moves at constant speed in a circular orbit, its acceleration is *centripetal*, meaning that it points to the center of the circle. The speed of a body in uniform circular motion remains constant while its direction continually changes. Notice that the velocity, if it is averaged over a full period, vanishes because the body has come back to its starting point.

EXAMPLE 2.7
The Circular Motion of Earth

Earth's orbit is actually an ellipse, but it is close to being a circle. In Figure 2.15, we treat it as a circle with a radius $R \simeq 1.5 \times 10^{11}$ m. Its period is one year, or $\sim 3.2 \times 10^7$ s. What is its centripetal acceleration?

Solution

First we determine the speed of Earth. In the course of a year, having traversed the $2\pi R$ circumference of its orbit, Earth is back where it began. Its speed is $V \simeq 3.0 \times 10^4$ m/s. From equation 10, we find for its acceleration, $a \simeq 6 \times 10^{-3}$ m/s^2. ∎

FIGURE 2.15
The velocity of Earth is always tangent to its orbit, which is very nearly a perfect circle of radius $R = 1.5 \times 10^{11}$ m. Its acceleration points toward the Sun.

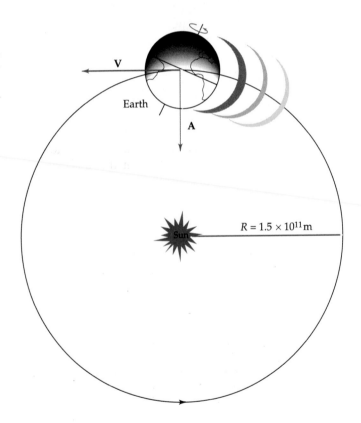

Earth

V

A

$R = 1.5 \times 10^{11}$ m

Sun

Exercises

7. Using the data in Example 2.4, find the distance between Albuquerque and Amarillo.

8. A Camaro accelerates uniformly from rest for 30 s, covering a distance of 1.5 km. What is its acceleration? What is its final velocity?

9. Starting from rest at $t = 0$, a train accelerates at a uniform acceleration of 0.8 m/s².
    a. What is its velocity at $t = 5$ s?
    b. What is its velocity at $t = 10$ s?
    c. What is its average velocity in the interval 0 s $< t < 10$ s?
    d. What is its average velocity in the interval 5 s $< t < 10$ s?
    e. How far has the train traveled in the first 5 s?
    f. How far has the train traveled in the first 10 s?

10. A train starts from rest at $t = 0$ with constant acceleration a in m/s².
    After 8 s, it has traveled a distance of 40 m.
    a. What is its average velocity over the first 8 s of its voyage?
    b. What is its velocity at $t = 8$ s?
    c. What is its acceleration a?
    d. What is its velocity at $t = 12$ s?
    e. What is its average velocity over the first 12 s?
    f. How far has it traveled in the first 12 s?

11. A spaceship begins at rest relative to Earth. It then moves in a straight line with a constant acceleration equal to that of gravity on Earth, 10 m/s². Show that its velocity, after one year of travel, is approximately the speed of light, about 3×10^8 m/s.

12. A toy train traverses a circular track 10 m in circumference at a constant speed of 1 m/s.
    a. How long does it take the train to complete a circuit? What is its average velocity over this interval?
    b. What is the magnitude of its average velocity over any 5-s time period?
    c. What is the magnitude of its centripetal acceleration?

13. A body moves at a constant speed in a circular orbit with a 10-m radius. It takes 30 s for the body to return to its starting point.
    a. What is the magnitude of its average velocity over a 15-s time interval?
    b. What is the speed of the body?
    c. What is the magnitude of its centripetal acceleration?

14. Analogs to velocity and acceleration are commonplace. The following table lists the population of small but growing Mudville in six consecutive years. The population is growing, but its rate of growth is slowing. The population of the town is like a position and its annual increase like a velocity:

Year	Population	Annual Increase
1970	10,000	n.a.
1971	11,400	1,400
1972	12,600	1,200
1973	13,600	1,000
1974	14,400	800
1975	15,000	600

a. What is the average annual population increase in the 5-yr period from 1970 to 1975? Show that the population increase over the prior year is given by $V(t) = 1600 - 200t$, where t is time in years since 1970. If we regard V as an annual average velocity, what is the analog of acceleration? Show that the population of Mudville, $P(t)$, is described by the formula

$$P(t) = 10,000 + 1500t - 100t^2$$

For the remaining parts of this problem, assume that this formula applies throughout the life of Mudville.

b. In what year was Mudville founded?
c. In what year was Mudville abandoned?
d. In what year did Mudville's population peak?

15. An elevator is at rest in the lobby of a skyscraper. It accelerates uniformly upward for 6 s at an acceleration of 3 m/s^2. It then coasts at constant speed for 10 s. At that time, it decelerates at -2 m/s^2 for 9 s. It is then at rest. Plot the acceleration of the elevator as a function of time. Plot the velocity and the position. What height above the lobby has it reached?

2.3 Copernicus and Kepler

In Section 2.2, we examined the description of motion. We learned how far and how fast a body moves if its acceleration is constant. We learned that uniform circular motion involves an acceleration that is constant in magnitude but centripetal. We did not learn about the causes of acceleration: of the forces bodies exert on one another by contact or at a distance. Nor did we learn how to determine the motion of a body from the forces that act on it. We begin our examination of these subjects in this section.

The creation of a science of mechanics depended on many developments in both astronomy and earthly physics. Precise astronomical observations led Copernicus and Kepler to ascertain the true nature of planetary motions; Galileo discovered the rules governing the motion of falling or thrown bodies; and Newton put it all together and constructed a unified theory of celestial and terrestrial mechanics.

Although it may seem obvious to us now that Earth and the planets circle the Sun, this was not always the case. Many ancient societies believed that the Sun, the Moon, and the stars move in circles about a stationary Earth. Just as we do, they saw these bodies traverse the heavens in concentric circles, as if they were embedded on rotating celestial spheres.

Rather than simply circling us, planets occasionally display an apparent *retrograde motion*, turning about and moving backward across the sky,

FIGURE 2.16
An example of the retrograde motion of Venus. The stars move across the night sky as if they were embedded in a celestial sphere. Planetary motion is more complex because the planets and the Earth move around the Sun. Planets sometimes display retrograde motion and move in loops relative to the stars. This drawing shows the apparent trajectory of Venus in 1983.

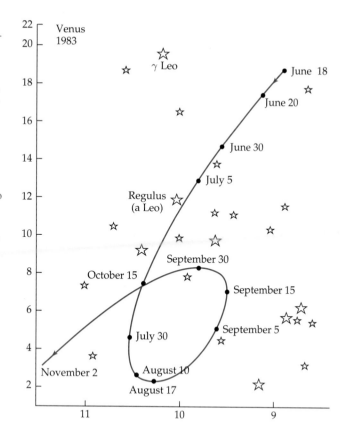

as shown in Figure 2.16. The word *planet* stems from Greek for wanderer and planets are indeed wanderers. To explain retrograde motion, Greek astronomers devised ingenious tricks to preserve the mandatory notion of circular motion. They showed how the paths of the planets could be described by circles within circles (epicycles), circles not centered about the stationary Earth (eccentrics), and circles traversed at varying speeds (equants). Eudoxus, in the fourth century B.C., postulated the existence of 27 crystalline spheres to describe the planets' behavior. Ptolemy (A.D. 100–170), in *The Almagest*, codified the work of the Greek astronomers in a system that survived for over a millenium.

Earth is immobile at the center of the Ptolemaic universe. Past and future planetary configurations could be anticipated and the times and dates of eclipses predicted. Occasional discrepancies between prediction and observation were resolved through tiny adjustments of the 70 simultaneous and independent circular motions of the seven heavenly bodies. Ptolemy's system was accepted even though it was intricate, contrived, and understood by very few. It worked, and it put Earth smack in the middle, where everyone knew it belonged. When the Dark Ages ended, Arab guardians of this ancient wisdom returned it to Europe, and the Ptolemaic system became part and parcel of Christian dogma.

The Contribution of Copernicus

Not until the late fifteenth century did anyone seriously question the Ptolemaic system. Nicolaus Copernicus (1473–1543) was a devout scholar and a canon of the Church. A firm believer in the primacy of circular motion, he was nonetheless bothered by the complexity and arbitrariness of the Ptolemaic system:

> The planetary theories of Ptolemy and most other astronomers, although consistent with numerical data seemed to present no small difficulty. Having become aware of these defects, I often considered whether there could perhaps be found a more reasonable arrangement of circles, in which everything would move uniformly about its proper center as the rule of absolute motion requires.

Nicolaus Copernicus (1473–1543) put forward the hypothesis that the Earth moves around the Sun. Most others at that time held the converse view and believed that Earth is at rest at the center of the universe. *Source:* Historical Pictures/Stock Montage, Inc.

Copernicus knew that the price for a simpler system was to put the Sun at center stage and have a rotating Earth be just another planet revolving around it. Instead of putting forth such a radical notion himself, however, he applied to authority:

> According to Cicero, Nicetus had thought the earth moved; according to Plutarch, certain others had held the same opinion. I myself also began to meditate upon the mobility of the earth. And although it seemed an absurd opinion, yet, because I knew that others before me had been granted the liberty of supposing whatever circles they chose in order to demonstrate the observations concerning the celestial bodies, I considered that I too might well be allowed to try whether sounder demonstrations of the revolutions of the heavenly orbs might be discovered by supposing some motion of the earth.

The Copernican system was as predictive as Ptolemy's, but less contrived. In his theory, each planet, Earth included, moves on a sphere centered about a fixed Sun (although a few small epicycles and an eccentric or two were added to give quantitative precision). This model of the solar system is heliocentric (Sun-centered), not geocentric (Earth-centered). Copernicus's vision of the heavens was consistent with his firm religious beliefs. He regarded his theory as a reflection of the mind of God, its greater simplicity a reaffirmation of the principles of Aristotle. Yet, Copernicus made few converts. Both the Roman Catholic church and the Reformed Protestant church were committed to the geocentric view as it is established in scripture:

> Then spoke Joshua to the Lord in the day when the Lord delivered up the Amorites before the children of Israel, and he said in the sight of Israel, Sun, stand thou still upon Gibeon; and thou moon in the valley of Ajalon. And the sun stood still, and the moon stayed, until the people had avenged themselves.

If God could stop the Sun, then surely it must move! The Copernican system was a useful computational tool (to set the dates for Easter and other holy days), but it was heresy to take it literally. Copernicus's writings were put on the index of forbidden books as "false and altogether opposed to Holy Scriptures," where they remained until 1835.

Kepler's Laws   Johannes Kepler (1571–1630) was a convinced Copernican whose aim was to perfect the heliocentric theory. Even more than Copernicus, he believed in the simplicity of natural law. He wondered why there were precisely six planets, no more and no less. Could the structure of the universe result from basic mathematical principles? Kepler knew there were just five regular geometric solids: tetrahedron, cube, octahedron, dodecahedron and icosahedron. If these are inscribed within a series of six nested spheres, as shown in the photograph, the radii of the spheres are approximately proportional to the solar distances of the planets. Although we now know that there are nine planets in the solar system and that Kepler's cosmology merely reflects an accidental coincidence, the idea reflects the boldness of his imagination, which, later in his life, yielded his three great laws of planetary motion.

Telescopes were not yet invented when Kepler searched for the laws of nature. But the quality of astronomical data available to him was far superior to what had been available to Copernicus. This was because of the painstaking work of the Danish astronomer Tycho Brahe (1546–1601), who devoted his life to measuring, night after night and year after year, the positions of stars and planets in the night sky. He designed and constructed the best instruments possible at that time to augment his naked–eye observations and collected a vast body of very precise data. Measurement errors in the positions of stars and planets in the sky were reduced to a tenth of what they had been. Kepler, who had been Brahe's assistant, tried desperately to fit the new data on Mars's orbit to his beloved Copernican theory. After four years of study, however, he found that the deed could not be done—the theory of Copernicus was wrong. *"The matter is obviously this,"* said Kepler, *"the planetary orbit is no circle; to both sides it goes inward and then outward. Such a figure is called an oval."*

Kepler's faith in the beauty and simplicity of nature was vindicated. Nonetheless, he proved that the orbits of planets are ellipses. An *ellipse* is the locus of all points the sum of whose distances from two fixed points

Johannes Kepler (1572–1630) proposed a mystical model of the universe in 1597. The outermost sphere, on which Saturn lies, circumscribes a cube in which Jupiter's sphere is inscribed. This bounds a tetrahedron in which Mars' sphere is placed, and so on. A decade later, after careful analysis of astronomical data, he discovered two of his three laws of planetary motion. He found the third law in 1618. *Source:* Courtesy Meyers Photo-Art.

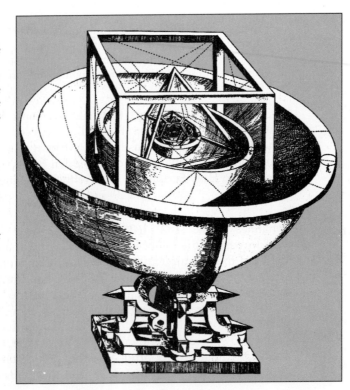

A BIT MORE ABOUT
Kepler and Brahe

Their lives preceded the institution of the Nobel Prize by three centuries, and there is no prize for astronomy anyway. Yet the rewards Johannes Kepler and Tycho Brahe received were even greater. On November 11, 1572, Brahe saw a brilliant new star appear in the constellation Cassiopeia. A few years later, on September 30, 1604, Kepler saw his own new star in the heavens. Both stars have since faded, but they are forever associated with these great astronomers.

The new stars were *supernovæ*, events in which a normal star suddenly becomes billions of times brighter than it was, only to fade away in a few weeks. These titanic stellar calamities are rare occurrences. A large star becomes a supernova when it runs out of nuclear fuel and can no longer resist the force of gravity. Its catastrophic implosion is followed by an explosive bounce. One of the most spectacular supernovæ was seen in A.D. 1054 by Chinese observers, who wrote that the star was bright enough to be seen all day and to rival the Moon at night. Although Indians of the American Southwest may have celebrated the event in their art, no European report of this event survives: Nary a soul in medieval Christendom dared admit the heavens could change.

Brahe's and Kepler's new stars are the most recent supernovæ in our galaxy, but in 1987 a new supernova burst forth in the Larger Magellanic Cloud, an appendix to our galaxy. A faint blur in a large telescope suddenly became a brilliant new star, visible to the naked eye. The photographs show how it appeared through a telescope before and after. The new supernova was easily seen in the southern hemisphere, but it was also detected as a burst of *neutrinos* in Japan and Ohio. Kepler, who was as much an astrologer as an astronomer, attributed his new star to the occurrence of a rare triple conjunction of Mars, Jupiter, and Saturn. We have come a long way since then, and we will learn more about supernovæ and neutrinos in Chapter 14.

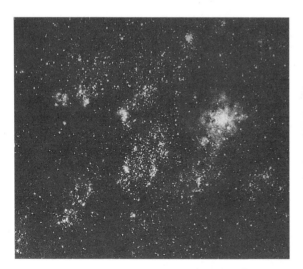 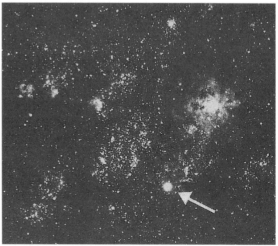

The appearance of the supernova in 1987 in the Large Magellanic Cloud (visible only from the Southern Hemisphere). The photo on the right was taken after the photo on the left. *Source:* Courtesy Drs. Barry F. Madore and Wendy L. Freedman, Observatories of the Carnegie Institution of Washington.

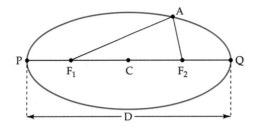

FIGURE 2.17
An ellipse is the set of all points (such as A), the sum of whose distances from two foci (F_1 and F_2) is a constant. In the special case where the foci coincide, the ellipse simplifies to a circle of diameter D. The value of D is the distance between the points P and Q. It is known as the major axis of the ellipse. The distance from P or Q to the center of the ellipse C is called its semimajor axis.

(F_1 and F_2 in Figure 2.17) is a constant. The fixed points are called the *foci* of the ellipse. The line from P to Q passing through both foci and the center of the ellipse C is called its *major axis*. The length of the line extending from C to either P or Q is often referred to as the *semimajor axis* of the ellipse.

Kepler's first law states that every planet describes an ellipse, with the Sun at one focus (Figure 2.18). No Ptolemaic tinkering is needed: Eccentrics and epicycles joined equants in the great garbage heap of discarded physical theories. This law replaced Aristotle's principle of circular motion. Kepler found no conflict between his religious faith and his science because he believed that the planets are imperfect material bodies, like Earth, and are permitted to move in imperfect ellipses rather than in perfect circles.

Kepler demolished another of Aristotle's principles as well. A planet does not travel at a constant speed in its orbit, but instead moves faster when closer to the Sun and slower when further away. More specifically, Kepler showed that a planet moves at such a speed that an imaginary line joining the planet to the Sun sweeps out equal areas in equal times, as shown in Figure 2.18. This result is known as *Kepler's second law*. Kepler's second law is a special case of the law of conservation of rotary

FIGURE 2.18
Kepler's first law says that a planet moves along an ellipse with the Sun at a focus. His second law says that the planet sweeps out equal areas in equal times. It takes the planet the same time to travel from A to B as from C to D. The two shaded regions have the same area.

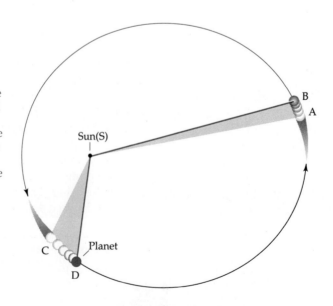

motion or angular momentum, something we shall learn more about in Chapter 3.

Kepler's Third Law

Kepler's third law is the most remarkable of his achievements. It has no Aristotelian precedent. Unlike his first two laws, which are concerned with the orbit of any one planet, the third law links the orbital properties of the different planets to one another. He came upon the law "after unceasing toil through a long period of time" until:

> At last the true relation overcame by storm the shadows of my mind, with such fullness of agreement between my seventeen years' labor on the observations of Brahe and this present study of mine that I at first believed that I was dreaming.

His stunningly simple result relates the size of a planet's orbit to its period. Kepler determined the semimajor axis R and the period P for each of the planets he knew of: Mercury, Venus, Earth, Mars, Jupiter, and Saturn.* In each case, he found that the ratio of the cube of the semimajor axis to the square of the period is the same number. This law is simply put as a formula:

$$R^3/P^2 = K_\odot \qquad (11)$$

(The subscript $\odot$ indicates the Sun, about which all the planets orbit.) The constant $K_\odot$ is the same for every planet. Mars is about four times further from the Sun than Mercury. Its period (the Martian year) is about eight

---

DO YOU KNOW

Why Summer Is Longer than Winter?

Spring begins at the spring equinox, which occurs on or about March 20. Fall begins at the fall equinox, around September 22. (The exact dates and times of the equinoxes vary from year to year.) The portion of the year consisting of the warmer seasons (spring and summer) is about a week longer than the remaining part of the year! How can this be?

If the orbit of Earth were a perfect circle, each of the four seasons would be exactly equal in length. However, its orbit is not a circle but an ellipse. Earth is about 3 percent closer to the Sun in January than it is in July. Kepler's second law tells us that Earth's orbital speed is about 3 percent swifter in midwinter than in midsummer. That's why the time between the fall equinox and the next spring equinox is shorter than the time between the spring equinox and the next fall equinox.

---

*Every planetary orbit, except for those of Mercury and Pluto, is almost a circle, with a radius nearly equal to its semimajor axis.

times greater than Mercury's. In discussing planetary orbits, astronomers often use the sidereal year as a measure of time and the semimajor axis of Earth's orbit as a measure of distance: 1 AU (astronomical unit) $\simeq$ 1.50×10^{11} m. The value of $K_{\odot}$ may be specified in these units, or it may be converted to SI units:

$$K_{\odot} = 1 \text{ AU}^3/\text{y}^2 \simeq 3.36 \times 10^{18} \text{ m}^3/\text{s}^2 \tag{12}$$

Kepler presented his three laws of planetary motion in a book entitled *The Harmony of the World.* It ends with a flourish:

> I write my book to be read either by present–day or by future readers, what does it matter? It may wait 100 years for its reader, since God Himself has been waiting 6000 years for one who penetrated His work.

Kepler did not deduce his laws from a consistent theoretical framework— this would be done a century later by Newton. Rather, he searched for and found his relationships by careful analyses of observational data and inspired guesswork. He was the among the most brilliant theoretical astronomers of any age, just as Brahe was among the greatest of observational astronomers.

Other Orbiting Bodies

Kepler's laws pertain to every object that orbits the Sun: the nine planets, the hundreds of asteroids and comets, and the innumerable rocks and

---

A BIT MORE About Planets

The seventh planet was discovered by William Herschel, the son of an oboist in the Hanoverian Footguards Band and an accomplished musician himself. Herschel built the best telescopes of his day and with them scanned the heavens. In 1781, he saw what seemed to be a comet, a small intruder into the solar system. Within months, its planetary nature became clear. Herschel called his new planet *The Georgian* to honor his king and patron, and so it was known in Britain. The French called it *Herschel*, after its discoverer. Swedes and Russians suggested Neptune, but a careful reading of mythology revealed a better choice: Uranus, god of the sky and husband to Earth, was father to Sat-

urn and grandfather to Jupiter, who begat Mars, Venus, Mercury, and Apollo (or the Sun). Today, Uranus is the name by which Herschel's planet is known.

Uranus was found by chance, but Neptune's discovery was a triumph of theory. Careful observations in the early nineteenth century revealed irregularities in Uranus's orbit. Two theoretical astronomers attacked the problem. They predicted an eighth planet whose gravitational force changed the Uranian orbit. The paper John Couch Adams left at the Royal Observatory on October 21, 1845, began:

> According to my calculations, the observed irregularities in the motion of Uranus may be accounted for by supposing the existence of an exterior planet, the mass and orbit of which are as follows . . .

dust motes that roam the solar system. The orbit of each of these bodies is an ellipse, with the Sun at one focus, in accord with the first law. A hypothetical line connecting each body to the Sun sweeps out equal areas in equal times, in accord with the second law. For each of the orbits, the ratio of the cube of its semimajor axis to the square of its period has the value $K_\odot$, given in equation 12.

Kepler's laws have an even wider range of applicability, however. They describe the motion of any body that is gravitationally bound to a much more massive primary body. For example, they describe objects that orbit Earth. The only such object that Kepler knew of was the Moon. Indeed, it travels in an ellipse with Earth at one focus and sweeps out equal areas in equal times. However, the ratio of the cube of the semimajor axis of its orbit (3.85×10^8 m) to the square of its sidereal period (2.36×10^6 s) does not equal $K_\odot$. For the Moon, we find:

$$R^3/P^2 = K_\oplus \simeq 1.02 \times 10^{13} \text{ m}^3/\text{s}^2 \tag{13}$$

(The symbol $\oplus$ is the astronomer's symbol for Earth.) This result applies to any body that is in orbit about Earth, including hundreds of artificial satellites. Some of these latter observe the weather, others gather intelligence for scientists and spies, and still others relay information from one part of the world to another. In each case, equation 13 relates the period of the satellite to its distance from Earth.

A few months later, Urbain Leverrier, in Paris, prepared a paper titled *"On the Planet which Produces the Observed Anomalies in the Motion of Uranus—A Determination of its Mass, Orbit and Position in the Sky."* Adams and Leverrier both explained where and when to search and what to look for.

On September 25, 1846, the director of the Berlin Observatory announced the discovery of the new planet, saying it was "the most outstanding proof of the validity of [Newton's] universal gravitation." French scientists at first called the new planet Leverrier, but this name was rejected elsewhere. Eventually, the name Neptune was agreed upon.

Small irregularities in the motions of the outer planets persisted. In 1915, Percival Lowell predicted a ninth planet, but because a powerful enough telescope was not yet available to detect this faint and distant object, he did not find it. Pluto was finally discovered by Clyde Tombaugh at the Lowell Observatory at Flagstaff, Arizona, in 1930, and announced on Lowell's birthday. Its symbol P is constructed from its champion's initials.

The discovery of Pluto ends our tour of the planets. The spectacular photos returned by the Voyager spacecraft of the outer planets and their many moons would have brought tears of ecstasy to Kepler's eyes. They do just that to lesser mortals.

EXAMPLE 2.8
Geosynchronous
Satellites

An artificial Earth satellite (and its orbit) is said to be geosynchronous if the satellite remains in the same position in the sky relative to an earthbound observer. A geosynchrous orbit must be circular; it must lie in the plane of the equator; and the satellite must move in the same direction that Earth turns about its axis (Figure 2.19). How far from the center of Earth is such a satellite?

FIGURE 2.19
A geosynchronous satellite hovers over the same point on Earth. It completes its equatorial orbit (P) in the same time that Earth turns about its axis. That is, its period equals a sidereal day. The satellite is at a distance R from Earth.

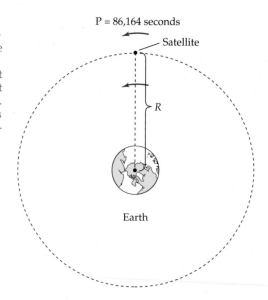

P = 86,164 seconds

Satellite

R

Earth

Solution

If the satellite is to remain fixed in the sky, its period P must equal the time it takes Earth to rotate once about its axis. This time interval, called the sidereal day, was shown in Chapter 1 to be about 86,164 seconds long. According to equation 13, the distance R from a geosynchronous satellite to the center of Earth is the cube root of the product of P^2 and $K_{\oplus}$. Carrying out the arithmetic, we obtain $R \simeq 42{,}300$ km. ∎

Kepler could not apply his laws to artificial satellites. But can you imagine his joy when he learned from his contemporary, the Italian scientist Galileo, of the existence of four moons circling the planet Jupiter? Careful observations of their motions showed that they satisfied his laws. What an astonishing triumph of theory in a new and unexpected domain! The Jovian moons that Galileo discovered are now called Io, Europa, Ganymede, and Callisto. They travel in elliptical orbits, with Jupiter at one focus, and sweep out equal areas in equal times. For each of their orbits, the ratio of the cube of the semimajor axis to the square of the period comes out to be the same constant:

$$R^3/P^2 = K_J \simeq 3.20 \times 10^{15} \text{ m}^3/\text{s}^2 \qquad (14)$$

The constant K_J appearing in this application of Kepler's third law lies between $K_\oplus$ and $K_\odot$. Newton later showed that the value of this constant (and, consequently, of R^3/P^2 for any particular orbiting body) is proportional to the mass of the heavy central primary body. If we denote the mass of the Sun by $M_\odot$, of Jupiter by M_J, and of Earth by $M_\oplus$, we learn from equations 12, 13, and 14 that

$$\frac{M_\odot}{M_J} = \frac{K_\odot}{K_J} \simeq 1000 \qquad \text{and} \qquad \frac{M_J}{M_\oplus} = \frac{K_J}{K_\oplus} \simeq 300 \qquad (15)$$

Kepler's third law (with a little help from Newton) tells us that the Sun is about 1000 times as massive as Jupiter, and that Jupiter itself is about 300 times as massive as Earth.

Why Kepler's laws describing the motions of planets (and, it turned out, the motions of planetary satellites) work could not be answered until Newton formulated his theories of motion and gravity. Kepler did not know that the natural state of motion of an undisturbed object, on Earth or in the heavens, is uniform, straight–line motion, or that the motions of bodies in the heavens and on Earth are controlled by the force of gravity.

Kepler's laws are not exact laws of physics. They are approximations to the truth, valid under special circumstances: The primary body must be much more massive than its satellites, and the satellites must be small enough, or far enough apart, that they do not much affect one another. The actual, precise motions of heavenly bodies are far more complex than Kepler could possibly have imagined. The orbit of any one body is ever so slightly affected by every other body in the solar system. These effects are tiny for the planets because the Sun is so massive. Newton's theory of mechanics allows scientists to take these effects, called *perturbations*, into account in their calculations. On several occasions, such as at the discovery of Neptune, the appearance of unexpected perturbations led to astonishing discoveries. Kepler's laws are important to the history of physics for three reasons: (1) They approximately describe the motions of many bodies in the heavens, (2) they do not describe them perfectly, and (3) they begged for a theoretical explanation based on fundamental precepts.

Exercises

The information in the following tables is used in the exercises for this section. Table 2.1 gives the orbital properties of the planets to three significant figures. Their periods are given in years. The semimajor axes of their orbits are given in astronomical units (AU) of $\sim 1.5 \times 10^{11}$ m. The AU is defined to be the semimajor axis of Earth's orbit. Except for Mercury and Pluto, the orbits of all the planets are very nearly circular. Table 2.2 gives the orbital properties of the four largest moons of Jupiter. Their orbits are also nearly circular.

16. Using the data in Table 2.1, calculate the value of R^3/P^2 in units of AU^3/y^2 for the first eight planets, and find the period of Pluto's orbit.

17. A hypothetical planet circles the Sun in 1000 y. How far is it from the Sun in AU?

18. Using the data in Table 2.2, compute the values of R^3/P^2 for Io, Europa, and Ganymede, and find the period of Callisto's orbit.

DO YOU KNOW

What Kepler's Third Law Says about the Universe?

The stars you see at night are a few of the billions that make up our galaxy: the Milky Way. There are billions of other galaxies in the universe. The stars and clouds of gas in a galaxy move in roughly circular orbits about the center of their galaxy, just as the planets of the solar system move around the Sun. Kepler's third law—suitably modified—applies to their motions as well.

The *rotation curve* of a system of orbiting bodies is a graph of orbital speeds V versus distances R from the center. Equation 11 may be recast to express V in terms of R: $V = 2\pi \sqrt{K_\odot/R}$. The rotational speeds of the planets are proportional to the reciprocal square roots of their distances from the Sun.

A galaxy differs from a solar system in structure as well as scale. Its mass is spread out over an enormous volume rather than concentrated at the very center. Its rotation curve, therefore, must differ from that of the solar system. Let $M(R)$ denote the amount of mass contained within a sphere of radius R about the center of a galaxy. Kepler's third law, for a spread out distribution of mass, says that

$$V \propto \sqrt{M(R)/R}$$

Starting at the center of the galaxy, the value of V was expected to increase from zero to a maximum value at a distance R_0 from the galactic center such that a sphere of that radius contains all of the stars in that galaxy. At larger distances, there are no more stars and only a tenuous cloud of orbiting gas molecules. At these distances, $V(R)$ was expected to fall off with $1/\sqrt{R}$. However, measurements of the orbital speeds of gas clouds do not agree with expectations, as you see in Figure 2.20. The rotation curve of a galaxy rises as the amount of mass interior to the body's orbit increases. The big surprise is the behavior of the rotation curve in the region far beyond its star-studded disk—it doesn't fall off at all with increasing distance.

Hardly any physicist or astronomer would abandon Kepler's law and sacrifice an otherwise triumphant theory of gravity. We need another explanation. And the only other explanation is that $M(R)$ increases directly with R to distances far beyond the visible galaxy. In other words, there is mass—and lots of it—where there is hardly any light. By applying Kepler's third law, astronomers found that 90 percent or more of the mass of a galaxy consists of a vast and invisible halo of dark matter that is not in the form of luminous stars. The dark matter does not seem to consist of stars, planets, asteroids, dust, gas, or golf balls. Elementary particle physicists have come to the rescue with more suggested solutions than you can shake a stick at and with names you never heard of. This wealth of so-called solutions, however, is an admission of failure: Nobody yet knows what the dark stuff is that dominates the mass of the universe.

TABLE 2.1 Planetary Orbits

Planet Name	Radius in AU	Period in Years
Mercury	0.387	0.241
Venus	0.723	0.615
Earth	1.00	1.00
Mars	1.52	1.87
Jupiter	5.21	11.9
Saturn	9.54	29.5
Uranus	19.2	84.0
Neptune	30.0	164
Pluto	39.4	?

TABLE 2.2 Orbits of Jupiter's Moons

Satellite Name	Radius in AU	Period in Years
Io	2.82×10^{-3}	4.85×10^{-3}
Europa	4.48×10^{-3}	9.72×10^{-3}
Ganymede	7.15×10^{-3}	19.6×10^{-3}
Callisto	12.57×10^{-3}	?

FIGURE 2.20
a. The rotation curve of a galaxy, showing the visible portion of the galactic disk, seen edge on. It consists of luminous stars that rotate about the center of the galaxy, so stars on the right recede from us while those on the left move toward us. The observed velocities of stars and gas clouds are shown as a solid curve. The dashed curve is what would be expected if most of the galaxy's mass were in the form of stars. Astronomers conclude that galaxies are surrounded by vast clouds of unseen dark matter. b. The rotation curve of the solar system. The orbital velocities of the planets are plotted versus their distances from the Sun. The velocity depends on one over the square root of the distance.

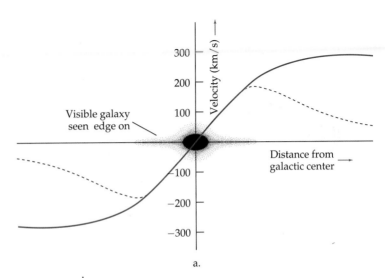

a.

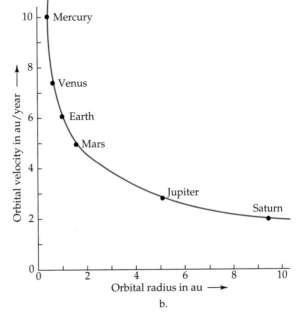

b.

19. Use equations 12 and 13 to find the ratio of the mass of the Sun to the mass of Earth.

20. Calculate the area swept out by a line connecting the center of Earth with the center of the Sun in 1 ms (microsecond). Compare your result with the area of Rhode Island.

21. What is the sidereal period in hours of an Earth satellite in a low equatorial orbit of radius 6600 km? If it travels in the same direction as Earth rotates, what is the time interval between successive sightings by an observer fixed on Earth?

22. Use Kepler's third law, equation 11, to express the speed of a planet in a circular orbit around the Sun in terms of its distance from the Sun.

2.4 Galileo Sets the Stage

Kepler's empirical laws work well, but why they work remained a mystery until scientists could understand such simple things as batted balls and falling bricks. Galileo Galilei (1564–1642) was both a physicist and an astronomer: His impact on both disciplines was enormous. His astronomical observations convinced the scientific world that Earth moves and Sun does not. His insights about falling bodies laid the groundwork for Newton's synthesis of the celestial with the mundane.

*Galileo and
Astronomy*

Galileo was not nearly so mathematically inclined as Kepler, and he never accepted his colleague's notion of ellipses. But, Galileo's telescopic observations confirmed his Copernican bent. In the early seventeenth century, a Dutch optician peered through a device consisting of two separated lenses and saw a magnified image of distant objects. When word of this discovery reached Galileo, he quickly built his own telescope. Within weeks, he discovered that the Milky Way (which is just that to the naked eye) consists of countless individual stars. He saw the figure of the planet Venus change from a crescent to a disk, just as the Moon does. He saw sunspots on the Sun, and of our nearest celestial neighbor, he wrote:

> The Moon is not smooth, uniform and precisely spherical as a great number of philosophers believe it (and the other heavenly bodies) to be, but is uneven, rough and full of cavities and prominences, being not unlike the face of the earth, relieved by chains of mountains and deep valleys.

Not spotted Sun nor craggy Moon were quite so perfect as faith decreed. When Galileo turned his telescope toward Jupiter, he discovered its four largest moons and verified that their motions obey Kepler's laws.

Galileo Galilei (1564–1642) was born in the year of Michelangelo's death. He was the first scientist to explore the heavens with a telescope and to understand falling motion on Earth. Because his scientific investigations were so controversial, he presented them as fiction, saying: "There are two sorts of poetic imagination: of those who invent fables, and of those who are disposed to believe them." *Source:* Historical Pictures/Stock Montage, Inc.

How could anyone with a small telescope, watching this miniature version of the solar system, doubt the heliocentric theory? It seems incredible to us today, but Galileo's observations were not generally accepted. A contemporary of his argued that what is seen through a telescope must be mere illusion since

There are seven windows in the head, two nostrils, two ears, two eyes and a mouth: so in the heavens there are two favorable stars, two unpropitious, two luminaries, and Mercury alone, undecided and indifferent. From which and many other phenomena of nature such as the seven metals, etc., which it were tedious to enumerate, we gather that the number of planets is necessarily seven. Besides, the Jews and other ancient nations, as well as modern Europeans, have adopted the division of the week into seven days, and have named them from the seven planets: now if we increase the number of planets, this whole system falls to the ground. Moreover, the satellites are invisible to the naked eye and therefore can have no influence on the earth and therefore would be useless and therefore do not exist.

The State responded differently. The telescope provided a valuable distant early warning system of approaching enemy ships. The military industrial complex of the Republic of Venice rewarded Galileo with a generous pension.

His telescopic observations convinced Galileo that Earth revolves about the Sun and rotates daily about its own axis. The Holy Office characterized both propositions as "absurd in philosophy and formally heretical, because they are expressly contrary to Holy Scripture." He was forbidden to "hold, teach or defend" the proscribed theories, but that is exactly what he did. In 1633, the Church reacted: Galileo was accused of heresy, threatened with torture, put under house arrest for the rest of his life, and forced to recant his belief in Earth's motion. Under his breath (so it is said) he muttered, "But it moves!"

Galileo was compelled to abandon his studies of the motions of heavenly bodies and turned his attention exclusively to the nature of motion on Earth. His reliance on precise and repeatable experiments, and his interpretations of the data, signaled the beginning of modern science.

Galileo and Mechanics

Mathematicians and theoretical physicists often do their best work by age 26. When Galileo was that age, he lived in Pisa. Legend has it that by dropping several unequal weights from its Leaning Tower he came to the astonishing conclusion that the speeds attained by two bodies of different masses in falling a fixed distance are the same. Much later, he wrote:

> I greatly doubt that Aristotle ever tested by experiment whether it be true that two stones, one weighing ten times as much as the other, if allowed to fall, at the same instant, from a height of 100 cubits [the height of Pisa's tower] would so differ in speed that when the heavier reached the ground, the other would not have fallen more than ten cubits.

Aristotle rarely stooped to observe nature at first hand. Instead, he accepted the view of his predecessors that a body falls at a constant speed proportional to its weight. Galileo consulted nature, not authority. He did experiments rather than read old books to find out what happens when a body falls. He showed that a small bullet and a cannonball, simultaneously dropped from the same height, reach the ground at virtually the same time and at the same speed. Decades later, Galileo realized that the rate at which bodies fall does not depend on a body's mass, its shape, or its chemical composition. In his book *Two New Sciences*, he set forth the basis of a theory of terrestrial motion, which we paraphrase:

© 1993 by Sidney Harris.

1. "Any velocity once imparted to a moving body will be rigidly maintained as long as the external causes of acceleration or retardation are removed, a condition which is found only in [frictionless] horizontal planes; for in the case of planes which slope downward there is already present a cause of acceleration, while on planes sloping upward there is retardation; from this it follows that motion along a horizontal plane is perpetual; for, if the velocity is uniform, it cannot be diminished, much less destroyed." This principle is sometimes called the law of inertia and was later adopted by Newton as the first of his three laws of motion.

2. In free fall through a vacuum, all objects of whatever weight, size and composition, fall a given distance in the same time. Galileo could not study falling bodies in a vacuum, but he realized that feathers and snowflakes fall more slowly because of the resistance offered by air.

3. The motion of a body in free fall (or when rolling down an inclined plane) is uniform acceleration; that is, the body's velocity increases by equal increments in equal times.

4. The laws of physics are the same when viewed in different, uniformly moving coordinate systems. Tennis, played on an ocean liner traveling across a calm sea at 30 knots, is no different from tennis played at Forest Hills (in a 30-knot wind). Galileo realized that we cannot readily detect the motion of Earth through space. To fulfill his promise to reject the Copernican heresy, he never applied this principle to the motions of heavenly bodies.

Falling Bodies— Motion in a Vertical Direction

Precise measurements of the velocities of falling bodies were difficult to make. For this reason, Galileo concocted the ingenious use of an inclined plane to slow things down. A ball rolling down such a surface can be seen to experience uniform acceleration. For freely falling bodies, such as a pile driver, he argued that the effect of the blow "will be greater and greater according to the height of the fall, that is, according as the velocity of the body becomes greater. From the quality and intensity of the blow, we are thus enabled accurately to estimate the speed of a falling body."

Galileo discovered that every freely falling body undergoes uniform acceleration. It does not fall at a constant speed, nor does its rate of descent depend on its size, mass, or nature. Whether it be thrown upward or released from rest, every freely moving body experiences a constant downward acceleration of about 9.8 m/s^2. This quantity is usually designated g, in Galileo's honor.

The acceleration of gravity is not a fundamental constant of nature, nor is it so invariable as Galileo believed or we might wish. It results from the gravitational force that our spinning and not quite spherical planet exerts upon all other bodies. Its attraction varies slightly with altitude, latitude, longitude, and even time. The acceleration of gravity is measurably lower in mile-high Denver than in Death Valley. Because Earth spins, and is not a perfect sphere, this acceleration is 9.88 m/s^2 at the poles and 9.78 at the equator. Because of the tides, it has a measurable daily variation. Underground deposits of dense ores produce tiny but detectable gravitational anomalies. Today's precise gravity meters, sensitive to a few parts per billion, are used to identify likely sites to find valuable mineral deposits.

For these reasons, the symbol g (which is known as the standard gravitational acceleration at sea level) is assigned the particular value of $g \equiv 9.80665 \text{ m/s}^2$. The difference between the measured acceleration of gravity at any time or place on Earth from this standard value is called the *gravitational anomaly*. In this book, we are not concerned with such precision and we often use the crude but adequate approximation $g \simeq 10 \text{ m/s}^2$.

Now you see the motivation for our earlier discussion of motion in one dimension with constant acceleration. The vertical motion of any freely moving object is an instance of such motion. Equations 7, 8 and 9 in Section 2.2 let us tackle such problems with ease by substituting g, the downward acceleration of any body due to Earth's gravity, for a. Example 2.9 should drive this point home.

EXAMPLE 2.9
A Ball Thrown
Upward

A ball is thrown straight upward at a speed of 30 m/s at $t = 0$. What is its speed at the times $t = 1, 1.5, 2, 3, 4$, and 6 seconds?

Solution

This is an instance of motion in one dimension with constant acceleration. Let the x-axis point up. We use equation 7 with $v(0) = +30$ m/s and $a = -g$. (The minus sign says that things fall down, not up.) Furthermore, we put $g = 10$ m/s². The velocity as a function of time is

$$v(t) = (+30 - 10t) \text{ m/s}$$

The initial velocity is $v(0) = +30$ m/s. Thereafter, the velocity decreases smoothly. One second later, its velocity is $v(1) = +20$ m/s. After 3 s, we find that $v(3) = 0$. The ball is instantaneously at rest at the highest point of its trajectory. Its velocity continues to decrease smoothly. It turns negative as the ball descends, becoming increasingly negative. Six seconds after it is thrown, the ball is caught on the ground having the same speed it started with, but the direction reversed, $v(6) = -30$ m/s.

■

At each of the specified times, how high above the ground is the ball? To answer this question, we must determine how far a body travels if its acceleration is a constant $a = -10$ m/s² and its initial velocity $v(0)$ is $+30$ m/s.

Solution

We've seen this kind of problem before. We use equation 9 to find: $d = (+30t - 5t^2)$ m. Note that $d = 0$ at $t = 0$ s (when the ball is thrown) and again at $t = 6$ s (when it is caught). At $t = 3$ s, the ball is at its highest point, 45 m up in the air. Table 2.3 shows our results.

TABLE 2.3   The Trajectory of a
Ball Thrown Upward

Time (s)	Velocity (m/s)	Height (m)
0	30	0
1	20	25
1.5	15	33.75
2	10	40
3	0	45
4	-10	40
6	-30	0

FIGURE 2.21
A ball is thrown up-
ward. Its height above
the ground is shown
versus time. The fig-
ure is a parabola. At its
highest point, the instantaneous velocity of the
ball vanishes.

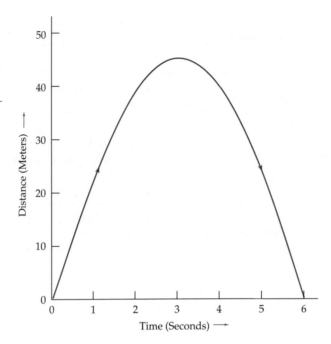

Or, we may display them in the form of the continuous graph, as in
Figure (2.21) of d versus t.   ■

Projectile Motion

Baseballs are rarely thrown or batted directly up or down, nor are they
simply dropped from rest. They often move both horizontally and vertically. Galileo showed how to deal with these motions. A projectile
(ball, bullet, or whatever) starts off at $t = 0$ at a given position and velocity. Afterward, it moves freely. Where is it, and what is its velocity,
at later times? We neglect the effect of air resistance in our discussion.
Curve balls and sliders form the subject of a more advanced treatise on
mechanics.

Position, velocity, and acceleration are vectors. To specify their components, we must choose a Cartesian coordinate system. Let the x-axis
point up, the y-axis point in a horizontal direction, and the z-axis point
in a direction that is perpendicular to both. Initially, the body is at a
position whose Cartesian coordinates are $x(0)$, $y(0)$, and $z(0)$. Its initial
velocity vector $\mathbf{v}(0)$ has components $v_x(0)$, $v_y(0)$, $v_z(0)$. To simplify the
problem, we choose the y-axis to be the direction of horizontal motion.
Hereafter, we put $z(0) = 0$ and $v_z(0) = 0$, where they remain throughout
the flight. Projectile motion is a problem in two spatial dimensions.

Horizontal and vertical motions are dealt with separately. The vertical motion of a freely moving body is uniform acceleration. The x

component of its velocity changes in accord with equation 7, and the vertical distance it traverses is given by equation 9. Thus, we find:

$$v_x(t) = v_x(0) - gt \qquad \text{and} \qquad x(t) = x(0) + v(0)t - \tfrac{1}{2}gt^2 \qquad (16)$$

If the effect of air resistance is ignored, the y component of its velocity remains constant. The horizontal motion of a projectile is uniform motion:

$$v_y(t) = v_y(0) \qquad \text{and} \qquad y(t) = y(0) + v_y(0)t \qquad (17)$$

EXAMPLE 2.10
A Baseball Problem

The pitcher hurls a baseball from a height above the ground of exactly 2 m (Figure 2.22). Initially, the pitch travels horizontally at a speed of 40 m/s. It is a strike, and is pocketed by the catcher who is 20 m from the pitcher. How high above the ground is the ball caught?

Solution

Choose a coordinate system with the origin at the mound. The x-axis points up, the y-axis points to the catcher. The action takes place in the x–y plane, with the initial position of the ball $x(0) = 2$ m and $y(0) = 0$ m. The components of the ball's initial velocity are: $v_x(0) = 0$, $v_y(0) = 30$ m/s. Equation 17 says that the ball reaches the catcher's mitt at $t = 0.5$ s. Equation 16 may then be used to tell us that $x(0.5) = 2 - 1.25$ m. The ball reaches the catcher's mitt when it is 0.75 m above the ground. ∎

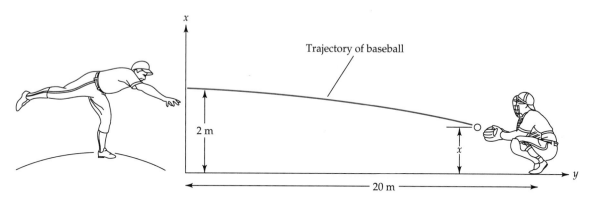

FIGURE 2.22   A pitch is thrown horizontally from a height of 2 m. If effects due to air resistance are ignored, the ball's trajectory is a parabola. When it reaches the catcher 20 m away, it has fallen to a height of 0.5 m. (In the real game, air resistance cannot be ignored and the behavior of pitched balls is more interesting.)

Reference Frames   Galileo considered the motion of a stone dropped from the mast of a moving ship:

> If the impetus with which the ship moves remains indelibly impressed in the stone after it is let fall from the mast; and if this motion brings no impediment or retardment to the motion directly downward natural to the stone, then there ought to ensue an effect of a very wonderful nature.

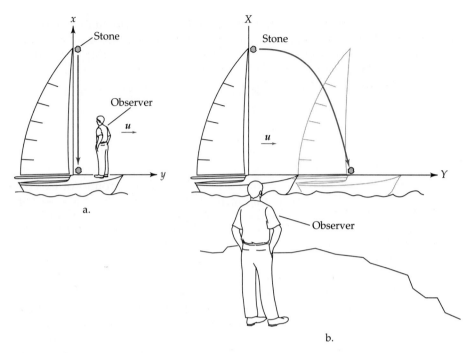

FIGURE 2.23 *a.* The (x–y) coordinate system used by an oberver aboard ship is fixed relative to the ship. The trajectory of a stone dropped from the top of a mast is a straight line. *b.* The (X, Y) coordinate system used by an observer on land is fixed relative to land. The trajectory of the stone is a parabola. The two descriptions of the stone's motion are equivalent to one another.

The wonderful effect is that the stone, when it is dropped from the top of the mast, falls to the base of the mast regardless of the speed of the ship. Aside from effects due to ambient air, no experiment performed aboard ship on a smooth sea can detect the stone's uniform motion. Let us examine the motion of the stone as it is seen by two different observers: A, who is aboard ship, and B, who is ashore (Figure 2.23).

A uses coordinates x and y, defined relative to the ship. The origin of A's coordinate system is the base of the mast, its x-axis points up and its y-axis points ahead. According to A, the initial position of the stone is $x(0) = h$ and $y(0) = 0$ and its initial velocity is zero: $v_x(0) = v_y(0) = 0$. The subsequent motion of the stone (until it hits the deck at $t = \sqrt{2h/g}$) is given by:

$$x(t) = h - \tfrac{1}{2}gt^2$$
$$y(t) = 0$$

(18)

According to A, the trajectory of the stone is a straight and vertical line extending from the top of the mast to its base.

B uses coordinates X and Y, defined relative to land. The origin of B's coordinate system is the location of the base of the mast at time $t = 0$.

The X-axis points up and the Y-axis points in the direction of motion of the ship. According to B, the initial position of the stone is $X(0) = h$ and $Y(0) = 0$. Its initial velocity is the velocity of the ship relative to land: $V_X(0) = 0$ and $V_Y(0) = u$, where u is the speed of the ship relative to land. The subsequent motion of the stone is given by:

$$X(t) = h - \tfrac{1}{2}gt^2$$
$$Y(t) = ut$$

(19)

where ut is also the Y coordinate of the moving mast.

According to B, the trajectory of the stone is a parabola extending from the top of the mast at $t = 0$ to its base at $t = \sqrt{2h/g}$. And, according to Galileo, the two descriptions of the stone's motion are equivalent to one another. In either case, the motion is described by the general equations of projectile motion, equations 16 and 17.

From the reference frame of A aboard ship, the motion of the stone is described by uniform downward acceleration. From the reference frame of B on land, the motion is also described by uniform downward acceleration. The only difference between the two analyses is in the initial conditions: To the sailor, the initial velocity of the stone is zero; to the landlubber, it is not zero.

To a caterpillar on the stone, the analysis of the stone's motion is fundamentally different. In its reference frame C, the stone remains at rest: It is not accelerated at all. However, as soon as the stone is dropped, the caterpillar sees everything else (the mast, the ship, the sea, and Earth) suddenly and inexplicably start accelerating *upward*. There is a moral to this tale. Certain reference frames, such as A and B, are singled out as the ones in which the laws of physics are simplest. They are called *inertial frames*. Any two inertial frames are in uniform motion relative to one another, and the laws of physics are the same in both. Reference frame C is an accelerated frame in which the laws of physics are not at all simple.

Galileo's principles of motion let us understand the motions of freely moving bodies at the surface of Earth, whether they are baseballs or ballistic missiles. He explained how the laws of physics are the same in all uniformly moving reference frames, and, implicitly, how Earth may move through space without its motion being noticed. Of his own scientific accomplishments, Galileo wrote:

> The theorems set forth in these brief discussions, if they come into the hands of other investigators, will continually lead to wonderful new knowledge. It is conceivable that in such a manner a worthy treatment may be gradually extended to all the realms of nature.

The hands that Galileo hypothesized were those of Isaac Newton, who was born in the year that Galileo died. He accomplished just what Galileo foresaw and admitted that "We see so far today because we stand on the shoulders of giants."

---◆··◆--------

Exercises

23. In Example 2.9, what is the average upward speed of the ball during its ascent (i.e., during the first three seconds of its flight)?

24. When a body is thrown upward at a speed v, show that it reaches a maximum height $v^2/(2g)$.

25. The acceleration of gravity on the Moon is about a sixth of what it is on Earth. If you can achieve a 1.5-m high jump on Earth, how high can you jump on the Moon? If you can achieve a 6-m broad jump on Earth, how far can you jump on the Moon? (For both events, assume that your initial velocity is the same on the Moon as on Earth.)

26. A ball is thrown horizontally at a speed of 30 m/s and from a height of 5 m. How far does it travel before it hits the ground?

27. An archer shoots an arrow at a coconut in a palm tree. He aims directly at the coconut, making no allowance for the vertical fall of the arrow.
    a. The arrow's flight time is half a second. Show that the archer misses his target by 1.25 m if the coconut stays put.
    b. He would have missed, except that he was lucky. At the very moment the shaft is released, the coconut drops from the tree! Show that the inept archer hits his target after all.

28. Galileo's example of a stone dropped from the mast of a moving ship, as viewed by an observer on land, illustrates the separation of horizontal and vertical motions. Suppose that the ship's speed is $u = 10$ m/s and $h = 20$ m. Using equation 19, make a table showing the values of X, Y and t for $t = 0, 0.4, 0.8, 1.2, 1.6$, and 2 s. Using these results, graph X (the height of the stone above the deck) versus Y, its horizontal position. The resulting parabola is the path of the stone as seen by the observer ashore.

2.5 The Newtonian Synthesis

Galileo's work was a good beginning but not yet a theory of motion. Galileo taught us that the horizontal motion of every moving body is maintained if it can be isolated from external causes of acceleration, and that its vertical motion is subject to a universal acceleration independent of its mass and composition. He proposed an early form of relativity which held that the laws of physics are the same on ships sailing at constant velocity as they are on land. (Centuries later, Einstein would invoke trains in uniform motion to make the same point in a modern context.) It follows that the absolute state of motion of a body cannot be

Isaac Newton (1642–1727) put it all together. His three laws of motion, together with his law of universal gravitational attraction, explain the motions of bodies on Earth and in the heavens. Physics was but one of his varied interests. Not only was he an ardent alchemist, mathematician, and biblical chronologist, but he became Master of the British mint. *Source:* Bettmann Archive.

determined, and, by inference, that the spinning Earth may move through the heavens without our being aware of it.

Galileo did not recognize force as the cause of acceleration. He could not work out the rules governing the collisions of balls on a billiard table, for example. And he did not identify the downward acceleration experienced by projectiles with the gravitational force exerted on them by Earth as a whole. Although he showed that Jupiter's moons satisfied Kepler's empirical laws, he never knew that they reflect the same principles that govern falling bodies. Galileo could not (or perhaps he dared not) identify the laws of ballistics with the laws of celestial mechanics.

The Rules of the Game

Isaac Newton (1642–1727) believed in the universality of physical law—that heavenly bodies obey exactly the same rules as balls and bricks on Earth. He proved that celestial mechanics and ballistics are one and the same discipline. Newton's hypothesis of a universal gravitational force explains the fall of apples and the validity of Kepler's laws. It describes the motions of planets and their moons, comets and asteroids, and, as well, ocean tides and the motions of bodies on Earth. Using Newton's laws and a computer, latter-day scientists and engineers plotted the trajectory astronauts would follow to land on the Moon and come safely home.

Newton published his mammoth *Principia*, setting forth the fundamental laws of motion, in 1687. Aside from position, velocity, and acceleration, these laws involve two additional mathematical constructs. *Mass*, loosely speaking, is a scalar quantity denoting the amount of matter in an object. *Force*, loosely speaking, is a vector quantity denoting the push or pull that causes the velocity of a body to change. Newton's three laws, describing the motions of bodies in response to the forces they experience, remain the logical basis to the science of mechanics. In a somewhat modern form, the laws are

1. Galileo's principle of inertia, generalized: A material body persists in its state of rest or uniform (unaccelerated) straight-line motion if and only if it is not under the influence of any external agency. Galileo recognized this principle for motion on horizontal frictionless planes. Newton extended it to describe the motions of real or imagined bodies on Earth or anywhere in space.

   Newton gives a precise meaning to the concept of an inertial frame, which was introduced in Section 2.4. An inertial frame is any reference frame in which the first law is satisfied. In the inertial frame that is at rest relative to the Sun, planets move in accord with Kepler's laws, and bodies that are beyond the gravitational influence of the Sun persist in a state of uniform motion. In the noninertial frame that is at rest relative to Earth, planets move in complex epicycles, and distant stars move in circles around Earth. Newton's three laws (and the rest of physics as well) are valid in any inertial frame, and only in inertial frames.*

2. The "external agency" of Newton's first law is given a name and a precise meaning by his second law. Force is that which causes velocity to change. This law is expressed as a formula,

$$\mathbf{a} = \mathbf{F}/m \qquad (20)$$

   It says that the instantaneous acceleration of a body is the ratio of the net external force acting on the body $\mathbf{F}$, to its mass m. The acceleration

---

*The starting point of Einstein's general theory of relativity was his postulate that the theory had to be valid in any reference frame, not just in inertial frames. It led him to a new theory of gravity but not to a unified theory of all of physics, which remains a sort of holy grail to many physicists.

of a body points in the same direction as the net force that acts on it. If the net force on a body vanishes, the body remains in a state of uniform motion, in accordance with the first law.

The second law involves the net force **F** that acts on a body. When two forces, **G** and **H**, act, their combined action is equivalent to that of their vector sum, $\mathbf{F} = \mathbf{G} + \mathbf{H}$. The net force on a body is the vector sum of all the forces acting on it.

3. The law of action and reaction, in Newton's words, is

> To every action there is always opposed an equal reaction: or, the mutual actions of two bodies upon each other are always equal, and directed to contrary parts. Whatever draws or presses another is as much drawn to or pressed by that other. If you press a stone with your finger, the finger is also pressed by the stone. If a horse draws a stone tied to a rope, the horse will be equally drawn back towards the stone; for the distended rope, by the same endeavor to relax itself, will draw the horse as much towards the stone as it does the stone towards the horse, and will obstruct the progress of the one as much as it advances that of the other.

Or, if hot gas is expelled from the rear of a rocket, the rocket is flung in the opposite direction, sometimes all the way to Mars. In general, if body A exerts a force **F** on body B, then body B exerts a force −**F** on body A.

All else is commentary, but there is a lot of it. Let's begin with the first law. It does not describe the behavior of freely moving bodies on Earth. We said before that bullets, balls, and cars with stalled engines soon come to rest. Planets move in ellipses. External forces, such as gravity, friction, and air resistance, are everywhere. Fortunately, nature is not capricious. The forces between bodies decrease with distance: Friction ceases when bodies are separated, air resistance vanishes where there is no air, and even the force of gravity falls away with the reciprocal square of the distance between the gravitating bodies. We may create circumstances where forces are negligible and bodies move at essentially constant velocities, or we may imagine bodies that are isolated from all forces—for example, perfect balls moving on idealized billiard tables.

EXAMPLE 2.11
Newton's Laws

A miniature cannon (with a mass of 2 kg when unloaded) lies at rest on a frictionless surface. It is armed with a 10-g cannonball and a charge of gunpowder (whose mass is small enough to be ignored). When the fuse is lit, the cannon fires (Figure 2.24, on page 94). The acceleration a experienced by the bullet is uniform throughout the $L = 20$-cm barrel. The speed of the bullet on leaving the gun is $v = 200$ m/s. Compute a and evaluate the force that acts upon the bullet while it is accelerating. How rapidly does the gun recoil?

Solution

The bullet starts at rest and experiences uniform acceleration over a distance of 20 cm. Its average speed while in the barrel is $v_{av} = \frac{1}{2}v$. It reaches the end of the barrel in a time L/v_{av}, or 2 ms. Its acceleration

FIGURE 2.24
a. A 2-kg mini-cannon is charged with gunpowder and loaded with a 10-g cannonball. *b.* When it is fired, a force *F* is exerted on the cannonball and an equal and opposite force is exerted on the cannon. *c.* The cannonball is accelerated to a velocity of 200 m/s and the cannon recoils with a velocity of −1 m/s.

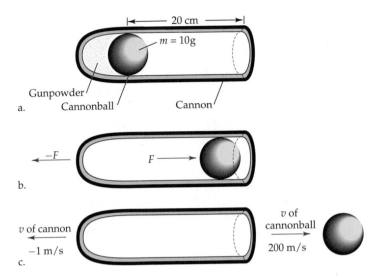

is the change of its velocity divided by this time, or $a = 1 \times 10^5$ m/s². According to Newton's second law, the force exerted by the gun on the bullet while it is being accelerated is *ma*, or $F = 1000$ N (newtons).

We appeal to Newton's third law to determine the recoil velocity of the cannon. It says that the force exerted by the cannonball on the cannon has the same magnitude as *F* but the opposite direction. The ratio of the backward acceleration of the cannon to the forward acceleration of the cannonball equals the ratio of the mass of the cannon to the mass of the cannonball. The cannon recoils backward with a velocity of −1 m/s.

The Apple Falls

Galileo discovered that falling bodies experience a downward acceleration *g*. Newton, by means of his second law of motion, attributed this acceleration to a force produced by Earth as a whole that acts on every body at or near the surface of Earth:

$$\mathbf{F} = m\mathbf{g} \qquad \text{on Earth}$$

The force due to Earth's gravity is directed downward. Its magnitude is proportional to the mass *m* of the body and is otherwise independent of its nature. No matter where we are on the globe, the force of gravity points to the center of Earth.

The Moon does not move in a straight line but in a nearly circular orbit. Its instantaneous acceleration points directly toward Earth. According to Newton's second law, the Moon's acceleration is caused by a force attracting it to Earth. Could it be, Newton asked, that Earth's gravity extends outward as far as the Moon, and beyond? Could a remnant of Earth's gravity provide the force that binds the Moon to us?

We have the tools to compute the acceleration of the Moon. Its distance from Earth is 384,400 km and its sidereal period is 2.36×10^6 s, a bit less than a month. Its speed in orbit follows from these data to be $\sim 1\,\text{km/s}$, and its centripetal acceleration v^2/R works out to be $\sim 2.7 \times 10^{-3}\,\text{m/s}^2$. The orbital acceleration of the Moon is much smaller than the acceleration of gravity on Earth. It is about $g/3600$. The force that is required to generate this acceleration is

$$\mathbf{F} \simeq \frac{M\mathbf{g}}{3600} \qquad \text{at the Moon}$$

where M is the mass of the Moon.

At the same time, the distance from the center of Earth to the Moon is about 60 (or the square root of 3600) times longer than the distance from the center of Earth to Earth's surface. Newton concluded that Earth's gravitational effect on a body must decrease with the inverse square of the distance between the body and the center of the Earth. In that case, the same mechanism—Earth's gravity—could explain the motions of bodies on Earth, the Moon's circular motion about Earth, and even the ocean tides.

Newton, however, was even more ambitious. If Earth's gravitational attraction causes the nearby Moon to revolve about us, could not the Sun's gravitational attraction be responsible for the motions of the planets? And could not Jupiter's gravity govern the motion of its moons? If this were so, gravity would not be a special feature of Earth but a common property of all massive bodies. It could explain the motions of all heavenly bodies.

From such reasoning, Newton formulated the universal law of gravitational attraction. Let M and M' be the masses of any two small or pointlike bodies that are a distance R apart. Newton postulated the existence of an attractive gravitational force between them of magnitude

$$F = \frac{GMM'}{R^2} \qquad (21)$$

In accordance with Newton's third law, each body attracts the other with the force F. The dimensional number G is a fundamental attribute of nature that is now known as Newton's gravitational constant. The attractive gravitational force is proportional to the product of the masses of the bodies and depends on the reciprocal of the square of the distance between them. Earth, however, is neither small nor pointlike. How can equation 21 explain the force of gravity that we experience? Your weight (the gravitational force that Earth exerts on you) results from the gravitational force exerted upon you by all the matter in Earth. Some of it lies just under your feet, some is in China, but most of it is deep underground. We must sum up the contributions to equation 21 of all of our planet's tiniest bits and pieces, as indicated in Figure 2.25. It would be a daunting task but for a marvellous mathematical tool: calculus.

Newton invented calculus. With it, he showed that the gravitational force exerted by a spherical Earth on a second body lying upon or outside it is identical to what the force would be if all Earth's mass were concentrated at its geometrical center. Call Earth's mass $M_\oplus$ and its ra-

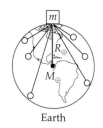

FIGURE 2.25
According to Newton's universal law of gravitation, every portion of Earth exerts a force on the mass m lying on its surface. Newton used calculus to prove that the net effect is as if all of Earth's mass $M_\oplus$ were concentrated at its center, at a distance $R_\oplus$ from m.

dius $R_{\oplus}$. As a consequence of Newton's theorem and the fact that Earth is roughly spherical, the gravitational force on an object of mass m lying at or near the surface of Earth is approximately:

$$F_{\oplus} = \frac{GM_{\oplus}m}{R_{\oplus}^2} \tag{22}$$

If the body is not subject to any other force, it will accelerate downward with an acceleration given by

$$g = \frac{GM_{\oplus}}{R_{\oplus}^2} \tag{23}$$

Scientists do not build their theories from the ground up. Rather, they add a stone or two to what was built by their predecessors. Newton added a whole new floor. His theory of gravity incorporates both Kepler's laws of planetary motion and Galileo's principles of earthly motion. In equation 23, which Newton deduced from scratch, he recaptured Galileo's discovery that all bodies fall at the same rate. His acceleration of gravity g is expressed in terms of his universal constant of gravity G and the mass and size of Earth.

Astronomical measurements alone cannot determine the mass of Earth or the Sun. This is because Newton's law of gravity, equation 21, involves the product of the mass of the gravitating body and Newton's constant G. Here is where physicists come to the aid of astronomers. Henry Cavendish, in 1798, succeeded in measuring the tiny gravitational attraction between two lead spheres in his laboratory. Since the masses of the spheres and their spatial separation were known, he was able to determine the value of G. His answer was correct to within 10 percent.* Today, G (and hence the mass of the Sun) is known much better. To an accuracy of .01 percent, it is

$$G = 6.673 \times 10^{-11} \text{m}^3/(\text{kg s}^2)$$

Using this result and equation 23, we may calculate the mass of Earth.

Mass and Weight   Physicists distinguish between the words *mass* and *weight*. Mass is the quantity of matter, and we measure it in kilograms. A kilogram brick has a mass of 1 kg here, on the Moon, or anywhere else. Weight, however, denotes the gravitational force that acts on a body, and is measured in newtons. It equals mg, where m is mass in kilograms and g is the local value of the acceleration of gravity in meters per second. Because Earth

---

*Earlier attempts had been made to measure G by determining the mass of Earth. In 1740, Pierre Bouguer carried out a series of experiments to this end in South America. He studied the effect of altitude on the rate of a pendulum's swing. By comparing the altitude effect at the Isle of Inca (at sea level) with data obtained on the high Andean Altiplano, he was able to estimate Earth's mass. Vicious weather made accurate measurements impossible. However, he was the first to demonstrate the possibility of such a measurement. The *Bouguer effect* remains important to today's geologists and prospectors.

spins about its axis and is also a bit lopsided, you weigh about 0.5 percent less at the North Pole than you do at home.

If you are the object to which equation 22 pertains, m is your mass and $F_\oplus$ is your weight on Earth. Armed with this result, it is possible to tell what your weight would be on the Moon. Its mass is $\sim 1.23 \times 10^{-2} M_\oplus$ and its radius is $\sim 0.272 R_\oplus$. Putting these values into equation 22 in place of Earth's mass and radius, we find that your lunar weight is about one-sixth of your weight on Earth. Of course, your mass on Earth or Moon is exactly the same: It may be changed by diet, but not by travel.

EXAMPLE 2.12   Most scales that read your weight are confusingly calibrated in units of mass such as kilograms or pounds. Ordinarily, this presents no difficulty because we spend little of our time in space, or in dive bombers and (with a rare exception) none of it on the Moon. Suppose that a woman has a mass of 50 kg. She brings her bathroom scale into the elevator of the Empire State Building and stands on it as the elevator moves upward with an acceleration of $a = 1.2$ m/s². What does the scale read?

Solution   Whether in the bathroom or in the elevator, the woman is subjected to two forces: a downward force mg (where m is her mass and g is the acceleration of gravity) and an upward force W exerted by the scale (Figure 2.26).

FIGURE 2.26
A woman weighs herself while she is accelerating upward at 12 percent of the acceleration of gravity. The scale reading is 12 percent higher than it normally would be.

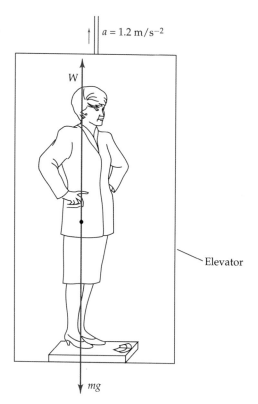

Newton's third law says W is also the downward force that the woman exerts on the scale: It determines the reading of the scale. The net force on the woman is $W - mg$. When she is at home, her acceleration is zero and Newton's second law says $W - mg = 0$ or $W = mg$. The scale reads 50 kg. In the elevator, her acceleration is a and the second laws says $W - mg = ma$ or $W = m(g + a)$. If we put $g = 9.8$ and $a = 1.2$ both in units of m/s^2, we find that $W = 1.12 \, mg$ when the woman is in the accelerating elevator. The scale reads about 56 kg.   ■

Kepler's Laws

The Sun's gravitational attraction determines the motions of the planets. Jupiter's gravity controls its satellites. Earth's gravity keeps our seas and atmosphere where they belong and our feet to the ground. It holds the Moon in thrall, which in turn produces the tides on Earth. In one bold stroke, equation 21 explains the reasons behind Kepler's laws of planetary motion and Galileo's universal acceleration of gravity on Earth. Celestial mechanics and terrestrial mechanics were unified in the year 1687.

The first of Kepler's laws—that planetary orbits can be ellipses as well as circles—is too technical for us to prove here. The second law, as we discuss in Chapter 3, is the statement of conservation of angular momentum. That leaves the third law, stating that R^3/P^2 is the same for all the planets. Newton deduced this result from equation 21, where R is identified as the semimajor axis of an elliptical planetary orbit. To prove this, we present a simplified treatment wherein all planetary orbits are approximated by circles. We begin by combining Newton's second law, equation 20, with his law of gravitation, equation 21, for the case of a planet of mass M circling the Sun at a distance R. The acceleration of the planet is

$$a = \frac{GM_\odot}{R^2} \tag{24}$$

The next step is to recall equation 10, the expression for the acceleration of a body in uniform circular motion, and to insert it into equation 24:

$$\frac{V^2}{R} = \frac{GM_\odot}{R^2}, \quad \text{or} \quad V^2 R = GM_\odot$$

Since V equals the circumference of the circle $2\pi R$ divided by the period of the orbit P, we obtain Kepler's third law:

$$\frac{R^3}{P^2} = \frac{GM_\odot}{4\pi^2} = K_\odot \tag{25}$$

The constant in Kepler's law is expressed in terms of Newton's constant and the mass of the Sun. For satellites of Earth or other planets, we now understand why R^3/P^2 is proportional to the mass of the primary body.

EXAMPLE 2.13
A Ball on a String

Suppose you swing a ball tied to a string in a circle about yourself. The radius of the circle is $R = 2$ m, the speed of the ball is $V = 12$ m/s, and its mass is $m = 250$ g. Imagine that you perform this act while in outer space, where no gravitational force acts on the ball to complicate matters (Figure 2.27). The circular motion of the ball results from a force that you exert on it. How large is the force?

Solution

The centripetal acceleration of the ball is given by equation 10 to be V^2/R, or $a = 72$ m/s^2. The force F that must be exerted on the body is given by equation 20 to be ma or 18 N. The force is supplied by the string, whose tension is 18 N. Ultimately, you are exerting that force on the ball.

Wait a second! Newton's third law says that the string pulls at you with a force of 18 N. You cannot remain fixed in space. The preceding analysis is correct only insofar as your mass is very much greater than the mass of the ball, so your own motion may be neglected. In this context, let's return to the controversy between Ptolemy's geocentric solar system and Copernicus's heliocentric model. In truth, they both got it wrong, although Copernicus was closer to the truth. Newton's theory says that the Sun doth move!

Planets move in ellipses because of gravitational forces exerted on them by the Sun. Newton's third law tells us that nine equal and opposite forces act on the Sun. The largest of these forces is due to Jupiter. The force is roughly constant in magnitude (since Jupiter's orbit is nearly circular), and its direction rotates in space with Jupiter's period. Consequently, the Sun moves in a circular orbit. The ratio of the Sun's orbital radius to that of Jupiter equals the ratio of Jupiter's mass to that of the Sun, a factor of about 1/1000. In each Jovian period of 12 years, the Sun

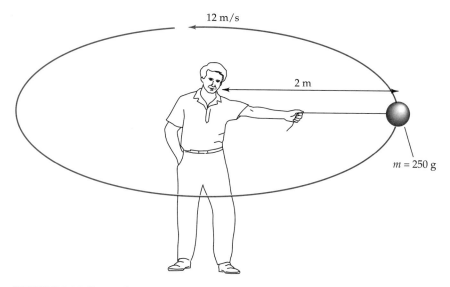

FIGURE 2.27 Somewhere in outer space, you are swinging a ball ($m = 250$ g) on a string in a circle of radius 0.5 m at a speed of 4 m/s.

FIGURE 2.28
The orbit of the Sun. Jupiter and the Sun have concentric circular orbits whose radii are inversely proportional to their masses. The radius of the solar orbit is only slightly larger than the radius of the Sun. (Smaller effects due to the other planets have been ignored.)

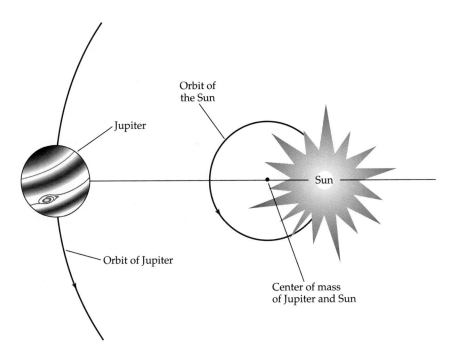

rotates about a fixed point lying on the line connecting the two bodies, as shown in Figure 2.28. The distance between this point and the center of the Sun happens to be slightly larger than the Sun's radius.   ∎

Newton's laws seemed to offer us an absolutely precise determination of planetary motions. Apparent discrepancies in the motions of the outer planets turned out to be confirmations of classical mechanics: They led to the discoveries of two new planets, Neptune and Pluto (see "A Little More About the Planets," in Section 2.3). On September 12, 1859, Leverrier, who had predicted the existence of Neptune some years before, reported an irregularity in the orbit of Mercury, the innermost planet. Its perihelion (the point on its orbit that is closest to the Sun) advances by 43 arcseconds per century. Leverrier suggested that the reason for this effect was "some as yet unknown action on which no light has been thrown. It is a grave difficulty worthy of attention by astronomers." Many suggestions were made to explain the anomaly. In the 1870s, some astronomers thought they had seen a planet closer to the Sun than Mercury. They called it Vulcan, but it turned out to be a mirage. Other astronomers suggested the possible existence of a flock of undiscovered intramercurian asteroids that caused the perturbation of Mercury's orbit.

The truth, however, was even more amazing: The inexplicable advance of the perihelion of Mercury is due to the failure of Newton's theory of gravity and of Euclid's understanding of geometry! Einstein, in 1915, published his general theory of relativity, in which he found a new and more profound meaning of the concept of mass. The mass of a body, according to Einstein, measures its effect on geometry: Space is intrinsically deformed, or curved, in the vicinity of a gravitating body. The force of gravity is not really a force at all but a reflection of the curvature of space. The impact on science of Einstein's general theory of relativity is similar to the impact caused by Newton's approximate theory. The general theory of relativity explains the tiny discrepancy in Mercury's orbit—but it does a lot more, besides. In Chapters 11 and 16 for example, we discuss how this theory is the logical basis upon which today's triumphant Big Bang picture of the expanding universe is based.

The moral is clear: However successful today's understanding of matter and the universe may be, tomorrow's will be even more so. Nonetheless, today's theoretical structures will not be abandoned. Newton's theory of gravity, although it is only an approximate realization of the more correct vision of Einstein, is as useful as it ever was. The successes of the past must be included within the science of the future.

Exercises

29. How fast must a plane fly west so that the sun appears fixed in the sky to its passengers? The route lies along the equator. Earth's radius is 6400 km, and your answer is to be in kilometers per hour.

30. Suppose that Earth's rotation is somehow stopped (gradually, so as not to cause flood and havoc). To accomplish this, the deceleration of a point on the equator must not exceed 1 m/s². What is the shortest time in which the deed may be done under this constraint? (*Hint:* First calculate the rotational speed of a point on the equator.)

31. By what multiplicative factor would your weight change if:
    a. Earth's mass were doubled and its radius kept constant?
    b. Earth's radius were doubled and its mass kept constant?
    c. Earth's mass were doubled and its mean density kept constant?
    d. Earth's mean density were doubled and its mass kept constant?
    e. Earth's radius were doubled and its mean density kept constant?
    f. Earth's mean density were doubled and its radius kept constant?

32. Earth is the best place to live in the whole solar system, but adventurers always seek new frontiers. Among the least inhospitable sites are our neighboring planets and two of Jupiter's moons. Calculate the ratio of your weight at each of these sites to your weight here. The masses and radii of these bodies are given here relative to Earth's mass $M_\oplus$ and radius $R_\oplus$:

Habitat	Radius	Mass
Mars	$0.53\ R_\oplus$	$0.11\ M_\oplus$
Venus	$0.95\ R_\oplus$	$0.82\ M_\oplus$
Ganymede	$0.41\ R_\oplus$	$0.025\ M_\oplus$
Europa	$0.25\ R_\oplus$	$0.008\ M_\oplus$

33. Normally, your bathroom scale (with you on it) reads 60 kg. If you are atop your scale in an elevator accelerating upward at 2 m/s^2, what does it read? What must your downward acceleration be for the scale to read 15 kg?

34. The Sun exerts a force on Earth that keeps it in its orbit. By the law of action and reaction, Earth exerts a precisely equal force on the Sun. The Sun is thereby accelerated, and its center describes a circular orbit. Show that the radius of the Sun's orbit is ~ 450 km, far smaller than its physical size. ($M_\oplus \simeq 3 \times 10^{-6} M_\odot$.) (*Hint:* The quantities mv^2/r are the same for each body. So are the periods of the two orbits, hence the quantities v/r. From these two facts, deduce that the orbital radii of the two bodies are inversely proportional to their masses.)

35. You are aboard a Ferris wheel of radius 15 m that is turning at a constant speed. You find yourself to be precisely weightless at the very top of the trajectory.
    a. What is the period of the wheel?
    b. When you stand on your bathroom scale, it ordinarily reads 65 kg. What would it read if you stood on it in the Ferris wheel at the moment that it was at the bottom of its trajectory?

Where We Are and Where We Are Going

Planets obey Kepler's precise mathematical laws, falling bodies those of Galileo. Newton deduced all these results from his laws of motion and his hypothesis of universal gravitational attraction. The gravitational force between massive bodies explains the motions of bodies on Earth and in the heavens. There are many other forces in nature. Some (such as friction, air resistance, and the impenetrability of solid bodies) impede motion, while others (those exerted by springs, motors, muscles, wind, tides, and so on) produce motion. Newton provided a general framework for the determination of the trajectories of moving bodies in terms of the forces that act upon them.

Motion may be described exclusively in terms of force, velocity, and acceleration. Nevertheless, we shall see that the further constructs of momentum and mechanical energy are not only useful and convenient but essential to link mechanics to the rest of physics. These quantities, and the conservation laws they satisfy, are introduced in Chapter 3, where they are applied to impact phenomena as well as planetary motion. Many of the examples and exercises in that chapter deal with cars and billiard balls, but the principles we learn will be applied, in later chapters, to molecules, atoms, and to their tiniest parts.

3 Energy and Momentum

My pocket dictionary says force is power, power is energy, and energy is force. Momentum is given as impetus, which in turn is force or energy. To physicists, however, these terms are precise and distinct. They bear different dimensions and are not interchangeable. We measure force in newtons, energy in joules, power in watts, and momentum in nameless kg m/s. Dimensions suggest, but cannot specify, their physical meanings.

In Hindu myth, Vishnu appears in nine avatars: fish, turtle, pig, monster, dwarf, Krishna, Buddha, and Rama the creator or the destroyer. As a white-winged horse, he will one day destroy Earth. Vishnu has much in common with energy. It, too, is an abstract quantity that is difficult to define because it appears in many guises. Its central property (shared with momentum and angular momentum) is conservation—energy can be neither made nor lost. Our primary source of energy is the Sun, but in a far off time it will fulfill Vishnu's prophecy: The Sun is destined to explode and engulf Earth.

Momentum and energy describe collisions, a familiar aspect of everyday life in games and on highways. Billiards was a subject of serious research by many famous scientists. Did Newton create mechanics as a by-product of time spent in pool halls? Perhaps we should teach physics in that salutary environment, since collision theory is basic to physics and chemistry. Billiard ball impacts mimic the countless elastic collisions of air molecules. Most of our knowledge of subatomic structure comes from the study of collisions of energetic particles. Particle physics is a kind of relativistic quantum billiards, played with twentieth-century rules.

Source: Historical Pictures/Stock Montage, Inc.

In the seventeenth century, Galileo showed how to calculate the motions of projectiles on Earth. A few decades later, Newton showed how to calculate the motions of planets in the heavens. In the centuries since, Newton's laws of motion and gravitation have evolved into a mathematical theory that can describe the motions of complex systems consisting of many interacting bodies. Some of these developments are described in this chapter. Section 3.1 describes momentum and two important forms of energy (kinetic and potential) in the context of motion in a line. Section 3.2 examines the roles played by energy and momentum in the collisons of moving bodies. Section 3.3 continues the study of vector algebra. The techniques we develop are applied to motion in the real three-dimensional world in Section 3.4.

3.1 Energy and Momentum in One Dimension

To introduce the notions of energy and momentum, and their relation to force, we return to the subject of uniform acceleration. Imagine a body in motion. We may be speaking of a falling body, or an accelerating rocket, or a stalled car being pushed by its driver. If we wish only to *describe* the motion, it is enough know the velocity from moment to moment. If we are to *explain* the motion in terms of the force acting on the body, we must also take the *mass* of the body into account. The velocity of a batted ball depends both on the force exerted by the batter during the impact and on the mass of the baseball.

An object of mass m is under the influence of a force F directed along the x-axis. Its acceleration is F/m. At time t_1, the position and velocity of the object are x and v (Figure 3.1). At the later time t_2, its position and velocity are x' and v'. If the force is constant, the velocity of the object changes at a fixed rate, so that

$$v' = v + (F/m)(t_2 - t_1) \tag{1}$$

Equation 1 may be rewritten as follows:

$$F\Delta t = mv' - mv \tag{2}$$

FIGURE 3.1
At time t_1, the position of the object is x and its velocity is v. A constant force F acts on the object. At time t_2, its position is x' and its velocity is v'.

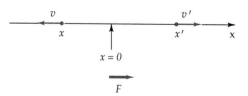

where $\Delta t = t_2 - t_1$ stands for the time interval over which the force has acted on the object. If the force is not constant, equation 2 may be used if the time interval Δt is sufficiently small that the change in F may be ignored.

If F is constant, the average velocity of the object during the time interval Δt is the average of its initial and final velocities, $v_{av} = \frac{1}{2}(v + v')$. The distance it has traveled is the product of the average velocity and the time interval:

$$x' - x = v_{av}(t_2 - t_1) = \tfrac{1}{2}(v + v')\,\Delta t \tag{3}$$

We may use equation 1 to eliminate v' from equation 3 so as to obtain an expression for x' in terms of the initial velocity v and position x and the time interval Δt:

$$x' = x + v\,\Delta t + \tfrac{1}{2}(F/m)\,(\Delta t)^2$$

This result was obtained in Chapter 2. It is useful if we wish to know where the object is after a given time has passed, but this is not our present focus. Here we use equation 3 to develop the essential concepts of momentum and kinetic energy.

Momentum

The quantity mv occurring in equation 2 is the product of the mass of the object and its instantaneous velocity. It is called its *momentum* p:

$$p = mv \tag{4}$$

Equation 2 says that the momentum imparted to a body by a constant force is the product of the force with the time over which the force acts. If the force is not constant, we take Δt to be tiny, and write equation 2 in the form

$$F = \frac{\Delta p}{\Delta t}$$

which is another way to interpret Newton's second law: The force acting on an object is the rate of change of its momentum.

To understand the significance of momentum, we must consider the cause of the force F. Call the object whose motion we have been studying A. Suppose that the force acting on A is exerted by another body B. (If A is a falling brick, B is Earth.) According to Newton's third law, the force F that B exerts on A is necessarily accompanied by a force $-F$ that A exerts on B. Since both forces act for the same time, the change of momentum of body A is equal and opposite to the change of momentum of body B. The effect of a force between two bodies is to transfer momentum from one body to the other, not to create it. The sum of the momenta of A and B is not affected by whatever forces they exert on one another.

Conservation of
Momentum

An *isolated system* is a collection of two or more bodies, all of which may exert forces on one another, but none of which are subject to any external force. Two colliding air molecules form an isolated system. So does the solar system. Consider an isolated system consisting of N bodies a, b, ... with masses m_a, m_b, ... The bodies are restricted to move along the x-axis, so we need not use vectors to describe their motions. Let their positions be $x_a(t)$, $x_b(t)$, ... and their velocities be $v_a(t)$, $v_b(t)$, ... The total momentum of this system is the sum of the momenta of all the bodies making up the system:

$$P_{\text{total}} = m_a v_a(t) + m_b v_b(t) + \cdots = p_a(t) + p_b(t) + \cdots \tag{5}$$

The force that any body (say, a) exerts on any other body (say, b) is equal and opposite to the force that b exerts on a. Momentum is transferred among the bodies, but the total momentum of the system does not change. It remains constant even though the motions of the individual bodies may be very intricate. *The total momentum of an isolated system is conserved.*

$$P_{\text{total}} = \text{Constant} \tag{6}$$

Consider the following linear combination of the positions of bodies:

$$X_{\text{cm}} = \frac{m_a x_a(t) + m_b x_b(t) + \cdots}{m_a + m_b + \cdots} \tag{7}$$

in which the denominator is the total mass of the system. The quantity X_{cm} has the dimension of length and is called the *center of mass* of the system. Its rate of change V_{cm} is called the *center-of-mass velocity:*

$$V_{\text{cm}} = \frac{m_a v_a(t) + m_b v_b(t) + \cdots}{m_a + m_b + \cdots}. \tag{8}$$

The numerator in equation 8 is the total momentum P_{total} as defined by equation 5. Its denominator is the total mass of the system, M_{total}. Thus, we find that

$$V_{\text{cm}} = \frac{P_{\text{total}}}{M_{\text{total}}} \tag{9}$$

The center-of-mass velocity of an isolated system is a constant equal to the ratio of its total momentum to its total mass. The center-of-mass of a system moves uniformly with the center-of-mass velocity.

EXAMPLE 3.1
Conservation
of Momentum

A 1000-kg automobile is moving at 50 m/s along a frictionless ice-covered road. At $t = 0$, it crashes into a 9000-kg truck that is initially at rest. The two vehicles stick together to form a single body (Figure 3.2). How fast does this wreck slip along the ice? Describe the motion of the center of mass of the car and truck.

FIGURE 3.2
a. A 1-ton (1000-kg) car approaches a stationary 9-ton (9000-kg) truck at a velocity of 50 m/s. *b.* The two vehicles collide on a frictionless road. *c.* The 10-ton wreck slides at a velocity of 5 m/s.

Solution   The law of conservation of momentum is so powerful that we can answer these questions without knowing any of the details of the collision or anything about the forces involved! Since the road is frictionless, the car and the truck form an isolated system whose total momentum P (the sum of the momenta of the automobile and the truck) does not change. Thus, P is the total momentum both before and after the collision. Prior to the accident, the momentum of the truck is zero. The total momentum is that of the automobile: $P = 5 \times 10^4$ kg m/s. It equals the final momentum of the wreck, which is the product of its mass (10,000 kg) and v, its unknown velocity: $10,000v = P$. Solving this equation for v with the known value of P, we find: $v = 5$ m/s. After the collision, the system consists of a single body whose position is its center of mass X_{cm}. Since the center-of-mass velocity of the system is unaffected by the collision, $X_{cm} = vt$ both before and after. The center of mass prior to the collision lies between the car and the truck at the point nine times further from the car than the truck.   ∎

Work and Kinetic Energy

The work done by a constant force that acts on an object is the product of the force F with the distance $\Delta x = x' - x$ that the object moves:

$$\text{Work done by } F = F \, \Delta x \qquad (10)$$

Equation 10 defines the work done by a constant force. If the force F is not constant, this definition may still be used if Δx is small enough that the change in F is negligible. Otherwise, Δx must be broken up into many tiny intervals, to each of which equation 10 must be applied. (This is the mathematical operation of *integration*.)

The kinetic energy of a body is defined to be one-half the product of its mass and the square of its velocity:

$$\text{Kinetic energy} = \tfrac{1}{2}mv^2 \tag{11}$$

These quantities are related to one another, as we now show. Equation 1 may be written $t_2 - t_1 = (v' - v)\,m/F$. With this result, we may eliminate t_1 and t_2 from equation 3 to obtain

$$F\,\Delta x = \tfrac{1}{2}mv'^2 - \tfrac{1}{2}mv^2 \tag{12}$$

The right-hand side of equation 12 is the change in kinetic energy of body A. The left-hand side is the work done by the force. Equation 12 says that the change of kinetic energy of a body is equal to the work done by the force acting on it. Work is a measure of what the force has accomplished.

EXAMPLE 3.2
The Work Done
by a Force

Part 1 Your car does not start. A friend pushes the initially stationary car with a constant horizontal force of 100 N. He continues to push until the car has moved 10 m. The mass of the car and driver is 2000 kg. What is the velocity v of the car if the effects of frictional forces are ignored?

Solution

A force of 100 N has been exerted on the car over a distance of 10 m. The work done by this force is 1000 J (newton-meters), which, by equation 12, is the change of the car's kinetic energy. Its initial kinetic energy is zero, so its final kinetic energy is given by $\tfrac{1}{2}mv^2 = 1000$ J, and hence, $v = 1$ m/s.

Part 2 Your friend continues to push the car until it reaches a speed of $V = 2$ m/s. How much further has the car moved?

Solution

Let d be the distance the car has moved beyond the 10-m point. Equation 12 tells us that $Fd = \tfrac{1}{2}mV^2 - \tfrac{1}{2}mv^2$, where $V = 2$ m/s, and $v = 1$ m/s. The initial kinetic energy of the car is 1000 J; its final kinetic energy is 4000 J. The difference is the work done by the 100-N force acting over the distance d: $100d = 3000$, or $d = 30$ m.

Part 3 The car starts and you drive away. Meanwhile, your friend continues his aerobic exercise by pushing at a parked car with a force of 100 N for another ten minutes. The parked car does not move because its brake is engaged. How much work has he done in this time?

Solution

Complex physiological processes take place within your friend's body to enable him to exert the force on the car. These processes consume metabolic energy. He feels as if he has done work, and he tires. However, the force he exerts on the car does no work at all because the car does not move. ∎

The total kinetic energy of a system of moving bodies is the sum of the kinetic energies of the individual bodies. The total kinetic energy of an isolated system, unlike its total momentum, is not necessarily conserved. In Example 3.1, the initial kinetic energy of the system is that of the car. Using equation 11, we find that it is 1.25×10^6 J. The final kinetic energy is that of the wreck. Again using equation 11, we find it is 1.25×10^5 J, which is only a tenth as large as the initial kinetic energy. Most of the initial kinetic energy of the car is transformed into other forms of energy and used to bend and break the two vehicles, and otherwise wreak havoc.

Potential Energy

Under certain circumstances—such as motion under a constant force—the kinetic energy of an object may change, but in a reversible fashion. If you throw a ball straight up with velocity v, it comes momentarily to rest at the top of its trajectory. Its kinetic energy disappears. It is not lost, but transformed into *gravitational potential energy*. When you catch the ball, its velocity has become $-v$. Its gravitational potential energy has changed back into kinetic energy. To understand this behavior mathematically, we rewrite equation 12 in the following fashion:

$$\tfrac{1}{2}mv^2 - Fx = \tfrac{1}{2}mv'^2 - Fx' \tag{13}$$

In equation 13, the terms on the left refer to the object at time t_1, and those on the right refer to the object at time t_2. What this means is that a certain quantity remains constant throughout the course of the motion:

$$E = \tfrac{1}{2}mv^2 - Fx \tag{14}$$

where v and x denote the velocity and position of the object at any time. E is the total energy of the object. It is the sum of two terms: $\tfrac{1}{2}mv^2$ is the kinetic energy of the object, and $-Fx$ is its *potential energy* referred to the point $x = 0$. As the object moves, both its kinetic and potential energy change, but its total energy remains constant.

The most familiar instance of a constant force is the downward force of gravity that acts on all objects on Earth. The magnitude of this force is mg, where m is the mass of an object and g is the acceleration of gravity. In this case, equation 14 becomes

$$E = \tfrac{1}{2}mv^2 + mgx \tag{15}$$

where x is the height of the object over the ground. The term mgx is the gravitational potential energy of the object relative to the ground.

Suppose that a brick of mass m, initially at rest, falls freely. Its initial kinetic energy is zero. When it has descended a vertical distance h, the force of gravity has done an amount of work on the brick equal to mgh. The potential energy of the brick decreases by this amount, while its kinetic energy increases by the same amount. Its total energy does not change. The kinetic energy of the brick, which has fallen from rest by a distance h, equals its loss of potential energy:

$$\tfrac{1}{2}mv^2 = mgh \quad \text{or} \quad v = \sqrt{2gh}$$

Conservation of energy offers a simple way to relate the speed of an object to the distance it has fallen.

We introduced the idea of potential energy in the context of a constant force acting in a fixed direction. However, the concept applies in many circumstances where the force is not constant, such as to the motions of heavenly bodies and ballistic missiles. (It does not apply to the force of friction, which turns kinetic energy into heat and other forms of energy.) The kinetic energy of an object is its energy of motion; its potential energy may be thought of as stored–up kinetic energy. There are many situations in which the total energy of an object is conserved throughout its motion, and there are many forms of potential energy—for example:

- *Gravitational potential energy* Imagine a pendulum swinging back and forth. If friction and air resistance were absent, the pendulum would continue to swing indefinitely. At the highest point of its swing, it is momentarily at rest and its kinetic energy vanishes. All of its energy is gravitational potential energy. At its low point, its kinetic energy is largest and its potential energy least. At every point in its path, the sum of the potential and kinetic energies of the pendulum is equal to its constant total energy.

- *The potential energy of a spring* Suppose you go bungee jumping off the Brooklyn Bridge using a perfectly frictionless bungee spring. If air resistance is neglected, you will bounce up and down forever. At the bottom of a bounce, when you have fallen a distance h, you are momentarily at rest. The energy $E = mgh$ is stored in the taut spring in the form of *mechanical potential energy*. At the top of a bounce, you are again at rest. The spring is slack and energy E is in the form of gravitational potential energy. At any point in between, your energy is E, but it is shared among kinetic energy, gravitational potential energy, and mechanical potential energy.

- *Electric and magnetic potential energy* Similar magnetic poles or charged bodies exert mutually repulsive forces on one another. Energy must be expended to place two such objects next to one another. The energy is not lost, but stored in the form of magnetic or electric potential energy. If the objects are released, their potential energy becomes kinetic energy as they spring apart.

EXAMPLE 3.3
A BB Gun

The BB is accelerated by the action of a coiled spring, which exerts a constant force of 2 N on the BB over a distance of 10 cm (Figure 3.3). If its mass is 1 g, how fast does the BB exit from the muzzle? Where did its energy come from?

Solution

The kinetic energy acquired by the BB is the work done by the spring—the force it exerts times the distance over which it acts, or 0.2 J. This is set equal to $\frac{1}{2}mv^2$ to obtain $v = 20$ m/s. The kinetic energy of the missile was previously stored in the spring as mechanical potential energy. It was supplied by the muscles of the person who cocked the gun.

FIGURE 3.3
a. Potential energy is stored in a spring in the BB gun barrel. *b.* When the trigger is pulled, the spring exerts a constant force of 2 N on the BB over a distance of 10 cm. The work done by the spring is 0.2 J. *c.* The potential energy of the spring becomes the kinetic energy of the 1-g BB, which has a velocity of 20 m/s.

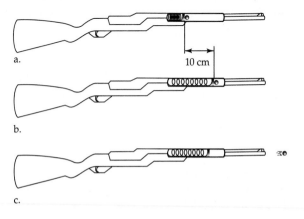

The Lever

Lift a stone of mass M to a height h. To counteract the force of gravity, the force exerted must equal the weight of the stone, Mg. The work done in the process is force times distance, or $W = Mgh$. If we raise the stone very slowly, kinetic energy plays no role in this process. The end result is not to set the stone into motion, but simply to raise it. What happens to the energy we have expended, or more precisely, the work we have done? It isn't lost because it reappears as kinetic energy if the stone falls down. The energy is stored in the form of gravitational potential energy. An object may be defined to have zero gravitational potential energy at sea level. At a height h above sea level, its potential energy is Mgh. When it falls, its potential energy becomes kinetic energy.

If the stone is too heavy to lift, we can use a lever, which is a simple tool that allows us to enhance the force of our muscles. In Figure 3.4, a downward force F is exerted on the lever at the point P, a distance H from its fulcrum. The other end of the lever is placed under the stone at point Q, a distance h from the fulcrum. The point P is pushed down a vertical distance D, and at the same time, the stone is raised by the distance d. The work done by the downward force F is FD. An upward force of Mg (the weight of the stone) must be applied to lift the stone.

FIGURE 3.4
FIGURE 3.4
The lever. A downward force F is applied to the lever at a horizontal distance H from its fulcrum. The stone of mass M is at a horizontal distance h from the fulcrum. The work done by F equals the work done on the stone: $FD = Mgd$. Since $D/H = d/h$, the force required to lift the stone is h/H times its weight.

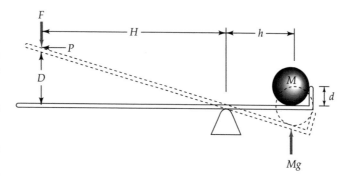

The work done by this force is mgd, and it equals the work done by the applied force: $FD = Mgd$. We also see from the geometry of the system that $d/D = h/H$. Thus, F is given by

$$F = (h/H)\, Mg \tag{16}$$

If H is much greater than h, the force F required to raise the stone is much smaller than Mg. The work done by this force becomes the gravitational potential energy of the stone.

The lever allows us to lift a weight larger than we normally can. The amplification of force is simply the ratio of lengths of the lever arms, or H/h. Of course, there's no free lunch: To pay for the amplification, we must exert the force over a distance that is longer by the same factor. In principle, we could lift an arbitrarily large weight with a long enough lever. Archimedes once said: "Give me a place to rest, and I will move the Earth." He would also need an impossibly long and strong lever.

Exercises

1. Calculate the kinetic energy in joules of:
   a. A 1.5-g bumblebee flying at 2 m/s.
   b. A 150-g baseball flying at 50 m/s.

2. A rubber band is used to project a 2-g spitball across the classroom. The stretched rubber band exerts a force of 0.25 N on the projectile over a distance of 10 cm.
   a. To what velocity is the spitball accelerated?
   b. What is the momentum of the spitball in flight?
   c. For what period of time was the spitball accelerated?

3. A stalled car with a mass of 1800 kg is being pushed by a good Samaritan. She exerts a constant force of 300 N on the car until it achieves a jump-startable velocity of 1.5 m/s. (Assume that frictional forces on the car are negligible.)
   a. For what distance does she push the car?
   b. For how long a time does she push the car?

4. A 50-lb child sits 2.4 m from the fulcrum of a seesaw. His father weighs 150 lb. Where should he sit for the seesaw to balance?

To raise the crate, three strands of rope connecting it to the pulley must be shortened. The tension on the rope is everywhere the same, so the force acting on the crate is three times the force F exerted by the worker.

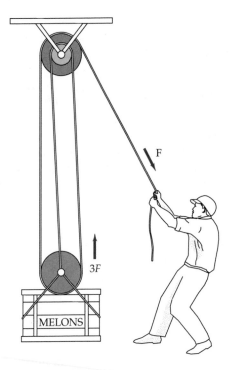

5. The pulley, like the lever, is a simple tool allowing us to raise heavy things, like the 240-kg crate of watermelons shown in the picture. What downward force in newtons must be applied to the free end of the pulley to lift the crate? (Notice that three strands of rope connect the crate to the overhead pulley. You must exert the force over 3 m of rope for every meter the crate is lifted.)

6. A ball is thrown up into the air. Which takes longer, its ascent or its descent? (*Hint*: Air resistance reduces the total energy of the ball.)

3.2 Collisions in One Dimension

The exquisite simplicity of the laws of physics often lies hidden within the countless complexities of nature. Newton and Galileo realized that an isolated body moves ceaselessly at a constant velocity. They made this discovery despite the fact that bodies on Earth or in the heavens do not act this way: They are subject to forces. Gravity, friction, air resistance, and so on are inescapable facts of life. Physicists must often imagine idealized circumstances in which they can exhibit the underlying principles of their discipline. We examine the collisions of two "balls," with masses m_1 and m_2, under the following hypothetical circumstances:

1. Their collisions are *elastic*. This means that no kinetic energy is lost when they collide. A tennis ball hitting the ground does not behave in this way. Much of its kinetic energy is lost in the bounce. It is turned to heat or used to strip away its fuzz. However, collisions made by air molecules are perfectly elastic, as are collisions of balls on a hypothetical frictionless billiard table.

2. Because we live in three-dimensional space, the description of collisions is rather complicated. Things are simpler on a billiard table, where motion is confined to a plane. Things are simpler yet in one dimension. In this section, we deal with the elastic collisions of structureless pointlike bodies that move along a straight line.

To Collide or Not to Collide?

At $t = 0$, an object is at $x = 0$ with velocity v_1 (Figure 3.5). A second object starts at the point $x = L$ with velocity v_2. Positive v means a body moves to the right and x increases; negative v means it moves to the left. Will there be a collision? When will it occur? Where will it occur?

Until a collision takes place, each ball moves uniformly with its initial velocity. Denoting the positions of the balls as the functions $x_1(t)$ and $x_2(t)$, we write

$$x_1(t) = v_1 t \qquad x_2(t) = L + v_2 t \tag{17}$$

A collision takes place if at some time t_c, the positions of the two balls coincide:

$$x_1(t_c) = x_2(t_c) \tag{18}$$

which, by the simple algebraic technique of inserting equation 17 into equation 18, becomes

$$t_c = \frac{L}{v_1 - v_2} \tag{19}$$

If the first ball is faster than the second, there will be a collision at a time t_c. The location of the collision point is determined by evaluating either

FIGURE 3.5
One object starts at $x = 0$ with velocity v_1. Another object starts at $x = L$ with the smaller velocity v_2. They collide when $t = L/(v_1 - v_2)$.

$x_1(t)$ or $x_2(t)$ at the time t_c:

$$x_1(t_c) = x_2(t_c) = \frac{Lv_1}{v_1 - v_2} \tag{20}$$

What happens when our idealized elastic collision takes place?

The Rules of Engagement

Two laws of physics come into play during a collision: the conservation of momentum and of kinetic energy. The first law is absolute. For any collision of two bodies, and in the absence of any external forces, the sum of the momenta of the two bodies does not change. The constant momentum of the two-body system is its initial momentum—the sum of the two initial momenta:

$$P = m_1 v_1 + m_2 v_2 \tag{21}$$

where m_1 is the mass of the first body and m_2 is that of the second. Since the motion of our balls is restricted to one dimension, there is no need to use vectors—all we need do is keep track of signs. A body with positive velocity has positive momentum, and a body with negative velocity has negative momentum. For motion along a line, vector addition is the same as ordinary addition.

The velocities of the two balls after the impact are v_1' and v_2' (see Figure 3.6). The total momentum before the collision must equal the total momentum afterward:

Momentum conservation: $m_1 v_1 + m_2 v_2 = m_1 v_1' + m_2 v_2' \tag{22}$

The kinetic energy of a body in motion is $\frac{1}{2}mv^2$ and that of a system of bodies is simply the sum of that of each body. Under idealized circumstances, the total kinetic energy of the two balls is unchanged by the collision. Such a collision, in which none of the kinetic energy is changed in form, is called an *elastic collision*. In such a case, the sum of the initial kinetic energies of the balls equals the sum of their final kinetic energies:

Kinetic energy conservation: $\frac{1}{2}m_1 v_1^2 + \frac{1}{2}m_2 v_2^2 = \frac{1}{2}m_1 v_1'^2 + \frac{1}{2}m_2 v_2'^2 \tag{23}$

Equations 22 and 23 can be solved for the final velocities v_1' and v_2' in terms of the initial velocities v_1 and v_2, but care must be taken to

FIGURE 3.6
Two balls collide with one another. Their initial velocities are v_1 and v_2. After the collision, their velocities are v_1' and v_2'. The motion takes place along a line.

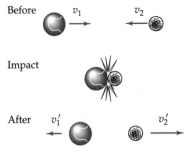

avert algebraic complexity—there are many ways to skin a cat, but some are easier than others. Let's rewrite the equations by bringing all terms involving m_1 to the left and all those involving m_2 to the right:

$$m_1(v_1 - v_1') = m_2(v_2' - v_2) \tag{24a}$$

$$m_1(v_1 - v_1')(v_1 + v_1') = m_2(v_2' - v_2)(v_2' + v_2) \tag{24b}$$

In equation 24b, we factored the expressions $v_1^2 - v_1'^2$ and $v_2'^2 - v_2^2$. These equations seem forbidding, but they are not difficult to deal with. If we divide each side of equation 24b by the corresponding side of equation 24a,* the masses drop out and we obtain a marvellously simple relation among the velocities: $v_1 + v_1' = v_2 + v_2'$. This formula is trying to tell us something. Rewritten as

$$V_{\text{rel}} = v_2 - v_1 = -(v_2' - v_1') \tag{25}$$

it says that the collision reverses the sign of the *relative velocity* of the two bodies (the difference between their velocities), but leaves its magnitude unchanged. After the collision, the balls separate from one another at the same rate that they approached each other beforehand.

Equations 24a and 25 are two linear equations in terms of the two unknowns v_1' and v_2'. They are readily solved. (Use equation 25 to express v_1' in terms of the other quantities. Insert the result into equation 24a to obtain a linear equation involving the single unknown quantity v_2'. Solve this equation to obtain the first of the following equations.) Here are the results:

$$\begin{aligned} v_1' &= \frac{\{m_1 v_1 + m_2 v_2\} + m_2(v_2 - v_1)}{m_1 + m_2} \\ v_2' &= \frac{\{m_1 v_1 + m_2 v_2\} + m_1(v_1 - v_2)}{m_1 + m_2} \end{aligned} \tag{26}$$

The combination of masses and velocities appearing within curly brackets is the total momentum. Divided by $m_1 + m_2$, it is the center-of-mass velocity defined by equation 8. Our results become more transparent when they are expressed in terms of the relative velocity V_{rel} and the center-of-mass velocity V_{cm}:

$$\begin{aligned} v_1' &= V_{\text{cm}} + \frac{m_2}{m_1 + m_2} V_{\text{rel}} \\ v_2' &= V_{\text{cm}} - \frac{m_1}{m_1 + m_2} V_{\text{rel}} \end{aligned} \tag{27}$$

Now that we have the answer, what do we do with it? For starters, suppose that the two balls have the same mass: $m_1 = m_2 = m$. In this case, equation 26 simplifies to yield $v_1' = v_2$ and $v_2' = v_1$. The velocity of one ball becomes that of the other, and vice versa. It is as if the

---

*This procedure is illegal if both sides of equation 24a vanish. That happens only when $v_1' = v_1$ and $v_2' = v_2$, corresponding to the trivial solution to equation 24a and 24b for which no collision takes place.

balls continue their uniform motion, passing through one another and swapping their identities!

EXAMPLE 3.4
Colliding Balls
of Equal Mass

At $t = 0$, a red ball is at $x = 0$ with a speed of 1.5 and a green ball is at $x = 1$ with a speed of 1. (The units of time, length, and velocity are seconds, meters, and meters per second.) Where and when does the collision take place? Where are the balls two seconds after the collision? (Treat the balls as if they were pointlike bodies.)

Solution

Until the collision, the position of the red ball is given by $r(t) = 1.5t$ and that of the green ball is $g(t) = t + 1$. They collide at such a time t_c that $r(t_c) = g(t_c)$, which is to say $1.5t_c = t_c + 1$, or $t_c = 2$. The collision takes place at the position $x = 3$. After the collision (for $t > 2$ s), the position of the red ball is given by $g(t)$ and that of the green ball by $r(t)$. At $t = 4$, the red ball is at $x = 5$, and the green ball is at $x = 6$. ∎

Consider another special case of two colliding bodies in which a moving ball (the projectile) approaches a stationary second body (the target). In this case, $v_1 \neq 0$ and $v_2 = 0$. We find from equation 26:

$$v_1' = \left(\frac{m_1 - m_2}{m_1 + m_2} \right) v_1 \qquad v_2' = \left(\frac{2m_1}{m_1 + m_2} \right) v_1 \qquad (28)$$

If the projectile is much lighter than the target, so that $m_1 << m_2$, it bounces backward. The velocity of the projectile after the collision is nearly its initial velocity v_1, but it is reversed in sign. The heavy target inches forward with speed $\sim 2(m_1/m_2)v_1$. (We may ignore the term m_1 in the denominator of equation 28 because it is so much smaller than m_2.)

If the projectile is much heavier than the target, $m_1 >> m_2$, it continues to move at nearly its initial velocity. (A bowling ball is not much affected by a collision with a Ping-Pong ball.) The target surges forward at practically twice the speed of the projectile.

You may find the preceding analysis to be heavy sledding. So did scientists of an earlier era. The following problem was considered by Gottfried Leibnitz and by the disciples of René Descartes.* A body with speed v and mass 4 kg collides with a stationary second body of mass 1 kg. If the collision is elastic, how rapidly does the second body recoil? Cartesians claimed the answer is $4v$. In 1685, Leibnitz argued for $2v$. Both were mistaken. Two years later, Newton published the correct solution in his world-shaking *Principia*. His answer, and ours, follows, directly from equation 28. It is 8/5 or 1.6 times v.

---

*Leibnitz and Descartes were both renowned philosophers and mathematicians. Descartes invented what we now call a Cartesian coordinate system (see Chapter 2). The question of who devised the methods of calculus, Leibnitz or Newton, raged throughout the eighteenth century. The controversy effectively split British mathematicians from their Continental colleagues. It was not resolved until the nineteenth century, when calculus evolved into a rigorous discipline.

A BIT MORE ABOUT
Elastic Collisions

Three identical elastic balls move along a line. What is the largest number of collisions they can make with one another?

Let's start with two balls on our one-dimensional table. Figure 3.7a is a graph of the balls' positions on the line as a function of time. Their initial relative velocity is negative, so they approach one another and collide. Both before and after the collision, each ball's trajectory is a straight line on the graph. When the collision takes place, the two balls interchange their velocities. Both straight lines continue, but the identity of the balls is switched. The collision is depicted as the intersection of two straight lines. There will be a collision if the ball on the left is initially moving rightward faster than the ball on the right is moving leftward.

If there are three balls, the maximum number of collisions is the number of times that three straight lines can intersect. Figure 3.7b shows that the answer is three. They will make three collisions if the ball on the left has the largest rightward velocity and the ball on the right has the least.

Now that you know the trick, you can solve this problem for any number of balls. How many collisions can four identical and elastic balls make with one another in one dimension? The answer is six, the largest number of intersections of four straight lines. The initial conditions may be chosen such as to yield any number of impacts from zero to six.

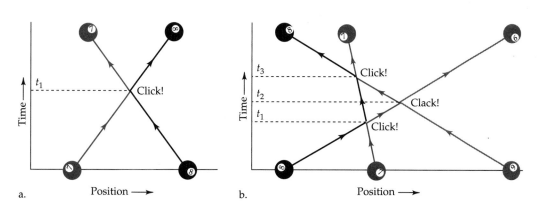

a.   b.

FIGURE 3.7 a. The positions of two balls on a line is shown as a function of time, which progresses upward. At $t = 0$ the two balls are approaching one another. At $t = t_1$, they collide. If they have the same mass and the collision is elastic, they simply exchange their velocities. At subsequent times, the left ball continues the initial trajectory of the right ball, and vice versa. The two balls move apart. b. Three balls of the same mass move along a line. Their posiitions are shown as a function of time. Initially, they are approaching one another. The left ball collides with the middle ball at t_1. Finally, the left ball collides once again with the middle ball. If the collisions are elastic, the three balls may make three collisions, but no more.

EXAMPLE 3.5
A Big Ball and
a Little Ball

Here is an example that you can try for yourself with a tennis ball and a basketball. Place the small ball a short distance above the center of the large ball. Drop both at the same time from a height of 1.8 m (Figure 3.8a). Describe what happens under the hypothesis that ball–ball and ball–ground collisions are perfectly elastic. This is an interesting and challenging problem. It would be best to do the deed before reading on.

Solution

In part b of Figure 3.8, the basketball touches ground in 0.6 s with a velocity of −6 m/s. At that time, the tennis ball has the same downward velocity. An instant later, the basketball bounces up and its velocity changes sign. It moves upward with velocity +6 m/s. In part c, it encounters the falling tennis ball, whose velocity is −6 m/s. Now, we come to the hard part: What happens when the two balls collide with equal and opposite velocities?

Let m_1 be the mass of the tennis ball and m_2 that of the basketball. Just before they collide, $v_1 = -6$ and $v_2 = +6$ in units of meters per second. We appeal to equation 26 to obtain the velocity of the tennis ball after the impact:

$$u_1 = \frac{18m_2 - 6m_1}{m_1 + m_2} \text{ m/s}$$

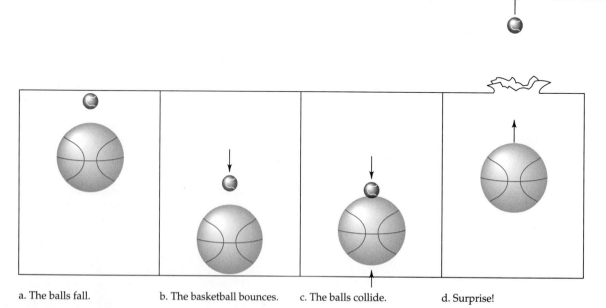

a. The balls fall.   b. The basketball bounces.   c. The balls collide.   d. Surprise!

FIGURE 3.8 In the first frame, a tennis ball is suspended above a basketball, and both are allowed to fall to the ground. The subsequent frames show what happens thereafter.

Basketballs are about five times heavier than tennis balls. Putting $m_2 = 5m_1$ into the above result, we find $u_1 = 14$ m/s. The result of the ball–ball collision is to increase the kinetic energy of the tennis ball by a factor of $(14/6)^2$ or ~ 5.44. It was dropped from a height of less than 2 m, but in part d, it bounces up to a height of nearly 10 m!   ∎

Inelastic Collisions

Most collisions we see about us are inelastic: Momentum is conserved but some kinetic energy is lost. Not even a superball bounces all the way back to the height from which it is dropped. Suppose that two balls with masses m_1 and m_2 move along a line and collide inelastically. As before, their initial velocities are v_1 and v_2, and their postimpact velocities are v_1' and v_2'. Conservation of momentum tells us that the total momentum is unchanged by the collision:

$$P = m_1 v_1' + m_2 v_2' = m_1 v_1 + m_2 v_2 \tag{29}$$

Suppose that the balls collide at $x = d$ and $t = 0$. Both before and after the collision, they move uniformly and we may write for their positions:

$$x_1(t) = \begin{cases} d + v_1 t & \text{for } t < 0 \\ d + v_1' t & \text{for } t > 0 \end{cases} \quad \text{and} \quad x_2(t) = \begin{cases} d + v_2 t & \text{for } t < 0 \\ d + v_2' t & \text{for } t > 0 \end{cases} \tag{30}$$

The center of mass $X_{cm}(t)$ of a system of objects was defined in equation 7. In this case, it is

$$X_{cm}(t) = \frac{m_1 x_1(t) + m_2 x_2(t)}{m_1 + m_2} \tag{31}$$

Substituting the expressions for $x_i(t)$ in equation 30 into equation 31, we find

$$X_{cm}(t) = d + \left(\frac{P}{m_1 + m_2} \right) t = d + Vt \tag{32}$$

where P is the total momentum and d is the location of the center-of-mass at $t = 0$, the moment of impact. Equation 32 holds true for all times, both before and after the collision. The center of mass of the two balls moves at the constant velocity $V = P/(m_1 + m_2)$.

EXAMPLE 3.6
A Game of Catch   John and Jane are 10 m apart on an icy, frictionless lake. Jane's mass is 40 kg exclusive of the 1-kg ball she holds. John's mass is 59 kg. Jane throws the ball to John with a horizontal velocity of 10 m/s. How far apart are Jane and John 10 s later? Show that their center of mass has not moved.

Solution   Let's start with Jane at $x = 0$ (Figure 3.9). John's initial position is $x = +10$. The center of mass of the isolated system consisting of John, Jane, and the ball is given by equation 31. It lies at a point between the players nearer to John than to Jane: $X_{cm} = +5.9$ m. We must analyze a sequence of two inelastic processes: the throw and the catch.

The throw: Jane throws the ball at time $t = 0$. She recoils backward with velocity v_1. The total momentum of the system consisting of Jane and the ball starts off at zero, and remains so. Thus, $40v_1 + 10 = 0$, and $v_1 = -0.25$ m/s. For $t > 0$, her position is given by $x_1(t) = -0.25t$. At $t = 10$, she is at $x_1 = -2.5$ m.

The catch: Before the catch, John is at rest at $x = 10$. The total momentum of the system consisting of John and the ball is that of the ball in flight, and so it remains. When John catches the ball at $t = 1$, he and the ball recoil with velocity v_2, where $60v_2 = 10$. Thus, $v_2 = +1/6$ m/s. After the catch, his position is $x_2(t) = 10 + v_2(t - 1) = 10 + (t - 1)/6$. At $t = 10$, he is at $x_2 = +11.5$ m. At this time, the players are 14 m apart. Their center of mass, according to equation 31, lies at $X_{cm} = (-2.5 \times 40 + 11.5 \times 60)/100 = +5.9$ m, just where it was at the beginning of the toss.

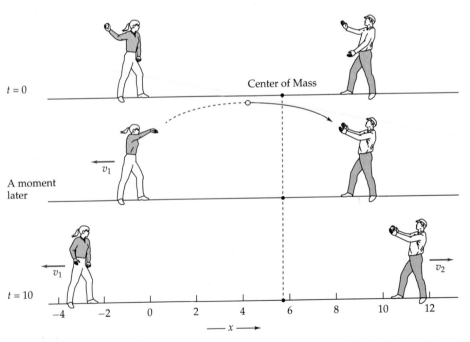

FIGURE 3.9 John and Jane are initially at rest 10 m apart on a frictionless surface. When Jane throws a ball to John, she recoils backward, acquiring a negative velocity v_1. When John catches the ball, he recoils backward and acquires a positive velocity v_2. The center of mass of Jane, John, and the ball remains fixed and is marked by an X.

The Coefficient of Restitution

Because momentum is conserved in an inelastic collision of two objects, we may determine the motion of their center of mass. However, kinetic energy is not conserved in such a collision. Equation 29 is insufficient to determine the final velocities of the objects in terms of their initial velocities. We must know something more to describe the outcome of an inelastic collision, something that tells us just how inelastic the collision is.

When two balls collide, whether elastically or inelastically, their center-of-mass velocity stays fixed, but their relative velocity changes. For an elastic collision in one dimension, equation 25 says that the relative velocity of the two bodies changes in sign, but not in magnitude. For an inelastic collision in one dimension, the relative velocity of the two bodies changes in sign, but it is also reduced in magnitude:

$$v_2' - v_1' = -\epsilon(v_2 - v_1) \tag{33}$$

The dimensionless factor ϵ lies between 0 and 1. It is called the *coefficient of restitution* or COR of the collision. Equation 33 replaces equation 25 in the case of an inelastic collision.

The value $\epsilon = 1$ corresponds to an elastic collision wherein no kinetic energy is lost. The value $\epsilon = 0$ corresponds to a inelastic collision wherein the largest possible amount of kinetic energy is lost. Example 3.1 illustrated such a collision. The car and the truck bodies remain in contact after the impact and move together as a single body with a center-of-mass velocity. Example 3.6 is another example of a collision with $\epsilon = 0$. Once John catches the ball, its velocity relative to him vanishes. In most familiar instances, as in the collision described in Figure 3.10, ϵ lies between these extremes.

Newton first introduced and named the coefficient of restitution. He found that the value of ϵ depends on the composition of the colliding bodies, but not on the velocity of the impact. For the collision of steel or

FIGURE 3.10
An inelastic collision. Two balls are shown *a.* before they collide, *b.* at the moment of collision, and *c.* after the impact. In this example, $m_1 = 3m_2$. Their initial velocities are $v_1 = 5$ and $v_2 = -3$. Their final velocities are $v_1' = 2$ and $v_2' = 6$. The center-of-mass velocity of the system of two balls is 3 m/s, both before and after the collision. Their relative velocity is -8 m/s beforehand, and $+4$ m/s afterward. The coefficient of restitution for this collision is $\epsilon = 0.5$.

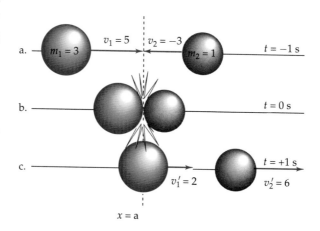

ivory balls, ϵ is much larger than it is for wooden balls. It is very small for balls fashioned from wet sponge or raw liver.

For the game of baseball, the major leagues specify the elastic properties of the ball. To be acceptable for play, its collisions must be neither too elastic nor too inelastic. Regulation balls thrown against a rigid ash wall at 58 mph must rebound at a speed lying between 30 and 33.5 mph. The officially permitted value of ϵ for bat–baseball impacts lies between 0.52 and 0.58.

The collisions of billiard balls are considerably more elastic than those of baseballs. Nonetheless, a significant portion of the kinetic energy is lost in a ball–ball collision. If one ball makes a head-on collision with another that is held firmly in place, the moving ball bounces back at ~ 80 percent of its initial speed. For billiard ball collisions, $\epsilon \simeq 0.8$.

In skeet shooting, a target is launched into the air and shot at. If a bullet lodges in the flying skeet, the two bodies thereafter move together. The collision is maximally inelastic and $\epsilon = 0$.

EXAMPLE 3.7
Croquet

The collisions of wooden balls, such as are used in the game of croquet, are not elastic. The appropriate coefficient of restitution may be assumed to be $\epsilon = 0.5$. If a croquet ball with a velocity of 6 m/s collides head-on with a stationary ball of the same mass, what are the velocities of the balls afterward?

Solution

The initial velocities are $v_1 = 6$ and $v_2 = 0$. Their final velocities are v_1' and v_2'. The center-of-mass velocity of two balls of the same mass is given by equation 8 to be $V_{cm} = \frac{1}{2}(v_1 + v_2) = 3$ m/s. Because it does not change as a result of the collision, $v_1' + v_2' = 6$ m/s. According to equation 33, $v_2' - v_1' = -0.5(v_2 - v_1)$. Solving this system of two simultaneous equations, we find $v_1' = 1.5$ m/s and $v_2' = 4.5$ m/s. ∎

Galilean Relativity and the Game of Billiards

Galileo claimed that the laws of physics are the same in any uniformly moving reference frames. Two observers, Norma and Ramon, decide to test the idea (Figure 3.11). Norma's coordinate system is at rest relative to a frictionless, one-dimensional billiard table. All billiard balls have the same mass m and the coefficient of restitution for ball–ball collisions is taken to be $\epsilon = 0.8$.

Norma sees two identical balls moving toward one another (Figure 3.11a). She measures their velocities both before and after the collision:

$$v_1 = +v \quad \text{and} \quad v_2 = -v \qquad \text{beforehand}$$

$$v_1' = -0.8v \quad \text{and} \quad v_2' = +0.8v \qquad \text{afterward}$$

FIGURE 3.11
Norma and Ramon observe an inelastic collision of two balls. Norma is at rest relative to the center of mass of the balls. *a.* The positions and velocities of the balls 1 s before and 1 s after the impact in Norma's reference frame. As she sees it, the ball on the left starts off moving to the right with $v_1 = v$, whereas the ball on the right starts off moving to the left with $v_2 - v$. Ramon is moving to the left with velocity $-v$ relative to Norma. *b.* The same data are shown in Ramon's reference frame. As he sees it, the ball on the right starts off at rest, whereas the ball on the left starts off with $\hat{v}_1 = +2v$.

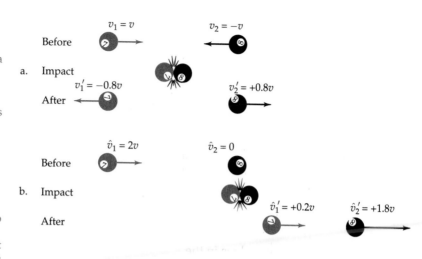

According to Norma, the center of mass of the two balls is stationary. Momentum is conserved, but the relative velocity afterward is 80 percent of what it was initially. The initial kinetic energy of the system of two balls is mv^2, while its final kinetic energy is $0.64\ mv^2$. The collision is not elastic and 36 percent of the initial kinetic energy is lost.

Ramon is moving leftward at velocity $-v$. The velocities that he measures are given little hats. Ramon observes the ball on the right to be at rest before the collision, $\hat{v}_2 = 0$. All of the velocities that Ramon sees are those seen by Norma, but increased by v due to his motion relative to her:

$$\hat{v}_1 = 2v \quad \text{and} \quad \hat{v}_2 = 0 \qquad \text{beforehand}$$
$$\hat{u}_1 = +0.2\ v \quad \text{and} \quad \hat{u}_2 = +1.8\ v \qquad \text{afterward.}$$

To Ramon, as to Norma, momentum is conserved and the relative velocity of the balls is reduced to 0.8 of its original magnitude. But what about energy?

According to Ramon, the initial kinetic energy is $\frac{1}{2}m(\hat{v}_1^2 + \hat{v}_2^2)$, or $2mv^2$. A similar calculation yields a final kinetic energy $1.64\ mv^2$. The collision is not elastic and 18 percent of the initial kinetic energy is lost. Norma concludes that the collision is 64 percent elastic; Ramon finds 82 percent. Which result is correct?

Neither observer erred. The percentage of the kinetic energy lost in the collision is different in the two reference frames. The physically significant quantity is not the percentage but the amount of kinetic energy turned into other forms. The loss of kinetic energy in Norma's frame is $(1 - 0.64)mv^2$. In Ramon's frame, it is $(2 - 1.64)mv^2$. In both cases, the kinetic energy lost in the collision is $0.36\ mv^2$.

A BIT MORE ABOUT
Reference Frames

Here's a puzzle that baffled many of my colleagues. Three observers (let them be Norma and Ramon again, along with their friend Armon) observe and interpret a physical process, but they come to wildly different conclusions. The process is simply this: A system consisting of a motorcycle and its rider (with a combined mass of 100 kg) starts at rest and accelerates to a ground speed of 100 m/s in a time T.

Armon (Figure 3.12a), who stands nearby, finds that the initial kinetic energy of the system is zero. When the action is completed, it is 5×10^5 J. Where did the kinetic energy come from? From the fuel, of course. Say that 10 ml (milliliters) of gasoline were consumed by the motorcycle engine. Armon computes the efficiency of the engine: its useful energy output per volume of fuel consumed. The energy output is the final kinetic energy minus the initial kinetic energy. The efficiency of the engine is 50 MJ/L (megajoules per liter).

Norma (Figure 3.12b) watches the action from a car moving at 100 m/s in the same direction as the motorcycle. Relative to her, the initial velocity of the motorcycle is −100 m/s. When it completes its acceleration, the speed of the motorcycle is zero relative to Norma. In her reference frame, the kinetic energy of the motorcycle starts off at 5×10^5 J and ends up zero. Yet, the same amount of fuel is used up. Norma deduces a negative efficiency for the motorcycle engine! Didn't Galileo tell us that every uniformly moving observer should come to the same physical conclusions?

Ramon (Figure 3.12c) offers a third horn to the dilemma. He cruises by in a car at a uni-

form 50 m/s. The initial velocity of the bike relative to Ramon is −50 m/s. Its final velocity is +50 m/s. The motorcycle ends up with the same kinetic energy as it started with. The efficiency of the engine, according to Ramon, is zero.

Armon's result for the efficiency is positive, Norma's is negative, and Ramon's is zero. Which observer is right and why are they all coming to different conclusions? Pause here and think for a moment, because I'm about to resolve the seeming paradox.

Armon is right, but it's a two-body problem! One body is the bike and its rider. The other is Earth itself. Earth's tiny recoil is entirely irrelevant to Armon's analysis, but it cannot be neglected by Norma or Ramon. All of the observers get the same answer for the efficiency of the engine when they take the recoil of Earth into account.

Call the mass of the bike and rider m and that of Earth M. According to Armon, the initial momentum of the two-body system is zero: They are at rest relative to her. Afterward, the momentum of the bike is $+mv$ and that of Earth is $-mv$. Earth recoils backward at the tiny speed mv/M, which for the case at hand is about the breadth of an atom in 1000 years. The kinetic energy of Earth relative to Armon is zero at first; afterward it is m/M times that of the bike. Armon may safely neglect Earth's recoil.

With respect to Norma, both the bike and Earth are moving with velocity $-v$ at the start. The initial total momentum is $-(m + M)v$, and the initial total kinetic energy is

$$E_i = (m + M)v^2/2$$

After the bike accelerates, it is at rest relative to Norma, with neither momentum nor kinetic energy. All of the initial momentum is carried by Earth, whose velocity afterward is $-v(1+m/M)$. Since the velocity of Earth has changed, so has its kinetic energy, which is the total kinetic energy of the system at the conclusion of the process:

$$E_f = Mv^2(1+m/M)^2/2 \simeq (M+2m)v^2/2$$

where I have neglected a tiny term proportional to m/M in the expression on the right. Notice that E_f is larger than E_i by the amount $mv^2/2$. The increase in kinetic energy of the two-body system is the same as what Armon found. Galileo was right after all! The proper treatment of this problem from Ramon's point of view is left to you.

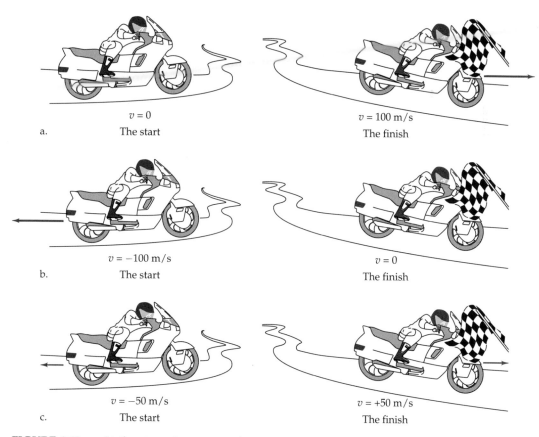

FIGURE 3.12 *a.* At the start, the motorcycle is at rest relative to Armon. At the finish, its velocity is 100 m/s. *b.* At the start, the velocity of the motorcycle is -100 m/s relative to Norma, whose ground velocity is $+100$ m/s. At the finish, the motorcycle is at rest in her frame. *c.* Ramon is moving relative to the ground at a velocity of 50 m/s. In his frame, the motorcycle starts with a velocity of -50 m/s and finishes with a velocity of $+50$ m/s.

Exercises

7. One train departs the station at noon and travels westward at 50 mph. A second train departs from the same station at 3 P.M., and travels westward along the same track at 70 mph. Where and when will an exceedingly inelastic collision take place?

8. Standing on a frictionless surface, you catch a 250-g ball moving at 40 m/s. If your mass is 79.75 kg, how fast do you recoil backward?

9. A 3-kg ball, initially at rest, is struck by a 1-kg ball moving to the right at 80 m/s. Subsequently, the 3-kg ball (now in motion) is struck by a second 1-kg mass moving to the left at 80 m/s. After both of these elastic collisions, in which direction (right or left) does the heavier ball move, and with what speed?

10. A 4-kg body moving at 8 m/s collides with a 1-kg body at rest. No external forces act on the bodies. The impact is elastic and the motion takes place along a line.
   a. What is the the kinetic energy of the system?
   b. What are the velocities of the two balls after the collision?
   c. What is the kinetic energy of the struck ball after the collision?

11. Suppose the impact described in exercise 10 is not elastic and the two bodies stick together after colliding. In this case, what is the kinetic energy of the system after the collision?

12. A 3-kg body moving at 4 m/s collides with a 1-kg body moving in the opposite direction at the same speed. No external forces act on the bodies. The impact is elastic and the motion takes place along a line.
   a. What are the kinetic energies of the balls before the collision?
   b. What are the velocities of the balls after the collision?
   c. What are the kinetic energies of the balls after the collision?

13. A, B, and C are three balls with masses of 1 kg, 3 kg, and 9 kg, respectively. Initially, A moves at 16 m/s while both heavier balls are at rest. A strikes B, which then strikes C. The collisions are elastic and all motion takes place along a line. What is the velocity of C after it is struck?

14. A pitched ball traveling at 80 mph encounters a swinging bat moving toward the mound at 70 mph. Assume that the mass of the ball is negligible compared to that of the bat, and that a baseball rebounds at half its speed from a stationary bat. What is the speed of the hit ball relative to the ground?

15. When a Ping-Pong ball is dropped to the ground from a height of 1 m, it bounces back to a height of 64 cm. What is the COR of the impact?

16. A car of mass m_1 travels north at speed V. A car of mass m_2 travels south at the same speed. They collide, forming a stuck-together wreck.
    a. What is the velocity of the wreck on the icy road?
    b. What is the ratio of the wreck's kinetic energy to the initial kinetic energy of the two cars?

17. The COR for the collision of croquet balls is one-half. If a ball moving at 4 m/s makes a head-on collision with an identical ball at rest, what speed does the struck ball acquire?

18. A 5-g bullet traveling horizontally at 500 m/s smashes through a 5-kg computer terminal sitting on a frictionless floor and emerges at a speed of 300 m/s. After the incident, what is the speed of the computer terminal?

19. Consider the puzzle illustrated in Figure 3.12. In Ramon's reference frame, the initial and final kinetic energies of the motorcycle are the same. Take Earth's recoil into account to show that the kinetic energy of the Earth has increased by $\frac{1}{2}mv^2$ in Ramon's reference frame.

20. A train is moving at 10 m/s. A 15-kg boy wearing frictionless roller skates stands at rest in the train. He catches a 1-kg ball moving at 8 m/s relative to the train and in the same direction. In the reference frame of the train, calculate the initial and final kinetic energies of the isolated system consisting of the boy and the ball. Perform the same calculations in the reference frame of an observer on the ground. Confirm that the amount of kinetic energy lost is the same in the two cases.

3.3 Multiplying Vectors

The histories of science and mathematics are linked. Newton and Einstein had to invent new mathematics to extend the frontiers of physical knowledge. Maxwell could not have formulated the laws of electrodynamics without differential equations. Today's theory of elementary particles is founded on what was regarded as pure mathematics only a generation ago. As I describe the history and substance of physics, I must translate intricate mathematical arguments into algebra, geometry, and ordinary English. No scientific accomplishment is worth its salt unless it can be described to the layperson. But beware! Any act of translation loses much of the flavor and beauty of the original. Having said this, I now discuss some more properties of vectors, since they are one branch of mathematics that physicists simply cannot do without.

In Chapter 2, you were told how to add and subtract vectors, and how to multiply them by ordinary numbers or scalars. Vectors can also be multiplied together. In fact, there are two distinct operations of vector multiplication, called the *dot product* (or scalar product) and the *cross product* (or vector product). Both operations are much used in physics. Let **A** and **B** be vectors to be multiplied together. Their magnitudes are A and B and the angle between them is θ (Figure 3.13).

FIGURE 3.13
The vectors **A** and **B** are shown. For simplicity, they are taken to lie in the x-y plane. Their dot product, **A** · **B**, is the number $AB \cos\theta$, where θ is the angle between them. The dot product may also be expressed in terms of the components of the vectors. **A** · **B** = $A_x B_x + A_y B_y$. For the vectors shown, the result is $A \cdot B = -2$. Their cross product, **A** × **B**, is a vector pointing in the z direction, up and out of the page. Its magnitude is $AB \sin\theta$. Since $(A \times B)_z = A_x B_y - A_y B_x$, its magnitude is $+6$.

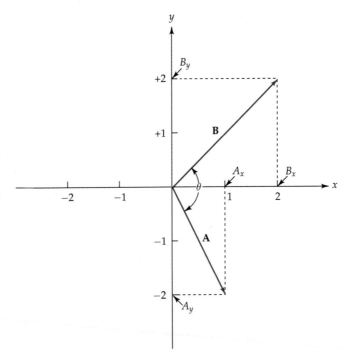

The Dot Product

The dot product of **A** and **B** is indicated by a dot between the factors. **A** · **B** is an undirected number, or scalar, whose value depends on the magnitudes of the factors and on the angle between them:

$$\mathbf{A} \cdot \mathbf{B} = AB \cos\theta \tag{34}$$

If **A** and **B** point in the same direction, their dot product is the ordinary product of their magnitudes. If they are perpendicular to one another, their dot product is zero. If they point in opposite directions, their dot product is minus the product of their magnitudes. The dot product shares many properties with ordinary multiplication. If **A**, **B**, and **C** are any three vectors, and p is a scalar, then:

$$\mathbf{A} \cdot \mathbf{B} = \mathbf{B} \cdot \mathbf{A} \tag{35a}$$

$$\mathbf{A} \cdot (\mathbf{B} + \mathbf{C}) = \mathbf{A} \cdot \mathbf{B} + \mathbf{A} \cdot \mathbf{C} \tag{35b}$$

$$\mathbf{A} \cdot (p\mathbf{B}) = p(\mathbf{A} \cdot \mathbf{B}) \tag{35c}$$

Equation 35a says that the order of the factors in a dot product doesn't matter: Dot multiplication is commutative. Equation 35b says that dot multiplication and vector addition satisfy the distributive law of arithmetic. Equation 35c says that the dot product of a vector with the product of a scalar and a vector is the ordinary product of the scalar with the dot product of the two vectors.

The dot product of two vectors may be expressed in terms of the components of the factors. The following result can be deduced from equation 34:

$$\mathbf{A} \cdot \mathbf{B} = A_x B_x + A_y B_y + A_z B_z \tag{36}$$

Equation 36 implies that the dot product of a vector with itself is the square of its magnitude.

The Cross Product

The dot product of any two vectors is a scalar. On the other hand, the cross product of $\mathbf{A}$ and $\mathbf{B}$ is another vector, which we write as $\mathbf{A} \times \mathbf{B}$. It has both a direction and a magnitude. Its magnitude $|\mathbf{A} \times \mathbf{B}|$ depends both on the magnitudes of the vectors and on the angle between them:

$$|\mathbf{A} \times \mathbf{B}| = AB |\sin\theta| \tag{37}$$

If $\mathbf{A}$ and $\mathbf{B}$ point in the same or opposite directions, their cross product is zero. If they are perpendicular to one another, the magnitude of their cross product is the ordinary product of their magnitudes. In general, the magnitude of the cross product of two vectors is the area of the parallelogram of which the vectors form two sides.

The direction of $\mathbf{A} \times \mathbf{B}$ is perpendicular to both $\mathbf{A}$ and $\mathbf{B}$. That is, if two vectors lie in the plane of this page, their cross product points either into the page or out of the page. To determine which, we appeal to the *right-hand rule*. Point the index finger of your right hand along $\mathbf{A}$. Then align the middle finger with $\mathbf{B}$. Your raised thumb points in the direction of $\mathbf{A} \times \mathbf{B}$.* Here are some properties of the cross product. If $\mathbf{A}$, $\mathbf{B}$, and $\mathbf{C}$ are any three vectors, and p is a scalar, then:

$$\mathbf{A} \times \mathbf{B} = -\mathbf{B} \times \mathbf{A} \tag{38a}$$

$$\mathbf{A} \times (\mathbf{B} + \mathbf{C}) = \mathbf{A} \times \mathbf{B} + \mathbf{A} \times \mathbf{C} \tag{38b}$$

$$\mathbf{A} \times (p\mathbf{B}) = p(\mathbf{A} \times \mathbf{B}) \tag{38c}$$

Equation 38a says that the order of the factors does matter: The operation of taking a cross product is not commutative. Equation 38b says that the cross product and vector addition satisfy the distributive law of arithmetic. Equation 38c says that the cross product of a vector with the product of a scalar and a vector is the ordinary product of the scalar with the cross product of the two vectors.

The components of the cross product of two vectors may be expressed in terms of the components of its factors. The result depends on the choice of the coordinate system. A *right-handed coordinate system* has

---

*For instances in which this feat is impossible, point your right index finger along $\mathbf{B}$ and your middle finger along $\mathbf{A}$. Your thumb now indicates the direction opposite to that of $\mathbf{A} \times \mathbf{B}$.

the following property: If you point the index finger of your right hand along the x-axis and your middle finger along the y-axis, your thumb shows the direction of the z-axis. The following equation applies in any right-handed coordinate system:

$$(\mathbf{A} \times \mathbf{B})_x = A_y B_z - A_z B_y$$

$$(\mathbf{A} \times \mathbf{B})_y = A_z B_x - A_x B_z \qquad (39)$$

$$(\mathbf{A} \times \mathbf{B})_z = A_x B_y - A_y B_x$$

This complicated formula is presented for the sake of logical completeness and to show that the cross product can be evaluated in terms of components. The cross product is essential for the description of such things as angular momentum and magnetic forces. However, we shall make use of it rarely, and only when necessary. Notice that equation 39 implies that the cross product of two parallel vectors is zero.

EXAMPLE 3.8
The Dot Product

The vector **F** has magnitude 10 and points 30° north of east. **G** has magnitude 6 and points 30° north of west (Figure 3.14). Evaluate $\mathbf{F} \cdot \mathbf{G}$.

Solution

Solution 1 The angle between the two vectors is 120°. Their dot product is obtained from equation 34, with $\cos 120° = -0.5$, and the magnitudes set equal to 10 and 6. The result is $\mathbf{F} \cdot \mathbf{G} = -30$.

FIGURE 3.14
F is a vector of magnitude 10 pointing 30° east of north. Its components are $(+5\sqrt{3}, +2)$. **G** is a vector of magnitude 6 pointing 30° west of north. Its components are $(-3\sqrt{3}, +3)$.

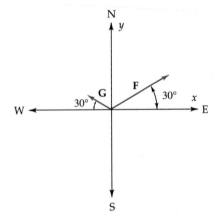

Solution 2 We may choose a coordinate system and use equation 36 to evaluate the dot product. Pick the x-axis to point east and the y-axis north. In this coordinate system, both vectors lie in the x-y plane and their components may be determined trigonometrically:

$$F_x = 5\sqrt{3}, \quad F_y = 5 \quad \text{and} \quad G_x = -3\sqrt{3}, \quad G_y = 3$$

The desired dot product is $F_x G_x + F_y G_y$ or $-45 + 15 = -30$. ■

EXAMPLE 3.9
The Cross Product

What is the cross product of the vectors **F** and **G** defined in Example 3.8?

Solution

Solution 1 The cross product is obtained from equation 37, with $|\sin 120°| = \sqrt{3}/2$ and the magnitudes as before. The magnitude of **F** × **G** is $30\sqrt{3}$. The right-hand rule says that the direction of **F** × **G** is up and out of the page.

Solution 2 If we adopt a right-handed coordinate system, whose z-axis points up, we can use equation 39 to evaluate the cross product. Since **F** and **G** lie on the x-y plane, their z components vanish, $F_z = G_z = 0$. Consequently, the x and y components of **F** × **G** are zero. For its z component, we obtain $F_x G_y - F_y G_x = (5\sqrt{3})3 - 5(-3\sqrt{3})$. The only nonzero component of the cross product is $(F × G)_z = 30\sqrt{3}$. ∎

Exercises

21. Consider a Cartesian coordinate system whose x-axis points North, y-axis points East, and z-axis points up. Is such a coordinate system right-handed or left-handed?

22. A 1-kg mass moves at a speed of 7 m/s. A 4-kg mass moves at a speed of 2 m/s in a direction 60° different from that of the first mass. Compute the magnitude of the sum of the two momenta in kilogram meters per second. (Use the component method to add vectors.)

23. Suppose that a clock has a 5-in. hour hand and an 8-in. minute hand. For the following questions, treat the hands as the vectors **h** and **m**, respectively. Evaluate **h** · **m** and **h** × **m**:
    a. At noon
    b. At 3:00
    c. At 4:30

24. Let **C** be a vector with magnitude $C = 4$ pointing 30° west of south, and **D** be a vector with magnitude $D = 3$ pointing 30° east of south.
    a. Compute the value of **C** · **D**.
    b. Determine the direction and magnitude of **C** × **D**.

25. Let **X** and **Y** be two vectors drawn from the same origin. Show that the magnitude of their cross product is twice the area of the triangle whose vertices are their tips and the origin.

26. Show that **A** · (**A** × **B**) = 0 for any two vectors **A** and **B**. Using equations 36 and 39, prove that the relation **A** · (**B** × **C**) = **B** · (**C** × **A**) holds for any three vectors.

3.4 Motion in Three-Dimensional Space

We can readily extend what we have learned about force, momentum, and energy to the case of motion in three dimensions. Let's return to the study of the motion of an object in response to a constant force **F**.

At the time t_1, the position and velocity of the object are **r** and **v**. At the later time t_2, its position and velocity are **r**$'$ and **v**$'$ (Figure 3.15). The acceleration of the object **a**, according to Newton's second law, is **F**$/m$. It points in the same direction as the force. If the force is constant, the velocity of the object changes at a fixed rate, so that:

$$\mathbf{v}' = \mathbf{v} + (\mathbf{F}/m)(t_2 - t_1) \tag{40}$$

Equation 40 may be rewritten as follows:

$$\mathbf{F}\,\Delta t = m\mathbf{v}' - m\mathbf{v} \tag{41}$$

where $\Delta t = t_2 - t_1$. The quantity $\mathbf{p} = m\mathbf{v}$ is the momentum of the object, and equation 41 says that the product of the force and the time over which it acts is the change in momentum of the object.

If the force is not constant, this result may still be used if the time interval Δt is small enough that the change in the force may be ignored. In that case, equation 41 takes the form:

$$\mathbf{F} = \frac{\Delta \mathbf{p}}{\Delta t} \tag{42}$$

and says that force is the instantaneous rate of change of momentum.

For a constant force **F**, the average velocity of the object during the time interval Δt is the average of its initial and final velocities, $\frac{1}{2}(\mathbf{v} + \mathbf{v}')$. The vector distance it has traveled is the product of its average velocity and the time interval:

$$\mathbf{r}' - \mathbf{r} = \tfrac{1}{2}(\mathbf{v} + \mathbf{v}')\,\Delta t \tag{43}$$

Equation 40 can be used to eliminate **v**$'$ from equation 43 to obtain an expression for **r**$'$ in terms of the initial velocity **v**, the position **r**, and the time interval Δt:

$$\mathbf{r}' = \mathbf{r} + \mathbf{v}\Delta t + \tfrac{1}{2}(\mathbf{F}/m)(\Delta t)^2 \tag{44}$$

This result allows us to study the motion of an object in response to the constant force of gravity—projectile motion, as described in Chapter 2—but in vector notation. Choose a coordinate system lying in the vertical plane and containing the initial velocity **v**. The x-axis points up and the y-axis points horizontally. The third direction in space plays no role. The components of the position **r** of the projectile are x and y, those of its velocity **v** are v_x and v_y, and those of the force of gravity **F** are $F_x = -mg$ and $F_y = 0$. In terms of its components, equation 44 reads

$$x' = x + v_x\,\Delta t - \tfrac{1}{2}g(\Delta t)^2$$
$$y' = y + v_y\,\Delta t \tag{45}$$

In the x direction, the object undergoes constant acceleration, as in equation 2.16. In the y direction, the object travels in uniform motion, as in equation 2.17. What vectors have done for us is to express these results compactly in equation 44.

The total momentum of a system of objects is the vector sum of the momenta of all the bodies making up the system:

$$\mathbf{P}_{total} = m_a\mathbf{v}_a(t) + m_b\mathbf{v}_b(t) + \cdots + = \mathbf{p}_a(t) + \mathbf{p}_b(t) + \cdots \tag{46}$$

where $\mathbf{p}_a = m_a\mathbf{v}_a$, $\mathbf{p}_b = m_b\mathbf{v}_b$, ... If the system is isolated from external forces, $\mathbf{P}_{total}$ does not change. This result follows from Newton's law of action and reaction: The force that object a exerts on object b is equal and opposite to the force that b exerts on a. Thus, momentum may be transferred among the objects, but it is neither created nor destroyed. For any isolated system consisting of bodies interacting with one another by whatever mechanism, the sum of the momenta of the constituent particles can never change. Each of the three components of $\mathbf{P}_{total}$ of such a system is said to be a constant of the motion.

The center of mass of a system of objects in three-dimensional space is defined in analogy with equation 7:

$$\mathbf{X}_{cm} = \frac{m_A\mathbf{r}_A(t) + m_B\mathbf{r}_B(t) + \cdots}{m_A + m_B + \cdots} \tag{47}$$

If the system is isolated, its center-of-mass velocity is constant,

$$\mathbf{V}_{cm} = \frac{\mathbf{P}_{total}}{M_{total}} \tag{48}$$

and the center-of-mass position uniformly:

$$\mathbf{X}_{cm} = \mathbf{V}_{cm}t + \mathbf{D}$$

FIGURE 3.15
An object moves under the influence of a constant force $\mathbf{F}$ pointing in the x direction. At time t_1, it is at the vector position $\mathbf{r}$ with velocity $\mathbf{v}$. At the later time t_2, the object is at $\mathbf{r}'$ with velocity $\mathbf{v}'$. In the time interval $\Delta t = t_2 - t_1$, it has moved a vector distance $\Delta\mathbf{r} = \mathbf{r}' - \mathbf{r}$.

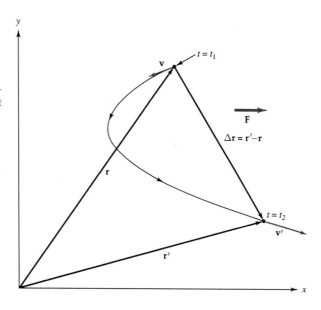

where **D** is the vector position of the center of mass at $t = 0$.

Since the Moon is about 1.2 percent as massive as Earth, the center of mass of the Earth–Moon system (the point P in Figure 3.16) lies about 1.2 percent of the way from the center of Earth to the Moon, and 1500 km under the surface of Earth. If we forget about the Sun, the point P would persist forever in a state of uniform motion, with Earth and the Moon rotating in concentric circles about it. But the Earth–Moon system is not isolated: The Sun attracts both bodies. The net effect of the Sun's attraction is as if all of the mass of Earth and the Moon were concentrated at P, which is the point that moves about the Sun in an elliptical orbit and satisfies Kepler's laws.

FIGURE 3.16
The center of the Moon M and the center of Earth E describe concentric circular orbits about P, their center of mass, which lies about 1500 km under Earth's surface. In about one month, the center of the Moon moves to M', while the center of the Earth moves to E'. The diagram is not to scale. In fact, the distance from P to M is about 80 times the distance from P to E.

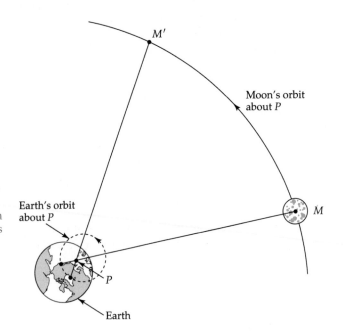

Work and Energy in Space

A little bit of mathematics will reveal the power of vector methods. If we take the dot product of each side of equation 43 by **F**, we obtain:

$$\mathbf{F} \cdot \Delta \mathbf{r} = \tfrac{1}{2}\mathbf{F} \cdot (\mathbf{v} + \mathbf{v}')\, \Delta t$$

where $\Delta \mathbf{r} = \mathbf{r}' - \mathbf{r}$. We use equation 41 to eliminate **F** from the right-hand side to find

$$\mathbf{F} \cdot \Delta \mathbf{r} = \tfrac{1}{2}m(\mathbf{v}' - \mathbf{v}) \cdot (\mathbf{v}' + \mathbf{v})$$
$$= \tfrac{1}{2}mv'^2 - \tfrac{1}{2}mv^2 \tag{49}$$

To get this result, we used the facts that $\mathbf{v} \cdot \mathbf{v} = v^2$, $\mathbf{v}' \cdot \mathbf{v}' = v'^2$, and $\mathbf{v} \cdot \mathbf{v}' = \mathbf{v}' \cdot \mathbf{v}$. The right-hand side of equation 49 is the change of the object's kinetic energy. The left-hand side is the work done by the force in moving the object the vector distance $\Delta\mathbf{r}$:

$$\text{Work done by a force} = \mathbf{F} \cdot \Delta\mathbf{r} \qquad (50)$$

If the force is not constant, $\Delta\mathbf{r}$ must be sufficiently small that the change in the force is negligible. Otherwise, $\Delta\mathbf{r}$ must be broken up into many tiny intervals, to each of which equation 50 is applied.

Equation 49 may be written as:

$$E = \tfrac{1}{2}mv'^2 - \mathbf{F} \cdot \mathbf{r}' = \tfrac{1}{2}mv^2 - \mathbf{F} \cdot \mathbf{r} \qquad (51)$$

where E is the total energy of the object. Its potential energy at any point $\mathbf{r}$ is $-\mathbf{F} \cdot \mathbf{r}$, relative to the origin. If the force is Earth's gravity and the origin is at the ground, the potential energy of an object is mgh, where h is its height above the ground.

Figure 3.17 is a schematic drawing of an ideal and frictionless roller coaster. The carriage is constrained to move along a track, but otherwise it is in free fall. The carriage reaches the point A, at height h, with velocity $\mathbf{v}$. Later, it reaches the point B at height h', with velocity $\mathbf{v}'$. At each point in its path, the carriage is subject to two forces. One is the vertical force of gravity. Another force $\mathbf{T}$ is exerted by the track on the carriage to keep the carriage on the track. If the track is frictionless, $\mathbf{T}$ is perpendicular to the instantaneous velocity of the carriage and does no work. We may apply equation 49 without taking $\mathbf{T}$ into account. Since gravity acts in the vertical direction, $\mathbf{F} \cdot \Delta\mathbf{r} = -mg(h' - h)$ and

$$\tfrac{1}{2}mv'^2 - \tfrac{1}{2}mv^2 = -mg(h' - h) \qquad \text{or} \qquad v' = \sqrt{v^2 + g(h - h')}$$

The final velocity of the carriage is determined by its initial velocity and the vertical distance it has fallen. The shape of the track plays no role.

FIGURE 3.17
The carriage reaches the point A, at a height h, with a velocity of $\mathbf{v}$. It later reaches the higher point B, at the height h', with velocity $\mathbf{v}'$. The change in kinetic energy of the carriage is $\mathbf{F} \cdot \Delta\mathbf{r}$. In the case shown, the kinetic energy of the carriage decreases by the quantity $mg(h' - h)$, where m is the carriage mass.

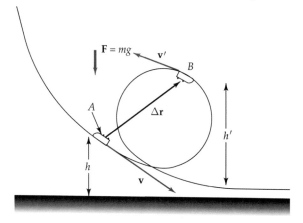

Collisions in Space

Let's begin our discussion of collisions in space with the elastic collision of two balls of the same mass m moving on a frictionless billiard table. Originally, the cue ball moves at velocity $\mathbf{w}$ and the target ball is at rest. Their collision is not central, and the subsequent motion of the balls is not confined to a line. After the impact, the cue ball velocity is **u** and the target ball velocity is **v**. The law of conservation of momentum yields the simple vector relation

$$m\mathbf{w} = m\mathbf{u} + m\mathbf{v} \qquad \text{or} \qquad \mathbf{w} = \mathbf{u} + \mathbf{v} \tag{52}$$

Equation 52 says that the three velocity vectors form a triangle, as shown in Figure 3.18.

Because the collision is elastic, kinetic energy is conserved. The initial kinetic energy of the cue ball is shared between the two balls. The conservation of kinetic energy says

$$\tfrac{1}{2}mw^2 = \tfrac{1}{2}mu^2 + \tfrac{1}{2}mv^2 \qquad \text{or} \qquad w^2 = u^2 + v^2 \tag{53}$$

where u^2, v^2, w^2 are the squares of the magnitudes of the corresponding vectors. The square of the length of one side of triangle formed by **u**, **v**, and **w** is the sum of the squares of the lengths of the others. Thus, the triangle formed by the three velocities is a right triangle whose hypotenuse is **w**. We have learned that the velocities of the two balls after the collision are mutually perpendicular.

If the collision is peripheral, the cue ball loses very little energy and is only slightly deflected. The target ball moves slowly in a perpendicular direction. In the case of an almost central collision, the situation is reversed, with the cue ball emerging almost perpendicular to its original direction. At a certain point between these extremes, the two balls share the energy equally. Each ball is deflected by 45° in opposite directions.

What happens when two moving bodies with *different* masses collide with one another *inelastically* in three-dimensional space? We may choose any inertial frame from which to study this problem. It's easiest to look at the collision from the point of view of an observer who is moving along with the center of mass. In this frame, the total momentum of the

FIGURE 3.18
An elastic collision of balls of equal mass. Initially, the cue ball moves with velocity **w**. It strikes a second ball off center and moves in another direction with velocity **u**. The struck ball acquires velocity **v**. The three velocity vectors form a right triangle, as shown in the insert.

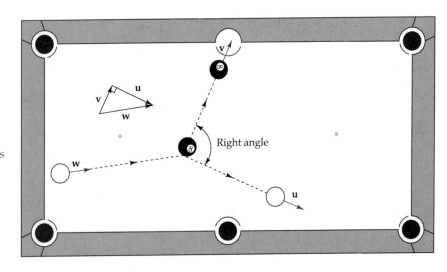

system of two balls is zero. Their center of mass remains fixed at the origin (Figure 3.19).

The masses of the balls are m_1 and m_2 and their initial velocities are $\mathbf{v}_1$ and $\mathbf{v}_2$. Since the total momentum of the system of two balls is zero, we know that

$$m_1\mathbf{v}_1 + m_2\mathbf{v}_2 = 0 \quad \text{or} \quad m_2\mathbf{v}_2 = -m_1\mathbf{v}_1 \tag{54}$$

The initial kinetic energy T_{initial} of the system is

$$T_{\text{initial}} = \tfrac{1}{2}m_1 v_1^2 + \tfrac{1}{2}m_2 v_2^2 = \frac{m_1(m_1 + m_2)}{2m_2} v_1^2 \tag{55}$$

where we have used equation 54 to express T_{initial} in terms of v_1.

Suppose that the velocities of the balls after the collision are $\mathbf{v}_1'$ and $\mathbf{v}_2'$. The law of conservation of momentum says that the total momentum of the two balls, which was initially zero, remains zero:

$$m_1\mathbf{v}_1' + m_2\mathbf{v}_2' = 0 \quad \text{or} \quad m_2\mathbf{v}_2' = -m_1\mathbf{v}_1' \tag{56}$$

The final kinetic energy T_{final} of the system is

$$T_{\text{final}} = \tfrac{1}{2}m_1 v_1'^2 + \tfrac{1}{2}m_2 v_2'^2 = \frac{m_1(m_1 + m_2)}{2m_2} v_1'^2 \tag{57}$$

If the collision were elastic, we would set T_{final} equal to T_{initial}. In the case of an inelastic collision with a coefficient of restitution of ϵ, the proper relation is

$$T_{\text{final}} = \epsilon^2 T_{\text{initial}}$$

When we use equations 55 and 57, we obtain a simple result:

$$v_1' = \epsilon v_1 \quad \text{and} \quad v_2' = \epsilon v_2 \tag{58}$$

FIGURE 3.19
An inelastic collision of balls of different mass, seen in the center-of-mass reference frame. The initial velocities are back to back, as are the final velocities, but their direction is changed by the angle θ. The final speed of each body is reduced by the factor ϵ from its initial speed.

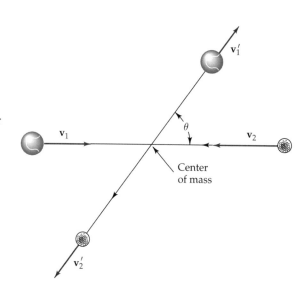

In the center-of-mass frame, the balls move away from one another in opposite directions. The *speed* of each ball is reduced by the factor ϵ from its initial value. The conservation equations determine the speeds of the balls after the collision. The particular direction at which they scatter from one another depends on the details of the impact.

Planetary Motion

According to Galileo, the acceleration of gravity on Earth is constant. Newton showed that Galileo's result does not apply to objects far from Earth. The gravitational force **F** that Earth exerts on an object is constant neither in magnitude nor direction. Its magnitude depends on the distance between the object and the center of Earth; it points from the object to Earth. The same is true for the gravitational force exerted by the Sun on a planet. It points from the planet to the Sun, and its magnitude depends on the distance between the two bodies. Kepler showed that each planet travels in an elliptical orbit with the Sun at one focus. To sweep out equal areas in equal times, a planet travels more swiftly when it is closer to the Sun. Its kinetic energy changes during the course of its orbit. Its total energy E, which is the sum of its kinetic and potential energies, remains constant.

Because the Sun is much heavier than any planet, we may ignore its motion and place it firmly at the origin. Imagine a planet of mass m moving under the influence of the Sun's gravity. Let $R(t)$ be the distance between the two bodies and $v(t)$ be the speed of the planet. Both quantities may change with time. Yet Newton's laws of motion and his expression for the gravitational force imply that the total energy E of the planet remains constant throughout its motion. Our analysis of constant forces is not up to the task of expressing E in terms of $v(t)$ and $R(t)$. Without further ado, I shall tell you that the total energy of the planet is

$$E = \frac{mv^2(t)}{2} - \frac{GM_\odot m}{R(t)} \tag{59}$$

where G is Newton's constant and $M_\odot$ is the mass of the Sun. The planet's total energy E consists of two parts, its kinetic energy and its gravitational potential energy:

$$E_{\text{kin}} = mv^2/2 \qquad E_{\text{pot}} = -GMm/R \tag{60}$$

Both contributions to E change with time except in the special case of a circular orbit, where both are constant.

The orbit of Halley's comet is an elongated ellipse (Figure 3.20). When the comet is at its closest point to the Sun, its potential energy is most negative and its speed is greatest. When it is furthest from the Sun, its speed is least. As time passes, kinetic energy is transformed into potential energy and vice versa. However, the total energy of the comet, the sum of its kinetic and potential energies, does not change at all. These remarks apply to more complex systems as well. In the case of a galaxy consisting of billions of stars, energy is continually being exchanged among the kinetic and potential energies of each of the constituent bodies. However, the sum of all these quantities, the total energy of the system, remains forever the same.

Suppose that the velocity of a body and its distance to the Sun are such that its total energy E is negative. Define the positive quantity W by $W = -E$. We may rewrite equation 59 as

$$\frac{GMm}{R(t)} = W + \frac{mv^2(t)}{2}$$

Because the kinetic energy cannot be less than zero, it follows that $GMm/R(t)$ cannot ever be less than W. This is equivalent to the constraint

$$R(t) \leq \frac{GMm}{W} = R_{\max}$$

In other words, if E is negative, the object can never travel further from the Sun than $R_{\max}$. The planet, comet, or asteroid is irrevocably bound to the Sun. It moves in an elliptical or circular orbit with a *binding energy* of W.

Suppose that the velocity of a body and its distance to the Sun are such that its total energy E is positive. It is not bound to the Sun. The body will eventually escape from the Sun's gravitational influence and

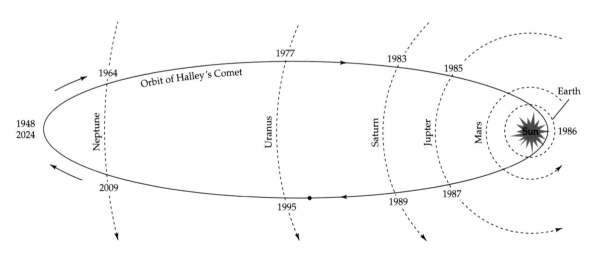

FIGURE 3.20 The orbit of Halley's comet is an elongated ellipse. The dates indicate when it passes several planetary orbits. Notice that it spends 30 years outside of Neptune's orbit, but a mere 2 years inside that of Jupiter. Its speed increases as it approaches the Sun.

head for the stars. Its orbit is an open curve in the shape of a hyperbola, as shown in Figure 3.21.

Throughout the study of physics, systems of particles are encountered that are held together by forces. The hydrogen atom consists of an electron and a much heavier proton. The total energy of the electron includes a positive kinetic energy and a negative electric potential energy. The sum is negative and the electron is bound to the proton. Energy must be supplied from outside the atom to remove the electron (or *ionize* the atom). In the same way, a molecule is a bound state of several atoms and an atomic nucleus is a bound state of neutrons and protons. The binding energy of any composite system (whether it be a galaxy, a solar system, an atom, a molecule, or an atomic nucleus) is the energy required to take it apart.

FIGURE 3.21
An object moves under the influence of the Sun's gravity. If its total energy is negative, it is forever bound to the Sun. It moves in an elliptical orbit such as that labeled *a*. If its total energy is positive, it will escape from the Sun. Its orbit is a hyperbola, such as that labeled *b*.

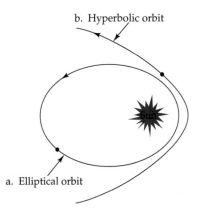

b. Hyperbolic orbit

a. Elliptical orbit

EXAMPLE 3.10
Escape Velocity from Earth

With what speed must a body be hurled from Earth if it is to escape into interplanetary space?

Solution

To achieve an unbound orbit about Earth, the sum of an object's kinetic energy and its gravitational potential energy due to Earth's gravity must be positive. Its kinetic energy $\frac{1}{2}mv^2$ must exceed minus its gravitational potential energy, or $GmM_{\oplus}/R_{\oplus}$. The necessary condition is

$$\frac{v^2}{2} \geq \frac{GM_{\oplus}}{R_{\oplus}}$$

Recall from Chapter 2 that g, the acceleration of gravity on Earth's surface, is equal to $GM_{\oplus}/R_{\oplus}^2$. We use this result to rewrite the above formula as $v \geq \sqrt{2gR_{\oplus}}$. Inserting numerical values of $g \simeq 10$ m/s and $R_{\oplus} \simeq 6.4 \times 10^6$ m, we find that a launch speed of at least ~ 11 km/s is required to escape from Earth. However, a considerably larger speed is required if the body is to escape the Sun's gravitational tug and so escape from the solar system. ∎

The Acceleration of Gravity on Earth

Let us return from space to apply what we just learned about gravitational potential energy to motion on Earth. Figure 3.22 shows an object of mass m located at a height h above the the surface of Earth. Its gravitational potential energy, according to equation 60, is

$$E_{pot} = -\frac{GmM_\oplus}{R_\oplus + h}$$

where $M_\oplus$ is the mass of Earth and $R_\oplus$ its radius. However, we know from Section 3.1 that the gravitational potential energy of the object relative to the ground is mgh. How are these results to be reconciled?

Under ordinary circumstances, h is much smaller than $R_\oplus$, and we may use the mathematical approximation

$$\frac{1}{R_\oplus + h} \simeq \frac{1}{R_\oplus} - \frac{h}{R_\oplus^2}$$

Jet planes generally cruise at an altitude of $h \simeq 10$ km, while $R_\oplus \simeq 6400$ km. For these values, the above formula is accurate to two parts per million. For motion taking place at or near Earth's surface, we may approximate the gravitational potential energy by

$$E_{pot} = -GmM_\oplus \left(\frac{1}{R_\oplus} - \frac{h}{R_\oplus^2} \right)$$

The first term in the parenthesis does not depend on h: It is a constant. If our concern is the motion of objects at or near Earth's surface, we may as well drop the constant and redefine the gravitational potential of a body so that it vanishes at sea level. What remains is simply

$$m \left(\frac{GM_\oplus}{R_\oplus^2} \right) mh = mgh$$

where we identify $mg = GmM_\oplus / R_\oplus^2$ as the force of gravity on an object of mass m near the ground, and mgh as its gravitational potential energy

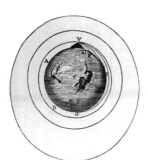

Newton, in 1710, drew this diagram showing bodies thrown horizontally from a mountain peak. If a body is thrown swiftly enough, it describes a circular orbit about Earth. If it is given twice as much kinetic energy, it will escape from Earth. *Source:* Courtesy Meyers Photo-Art.

FIGURE 3.22
An object of mass m is at a distance h above the surface of Earth. The gravitational attraction of Earth on the object depends on $R_\oplus + h$, the distance between the object and the center of Earth. Under ordinary circumstances, h is much smaller than $R_\oplus$, and the force of gravity is effectively constant.

at a height h above the ground. Galileo's principle of the universal acceleration of falling bodies follows as an approximate consequence of Newton's theory of gravity.

Angular Momentum

Consider a planet (or other body) of mass m orbiting the Sun (Figure 3.23). Its *angular momentum* is defined to be the vector

$$\mathbf{J} = \mathbf{r} \times \mathbf{p} \tag{61}$$

where $\mathbf{r}$ is the planet's instantaneous position relative to the Sun and $\mathbf{p} = m\mathbf{v}$ is its instantaneous momentum. A time Δt later, the momentum and the position of the planet have changed. The planet's position becomes $\mathbf{r} + \Delta \mathbf{r}$ and its momentum becomes $\mathbf{p} + \Delta \mathbf{p}$. Consequently, its angular momentum becomes $\mathbf{J} + \Delta \mathbf{J}$, where

$$\Delta \mathbf{J} = (\mathbf{r} + \Delta \mathbf{r}) \times (\mathbf{p} + \Delta \mathbf{p}) - \mathbf{J} = \Delta \mathbf{r} \times \mathbf{p} + \mathbf{r} \times \Delta \mathbf{p} + \Delta \mathbf{r} \times \Delta \mathbf{p}$$

The rate of change of $\mathbf{J}$ is the ratio of this expression to Δt:

$$\frac{\Delta \mathbf{J}}{\Delta t} = \left(\frac{\Delta \mathbf{r}}{\Delta t}\right) \times \mathbf{p} + \mathbf{r} \times \left(\frac{\Delta \mathbf{p}}{\Delta t}\right) + \left(\frac{\Delta \mathbf{r}}{\Delta t}\right) \times \Delta \mathbf{p}$$

If we consider smaller and smaller time intervals Δt, the ratio of $\Delta \mathbf{r}$ to Δt approaches the instantaneous velocity of the planet, $\mathbf{v}$. According to equation 42, the ratio of $\Delta \mathbf{p}$ to Δt becomes the force $\mathbf{F}$ acting on the planet. As Δt approaches zero, the rate of change of $\mathbf{J}$ becomes

$$\frac{\Delta \mathbf{J}}{\Delta t} = \mathbf{v} \times \mathbf{p} + \mathbf{r} \times \mathbf{F} + \mathbf{v} \times \Delta \mathbf{p}$$

The first term on the right-hand side vanishes because $\mathbf{v}$ points along $\mathbf{p}$ and their cross product is zero. The second term vanishes because $\mathbf{F}$ points diametrically opposite to $\mathbf{r}$. The third term vanishes as well, because $\Delta \mathbf{p}$ goes to zero along with Δt. Thus, the rate of change of

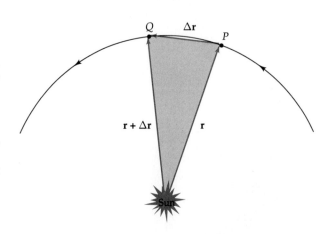

FIGURE 3.23
A portion of the orbit of a planet is shown. At a certain time it is at P, at a vector distance $\mathbf{r}$ from the Sun. In the time interval Δt, it moves to Q, a vector distance $\Delta \mathbf{r}$ from P. The magnitude of the vector $\Delta \mathbf{r} \times \mathbf{r}$ is twice the area of the shaded triangle. If Δt is small, it is the area swept out by the line from the Sun to the planet in that time.

the angular momentum of a planet is zero. The angular momentum of a planet in orbit about the Sun does not change. Each of its three components is a constant of the motion.

The direction of **J** is perpendicular both to the planet's position vector and its velocity vector. That is, it is perpendicular to the plane in which the motion takes place. Because the direction of **J** is fixed in space, the orbit of a planet lies in a fixed plane. The angular momentum of a planet divided by its mass is $\mathbf{r} \times \mathbf{v}$ or $(\mathbf{r} \times \Delta\mathbf{r})/\Delta t$. The magnitude of $\mathbf{r} \times \Delta\mathbf{r}$ is twice the area swept out by a line from the Sun to the planet in the time Δt. Because the magnitude of **J** is constant, Kepler's second law applies to the motion of any orbiting body.

EXAMPLE 3.11
Angular Momentum

Think of an ice skater spinning with arms outstretched on a frictionless surface. When she brings her hands inward, her rate of spin increases. Lo! Angular momentum is conserved. Figure 3.24 illustrates the same principle. Two satellites of mass m are tethered together by a massless cable of length $2L$ (Figure 3.24a). They are in deep space, away from any gravitational influences. Both satellites rotate about the same circle of radius L. The center of the circle is at the midpoint of the cable, which is the fixed center of mass of the satellites. At each moment, their velocities are equal and opposite. Astronauts within each satellite reel in the cable until its length is halved. The configuration shown in Figure 3.24b is produced. At what speeds do the satellites move in their diminished orbits?

Solution

The kinetic energy of the system of satellites changes because it takes energy to reel in the cables. However, the angular momentum of the system does not change. At first, the magnitude of the angular momentum of each satellite is mvL. The angular momenta of the two satellites point in the same direction, so that the total angular momentum is $J = 2mvL$. Let v' be the speed of the satellites after the cable is drawn in. Since the radii of their circular orbits has become $L/2$, we conclude that $J = mv'L$, and hence, that $v' = 2v$. Halving the radius doubles the speed.

FIGURE 3.24
a. Two satellites are tethered by a cable of length $2L$. The satellites move in circular orbits about the center of the cable with speeds v.
b. The cable is reeled in to half its original length, so that the orbital radius is halved. Conservation of angular momentum implies that the orbital speeds of the satellites are doubled, $v' = 2v$.

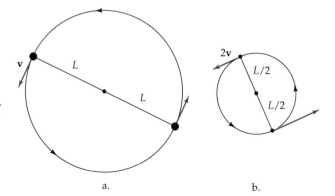

Conservation Laws

When isolated bodies collide or otherwise exert forces on each other, their speeds and directions change. However, certain combinations of their masses and their velocities do not. These are called constants of motion. The quantities that remain unchanged are said to satisfy conservation laws. Imagine that two bodies are originally moving freely. They interact with each other in some fashion, and then come apart, once again moving freely.

- *The conservation of mass* According to this law, the sum of the masses of the various bodies M cannot change as a result of a collision or other interaction. Let the masses of two bodies be m_1 and m_2 before their encounter and m_1' and m_2' afterward. Then,

$$\text{Total mass} = M = m_1 + m_2 = m_1' + m_2'$$

  The total mass of the system is a constant of the motion, even though the individual masses could change as bodies stick together or tear apart. (If the bride carries the groom over the threshold, their combined mass is the sum of their individual masses, no matter what chemical or physical processes ensue.)

- *The conservation of linear momentum* The momentum of a particle is the product of its mass and its velocity. It is a directed quantity, or a vector. According to this law, the sum of momenta of all the bodies cannot change. Again distinguishing initial and final values with a prime,

$$\text{Total momentum} = \mathbf{P} = m_1\mathbf{v}_1 + m_2\mathbf{v}_2 = m_1'\mathbf{v}_1' + m_2'\mathbf{v}_2'$$

  The total momentum of the system $\mathbf{P}$ is a constant of the motion even though the individual masses and velocities may change.

- *The conservation of kinetic energy* In certain cases, such as the collisions of air molecules (or of idealized bodies), all of the kinetic energy remains in the form of kinetic energy. For these *elastic* collisions, the sum of the individual kinetic energies does not change:

$$\text{Total kinetic energy} = E_{\text{kin}}$$

$$= \tfrac{1}{2}m_1v_1^2 + \tfrac{1}{2}m_2v_2^2 = \tfrac{1}{2}m_1'v_1'^2 + \tfrac{1}{2}m_2'v_2'^2$$

  and the total kinetic energy E_{kin} is a constant of the motion. For the motions of heavenly bodies or of falling bodies on Earth, gravitational potential energy must be included in the total energy as well. However, conservation of energy is not apparent on the billiard table or the ball field because frictional forces inexorably transform kinetic energy into heat energy and soon end the action.

- *The conservation of angular momentum* This is the last and most complicated of the kinematic conservation laws. The angular momentum of an object is the vector defined by equation 61. The total angular momentum of a system of objects is the vector sum of the angular momenta of each object. The total angular momentum of an isolated system is constant. In particular,

$$\text{Total angular momentum} = J_{\text{total}} =$$

$$m_1 \mathbf{r}_1 \times \mathbf{v}_1 + m_2 \mathbf{r}_2 \times \mathbf{v}_2 = m_1 \mathbf{r}_1' \times \mathbf{v}_1' + m_2 \mathbf{r}_2' \times \mathbf{v}_2'.$$

A detailed discussion of angular momentum is beyond the scope of this book. Its importance lies in the fact that it is one of a very few quantities, along with linear momentum and energy, that is rigorously conserved by all processes taking place in isolated systems, like molecules, atoms and elementary particles. The concept plays a vital role throughout all of the physical sciences, including astronomy, geology, physics, and chemistry. Here is an example: The attraction of the Moon on the matter of Earth produces tides. Because energy is conserved, tidal energy must come from somewhere. In fact, it is being extracted from the rotational energy of Earth. Earth is gradually spinning more slowly and the day is growing longer. However, the angular momentum of Earth cannot be lost. It is being transferred to the Moon, the radius of whose orbit is slowly increasing.

The conservation laws we have discussed have been refined, but not essentially altered in the three centuries since their discovery. Einstein's special theory of relativity taught us that mass must be regarded as a form of energy, and that the laws of energy and mass conservation must be put together into a single law. The two laws have become one. The conservation laws for mass–energy, and for linear and angular momenta apply not only to the science of mechanics, but to all aspects of nature. However elegant and useful they are, they have a more profound significance. Within the context of modern physical science, they indicate subtle truths about our universe and its relationship to space and time.

Students of physics learn that the conservation of linear momentum is equivalent to the statement that the laws of physics are the same everywhere. They are the same here on Earth as they are on the back of the Moon. An intelligent alien living in a distant galaxy would learn the same things about quarks and stars (and everything in between) as we have. Momentum is conserved because space is *homogeneous*.

The conservation of angular momentum is equivalent to the statement that space is *isotropic*. This means that no direction in space is singled out from any other direction. Of course, up is a very special direction to us, but it is so only because we live on the surface of a planet. Deep in space (or, with the effects of gravity taken into account) no direction is singled out from any another.

In a similar vein, the conservation of mass–energy reflects the fact that the laws of physics do not change with time. They are true everywhen. They were precisely the same in the Age of Dinosaurs, and they will remain valid long after *homo sapiens* is extinct.

27. How rapidly must a body be launched from Earth if it is to escape from the solar system? (*Hint*: The total gravitational potential energy of a mass m on Earth's surface is the sum of Earth's contribution $-GM_\oplus m/R_\oplus$ and the Sun's contribution $-GM_\odot m/1\,\text{AU}$.)

28. What is the binding energy of Earth to the Sun? That is, how much kinetic energy must be supplied to Earth to remove it from the solar system?

29. Suppose that a planet has two satellites, Noo and Foo. Both are in circular orbits and both have the same mass. However, Foo is twice as far from the planet as Noo. What is the ratio of the angular momentum of Foo to that of Noo?

30. A ball moving at 6.5 m/s collides elastically with an identical ball at rest on a frictionless table. The collision is glancing and the first ball proceeds with a speed of 6 m/s afterward. What speed does the struck ball acquire. Roughly describe its direction.

31. A body slips on a frictionless plane surface that is tilted relative to the horizontal by an angle θ. It is subject to a gravitational acceleration along the plane equal to $g\sin\theta$. Show that for $\theta = 30°$, the time required for a body to descend a given height from rest is twice what it would be in free fall. (*Hint*: The velocity of the body, which is directed parallel to the plane, is $gt\sin\theta$. Its component in the vertical direction, which determines how fast it descends, is $gt\sin^2\theta$.)

32. In Example 3.11, the tension in the cable provides the centripetal acceleration of the tethered satellites.
    a. What is the tension before and after the cable is shortened. Express your answers in terms of m, v, and L.
    b. Before the cable was shortened, the acceleration of each satellite was equal to g, the acceleration of gravity on Earth. Those aboard the satellites perceived the acceleration as an apparent *centrifugal force* driving them outward. They "weighed" the same as they did at home. How did their weight change when the tether was shortened?
    c. Before the cable was shortened, the period of the orbit was one day. What was its period when the tether was shortened.

Where We Are and Where We Are Going

In the next chapter, our focus turns from the abstract study of motion to an apparently unrelated subject: the investigation of the properties of gases. It is less of a logical leap than it seems. Of the three familiar forms of bulk matter (gas, liquid, and solid), the gaseous state is the simplest. In liquids and solids, molecules touch one another and are in continual interaction. However, the molecules of a gas are in erratic motion and only rarely collide with one another.

The elastic impacts of gas molecules against any solid surface transfer momentum to it and thereby exert a force. This effect is interpreted as a property of a gas called its pressure, and its propagating fluctuations are what we hear as sound. The average kinetic energy of the molecules of a gas is a measure of another property of a gas: its temperature. Those are some of the reasons that we studied force, momentum and energy before turning to the properties of matter. Of course, none of the microphysics underlying macroscopic phenomena was known when the quantitative study of gases began in the seventeenth century. It was largely by means of these endeavors that the nature of heat was discovered, the atomic hypothesis vindicated, and combustion shown to be oxidation. Modern chemistry and physics emerged from the study of gases.

4 The Behavior of Gases

We have seen that atmospheric air is composed of two gases, one of which is capable, by respiration, of contributing to animal life, and in which metals are calcinable, and combustible bodies may burn; the other, on the contrary, is endowed with directly contrary properties; it cannot be breathed by animals, neither will it admit of the combustion of inflammable bodies, nor of the calcination of metals. We have given to the former, or respirable portion of the air, the name oxygen. The chemical properties of the noxious part of the atmosphere, being hitherto little known, we have been satisfied to derive its name from its known quality of killing such animals as are forced to breathe it, giving it the name azote. [From Greek for lifeless, *azote* is still French for nitrogen.] Water is composed of oxygen combined with an inflammable gas in the proportions of 85 parts by weight of the former, to 15 parts of the latter. Thus water, besides the oxygen, which is one of its elements in common with many other substances, contains another element for which we must find an appropriate term. None that we can think of seems better adapted than the word hydrogen

A. L. Lavoisier, French scientist

We live under an ocean of air. *Source:* NASA.

The history of physical science is linked to the study of air and other gases. Of the three forms of matter—gas, liquid, and solid—the gaseous state is by far the simplest to understand. The molecules in a gas are far apart and rarely encounter one another. Section 4.1 opens with an introduction to some of the earliest speculations about the nature of matter. Afterward, we turn to the discovery, in the seventeenth century, that air is a substance subject to scientific scrutiny. Section 4.2 is devoted to the physical properties of gases and to Robert Boyle, who put forward the first precise law governing the behavior of matter: a quantitative relation between the pressure and volume of a gas. Although the physical properties of gases would ultimately be explained in terms of the chaotic motions of molecules, it was through the study of chemistry, and especially the chemistry of gases, that scientists were driven toward the atomic hypothesis. The final section discusses the discoveries of many different gases with different chemical properties, and especially to the realization that combustion is a chemical reaction requiring oxygen and that water is formed by the union of two gases: hydrogen and oxygen.

4.1 From the Greeks to Torricelli

Egyptian technology and Babylonian astrology, two streams of thought that influenced one another for millenia, converged in ancient Greece to nurture the seeds of modern science. Technology had advanced far enough to serve the nation's needs—the mysteries of the priesthood had become the tools of the artisan. An intelligentsia emerged with the freedom to speculate about the ultimate nature of reality: Science arose from the leisure of the theory class. These Greek philosophers were the first scholars of pure science. They developed surprisingly modern notions of atoms and elements. However, the pendulum of their science swung from the empirical to the purely cerebral. Their high-flying thoughts became far removed from the phenomena they sought to explain. They would frame glorious hypotheses about nature, but rarely would they stoop to test them by mundane experiments. So began the wild Greek strivings to explain all the properties of matter in terms of its ultimate constituents.

Thales (640–546 B.C.) is the first Greek philosopher whose name survives. He surmised that the myriad forms of matter we see about us are manifestations of a single fundamental substance, water, which evaporates to form a vapor and freezes to form solid matter. Anaximedes chose air as his ultimate material, believing that it yielded fire on rarefaction and water, earth, and stones on condensation. Heraclitos opted for fire, which he hypothesized could be condensed to form to air, water, and earth.*

In the end, we may blame Empedocles for the Greek model with its four so-called elements—earth, water, air, and fire—playing equal roles. Each element was supposed to be made up of minute and unchanging particles called atoms, meaning uncuttable. Every atom of a given element was to be identical to every other atom. All of the matter on Earth was to be made of complex combinations of these simple elements bound together by Love and caused to separate by the opposing force of Strife. The Greeks gave us our first list of hypothetical fundamental particles and of the basic forces governing their behavior. This pattern repeats itself several times over the centuries-long quest for the ultimate building blocks of which matter is made. Today's vision of the structure of matter makes use of six quarks and six leptons, which interact through four primal forces.

---

*The Greeks confused the three states of matter with elements. Ice, liquid water, and steam represented earth, water, and air. Condensation led from air, to water, and then to earth. Rarefaction was the inverse process: Ice melts to form water, which boils to form steam, which the Greeks did not distinguish from air. Because flames go up, it seemed to them that fire was a further rarefaction of air.

"The periodic table." © 1993 by Sidney Harris.

Greek atomic theory culminated with the work of Demokritos (460–370 B.C.). His atoms of earth, air, fire, and water were in continual and random motion within an otherwise empty space or void. His principles of natural science are almost acceptable today:

1. From nothing comes nothing. Nothing that exists can be destroyed. All changes are due to the combination and separation of molecules.
2. Nothing happens by chance. Every occurrence has its cause from which it follows by necessity.

3. The only real things are atoms and empty space; all else is mere opinion.

4. Atoms are infinite in number and various in form. Their collisions, lateral motions and whirlings are the beginnings of worlds.

5. The varieties of all things depend upon the varieties of their atoms, in number, size and aggregation.

6. The soul consists of fine, smooth round atoms like those of fire. They are the most mobile of all. They interpenetrate the whole body and in their motions the phenomena of life arise.

Demokritos's vision resembles modern atomic theory, with its ordinarily unchangeable atoms moving randomly and ceaselessly through a void. His primitive, but basically correct, model was proposed millennia before its time, but his ideas were swallowed up by the work of later Greek philosophers, including Plato and Aristotle, whose thoughts were much further from the mark.

Archimedes, Density, and Buoyancy

Greek science did not end with Aristotle, although he is often considered the last of the great Greek thinkers. Mathematics thrived elsewhere with the works of Euclid (in Egypt) and Archimedes (in Sicily). Euclid's rigorous geometrical arguments make a better textbook than much of what is offered to pupils today. Archimedes was principally a mathematician, but he made many practical discoveries as well, one of which is the principle of buoyancy.

Archimedes was challenged to find out whether a crown made for the king was made of pure gold. The problem was to determine this

Archimedes in his bath, from Walter Ryff, Bawkunst. Basel, 1582. When Archimedes sat in his bathtub, he noticed that the water level rose. This phenomenon suggested a method by which he could measure the volume of any object, in particular the king's crown. *Source:* Historical Pictures/Stock Montage, Inc.

without destroying the crown.   As the story goes, inspiration hit Archimedes as he stepped into the bath and saw the water displaced. "Eureka!" he cried, and ran off, nude and dripping, into the streets of Syracuse.

Archimedes' solution was based on the fact that most solid objects (like corks and crowns, but not sponges) are impermeable. When such a body is completely immersed in a liquid, it displaces a volume of liquid equal to its own. Archimedes plunged the crown into a graduated container of water and measured the change of its level. Thereby, he learned the volume of the crown. Taking the ratio of its mass to its volume, he determined its density.* Archimedes found that the density of the king's crown was less than that of pure gold, and, therefore, that it consisted of an alloy of gold with a cheaper and less dense metal.

Archimedes' insight led him to the principle of *buoyancy*. When a body is totally immersed in a fluid, it is subject to an upward buoyant force equal in magnitude to the weight of an equivalent volume of the fluid. Figure 4.1a shows a body with volume V and density ρ immersed in a fluid of ρ_0. The net force on it is the sum of its weight $W = -\rho V g$ (pointing down) and the buoyant force $F = +\rho_0 V g$ (pointing up). If the body is denser than the fluid, the sum is negative and the body sinks. If the body is less dense than the fluid, the sum is positive and the body rises. Figure 4.1b shows a body floating at the fluid surface and only partly submerged. The volume V_0 of its submerged portion is such that the buoyant force on it exactly cancels the weight of the body and $\rho_0 V_0 = \rho V$.

From Archimedes' principle, we may determine how much lower in the water a ship sinks when it is loaded and what payload a helium-filled balloon can lift. Years after Archimedes' discovery, while he was constructing a geometrical proof by drawing figures in the sand, Archimedes was murdered by a Roman soldier, a proud conqueror of Syracuse.

FIGURE 4.1
a. An object of density ρ and volume V is immersed in a liquid of density ρ_0. If the buoyant force on it, $F = g\rho_0 V$ exceeds its weight, $W = g\rho V$, the object rises. Otherwise, it sinks. *b.* In this case, the object is floating. The volume V_0 of its submerged portion determines the buoyant force to be $F = g\rho V_0$, and it is equal to the object's weight.

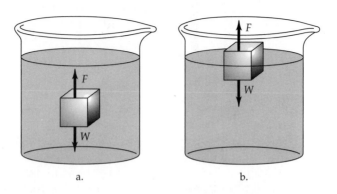

a.   b.

*The density, ρ, of an object is defined to be its mass M divided by its volume V: $\rho = M/V$.

EXAMPLE 4.1
The Density
of an Alloy

A solid object of mass M and volume V is made from an alloy of two different metals. The object contains the mass m_1 of a metal whose density is ρ_1. Its other constituent is the mass m_2 of another metal whose density is ρ_2. Thus, $M = m_1 + m_2$. The constituents, if they were extracted, would have volumes $v_1 = m_1/\rho_1$ and $v_2 = m_2/\rho_2$. The volume of the object is *nearly* equal to the sum of these individual volumes. The equality is not exact because the atoms of the pure metals rearrange themselves in the alloy. For the following problem, we neglect this small effect and set $V = v_1 + v_2$.

   A 1-kg crown is 75 percent gold and 25 percent copper by mass. The density of gold is 19.3 kg/L (or grams per cubic centimeter, g/cm^3); that of copper is 9 kg/L. How much water does the crown displace? How much water would the crown displace if it were made of pure gold? What is the density of the crown?

Solution

   An object displaces a volume of water equal to its own volume. The crown contains 0.25 kg of copper of density 9 kg/L, whose volume is $v_1 = 27.8$ mL (milliliters). The crown contains 0.75 kg of gold, whose volume is $v_2 = 38.9$ mL. The volume of the crown is $v_1 + v_2 = 66.6$ mL. (Were the crown pure gold, its volume would have been 51.8 mL.) The density of the crown is the ratio of its mass to its volume: $\rho = M/V$, or 15.0 kg/L. ∎

EXAMPLE 4.2
Buoyant Forces

A wooden plank weighing 12 kg floats in a lake. How many 1-kg bricks can be piled on the plank before it sinks? The density of the plank is 0.5, that of water is 1, and that of the brick is 2.5, all in units of kg/L.

Solution

   The laden raft has a total mass M, which is the sum of the masses of the plank and the bricks upon it. Its total volume V is the sum of the volumes of the plank and the bricks. The raft sinks when M/V is greater than 1 kg/L, the density of water. From the stated densities, the volume of the plank is 24 L and that of a brick of 0.4 L. If N bricks are atop the plank, the total mass of the raft is $M = 12 + N$ kg and its total volume is $V = 24 + 0.4N$ L. We set $M/V = 1$ to obtain: $12 + N = 24 + 0.4N$, or $N = 20$. No more than 20 bricks may be floated on the plank. ∎

Air Is Matter

One of the next big discoveries took place a few hundred years later, in the first century A.D. Hero of Alexandria invented many machines based on ideas that approach modern thought (Figure 4.2, on page 158).

   He invented a mechanical slot machine that could open the temple doors or deliver a dollop of holy water in exchange for a coin. Many of his devices were powered by steam, including a primitive engine. He wrote from professional experience:

   Vessels which seem to most men empty are not empty, as they suppose, but full of air. Now air, as those who have treated of physics are agreed, is composed of particles minute and light, and for the most part invisible. If then, we pour water into an apparently empty vessel, air will leave the vessel proportional in quantity to the water which enters it. Hence, it must be assumed that air is matter.

Hero went on to describe how air may be blown into a globe with a single aperture and be compressed (or, sucked out of a globe and be rarefied). When the pressure is released, air rushes out of the globe. He came to the unorthodox conclusion that air consists of atoms moving through a void, and could thereby be compressed or rarified. Hero's contemporaries favored the conventional wisdom that matter and space were indivisible, that the concept of space without matter—a vacuum— was unthinkable. Hero could not prove that his colleagues were wrong. All he could do was to point out that if experiments could demonstrate "that there is such a thing as a vacuum, they will no longer be able to hold their ground." Hero approached the modern atomic view, but his ideas sank along with those of Demokritos. Not until the Renaissance were the works of the Greek atomists translated and distributed: those of Demokritos in 1473, of Hero in 1575. That is why we now make an enormous leap in time to 1638, to Galileo in his last years. Not a great deal had happened in the interim as far as our understanding of air is concerned. The Dark Ages were indeed dark.

As we saw in Chapter 2, Galileo spent his later years studying the motions of falling bodies. He proved, by means of careful experiments, that all dense bodies fall in the same fashion. He also knew that tenuous things, such as feathers and leaves, float slowly to the ground. He at-

FIGURE 4.2
Two of Hero's inventions. *a.* As the thumb is pressed on or released from the hole in its handle, the magic jug will pour or not. *b.* A primitive steam engine, by which the reaction force due to the steam jets causes the globe to spin.

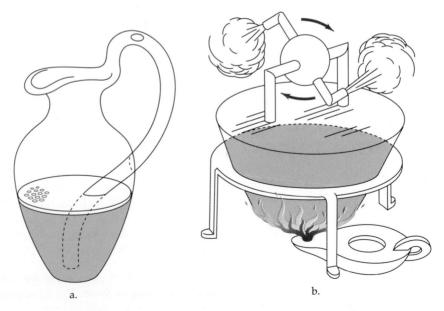

a.   b.

tributed this effect to air resistance. In the reference frame of the falling body, the ambient air moves upward and acts to retard its fall. Galileo decided to test this notion. If air offers resistance to motion, must it not be a substance? Fitting a large flask with a valve "through which by means of a syringe," Galileo "forced into the flask a great quantity of air," he weighed the flask, released the valve and weighed the flask again. He found that air has weight and measured its density to be about 1/500 that of water. (The correct answer, at sea level, is 1/816. But because there were few precision instruments available in Galileo's day, his result, although it was half again too high, was nonetheless a remarkable achievement.)

Curiously enough, the understanding of air was furthered by a problem that Galileo had with his plumbing. His upstairs water pump didn't work when the level in his outdoor water storage tank fell too low. In his *Dialogs Concerning Two New Sciences*, he admits to being enlightened by a common plumber:

> *Sagredo*: 'I have learned the cause of a certain effect which I have long wondered at and despaired of understanding. I once saw a cistern which had been provided with a pump. The pump worked perfectly so long as the water in the cistern stood above a certain level; but below this level the pump failed to work. The workman whom I called in to repair it told me the defect was not in the pump but in the water, which had fallen too low to be raised through such a height and he added that it was not possible, either by a pump or by any machine working on the principle of attraction, to lift water hair's breadth above 18 cubits [33 feet]; whether the pump be large or small this is the extreme limit of the lift. Up to this time I had been

---

DO YOU KNOW
How Fast Raindrops Fall?

Suppose that a raindrop fell from a cloud $h = 1000$ m high. If its acceleration were constant and equal to g, its velocity when it hit you would be $\sqrt{2gh} \simeq 144$ m/s, or 324 mph! Rain would be a deadly hazard.

Air resistance saves the day! Imagine an object of mass m falling through the air. It is acted on by two forces. The gravitational attraction of Earth is a constant downward force of magnitude mg. The upward force of air resistance F depends in a complicated way on the size and shape of the falling object. It also depends on the velocity v of the object relative to the air about it. In particular, F increases as the body accelerates. When the object begins to fall, F is very small. As the object accelerates, F becomes larger. Consequently, the acceleration of the falling object decreases. At a certain point, when the object's velocity is such that $F = mg$, the upward and downward forces acting on it balance each other. The falling body continues to fall, but at a constant velocity. The value of this velocity is called the *terminal velocity* of the object. For a skydiver, it is about 100 m/s, but for a raindrop it is a mere 2 m/s. Rain won't hurt you.

so thoughtless that, although I know a rope, or rod of wood or of iron, if sufficiently long, would break by its own weight when held by the upper end, it never occurred to me that the same thing would happen, only much more easily, to a column of water attached at the upper end stretched more and more until finally a point is reached where it breaks, like a rope, on account of its own weight.'

Salviati: 'That is precisely the way it works; this fixed elevation of 18 cubits is true for any quantity of water whatever, be the pump large or small or even as fine as straw. We may therefore say that, on weighing the water contained in a tube 18 cubits long, no matter what the diameter, we shall obtain the value of the resistance of the vacuum in a cylinder of any solid material having a bore of the same diameter.'

Aristotle taught that the working of a pump resulted from the principle that "nature abhors a vacuum." Galileo saw that nature does so only up to a point, a most unsatisfying result. A rational explanation of the pump's behavior was called for—one that could put an end to the doctrine of the abhorrence of the vacuum and facilitate the rebirth of atomic theory. Evangelista Torricelli found it. My junior high school science teacher, in trying to explain Torricelli's discovery, once said to me, "It's the air pressure that pushes the Coke up the straw, dummy!"

The Pressure of Air

Evangelista Torricelli (1608–1647) was a young and promising Italian scientist whose dream was to work with Galileo. In his letter of introduction to the master, Torricelli described himself as "a Copernican by conviction, by profession and sect Galilean." He became Galileo's assistant in October 1641, three months before the great man's death.

Galileo thought that a suction pump worked because the water rose up in the pipe to avoid the formation of a vacuum at the top. He believed that the height limit to the operation of a such a pump—which is about 10.3 m—corresponded to the maximum force the vacuum could exert, or perhaps, to the breaking point of a column of water. Torricelli suspected otherwise. He believed that the water was forced up the pipe by the pressure of the air on the surface of the water in the cistern below (see the illustration). Torricelli filled long tubes, sealed at one end, with different liquids and lashed them upright to ships' masts. He found that the height of the column varied inversely with the density of the fluid within. A column of wine (which is less dense than water) was higher than a column of honey (which is more dense).

Torricelli had a brainstorm: Why not use mercury, the densest known fluid, in his column? If his idea were correct, mercury, whose density is 13.6 times that of water, should fill a column 1/13.6 as high as the the maximum height of a water column: 76 cm rather than 10.3 m. In 1644, Torricelli prepared a meter-long glass tube that was sealed at one end. It was entirely filled with mercury, plugged with a finger, inverted in a mercury bath, then unplugged. As Torricelli expected, the level of

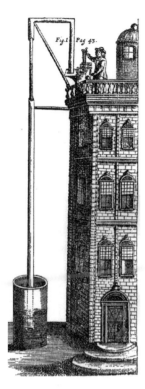

Galileo's suction pump was at the top of a pipe leading to a water supply below. When the distance between the pump and the level of water in the cistern exceeded 10 m, the pump failed to work. (In fact, this illustration shows a later demonstration prepared by Robert Boyle.) *Source:* Historical Pictures/Stock Montage, Inc.

mercury in the tube immediately plunged to a height of about 76 cm above the level in the bath, as shown in Figure 4.3. No matter what size or shape tube Torricelli used, the mercury always fell to the same level.

What could be in the space above the mercury but a vacuum? Hero's ancient challenge had been met and Torricelli's interpretation of his experimental result was inspired. The mercury is not sucked up by the vacuum. The vacuum doesn't affect the mercury at all. It is the pressure of the atmosphere pushing on the reservoir that drives the mercury up the tube. Torricelli invented the notion of air pressure. Galileo's pump and the mercury-filled tube behaved as they did because, in Torricelli's words,

> We live submerged at the bottom of an ocean of the element air, which by unquestioned experiments is known to have weight.

The attentive student will surely ask how the weight of the atmosphere (a downward force) can generate a pressure to push the mercury upward? Long ago, Torricelli attempted to explain the nature of pressure:

> There once was a philosopher who, seeing his old servant put a faucet on a barrel, scoffed at him, saying that the wine would never come out because the nature of weight is to press downward and not horizontally from the sides; but the servant made him see with his own eyes that although by nature liquids gravitate downward, they press and spout in every direction, even upward, as long as they find places to reach, that is places which resist with less force than their own. Sink a pitcher entirely in water with its mouth downward, then make a hole through the bottom so that the air can come out, and you will see with what impetus the water moves up from below to fill it. You try it for yourself, for I shall not annoy you further.

Torricelli solved the puzzle of the pump, produced the world's first artificial vacuum, discovered air pressure, and computed the height of Earth's atmosphere. His stated intent was "not simply to produce a

FIGURE 4.3
Torricelli's mercury columns. Whatever the shape of the tube, the height of the mercury column is the same. The space above the column is a vacuum.

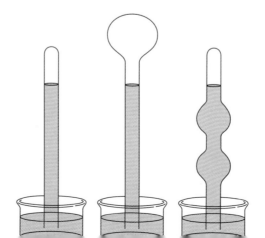

vacuum, but to make an instrument which would show the changes in the air, which is at times heavier and thicker and at times lighter and more rarefied." Thus, he invented the barometer as well.

Blaise Pascal (1623–1662) was a French prodigy, an accomplished mathematician at 16, an inventor, physicist, and renowned philosopher of religion. Upon learning of Torricelli's work, he set out to prove for himself that air pressure sustains the Torricellian column. He began his experiments in 1647, the very year that Torricelli died, with a repetition of the original experiment.

If air pressure were responsible for holding up the mercury in a barometer, Pascal reasoned, such a device, if it were placed in a vacuum, should not support any mercury at all. However, there was no way to perform the experiment in a vacuum, neither above the atmosphere nor in an evacuated room. Nevertheless, Pascal invented an ingenious device that could produce a vacuum within a vacuum. It consisted of the bent glass tube shown in Figure 4.4. With the stopper in place (Figure 4.4a), the entire device was filled with mercury and inverted in a mercury bath. The mercury distributed itself as shown. The mercury level in the left-hand part of the tube C is 76 cm above the level in the reservoir D. Because there is air nowhere in the device, the mercury levels at A and B are the same. By this means, Pascal showed that Torricelli's column is not sustained in a vacuum. When the stopper is removed from Pascal's device (Figure 4.4b), air enters the column and level C falls to the level of the reservoir. At the same time, level A rises 76 cm above level B, thus offering a further confirmation of Torricelli's theory of air pressure.

A decisive proof that a barometer measures the weight of air above could be obtained if such a device were brought to a mountaintop, where there is less air above and atmospheric pressure is distinctly less. Frail, ill,

FIGURE 4.4
Pascal's vacuum within a vacuum. *a*. The tube is filled with mercury and inverted with the stopper in place. A column of mercury is trapped between two vacua. Its level at A is equal to its level at B. The mercury level at C, however, is about 76 cm above the reservoir level D due to air pressure in the laboratory. *b*. When the stopper is removed, air enters the tube and the mercury column on the left collapses. At the same time, air pressure forces the mercury level down at B, and it consequently rises at A. The difference in height between B and A becomes 76 cm.

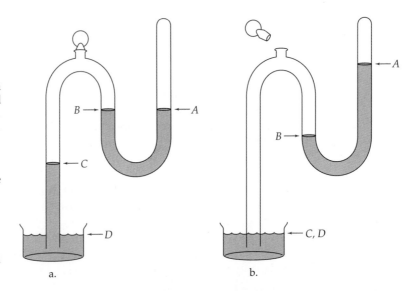

a.   b.

acrophobic, and lacking a nearby Parisian mountain, Pascal proposed the experiment to his brother-in-law, Florin Perier who lived in the mountainous Auvergne region of France. The deed was done on September 19, 1648. As expected, the barometric reading at the summit of the Puy-de-Dôme was a few inches less than it was at the foot of the mountain. Perier wrote, "We were so carried away with wonder and delight, and our surprise was so great that we wished, for our own satisfaction, to repeat the experiment. So I carried it out with the greatest care five times more." These athletic feats confirmed Pascal's belief that air is an elastic substance, whose pressure and density decrease with altitude.

Pascal realized that a column of mercury could function as an altimeter as well as a barometer. It could tell "whether two places are at the same altitude, or which of the two is higher, however far apart they may be." Pascal concluded his study of air pressure with an impressive list of familiar phenomena that are explained by Torricelli's discovery. According to Pascal—and quite correctly—air pressure:

1. Causes the difficulty in opening a sealed bellows.
2. Causes the difficulty one feels in separating two polished bodies in close contact.
3. Causes of the rise of water in syringes and pumps.
4. Causes water to remain in tubes sealed at the top.
5. Causes water to rise in siphons.
6. Causes the flesh to rise in the process of cupping.
7. Causes the attraction produced by sucking.
8. Causes the flow of milk to nursing infants.
9. Causes the in-drawing of air which occurs in breathing.

The story that began in Italy and continued in France, now moves to Germany. Otto von Guericke was the mayor of the town of Magdeburg and an accomplished amateur scientist. He invented an ingenious and powerful air pump with which he could produce a vacuum in a sizable volume. Von Guericke confirmed that air is a material substance by showing that all objects, even feathers, fall equally fast in a vacuum. By pumping the air out of a bottle containing small animals, he showed that air is essential to life.

As we describe the history of physics, we shall find that the scientists of earlier times, like those of today, were often showmen as well as scholars. Just as Galileo dropped weights from the Leaning Tower of Pisa to convince the public of his theory, so did von Guericke offer a dramatic demonstration of the reality and strength of atmospheric pressure. He built two hollow copper hemispheres, each about 55 cm in diameter. They were fitted together and most of the air within the sphere was removed by the pump. The pressure of the surrounding air was so great that two teams of eight horses could not pull the hemispheres apart!

In the mid-seventeenth century, Torricelli conjectured and Pascal proved that we live under an ocean of air. Von Guericke's air pump made it possible to explore the properties of air in the laboratory. Air is

Otto von Guericke. the mayor of Magdeburg, was also an amateur scientist. To demonstrate the immense power of air pressure, he pumped the air from two brass hemispheres, that were fitted togther. When he attached teams of eight horses to rings on the hemispheres, their combined force could not separate them. Yet, when von Guericke opened the stopcock, the hemispheres fell apart. *Source:* Historical Pictures/Stock Montage, Inc.

a form of matter whose attributes, like those of any other form of matter, can be determined empirically and explained by scientific laws. Before we begin to describe these laws and how they were established, we must explain the nature of pressure more carefully.

The Meaning of Pressure

As you dive into a swimming pool, the pressure of the water around you increases. You feel the effect on your eardrums, which are the body's only pressure detectors. In fact, though, the pressure of the water exerts a force on every part of your body that it touches. So does the atmosphere about us. The pressure of a fluid (a gas or a liquid) is the property by which it exerts a force on any surface, real or imagined. Figure 4.5 shows an element of surface. It might be a postage stamp, a pane of glass, or a hypothetical surface. The surface might be located in the air of this room, 10 m under the sea, or in any fluid that is under pressure. Whatever the orientation of the surface, the fluid exerts a force on each of its sides. The force is directed perpendicular to and toward the surface and its magnitude is given by

$$F = PA \tag{1}$$

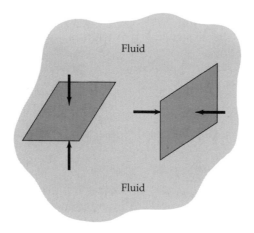

where P is the pressure of the fluid and A is the area of the surface. The surface is not moved by the force because the fluid exerts equal and opposite forces on both of its sides. The dimensions of pressure are determined by equation 1 to be force per unit area, for which the SI unit is N/m^2, also known as the *pascal* (pa).

Atmospheric pressure at sea level P_0 is about 10^5 pa, although its precise value varies by a few percent with the weather. The outward force exerted on a small window of area 10^{-2} m^2 by the air within a room is ~ 1000 N, which is roughly my weight! The window does not break because the air outside exerts an equal inward force. The air pressure indoors and outdoors is the same, so there is no net force on the window. This is not so for jet aircraft flying at high altitude, where atmospheric pressure is much lower than it is at sea level. The air pressure in the cabin is kept at nearly sea-level air pressure for the health and comfort of the passengers. The windows of jets are both small and strong in order to withstand the large outward force of the cabin pressure at high altitude.

Atmospheric pressure results from the weight of the air above us. Its value is the weight per unit area of a vertical column of air extending upward to the top of the atmosphere. The weight of this column of air is its mass m multiplied by the acceleration of gravity $g \simeq 10$ m/s². Thus, the mass of a column of the atmosphere with a cross-sectional area of 1 cm² is about 1 kg. The mass of a 10-m high column of water with the same cross-sectional area is also 1 kg. The pressure at its base due to its weight equals the ambient air pressure. Atmospheric pressure cannot support a higher column of water. This solves Galileo's plumbing puzzle and explains the height of a Torricellian water column. It also explains why the level in a mercury barometer is about 76 cm. Because there is no pressure in the vacuum above the mercury, the height of the column adjusts itself until the pressure at its base equals the pressure of the ambient air. In general, the height h_0 of a barometric column satisfies the equation

$$P_0 = \rho g h_0 \qquad (2)$$

where ρ is the density of the liquid in the column. If a storm is brewing and the P_0 falls, so does the height h_0 of the column.

EXAMPLE 4.3
A Barometer Problem

Suppose that a quantity of water is trapped within a mercury barometer. The water extends upward from the mercury level for a distance of 27.2 cm. The column of a water-free mercury barometer reads 76 cm. How high is the column of mercury in the watered barometer?

Solution

The total amount of fluid in the vertical column must weigh the same in both cases. Mercury's density is 13.6 times that of water, so the 27.2 cm of water weighs as much as 2 cm of mercury. Therefore, the water sits atop a mercury column 74 cm high. ∎

Let's consider the operation of a barometer more carefully. Let $P(h)$ denote the pressure of the mercury at a height h above the reservoir level. The pressure of the fluid is not the same everywhere in the column. For example, $P(0)$ equals the air pressure P_0. At the interface of the fluid with the vacuum above, the pressure is zero, $P(h_0) = 0$. At any point between, $P(h)$ equals the weight per unit area of the fluid lying above that point, which is $\rho g(h_0 - h)$. Thus, we obtain the relation

$$P(h) = \rho g(h_0 - h) \qquad \text{or} \qquad P(h) = P(0) - \rho g h \qquad (3)$$

This result has wider implications. Consider any device containing an incompressible fluid under pressure.* According to equation 3, the pressure is the same at all points in the fluid that are at the same height. Let A and B be two distant points in a horizontal water pipe. If the pressure at A is increased by a pump, this increase is rapidly communicated to point B. For example, the water pressure in your garden hose may be supplied by a pump at the waterworks. If the pump fails, you will soon notice the drop in pressure.

---

*If the fluid can be compressed, its density is not constant, and equation 3 cannot be used.

DO YOU KNOW
What Air Pressure Can Do?

Many curious effects of air pressure are easily seen without the need for horses and hemispheres. Here are two do-it-yourself demonstrations.

1. Fill a tumbler with water to its brim and slide a plastic playing card over the top. Holding the card in place with a finger, invert the tumbler. Now remove your finger. The water remains in the tumbler and the card stays in place! Air pressure exerts an upward force sufficient to counteract the weight of the water. This trick would fail on the Moon, where there is no atmosphere and no atmospheric pressure.

2. For this demonstration, you will need two large plastic soda bottles, a balloon, a sharp knife and some quick-setting glue. Create a two-necked bottle by cutting a hole in the side of one bottle and gluing on the severed neck of the other. Insert a balloon into the bottle with its end wrapped about and glued to the original bottle neck. Suck out the air from the other neck. As the interior pressure is reduced, the pressure of the air outside causes the balloon to inflate within the bottle. Cap the second neck, and you can enjoy the unusual spectacle of an inflated balloon which is open to the air!

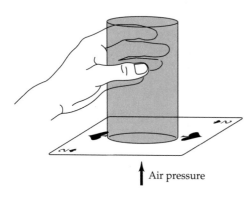

Air pressure

A playing card covers a glass of water that is full to the brim. When the glass is carefully turned over, the water does not spill out. The upward force of air pressure exceeds the weight of the water.

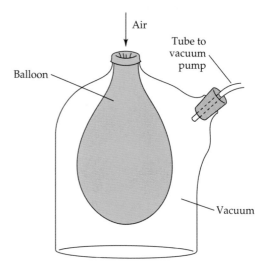

A balloon is inserted into one of the necks of a two-necked bottle. When the air in the bottle is pumped out, the balloon expands and fills the bottle.

Exercises

Use for air pressure $P_0 = 10^5$ N/m^2; for the acceleration of gravity, $g = 9.8$ m/s^2; for the density of water, 1 kg/L; for the density of gold, 19.3 kg/L; for the density of copper, 9 kg/L; and for the density of mercury, 13.6 kg/L.

1. A 10-kg solid metal cube consists half of gold and half of copper by mass.
   a. What is the density of the cube?
   b. What is the length of a side of the cube?

2. A bottle has an interior volume of 2 L. It is made of glass with a density of 2.5 kg/L, and when empty its mass is 0.5 kg.
   a. What is the volume of the glass from which the bottle is made?
   b. The empty bottle is tightly sealed. What is the magnitude of the downward force required to keep it completely submerged under water? (Neglect the mass of the stopper, but remember that the total volume of the bottle is larger than its interior volume.)
   c. While it is submerged, the bottle develops a leak and fills completely with water. What is the magnitude of the upward force required to keep it from sinking further?

3. A balloon is filled with 10 m^3 of helium gas. What is the mass of the empty balloon and its payload such that it floats in the air, neither rising nor falling. (A cubic meter of air has a mass of 1.3 kg while one of helium has a mass of 0.18 kg. Neglect the volume of the payload.)

4. A perfect vacuum within causes two hollow hemispheres of diameter 1 m to bind together. What is the force in newtons that is needed to part them. (*Hint*: Air exerts a net force on each hemisphere equal to what it would exert on a disk of the same radius—that is, the product of air pressure with the area of the disk.)

5. Explain how the two-hemisphere demonstration could have been done in a more impoverished land with just one team of horses.

6. Two highly polished metal plates each have an area of 100 cm^2. One is securely attached to the ceiling of a cathedral. The second plate is placed flush underneath the first, and is bound to it by air pressure. A daredevil hangs by his teeth from a hook attached to the second metal plate. What is the mass of the fattest daredevil who could possibly survive this rash stunt?

7. A long Torricellian barometer is put underwater so that the surface of the mercury in the reservoir is at a depth of 10 m (the equivalent of 75 cm of mercury, which is the reading of the barometer before it is submerged). What is the reading of the barometer while it is submerged?

8. If a barometer reads 76 cm of mercury, what is the precise value of the air pressure in N/m^2?

9. The surface area of Earth is $\sim 5 \times 10^{14}$ m^2. From the values of air pressure at sea level and the acceleration of gravity, calculate the total mass of Earth's atmosphere.

10. At its deepest point, the ocean is ~ 10 km deep. What is the pressure in pascals at that depth? About how many times greater is it than atmospheric pressure at sea level?

11. Pascal listed nine consequences of air pressure. Choose any three of them and discuss more fully the physical principles involved.

FIGURE 4.6
The oil pressure at the top of the large tube exceeds air pressure by the weight of the large object divided by the area of the large tube. The pressure at the top of the small tube exceeds air pressure by the weight of the small object divided by the area of the small tube. Since the two weights are at the same height, Pascal's law says that the two pressures are equal.

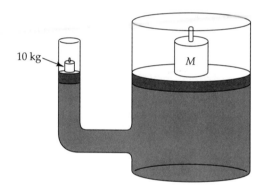

12. The hydraulic press takes advantage of Pascal's law, which states that the pressure applied to a confined fluid acts uniformly throughout the fluid. A hydraulic seesaw consists of two pistons of area 0.05 m^2 and 1.0 m^2 enclosing a reservoir of oil, as shown in Figure 4.6. What mass must be placed on the larger piston so as to balance a 10-kg mass placed on the smaller piston?

4.2 The Physical Properties of Gases

The differences among the three states of matter are easy to describe. A solid maintains its shape: A bottle full of marbles contains marbles, which retain their shape as little spheres, with air filling in the remaining space among them. A liquid maintains its volume but not its shape: A pint of milk in a quart bottle assumes the shape of the bottle, but fills only half of it, the rest being filled with air. A gas, the most fascinating state of matter, has neither a fixed shape nor a fixed volume: Most of the air may be pumped out of a bottle, but what is left fills the bottle. When von Guericke discovered the principle of the air pump, he made it possible for his successors to study the curious properties of gases.

Robert Boyle (1627–1691) made several great leaps forward as soon as he heard of the works of Torricelli, Pascal, and von Guericke. Boyle was the fourteenth child of the Earl of Cork, who was one of the richest and most powerful men in Ireland. After being educated at home and abroad, the young man joined a group of Utopian intellectuals who called themselves the *Invisible College.* The group met in secret because of the political turbulence then raging in England. Years later, it became famous as the *Royal Society.* In 1653, Boyle moved to Oxford as a full-time scientist. He had his assistants build the world's most powerful air pump with which he might study the properties of air, which was then the only known gas. With this pump, Boyle first investigated the behavior of the mercury barometer in a vacuum. A diagram of his apparatus is shown in Figure 4.7.

Boyle knew that the column of mercury is supported by air pressure. If the air were pumped out, the column could not be sustained. He wrote, "So if we could perfectly draw the air out of the receiver, it would conduce as well to our purpose, as if we were allowed to try the experiment beyond the atmosphere." As the pump removed the air from the glass globe, the mercury level fell. This was a direct proof of Torricelli's assertion that air pressure supports the mercury column in a barometer.

Boyle went on to show that air is essential for the transmission of sound.* Later, he turned to the intrinsic elasticity of air itself, a property

FIGURE 4.7
As the air is pumped out of the bell jar, the air pressure is reduced and the level of mercury in the barometer falls.

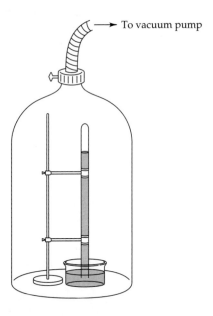

To vacuum pump

*Sound is a wave consisting of alternate regions of higher and lower pressure. Its existence depends on the presence of a material medium. That is why outer space is silent, and why a ringing bell in a jar from which the air has been removed cannot be heard.

that underlies the transmission of sound. He took "a lamb's bladder large, well-dried, and very limber, and leaving in it about half as much air as it could contain, we caused the neck of it to be strongly tied." He placed the half-filled bladder in an evacuated vessel and observed, "that before we had exhausted the receiver near so much as we could, the bladder appeared as full and stretched." In another experiment, Boyle began with a fully inflated bladder placed in the vessel,

> and upon drawing out the ambient air that pressed on the bladder, the internal air not finding the wanted resistance first swelled and distended the bladder, and then broke it, with so wide and crooked a rent, as if it had been forcibly torn asunder.

Finally, he placed an almost empty bladder in his device and pumped out the surrounding air. The bladder increased in volume by a factor of 10. At this point Boyle was close to his greatest discovery: the law bearing his name and governing the elasticity of air.

Boyle's Law

Having shown that air in a bladder expands when the ambient pressure is reduced, Boyle began to investigate whether air could be compressed. Using his pump, he showed that, indeed, a large quantity of air could be forced into a small glass chamber. When the chamber was sealed, the pressure remained undiminished for a long time. He concluded that "there is a spring or elastic power to the air we live in." He then went on to demonstrate the inherent relationship between the pressure and the volume of an enclosed gas.

Boyle used a J-shaped glass tube whose shorter leg was sealed and whose longer leg was open. He filled the tube with mercury so that the mercury level was the same on each side (Figure 4.8). The pressure of the air trapped in the short end was equal to the air pressure in the room.

FIGURE 4.8
How Boyle found his law. *a.* Air is trapped in a J-shaped tube that is partly filled with mercury. The level of mercury on either side is the same. The pressure of the trapped air is equal to the air pressure, which corresponds to 76 cm of mercury in a barometer. *b.* Mercury is added to the open end of the tube until the mercury level on the left is 76 cm higher than the level in the short end of the tube. At that point, the pressure of the trapped air has been doubled and its volume halved.

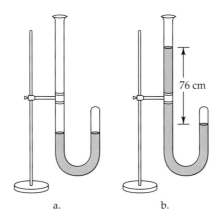

a.   b.

This done, we begin to pour quicksilver into the longer leg of the siphon, which by its weight pressing up that in the shorter leg, did by degrees straighten the included air: and continuing the pouring in of quicksilver till the air in the shorter leg was by condensation reduced to take up but half of the space it possessed before; we cast our eyes upon the longer leg of the glass, on which was likewise pasted a list of paper carefully divided into inches and parts, and we observed, not without delight and satisfaction, that the quicksilver in the longer part of the tube was 28 inches higher than the other.

In short, as mercury was poured into the open end of the tube, the air in the short leg was compressed. When the left-hand column was about 76 cm higher than the right-hand column, the pressure of the trapped air became twice what it had been. Boyle observed that the height of the column of trapped air (and consequently, its volume) had been halved. By means of a detailed series of experiments over a wide range of pressures, he established a precise quantitative law.

Suppose that the pressure of a quantity of gas enclosed within a container of volume V_1 is P_1. The volume of the container is then changed to V_2, but no gas is allowed to enter or leave. The pressure of the gas changes to P_2, and these quantities satisfy the equation:

$$P_1 V_1 = P_2 V_2 \tag{4}$$

This may be said more generally: The volume V of a fixed sample of air that is kept at a constant temperature varies inversely with its pressure P:

$$\text{Boyle's law:} \quad PV = \text{Constant} \tag{5}$$

where by "constant" is meant something depending on the quantity of gas and its temperature, but independent of its pressure. In this fashion, Boyle established the first quantitative law describing the measurable properties of matter. He was certain that there were many more such laws. Boyle was led to a philosophy in which the universe, once created, simply follows the laws ordained by its creator—the clockwork universe. In view of the importance of Boyle's law to both physics and philosophy, it behooves us to depart from our narrative to examine several specific instances of its application.

EXAMPLE 4.4
A Cylinder
and a Piston

Imagine the device shown in Figure 4.9. A cylinder, closed at the bottom, has a cross-sectional area of 100 cm^2. Its top is sealed with a freely moveable piston of negligible weight. One liter of air is enclosed. (Thus, the distance between the piston and the bottom of the cylinder is 10 cm.) Atmospheric pressure exerts a downward force on the piston of 1000 N. The piston does not move because this force is exactly compensated by the upward force exerted by the trapped air. The pressure is the same inside and out. What does the volume of the enclosed air become if a 50-kg mass is placed atop the piston?

FIGURE 4.9
a. A massless piston confines 1 L of air in a cylinder. The pressures inside and outside are equal to atmospheric pressure. Equal and opposite forces of 1000 N act on the two sides of the piston. *b.* A 50-kg weight is placed on the piston, which increases the downward force to 1500 N. The piston falls so as to reduce the volume of the trapped air to two-thirds its initial value, and to increase its pressure to three-halves its initial value.

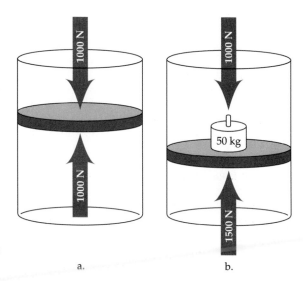

a.   b.

Solution   Call the initial pressure and volume P_1 and V_1, and the final pressure and volume P_2 and V_2. When the weight is placed on the piston, the downward force on it is increased by 500 N, or by a factor of 1.5. To balance this force, the pressure of the enclosed gas increases such that $P_2 = 1.5\,P_1$. Equation 4 says that its initial volume of 1 L decreases to 2/3 L. If the mass on top were increased to 100 kg, the volume of the gas would be halved. If there were no mass on the piston, but the device were submerged under 10 m of water, the volume of the trapped gas would likewise be halved.   ■

EXAMPLE 4.5
Fun with a
J-Shaped Tube

Let us return to Boyle's experiment with the J-shaped tube. Let its cross–sectional area be 1 in². On Monday (when the ambient air pressure corresponds to 28 in of mercury), the tube is partly filled with mercury so that the level is the same on either side (Figure 4.10, on page 174). That is, the pressure of the trapped air is the same as the pressure outside. The sealed end of the tube extends 15 in above the level of mercury. The volume of the cylinder of trapped air is its height times its area, or 15 in³.

On Tuesday, the atmospheric pressure rises and a mercury barometer reads 32 in. The mercury level in the open end of the tube falls by a distance x (Figure 4.10). The level in the sealed end rises by the same distance because the total quantity of mercury in the tube is unchanged. What is the value of x?

Solution   The length (and hence, the volume) of the column of trapped air is reduced by the factor $(15 - x)/15$. Since the product of pressure and volume stays constant, its pressure is increased from 28 to $28 \times 15/(15-x)$ in of mercury. On the other hand, the mercury level on the sealed end

FIGURE 4.10
a. On Monday, the pressure of the trapped air is the same as the barometric pressure, which corresponds to 28 in of mercury. *b.* On Tuesday, the barometric pressure has risen to 32 in of mercury, thereby depressing the mercury level in the open end of the tube by a distance *x*. In the text, it is shown that $x = 1$ in.

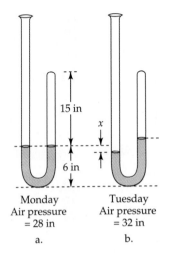

15 in

x

6 in

Monday
Air pressure
= 28 in

Tuesday
Air pressure
= 32 in

a.

b.

is $2x$ in higher than it is in the open end. Tuesday's air pressure of 32 in must equal the sum of these quantities:

$$28\left(\frac{15}{15 - x}\right) + 2x = 32$$

A BIT MORE ABOUT

Thermometers

A thermometer shows the intensity of heat, or temperature, of a body. Temperature indicates the ability to communicate heat: Heat flows from a hotter body to a cooler one, but not the other way. The first crude thermometer was invented by Galileo in 1597. It consisted of a glass tube with a large air-filled bulb on top, the lower end dipping into a vessel (Figure 4.11). The vessel was open to the atmosphere and contained a colored fluid. The air trapped in the bulb expands when the temperature rises and contracts when it falls. Thus, the height of the liquid in the tube depends on the ambient temperature. However, the height of the liquid is also affected by changes in atmospheric pressure. Galileo's instrument, which he called an *air thermoscope*, was a combination of a barometer and a thermometer.

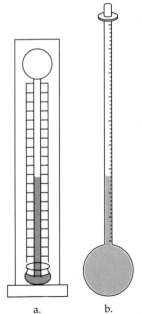

FIGURE 4.11
a. Galileo's device was an air-filled glass globe with a tube leading to a reservoir of liquid. If the temperature rises, the air expands, driving the fluid level lower. If the temperature falls, the fluid level rises. This device is also affected by changes in air pressure. *b.* A late seventeenth century thermometer. A reservoir of liquid at the bottom leads to a sealed glass tube. If the temperature rises, the liquid expands driving the fluid level higher, and vice versa. Because the tube is sealed, this improved thermometer is not affected by changes in air pressure.

a.

b.

This may seem to be a complicated equation, but look at it: It is satisfied by $x = 1$, which is the answer to this problem. The mercury level falls by 1 in on the left and rises by 1 in on the right. ∎

EXAMPLE 4.6
More Fun With a J-Shaped Tube

The initial setup in this example is the same as in Example 4.5. Mercury is poured into the open end of the tube until the length of the column of trapped air is reduced to 10 in (and its volume is decreased to two-thirds of its original volume). How much mercury (in cubic inches) was added?

Solution

Let the initial height of the mercury column on either side be d inches. (In Figure 4.10a, $d = 6$ in.) After the mercury is added, the height of the column in the short leg is $d + 5$ in. The pressure of the trapped air has been increased by a factor of three-halves because of Boyle's law. It has become 42 in of mercury, or 14 in more than the ambient air pressure. It follows that the level of mercury in the open leg of the tube has become $d + 19$ in. All told, 24 in^3 of mercury were added. ∎

Several complications were omitted in our discussion of barometers, having to do with the thermal expansion and vapor pressure of fluids.

By 1667, greatly improved thermometers made by the glassworkers of Florence were in common use. They consisted of sealed tubes, with a bulb at the lower end containing colored alcohol or oil. In this case, the thermal expansion or contraction of the liquid in the bulb affects the level in the tube. Since this device is not open to the air, it responds to changes in the temperature of the atmosphere, but not changes in its pressure. An arbitrary scale was provided marking the highest temperature of summer and the lowest of winter. In the eighteenth century, many attempts were made to introduce a standard and reproducible measure of temperature:

• In 1701, Newton defined the reading of a thermometer that was placed in a mixture of ice and water to be 0°. Its reading when he grasped it firmly in his hand was to be 12°. Intermediate temperatures from 1 to 11 were equally spaced between these readings.

• In 1714, the German physicist Gabriel Fahrenheit adopted mercury as the fluid of choice for thermometers. He introduced a system in which the temperature of a mixture of ice and salt is 0° and body temperature is 100°. This scale is no longer widely used, except in the United States.

• In 1731, the French mathematician and meteorologist René Réaumur picked 0° for the freezing point of water and 80° for its boiling point. The French were, and still are, addicted to counting by 20s.

• In 1742, the Swedish astronomer Anders Celsius introduced a scale in which 0° is the freezing point of water and 100° is its boiling point. The Celsius scale of temperature is part of the international SI system of units.

The empty space at the top of a Torricellian column is never a true vacuum. It contains water (or mercury) vapor at a pressure equal to the *vapor pressure* of the fluid, one of its intrinsic physical characteristics. The weight per unit area of the column plus the vapor pressure of the fluid equals the ambient air pressure. If the temperature varies, the density of the liquid changes as the fluid expands or contracts, and so does its vapor pressure. Because of these thermal effects, the height of the column can change, even though the atmospheric pressure may not have changed at all. Fortunately, both effects are small in the case of a mercury column, small enough to have escaped the notice of seventeenth-century scientists. If it were otherwise, the history of physics might have been different.

Torricelli's discovery of air pressure and Boyle's formulation of his law were essential first steps in our understanding of the nature of matter, but they posed as many puzzles as they solved. Why is there such a simple relation between the pressure and volume of a gas? What is the underlying mechanism by which a gas exerts forces on a containing vessel: an upward force on the top, a downward force on the bottom, and sidewise forces on the sides? Newton, flushed with the success of his gravitational theory, offered an explanation in terms of a hypothetical repulsion between air molecules. He was wrong. The pressure of a gas arises from the random collisions of its molecules upon the surface of the container.

Gay-Lussac's Law

The volume of a given amount of gas depends upon its temperature as well as its pressure: Gas expands when heated. Most things expand when they are heated. (But not all! Water contracts when it is heated from 0° C to 4° C.) The rails on which a train rides do not quite touch one another. A gap between each section allows for thermal expansion in hot weather. Similarly, the mercury (or alcohol) in a thermometer expands as it warms, letting us measure the temperature. Unlike other forms of matter, gases satisfy a remarkably simple law of thermal expansion, and one that may be readily deduced from the atomic hypothesis.

Soon after a reliable and reproducible method to measure temperature was developed, quantitative research on the nature of heat began. Boyle's law described how the volume of a gas depends on its pressure. The first detailed study of the effect of temperature changes on gases was done by Jacques Alexandre César Charles in 1787, but these studies were carelessly carried out and never published. The law governing the thermal expansion of gases is usually and correctly attributed to Gay-Lussac.* He carefully studied many different gases and found an astonishing result. If its pressure is kept fixed, every gas expands by about 1/300 in volume for each degree Celsius it is heated. His law is

---

*In his discovery paper, Gay-Lussac candidly remarked, "Charles had remarked on these properties of gases 15 years ago; but not having published his results, I have the great good luck to make them known."

FIGURE 4.12
A sample of gas re-
mains at a fixed pressure
within a horizontal grad-
uated tube immersed in
a beaker of water. A
freely moving plug al-
lows the volume of the
gas to change without
offering resistance. As
the water is heated, the
volume of the gas and
its temperature are mon-
itored. The data may be
presented in the form of
a graph showing the gas
volume vs. its tempera-
ture.

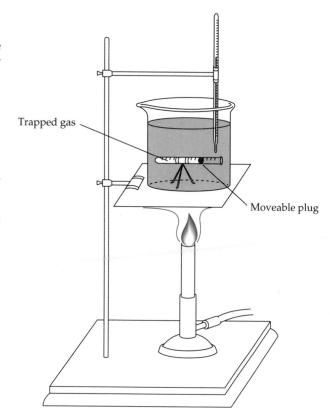

Trapped gas

Moveable plug

universal: The law of expansion is independent of the chemical composition of the gas. It is the same for all gases. The experiment he designed
to obtain this result is shown in Figure 4.12.

Gay-Lussac's findings may be presented graphically by plotting the
volume occupied by a fixed quantity of gas against its Celsius temperature. The result is a straight line, as shown in Figure 4.13 (on page
178). Although the figure displays data for temperatures lying between
the freezing point and boiling point of water, the volume of a gas will
continue to rise at higher temperatures and shrink at lower ones.

Gay-Lussac discovered that the volume V occupied by a sample of
gas (at a fixed pressure) is a linear function of its temperature T. Any
such relation may be expressed as an equation:

$$V = c\,(T - b)$$

where c and b do not depend on T. The parameter c is not a constant.
Its value depends both on the quantity of the gas in the sample and its
pressure. However, Gay-Lussac found that the parameter b is truly a
constant: *b has the same value for any size sample of any gas at any pressure.*
A result of such generality demands further comment.

FIGURE 4.13
Gay-Lussac showed that the volume of a gas kept at constant pressure increases linearly with the temperature. Extrapolated to negative Celsius temperatures, the volume of the gas would vanish at $T = -273°$ C.

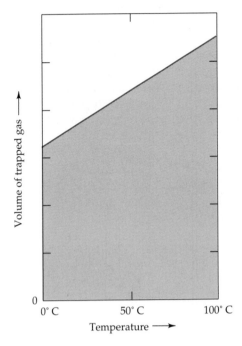

Notice that b has the dimensions of temperature. Let's call it a temperature, and rewrite Gay-Lussac's law in the form

$$V = c(T - T_0) \qquad (6)$$

where $T_0 = b$. Equation 6 says that a gas expands when it is heated and contracts when it cools. Gay-Lussac could have determined the value of T_0 by extrapolating his data to very low temperatures. He would have discovered that T_0 is approximately equal to $-273°$ C, a temperature very far below anything he could produce in his laboratory.

According to equation 6, the volume of any gas should approach zero as its temperature descends toward $-273°$ C. Indeed, at a temperature equal to T_0, its volume should vanish entirely! In fact, this point is never reached. At a certain temperature, which depends on the identity of the gas, its molecules touch one another and the gas becomes a liquid or a solid. Oxygen liquefies at $-133°$ C. At even lower temperatures, it becomes a solid. Carbon dioxide, when cooled sufficiently at atmospheric pressure, turns directly into dry ice. Helium survives as a gas until it is cooled to $-269°$ C, while steam gives up the ghost at $100°$ C. Gay-Lussac's law applies to any gas, but not to liquids or solids for which there is no simple law of thermal expansion.

The relation between the volume and temperature of a gas looks neater if we adopt what is called an *absolute temperature scale.* The Scottish physicist William Thomson (1824–1907), also known as Lord Kelvin, introduced such a scale in 1847. The degrees of the Kelvin scale are equal to Celsius degrees, but the zero point is chosen to be the universal temperature at which the volume of any gas would vanish if it were to continue to contract when cooled. (Of course, real gases do no such thing.) Because every gas, independent of its composition, expands at the same rate with temperature, the zero point of our new scale is unique and unambiguous. The zero point of the Kelvin scale is defined to be the temperature T_0, which is to say $\sim -273°$ C. It is known as *absolute zero*. In the Kelvin scale, ice melts at 273 K and water boils at 373 K. In general, Kelvin temperature equals Celsius temperature plus 273:

$$T_{Kelvin} = T_{Celsius} + 273$$

In terms of the Kelvin temperature, Gay-Lussac's law takes a simple form:

$$\text{Gay-Lussac's law:} \qquad V = \text{Constant} \times T \qquad (7)$$

where by "constant" is meant something depending on the quantity of the gas and its pressure, but not on its temperature. Now we see that T_0 is not so much a universal constant as a natural definition of the zero point of our temperature scale, a definition in terms of which the laws of physics (such Gay-Lussac's law) assume their simplest forms.

Boyle's law, equation 5, says that the pressure P of a given sample of gas is proportional to its volume V, if its temperature does not change. Gay-Lussac's law, equation 7, says that the volume of the sample is proportional to its temperature, if its pressure does not change. The two laws can be combined together into one simple formula:

$$\text{The combined gas law:} \qquad PV = aT \qquad (8)$$

where a is a constant for any particular sample of gas. Equation 8 yields Boyle's law if the temperature is held constant. It yields Gay-Lussac's law if the pressure is held constant. However, equation 8 represents more than the sum of its parts. For example, it tells us that the pressure of a gas that is confined to a fixed volume is proportional to its absolute temperature. We shall see in Chapter 5 that a is determined by the mass of the sample of gas, its molecular weight, and a truly universal constant of nature.

Physics at Low Temperature

Absolute zero is the lowest point of the temperature scale, the point at which essentially all molecular motions cease. Although it may never be attained, physicists have succeeded in bringing substances very close to absolute zero. Just as particle physicists seek higher and higher particle energies, low-temperature physicists drive toward ever lower temperatures. Millidegrees Kelvin are commonly reached at any respectable laboratory. The lowest temperature ever reached is measured in nanodegrees Kelvin. Several important Kelvin temperatures are shown in Table 4.1.

TABLE 4.1   Some Kelvin Temperatures

Absolute zero..........	0	Water boils.......	373
Lowest lab temperature	10^{-7}	Iron demagnetizes	1000
Helium liquefies.......	4	Iron melts........	1800
Temperature of Uranus	58	Iron boils.........	3000
Nitrogen liquefies	77	Solar surface	6000
Dry ice	196	Solar center	10^7
Water freezes..........	273	Red giant center..	10^8
Room temperature.....	293	Supernova center.	10^{11}

Quantum mechanical effects, which ordinarily come into play only for tiny things like atoms and elementary particles, become of paramount importance at very low temperatures, where the molecules of bulk matter move very slowly. Under these conditions, weird and wonderful things happen. Some materials become *superfluids* and creep up the walls and out of any containing vessel. Others become *superconductors* and offer no resistance to the passage of electric current. They promise lossless storage and transmission of electric power, and speedy and noiseless magnetically levitated trains, among many other marvels. Today's superconducting devices operate at liquid helium temperatures (~ 4 K), which are difficult to reach and costly to maintain. In 1987, physicists discovered new materials that become superconducting at readily accessible liquid nitrogen temperatures and above. These new materials may lead to revolutionary technological developments.

EXAMPLE 4.7
Using the
Combined Gas Law

A bottle at a temperature of 300 K is almost filled with 0.75 L of wine. It is tightly sealed, leaving a bubble of air within at atmospheric pressure and with a volume of 0.09 L. The bottle is placed in a freezer and brought to a temperature of 250 K. The wine freezes, and in so doing, expands to occupy a volume of 0.79 L. Assume that the volume of the bottle in unchanged. What is the pressure of the bubble now that it is squeezed into a volume of 0.05 L?

Solution

Let the initial pressure, volume, and temperature of the air trapped in the bottle be denoted with the subscript i, their final values with the subscript f. The combined gas law, equation 8, says that the product of the pressure of a fixed sample of gas with its volume, divided by its Kelvin temperature, cannot change. That is:

$$P_f V_f / T_f = P_i V_i / T_i$$

or $P_f = (V_i/V_f)(T_f/T_i) \times 1$ atm (atmosphere) for the case at hand. The ratio of the initial gas volume to its final volume is 9/5. The ratio of its final temperature to its initial temperature is 5/6. Thus, the pressure in the bottle has increased to 1.5 atm. ∎

Exercises

13. Describe a thought experiment to show that the pressure of a confined gas increases in proportion to its absolute temperature.

14. Suppose that Boyle's J-shaped tube were initially filled with mercury, leaving no air space at the sealed end. If it were sufficiently long, it would then act as a barometer, with the height of the mercury in the sealed leg being 28 in above the level in the open leg. If the air pressure rises to 32 in, by how many inches does the mercury level rise in the sealed end of the tube?

15. Suppose that a scuba tank has sufficient air for a person to breathe for an hour at sea level. However, the diver is at a depth of 20 m where the ambient pressure is 3 atm. The air in his lungs is at that same pressure. If he breathes at the same rate (in liters per second) as he would at sea level, how long will his air supply last?

16. In the preceding problem, the air tank runs out! Untrained in the art of scuba diving, the diver rises to the surface while holding his breath. Explain why he has made a disastrous (and perhaps fatal) error.

17. A balloon is inflated at sea level to a diameter of 30 cm. Neglect the force that the rubber exerts on the gas within. The pressure inside equals the pressure outside. What is the diameter of the balloon when it is brought to a mountaintop where air pressure is half what it is at sea level?

18. A Torricellian barometer indicates an air pressure of 10^5 N/m^2. A valve attached to the top of the column is opened and a small amount of air is admitted. The mercury level falls to a fourth of its initial height. What is the pressure of the air trapped above the mercury?

19. A noxious gas is stored in a steel cannister at a pressure of 300 atm and a temperature of 27° C. The tank will burst when the internal pressure is 400 atm. At what Celsius temperature will the tank explode?

20. If the temperature of the tank in exercise 19 is increased to 127° C, what will the internal pressure be?

21. A quantity of gas occupying 2.24 m^3 at a pressure of 10^5 N/m^2 and a temperature of 273 K is compressed to half its volume and heated to 819 K. What does its pressure become?

4.3 The Chemical Properties of Gases

Two great controversies arose during the seventeenth century that were not settled until the following century. One was confusion concerning the nature of heat and fire. We reserve this fascinating tale for the next chapter. The other resulted from the stubborn insistence of some to regard air as a fundamental element and not to recognize the existence of other types of gases with chemical properties different from those of air. Jan van Helmont, a wealthy Belgian physician of the early seventeenth century, observed that gases are generated by a wide variety of chemical and biological processes. Although he knew that these gases (carbon dioxide and impure samples of sulfur dioxide, hydrogen, and oxides of nitrogen) had different properties, he did not regard this fact as important. He is remembered today for contributing a new word to the scientific nomenclature:

> I call this spirit, unknown hitherto, by the new name of Gas, which can neither be contained by vessels, nor reduced into a visible body. But bodies do contain this spirit, and do sometimes wholly depart into such a spirit, not indeed because it is actually in these bodies (for truly it could not be detained, yet the whole compound body should fly away at once), but it is a Spirit grown together. [The word *gas* derives from Greek for chaos.]

Gases, Gases, Everywhere!

The practitioners of science were not always so arbitrarily pigeonholed as they are today. That is why our narrative cannot exclude developments in chemistry, astronomy, mathematics, and biology. Boyle was not only a physicist and a chemist, but he was also an accomplished biologist. At the time, people knew that small animals would not survive for long in a sealed container, even though food and water were provided. Their deaths were attributed to the fouling of the air by the animals themselves, or in Boyle's words, to the "fulginous steams, from which exspiration discharges the lungs; and which may be suspected, for want of room, to stifle those animals that are closely penned up." Boyle tested his hypothesis by placing animals in a vessel from which he evacuated the air. Seeing that the animals sickened or died much more quickly than otherwise, he concluded, "that the death of the forementioned proceeded rather from the want of air, than that the air was overclogged by the steams of their bodies." Here was the first hint that air was not merely a passive medium but a chemically active substance that is vital to life.

Boyle suspected that air was a complex material, "that our atmospherical air may consist of three differing kinds of corpuscles." These were: (1) the exhalations that "ascend from the earth, water, minerals, vegetables and animals"; (2) the "innumerable particles that produce what we call light," and (3) the many "elastical" particles seemingly needed to explain the resilience of air. Of the final one, Boyle wrote, "One may think them to be like the springs of watches, coiled up, and still endeavoring to fly apart." Boyle is often credited with the modern definitions of element and compound, as the following extract from his *Sceptical Chymist* suggests:

> I now mean by elements certain primitive and simple, or perfectly unmingled bodies; which are not made of other bodies, or of one another, are the ingredients, of which all those called perfectly mixt bodies [compounds] are immediately compounded, and into which they are ultimately resolved.

Later on, however, he seems to come to the very opposite conclusion:

> I see not why we must needs believe that there are any primogeneal and simple bodies, of which, as of pre-existent elements, nature is obliged to compound all others. Nor do I see why we may not conceive, that she may produce the bodies accounted mixt out of one another, without resolving the matter into any such simple or homogeneous substances, as are pretended.

Chemistry had a long way to go. The modern discipline emerged in large measure from the study of air and other gases by Boyle's successors.

Stephen Hales (1677–1761), the sixth son of the baronet Sir Robert Hales, had a small but important part to play in this story. He invented a

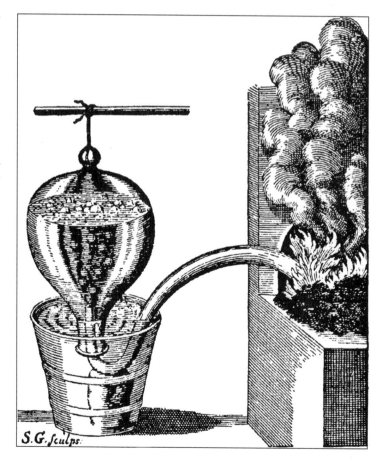

Hales' apparatus for collecting gas over water. The substance to be decomposed is placed in the sealed end of a gun barrel, whose other end leads into an upsidedown bottle in a pail of water. The sealed portion is placed in a fire, and the evolved gas is collected. *Source:* Historical Pictures/Stock Montage, Inc.

S.G. *Sculps.*

device for collecting gases over water. The illustration shows a bent gun barrel, the closed end of which, containing the material to be investigated, was placed in a furnace. The other end opened into an inverted water-filled vessel into which the evolved gas bubbled.

Hales collected the gases released by heating blood, tallow, horn, shells, wood, seeds, honey, beeswax, sugar, coal, tartar, kidney stones, chalk, pyrites, and saltpeter. He was concerned about the *quantity* of gas liberated in each instance, and not about its particular *quality*. Air was air, to Hales, and he missed the profound discovery that there exist different kinds of gases with different physical and chemical properties. Hales was famous in his day for quite another air-related endeavor. He designed and manufactured a ventilator with which fresh air could be introduced into jails, hospitals, mines, and ships' holds. After the installation of Hales's device at the Savoy prison in 1749, the mortality rate of prisoners fell from about 100 to merely 1 per year.

A Gas Different from Air!

Joseph Black (1728–1799), a Scottish physician and chemist, wrote his doctoral thesis on the use of milk of magnesia to counteract flatulence. From the pursuit of this engaging endeavor, he discovered the first gas whose properties are demonstrably different from those of common air. He called his new gas *fixed air*. We know it today as carbon dioxide, or in chemical terminology, CO_2.

Black showed, by examining a series of chemical reactions, that carbon dioxide plays a specific and quantitative role in chemical transformations. In one of his most famous demonstrations, he heated common chalk ($CaCO_3$) in a retort. As it decomposed to form quicklime (CaO), a large volume of fixed air (CO_2) was released and collected. One of his contemporaries, in the introduction to the posthumous publication of Black's lectures on chemistry, describes the wondrous nature of Black's discovery:

> He had discovered that a cubic inch of marble [which is chemically identical to chalk] consisted of about half its weight in pure lime and as much air as would fill a vessel holding six wine gallons! What could be more singular than to find so subtle a substance as air existing in the form of a hard stone, and its presence accompanied by such a change in the properties of the stone?

Black dissolved the resulting quicklime in water. (This, in itself, is remarkable since the original chalk or marble is totally insoluble in water.) The previously collected carbon dioxide was then made to bubble through the limewater. The fixed air reacted with the limewater to form chalk particles, which fell to the bottom of the container. The mass of the chalk that Black obtained equaled the mass of the chalk that he started with. Black demonstrated both the decomposition by heat of chalk into lime:

$$CaCO_3 \rightarrow CaO + CO_2$$

and the recombination of lime and carbon dioxide (in water solution) to form chalk:

$$CaO + CO_2 \rightarrow CaCO_3$$

Furthermore, he showed that potash (K_2CO_3, a mild alkali prepared from ashes and used for millennia to make soap) reacts with limewater to form chalk and caustic alkali, KOH, thus showing that potash contained a fixed percentage by weight of carbon dioxide as a chemical constituent.

Black's researches dispelled the belief that air was the only type of gas. The gas he called fixed air was evidently a chemically reactive substance that was entirely distinct from common air. It was produced by respiration (as one could readily see by blowing bubbles in limewater through a straw), by fermentation, and by the combustion of wood or coal. The science of pneumochemistry—the quantitative study of the chemical properties of gases—had begun.

Hydrogen, Nitrogen, and Oxygen

The gas we know as hydrogen was first prepared by the Russian poet-scientist M.V. Lomonosov. In 1745, he wrote: "On solution of any non-precious metal in acid, there emerges an inflammable vapor that is nothing else than phlogiston." (For a discussion of this historically important but nonexistent material, turn to Chapter 5.) Only a few decades later, scientists discovered that water is not an element and that hydrogen is one of its elemental constituents. In fact, hydrogen is far and away the most abundant chemical element in the universe.

Henry Cavendish (1731–1810) measured the densities of the new gases.* He found that carbon dioxide is about 1.5 times as dense as air, while hydrogen is only about a tenth as dense. The existence of a gas lighter than air caused a considerable stir and inspired a remarkable dinner *chez* Black. He prepared a balloon (formed of the fetal membrane of a calf) filled with hydrogen, which rose to the ceiling when it was released. The guests were convinced that the effect was an illusion. They were amazed when no trace of trickery could be found.

Only a few years later, what had begun as a dinner entertainment evolved into the first instance of manned flight. On November 21, 1783, a balloon, designed and constructed by the Montgolfier brothers, carrying two passengers, was launched into the skies of Paris to fly for 25 minutes over a distance of some five miles. It was a hot air balloon rather than one filled with hydrogen. In response to news of the flight, Black said:

> The experiment with the bladder was so very obvious that any person may have thought of it; but I certainly never thought of making large artificial bladders [to] carry men up into the air. The idea being founded upon a

---

*Cavendish made many important contributions to physics and chemistry, although he rarely published his results. A brilliant and wealthy recluse, he was known as "the richest of the learned and the most learned of the rich." We shall encounter him again in connection with the discovery of argon and the elucidation of the inverse-square laws characterizing electric and gravitational forces.

The first flight of a hot air balloon with a live cargo took place on September 19, 1783, in Paris. It carried a lamb, a duck, and a rooster. Two months later, a manned flight was launched. *Source: The Bettmann Archive.*

principle which has long been known, which has no connection with Mr. Cavendish's discovery, it is only surprising that the Montgolfiers should not have put it sooner into practice. I suppose [they] never were roused into an operation for making the trial until others began to think of flying by means of inflammable air. Who first thought of the method I cannot tell, for I confess I did not read the history of the Experiments; they never interested me in the least.

Despite Black's modesty and lack of interest, humans learned to fly as a serendipidous result of his commitment to understand air. Here was an early instance of an unexpected technological development emerging as a spinoff from pure scientific research.

Neither Black's fixed air (carbon dioxide) nor Cavendish's inflammable air (hydrogen) are significant components of common air. Black's doctoral student, Daniel Rutherford, was assigned the task of isolating that part of air which remained after combustion takes place. The problem Rutherford faced was compounded by the fact that carbon dioxide is often produced in the process of combustion. Rutherford burned a piece of phosphorus in a sample of air trapped in a jar inverted in a pan of water. The fumes were readily absorbed by the water. When the process was complete, about three-fourths of the air remained. Rutherford had isolated nitrogen. He found that it did not support combustion or respiration (like fixed air), but did not turn limewater cloudy (unlike fixed air). Rutherford called his new gas *noxious air*. Nitrogen, the dominant elementary constituent of the atmosphere, was thus discovered in 1772. But the most important gas, oxygen, was still at large.

Joseph Priestley was the self-taught son of a tailor, a Unitarian minister and an outspoken supporter of the rebellious American colonists. His ministry was located conveniently near a brewery, where large quantities of fixed air (and good English beer) were readily available. He discovered how to make artificial soda water, thereby winning the prestigious Copley Medal of the Royal Society. Seltzer is serious stuff. Priestley improved Hales's pneumatic trough by replacing the water bath with a mercury bath. With this device, gases that dissolve in or react with water could be collected. And Priestley found scads of them: nitric oxide, nitrous oxide, hydrogen chloride, ammonia, sulfur dioxide, silicon tetrafluoride, and more. In 1774, he found the fairest gas of all—oxygen.

Priestley's discovery of oxygen has to do with a curious property of mercury. When it is heated in air, mercury forms a red oxide. Heated even more, mercuric oxide decomposes into its constituent elements. Priestley put mercuric oxide in a retort that led into an inverted mercury-filled flask (Figure 4.14). Sunlight was focused on the mercuric oxide to bring it to a high temperature. As it decomposed, oxygen collected in the flask. Priestley found his new gas to be fit to breathe and showed that it causes a candle flame to burn brighter than in common air. Mice in an oxygen-filled vessel were unusually frisky and survived about twice as long as they did in common air. Priestley concluded that his new gas

FIGURE 4.14
Schematic diagram indicating Priestley's discovery of oxygen. The intense heat produced by focused sunlight decomposes mercuric oxide. The evolving oxygen is collected in an inverted mercury-filled flask.

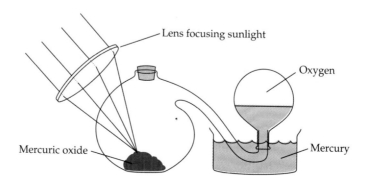

was "even purer" than common air. Priestley called it *dephlogisticated air* and said of it:

> After having ascertained the superior goodness of dephlogisticated air by mice living in it, I have had the curiosity of tasting it myself. The feeling of it to my lungs is not sensibly different from common air, but I fancied that my breath felt peculiarly light and easy for some time afterwards. Who can tell that, in time, this pure air may become a fashionable article of luxury? Hitherto, only two mice and myself have had the privelege of breathing it.

Oxygen instantly became a central focus of interest to chemists in England and on the Continent. Air was found to consist of about 80 percent nitrogen and 20 percent oxygen. When Cavendish extracted the oxygen (by burning) and the nitrogen (by sparking) from a sample of common air, he was left with a small and inexplicable remnant. Not until 1894 was it realized that Cavendish's bubble consisted of the inert gas argon, which makes up about 1 percent of common air.

Fire and Water

Cavendish noticed in 1766 that hydrogen burns, but he did not identify its combustion product. In 1781, he succeeded in igniting a mixture of common air and hydrogen gas by a spark within a sealed tube, only to discover, to his astonishment, that water was produced! His more quantitative studies, published in 1784, showed that two volumes of hydrogen gas react with five volumes of common air in such a way as to consume all of the hydrogen and to decrease the volume of common air by 20 percent.

Although the evidence pointed clearly to the fact that hydrogen and oxygen combine together to form water, Cavendish insisted on an explanation in terms of phlogiston: "There seems the utmost reason to think that dephlogisticated air [oxygen] is only water deprived of phlogiston, and that inflammable air [hydrogen], as before said, is either phlogisticated water, or else pure phlogiston, but in all probability the former." Cavendish was wedded to the false notion that water is a fundamental and indecomposable element. He regarded both hydrogen and oxygen as varieties of water with modified phlogiston content.

The chemistry of gases, thus far, had been a mostly British endeavor: from Hales, to Black, Rutherford, Cavendish, and Priestley. The next and most profound development took place across the Channel, in the laboratory of Antoine Laurent Lavoisier (1743–1794), who was truly the founder of modern chemistry. The well-bred son of a wealthy Parisian lawyer, he was elected to the French Royal Academy of Science at the age of 25, in 1768. At the same time, he purchased membership in the Ferme Générale, a private institution whose purpose was to collect taxes for the French government. His wife, Marie-Anne *née* Pierrette, was the beautiful and talented daughter of another member of the Ferme. We shall hear more about her in the next chapter. Their home became a center of international science, visited by Thomas Jefferson and Benjamin Franklin, as well as by Black and Priestley.

Lavoisier's experiment on the decomposition of water, as he drew it. Water, introduced at *A*, passes through a red-hot gun barrel at *B*, where it decomposes into oxygen (which unites with the iron) and hydrogen gas. Undecomposed steam is condensed and removed at *C*, and hydrogen gas is collected over water at *D*. *Source:* Historical Pictures/Stock Montage, Inc.

Months after the discovery of oxygen, and shortly after a visit by Priestley to Paris, Lavoisier prepared and studied oxygen at his own laboratory. Of its discovery, Lavoisier writes in 1789:

A taper burned in it with a dazzling splendor, and charcoal, instead of consuming quietly as it does in common air, burnt with a flame, attended with a decrepitating noise, like phosphorus and throughout such a brilliant light that the eyes could hardly endure it. This species of air was discovered at the same time by Mr. Priestley, Mr. Scheele, and myself. Mr. Priestley gave it the name of 'dephlogisticated air', Mr. Scheele called it 'empyreal air'. At first I named it 'highly respirable air', to which subsequently has been substituted the term 'vital air'. We shall presently see what we ought to think of these denominations.*

Lavoisier discovered the true nature of fire by showing that combustion is the chemical combination of oxygen with a burning substance. He showed that the metabolism of humans and of animals involved the "internal combustion" of carbon and hydrogen (in our foods) to exhaled CO_2 and H_2O. Cavendish's discovery of the burning of hydrogen to form water played a key role in Lavoisier's thinking. In the illustration, you can see how Lavoisier recovered hydrogen gas after passing steam through a red-hot gun barrel.

The title of his first great paper, published in 1783, is self-explanatory: *On the Nature of Water and on Experiments that appear to prove that this Substance is not properly speaking an Element, but can be decomposed and*

*Lavoisier coined the word *oxygen* from the Greek for acid-former because he believed (falsely!) that all acids contain oxygen. His name for inflammable gas, *hydrogen*, meaning water-former, makes more sense.

recombined. Water is a compound of the elements hydrogen and oxygen. All four of the Greek elements had by now bitten the dust: Air is a mixture of elements, water is a compound, fire is the process of oxidation, and earth comes in many chemically distinct varieties.

With the work of Lavoisier, chemistry became a mature science as opposed to an almost magical art. His notion of chemical elements as irreducible simple substances is almost the same as ours. This becomes clear from a perusal of Table 4.2, his list of chemical elements.

Lavoisier suspected (correctly!) that the earths were actually oxides of metallic elements that he could not isolate. The radicals posed a more serious problem. Humphry Davy, in the early nineteenth century, showed that muriatic acid is a compound of hydrogen and chlorine—the first known acid that does not contain oxygen. Lavoisier's muriatic radical is chlorine. Elemental fluorine was not prepared until 1886. Light and heat remained deep mysteries to Lavoisier, and would remain so for decades. The number of known chemical elements was to increase rapidly from Lavoisier's time onward, as shown by the graph in Figure 4.15.

Roughly speaking, the pace of discovery has been constant in time, with the last three elements (107, 108, and 109) created, atom by atom, by German scientists in the 1980s. The structure of the curve of known chemical elements reflects the emergence of new and more powerful techniques, such as electrochemistry and spectroscopy. The most recent spurt of discovery, beginning about 1940, results from the syntheses, by means of nuclear reactions, of about 20 chemical elements that do not ordinarily exist on Earth. We shall have more to say about the chemical elements in Chapter 6.

TABLE 4.2   Lavoisier's 33 Elements

Three Gases	Three Radicals
Hydrogen	Muriatic
Nitrogen	Boric
Oxygen	Fluoric
Fifteen Metals	**Five Non-Metals**
Bismuth	Antimony
Cobalt	Arsenic
Copper	Carbon
Gold	Phosphorus
Iron	Sulfur
Lead	**Five Earths**
Manganese	Calcia
Mercury	Magnesia
Molybdenum	Baria
Nickel	Alumina
Platinum	Silica
Silver	**Two Boo-Boos**
Tin	Light
Tungsten	Heat
Zinc	

Radicals are, in fact, compounds with hydrogen.
Earths are, in fact, compounds with oxygen.
Light and *Heat* are not chemical elements at all.

FIGURE 4.15
The number of known chemical elements vs. time. Lavoisier listed 31 elements, while today we know of 109, many of them artificial and unstable. Phosphorus is the first chemical element whose discoverer is known. It was found by Hennig Brand of Hamburg in 1676. Rhenium, which is used in turbine blades, is the last stable chemical element to be found. Ida Noddack of Berlin (and others) isolated it in 1925.

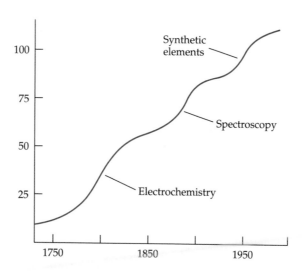

Exercises

22. The lifting power of a balloon of volume *V* equals the difference between the mass of that volume of air and the gas contained in the balloon. A hot air balloon rises because hot air is less dense than cool air. A helium-filled balloon rises because helium's density is only ∼ 14 percent that of air. Hydrogen gas is half as dense as helium. Does this mean that a hydrogen-filled balloon has twice the lifting power of a helium-filled balloon? Explain your answer.

23. A hot air balloon is filled with 400 m³ of air at a temperature of 127° C and at atmospheric pressure. The outside temperature is 27° C. What is the largest mass (including the balloon and its payload, but not the air within) that will fly? (To answer this, you must know that a cubic meter of air at 27° C and atmospheric pressure has a mass of about 1.3 kg.)

24. A flat raft and a canoe have the same mass and are made of the same kind of wood. Explain why the canoe can carry a larger load without foundering.

25. Many of the developments discussed in this chapter had to do with the curious chemical and physical properties of mercury. How could the secrets of air pressure and oxidation have been discovered if mercury did not exist?

26. Assume that air consists of 80 percent nitrogen gas and 20 percent oxygen gas *by volume*. If the density of pure nitrogen is 1.25 kg/L and that of pure oxygen is 1.43 kg/L, what is the density of air?

Where We Are and Where We Are Going

None of the scientists whose work we discussed in this chapter was *just* a physicist. Each of them was also a philosopher, physician, astronomer, mathematician, biologist, or chemist. Pascal is known today primarily as a mathematician and a philosopher of religion. We described Black's chemical investigations, but we shall learn in the next chapter that he also carried out the first quantitative studies of heat. Priestley discovered oxygen and found that it is consumed by animals but regenerated by green plants and seaweeds. His textbook on electricity first proposed the inverse-square law governing the force between electric charges. Lavoisier distinguished atmospheric oxygen from nitrogen chemically (by its role in combustion), biologically (by its ability to sustain animal life), and physically (by its greater density). Science, as we know it, was being invented from scratch, and its subdisciplines had not yet emerged.

The first hesitant steps on the way to modern science were taken in the seventeenth century when people realized that things are not as they seem: that Earth is not flat and that it moves; that the natural state of motion is perpetual motion; and—the starting point of this chapter—that we live under an ocean of air. In the eighteenth century, air was acknowledged to be a form of matter, and its physical and chemical properties were determined. Many different kinds of gas were discovered, and air was found to be a mixture consisting mostly of oxygen and nitrogen. Oxygen is required by living creatures and is consumed in the process of combustion. The science of chemistry had taken off. Any gas, no matter what its chemical composition, satisfies a precise physical law by which its pressure, volume, and temperature are related to one another. The science of physics had taken off.

However great were their contributions to the sciences, practically all scientists up to the early nineteenth century shared a flawed faith. They believed that heat is a material substance. Lavoisier called it caloric and went so far as to include it in his list of chemical elements. Chapter 5 describes how nineteenth-century scientists learned that heat is not a form of matter but a form of energy. It consists of the random motions of the individual molecules of which matter is made. For a gas, the simplest state of matter, temperature is a measure of the mean kinetic energy of its freely moving molecules. Once the nature of heat was understood, the two great laws of thermodynamics were formulated. Only then could modern science and technology develop.

5   Heat Is a Form of Motion

The idea that matter consists of tiny particles is one of the oldest in human history. What concerns us here is the transformation of atomism from speculative philosophy to testable physical theory.... It was not the belief in continuous matter which the kinetic theory had to displace, but rather an opposing doctrine about the nature of heat. The proponents of the caloric theory were also in many cases atomists, but they attributed the expansive power of a gas to the repulsion of its atoms rather than to their free motion. The kinetic theory could not flourish until heat as substance had been replaced by heat as atomic motion. Hence a thorough development of the kinetic theory had to wait until the second half of the 19th century....

A convenient starting point is the notion of air pressure which arose in the 17th century through the work of Torricelli, Pascal and Boyle. By a combination of experiments and theoretical reasoning, they persuaded other scientists that the earth is surrounded by a sea of air that exerts pressure in much the same way that water does, and that air pressure is responsible for the phenomena previously attributed to "nature's abhorrence of the vacuum." Instead of postulating occult forces and teleological principles to explain natural phenomena, scientists started to look for explanations based simply on matter and motion.

Stephen G. Brush, American historian of science

Distillation furnace of the fifteenth century. This device is not very different from the stills that modern Scots use to make whiskey. Our forebears may not have known the true nature of heat, but they used it effectively for thousands of years. *Source:* Courtesy Meyers Photo-Art.

The stories of heat and civilization are inextricably bound together. According to a Greek myth, Prometheus stole fire from the gods and presented it to mortals, who put it to good use: to cook food, keep warm, and extract metals with which to make tools and ornaments. Harnessed by steam engines, heat from fuels impelled the Industrial Revolution. Harnessed by turbines and internal combustion engines, heat lets people travel by air, land, and sea. Heat released by the combustion of fossil fuels, or by nuclear fission, makes modern society what it is.

What are heat and fire? Eighteenth-century scientists believed that heat was a fluid called *caloric.* We learned in Chapter 4 that fire was thought to be a process by which a burning body expelled another fluid called *phlogiston.* These were the devils whose exorcism was necessary before the mysteries of fire and heat could be solved. Section 5.1 describes the fall of the phlogiston theory and the discovery that the quantity of matter is unchanged by any process. Section 5.2 describes the fall of the caloric theory and the linked discoveries that heat is a form of energy and that the quantity of energy is unchanged by any process. Section 5.3 turns to the study of kinetic theory, which describes the thermal properties of gases in terms of their underlying molecular structure. The final section introduces the science of thermodynamics, the quantitative study of the relationship between heat and mechanical work.

5.1 The Law of Conservation of Mass

When we discussed mechanics in Chapter 3, we assumed—without presenting any empirical evidence—that the total mass does not change when objects collide or otherwise interact. But is it true that all matter now existing existed in the past and will continue to exist in the future? After all, things change their shapes, forms, colors, and chemical natures. Ancient Greek philosophers speculated that the overall quantity of matter cannot change. Their view was succinctly stated by the Roman poet Lucretius: "Things cannot be born from nothing, cannot when begotten be brought back to nothing." In the seventeenth century, Von Helmont put the hypothesis of matter conservation to a test. He dissolved a weighed amount of glass in a strong alkali.* Treating the solution with acid to reconstitute the silica, he showed that the recovered material weighed the same as the original glass. But the conservation of mass did not seem to apply to the most familiar type of chemical transformation, combustion.

Robert Boyle thought that fire consisted of innumerable small sharp particles moving so rapidly that they could pass through material bodies. He sealed bits of metal in glass flasks and heated them until the metal was transformed into what we today call their oxides. Opening the flasks and weighing the metal, he noted an increase in mass that he attributed to the penetration of fire particles through the glass and into the metal. In fact, the metal had chemically combined with the oxygen in the flasks. Wouldn't Boyle have been surprised if he had weighed the sealed flasks before and after heating them?

The Rise and Fall of Phlogiston

Combustion can be confusing. Oil, gas, and candle wax disappear into thin air when they burn, while wood, paper, and coal leave hardly any ash. However, when metals are intensely heated in air, they get heavier. George Ernst Stahl (1660–1734) devised an ingenious but false theory of combustion that remained chemical dogma until the work of Lavoisier. In his theory of phlogiston, elements were drawn to one another by a mysterious attraction: They never occurred in isolation, but only as constituents of material things.† Phlogiston (from the Greek to burn) was one of these hypothetical elements—it was the agent underlying combustion, the be-all of burning.

What Stahl did not know is that ashes left by burning wood are light because most of the reaction products escape as CO_2 and steam. He believed that burning bodies release phlogiston into the air. To him, air was just there—it played no active part in combustion. That metals became heavier when fired was not seen as an inconsistency because Stahl

*Von Helmont studied gas as well as glass, as we learned Chapter 4.

†In this respect, Stahl's elements had something in common with today's quarks, which cannot be isolated from the particles of which they form parts.

and his followers did not assume that quantity of matter is conserved under all circumstances.

Later chemists took to phlogiston like ducks to water. Black's fixed air (carbon dioxide) became known as phlogisticated air. According to Black, fixed air does not support fire because it is loaded with phlogiston and can accept no more: Animals die in it because they cannot rid themselves of phlogiston in an already phlogisticated atmosphere. Priestley's dephlogisticated air (oxygen) was supposed to be air from which phlogiston had been purged: Flames burn so brightly and animals are so frisky in oxygen because it can absorb phlogiston more readily than phlogiston-tainted common air. Lighter-than-air hydrogen was suspected to be phlogiston itself, thus hinting at an explanation for the gain in weight of metals when they burn.

They had it backwards. Combustion is the chemical combination of a fuel with oxygen. Well after Lavoisier found the truth, however, many scientists clung to the old ways. Priestley, driven from England because of his overt sympathy for the American Revolution, spent his last decade in rural Pennsylvania, a lifelong believer in phlogiston.

Lavoisier studied combustion by means of many ingenious experiments. He knew that diamonds burn, leaving no ash. Borrowing some from a well-known Parisian jeweller, he sealed them in airtight clay vessels and baked them. In the absence of air, the diamonds survived unscathed—to the delight of the scientist and the relief of the jeweller. He showed that oxygen is needed for combustion. "It was the glory of Lavoisier," wrote the British chemist Humphry Davy a generation later, "to lay the foundation of a sound logic in chemistry by showing that the existence of phlogiston, or of any other principles, should not be assumed where they could not be detected."

Conservation of Mass

Lavoisier's studies of combustion led him to formulate the law of matter conservation. If mercury is heated in air, it oxidizes to form red mercuric oxide. With the device shown in the illustration, Lavoisier showed that air is consumed as the mercury turns to a red powder. When heated further, the mercuric oxide decomposes into the original mercury. Like Priestley before him, Lavoisier collected the gas that was liberated and found it to be pure oxygen. In yet another experiment, he placed a weighed quantity of iron in a sealed tube filled with air. After heating the tube, he weighed the reaction product and measured the amount of air that was consumed. He wrote:

> The air will be found to be diminished in weight exactly to what the iron has gained. Having therefore burnt 100 grains of iron, which has acquired an additional weight of 35 grains, the diminution of air will be found exactly 70 cubical inches; and it will be found in the sequel, that the weight of vital air is pretty nearly half a grain for each cubical inch; so that, in effect, the augmentation of the weight of the one exactly coincides with the loss of it in the other.

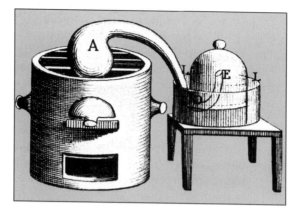

Lavoisier's device to oxidize mercury and measure the air consumed. The spout of a retort leads into a reservoir of air in an inverted bell jar. When the mercury in the retort is heated, flakes of red mercuric oxide appear on its surface. At the same time, air is consumed and the water level in the bell jar rises. Using this device, Lavoisier determined the oxygen content by weight of mercuric oxide and confirmed the law of mass conservation. *Source:* Courtesy Meyers Photo-Art.

Lavoisier proved that a gain in mass of one part of a closed system is always accompanied by a loss in mass of another. In 1789, he advanced the law of conservation of matter:

> We must lay it down as an incontestible axiom, that in all the operations of art and nature, nothing is created; an equal quantity of matter exists both before and after the experiment. Upon this principle, the whole art of performing chemical experiments depends.

Lavoisier may have been the greatest chemist of all time. He explained combustion as oxidation, identified water as a compound of hydrogen and oxygen, recognized what are and are not chemical elements, and proposed the law of conservation of mass. At the age of 50, he was beheaded because he was a tax collector for the *ancien régime.* An appeal on his behalf was met by the infamous statement: *"La République n'a pas besoin de savants"* (The state needs no scientists). On May 8, 1794, immediately after the execution of his father-in-law, Lavoisier took his turn at the guillotine. His colleague Legendre said of this murderous act that it took but a moment to cut off Lavoisier's head, but a century might not suffice to produce its equal.

Exercises

1. A quantity of mercury contained within a sealed flask is heated until it becomes a red powder. Is the mass of the powder more or less than the mass of the mercury? Is the pressure within the flask after it has cooled more or less than air pressure? Explain your answers.

2. Three kilograms of carbon, when it is burned, produces 11 kg of CO_2. What mass of oxygen is consumed in this reaction?

3. A small sample of a solid substance with mass m is placed in a 2.5-L flask, which is then tightly sealed. The flask is intensely heated, causing the oxygen in the air to combine chemically with the sample.

The density of air at standard conditions of temperature and pressure is 1.29 g/L, and it contains 23 percent oxygen by mass.

a. What is the mass of oxygen gas in the flask before it is heated? (Neglect the small volume of the sample in the flask.)

b. In the first experiment, the sample consists of 1 g of iron. The iron reacts with the air in the flask and is entirely converted into iron oxide, Fe_2O_3, with a mass of 1.43 g. How much oxygen remains in the flask after it is heated, but before the seal is cracked?

c. In the second experiment, the sample consists of 3 g of iron. All of the oxygen in the flask combines with the iron to form Fe_2O_3, but no other reaction takes place. What is the mass of the residue in the flask?

d. In the last experiment of the series, 1 g of sulfur is placed in the flask. All of the oxygen is consumed, leaving 0.355 g of sulfur in the flask. What is the mass of the gas that was produced by the union of sulfur and oxygen. What is its composition by mass?

4. Earth's atmosphere has a total mass of 5.1×10^{18} kg. Earth's oceans, lakes, and rivers have a mass of 1.7×10^{21} kg and water is 89 percent oxygen by mass. What is the ratio of the total mass of oxygen in water to that in air? (See Exercise 3.)

5. If all the oxygen in the atmosphere were burned to form CO_2, which remained in the atmosphere, by how much would the air pressure on Earth change? (*Hint*: Air pressure is the total weight of the atmosphere divided by the surface area of the Earth. By what fraction does the weight of the atmosphere change when each O_2 molecule is replaced by a heavier CO_2 molecule? See Exercises 2 and 4.)

5.2 The Law of Conservation of Energy

The law of conservation of energy has a richer history than the law of conservation of mass. It cannot be attributed to any one scientist, and it was not established until the middle of the nineteenth century. Lavoisier explained combustion, but left the subject of heat out in the cold. The words we use to describe heat recall the confusion of times gone by. *Thermometer* was coined in 1633 and means "heat measurer" in Latin. It was thought to show the amount of heat in a body. *Temperature* derives from temper and signified a mixture or proportion, as in: justice tempered with mercy, or a person's temper as a balance between good and evil fluids. The temperature of a substance once referred to the extent of its admixture with the hypothetical heat fluid, caloric. But heat is not a substance. It is the energy inherent in a substance in the form of molecular motions. Temperature indicates the magnitude of these motions. Not until these facts were known could scientists learn that the total amount of energy in an isolated system cannot change—that energy can neither be created nor destroyed.

The Measure of Heat

In the eighteenth century, the true nature of heat was unknown. Nonetheless, the effects of heat on matter were studied. Black and Lavoisier measured the heat content of objects by placing them in a container within a larger insulated container. The space between the inner and outer vessels was filled with ice. This contrivance, called an *ice calorimeter*, allowed an object's heat content to be reckoned by the mass of ice it would melt. Heat could be quantified without using a thermometer. That's the good news, because there was no commonly accepted temperature scale at the time. The bad news was that ice was hard to come by, because there were no refrigerators either.

In the nineteenth century, when reliable and standardized thermometers became available, calorimetry—the measurement of heat—became more sophisticated and several standard units of heat were defined. The British thermal unit (BTU) is the amount of heat required to raise the temperature of 1 lb of water by 1 Fahrenheit degree. On the Continent, the unit of choice became the amount of heat required to raise the temperature of 1 g of water by 1 Celsius degree. The latter unit is called the *calorie*.*

Specific Heat

Joseph Black, who discovered carbon dioxide, devoted much of the time in his Edinburgh laboratory from 1740 to 1760 to the study of heat. Black set out to determine how much heat must be added to an object to change its temperature. He introduced the concept of the *specific heat*, or *heat capacity*, of a substance. This is the quantity of heat that must be added to 1 g of that substance to increase its temperature by 1° C. The calorie had been defined so that the specific heat of water is 1 cal/g °C. The heat Q needed to raise a quantity of water from the temperature T to the temperature $T + \Delta T$ is $Q = m \Delta T$, with Q expressed in calories, m in grams, and ΔT in Celsius degrees.

Black found different substances to have different specific heats. When 1 calorie is added to 1 g of water, its temperature increases by 1° C. However, when 1 calorie is added to 1 g of mercury, its temperature increases by 30° C. Thus, the specific heat of mercury is 0.033 cal/g °C. In general, if an object is made of a substance with a specific heat σ, the heat Q that must be added to it to raise its temperature by ΔT is given by

$$Q = \sigma m \Delta T$$

The specific heats of ten common substances are listed in Table 5.1, along with their densities. Water has a higher specific heat than most other substances. Furthermore, it is usually, but not always, true that the denser a substance is, the smaller is its specific heat.

*When the unit is capitalized, as in Calorie, it denotes a nutritional calorie of 1000 calories—a kilocalorie, or the amount of heat needed to raise 1 kg of water by 1° C. Food energy is often given in Calories.

EXAMPLE 5.1
Antifreeze

The heat produced by an automobile engine is extracted by its cooling system. A liquid is pumped through the engine to be heated, and then through the radiator to be cooled. The cooling system of a Buick, garaged in sunny Florida, contains 10 L of pure water as its coolant. An Alaskan Chevy uses 10 L of methanol, whose freezing point is −96° C. How much added heat brings the the Buick's coolant from a temperature of 20° C to 100° C, the boiling point of water? How much heat will raise the coolant temperature of the Chevy from −14° C to the 66° C, the boiling point of methanol?

Solution

In both cases, we are asked to determine the quantity of heat needed to raise the temperature of a specified volume of fluid by 80° C. To obtain the answers, we use equation 1 and the specific heats and densities given in Table 5.1. The mass of the water in the Buick is 10 kg. Thus, 8×10^5 cal of engine heat will heat it by $\Delta T = 80°$ C and bring it to a boil. The mass of the methanol in the Chevy is 8 kg, and its specific heat is 0.6 cal/g °C. Consequently, a mere 3.64×10^5 cal brings the methanol to its boiling point. ∎

Before we proceed further, we must point out two everyday yet fundamental principles of heat:

A BIT MORE ABOUT

The Calorie

There are three questionable aspects to the definition of a calorie stated in the text. What is the pressure? What is the temperature? And, what is water? As its pressure changes, water's properties change as well. Thus, the standard calorie became the amount of heat required, at a pressure of 1 *standard atmosphere* (or 1.01325×10^5 N/m²), to raise the temperature of 1 g of water by 1° C. Since barometric pressure doesn't change very much from day to day, this refinement of the meaning of a calorie is ordinarily unimportant.

In 1870, the American physicist Henry Rowland showed that water's properties depend on temperature. Between its freezing point and boiling point, the heat needed to raise the temperature of a fixed amount of water by 1° C varies by 1 percent. The definition of the calorie had to include a specification of the starting temperature. This became the amount of heat required, at a pressure of 1 standard atmosphere, to raise the temperature of 1 g of water from 15° C to 16° C.

Water is made of hydrogen and oxygen, but hydrogen atoms are not all the same. Nor are oxygen atoms. We shall learn in Chapter 6 that atoms occur in various *isotopes* with different masses. Different samples of pure water differ in their *isotopic composition*, and in their physical properties. Fortunately, today's definition of the calorie has been divorced from the properties of water. In the SI system, 1 calorie is an amount of heat energy equal to 4.184 J, which is *about* enough to raise the temperature of 1 g of any sample of water by 1° C.

TABLE 5.1   Specific Heats and Densities of
Various Materials

Substance	Specific Heat in cal/g °C	Density in g/cm³
Ammonia (NH$_3$)	1.13	0.61
Water	1.00	1.00
Paraffin	0.70	0.90
Ice at −2° C	0.50	0.92
Methanol	0.60	0.80
Rock salt	0.22	2.20
Granite	0.19	2.70
Iron	0.11	7.90
Lead	0.031	11.0
Mercury	0.033	13.6

1. *The Principle of Heat Conservation*   The quantity of heat in a body or a system of bodies is conserved under very special circumstances. In order to apply this principle, no heat may be added or taken away. Thus, the system under consideration must be thermally isolated by placing it in an insulated container through which heat cannot pass. Neither chemical reactions nor mixings of dissimilar liquids may take place in the system because these processes generally produce or absorb heat. Vigorous stirring is also forbidden, since it would transform mechanical energy into added heat.

2. *The Principle of Thermal Equilibrium*   Temperature reflects the ability of matter to transfer heat to or from another body. When two bodies are placed in thermal contact, heat flows from the warmer body to the cooler one. The process continues until the two bodies come to the same intermediate temperature. They are then said to have come to thermal equilibrium.

Suppose that two objects, with masses m_1 and m_2 and specific heats σ_1 and σ_2, are originally at temperatures T_1 and T_2. They are placed in thermal contact in an insulated container and allowed to come to an equilibrium temperature T. If no chemical or physical reactions take place, the heat content of the first body changes by $Q_1 = m_1\sigma_1(T - T_1)$, while the heat content of the second body changes by $Q_2 = m_2\sigma_2(T - T_2)$. The total quantity of heat is conserved, so that

$$Q_1 + Q_2 = 0 \qquad (2)$$

If the masses and specific heats of the bodies are known, equation 2 may be used to express the equilibrium temperature T in terms of T_1 and T_2.

EXAMPLE 5.2
**Some Like
It Lukewarm**

An insulated bathtub is filled with a mixture of 100 kg of water at 90° C and 300 kg of water at 10° C. What is the equilibrium temperature T of the bathwater?

Solution

The change in the heat content of the hot water is $Q_h = (T - 90) \times 10^5$ cal and is negative. The change in the heat content of the cold water is

$Q_c = (T - 10) \times 3 \times 10^5$ calories. Equation 2 applies, so $Q_h + Q_c = 0$. We solve this linear equation in T to find $T = 30°$ C. ∎

EXAMPLE 5.3
How to Measure
Specific Heat

One kilogram of an unknown metal is heated to $100°$ C. It is plunged under 2 kg of water at $20°$ C in an insulated container, after which the system comes to a common temperature of $25°$ C. What is the specific heat σ of the metal?

Solution

As the metal cools by $75°$ C, its heat content changes by $Q_m = -7.5 \times 10^4 \sigma$, where σ is its unknown specific heat. Meanwhile, the heat content of the water changes by $Q_w = +10^4$ cal. Since $Q_m + Q_w = 0$, we conclude that $\sigma = 0.133$ cal/g °C. ∎

Latent Heat

Sometimes heat may be added to a substance without changing its temperature. A pot of boiling water on a stove remains at a constant temperature of $100°$ C—its boiling point—until it boils away. A glass of ice water in a warm room remains at a constant temperature of $0°$ C—the melting point of ice—until all the ice is gone. To describe these phenomena, Black introduced the concept of *latent heat*. In general, it is the amount of heat needed to change the state, but not the temperature, of 1 g of a substance.

A pot of boiling water turns to steam, but not all at once. Black found that about 540 calories must be added to 1 g of water at $100°$ C to turn it into steam:

$$1 \text{ g water at } 100° \text{ C} + 540 \text{ calories} \longleftrightarrow 1 \text{ g steam at } 100° \text{ C}$$

Notice that the arrow in this relation points both ways. When 1 g of steam condenses to water, it gives off 540 calories. That's what goes on in a steam radiator to keep you warm in winter.

The heat needed to transform 1 g of a particular liquid to its gaseous state is called its *heat of vaporization*. Like other latent heats, it is measured in cal/g. The heat of vaporization of a fluid depends on the nature of the liquid and on the ambient air pressure. For water at atmospheric pressure, it is about 540 cal/g. Black found that different liquids have different heats of vaporization. That of methanol is 260 cal/g, while that of mercury is 65 cal/g.

The ice in a glass of ice water turns to water eventually, but it takes about 80 calories to melt each gram of ice. As the ice melts, it remains at a constant $0°$ C:

$$1 \text{ g ice at } 0° \text{ C} + 80 \text{ calories} \longleftrightarrow 1 \text{ g water at } 0° \text{ C}$$

Once again, the arrow is two-sided: 80 calories of heat must be extracted from 1 g of water to freeze it.

The heat needed to transform 1 g of a particular solid to its liquid state is called its *heat of fusion*, and it depends on the composition of the solid. For water, it is about 80 cal/g, while for paraffin it is 35 cal/g, and for lead it is merely 6 cal/g. Suppose that the latent heat for the transition of a substance from state a to state b is σ_l. (The transition may be from solid to liquid, or liquid to gas, or from one crystalline form to another.) The heat in calories that must be added to the substance to cause its transformation is

$$Q_{a \to b} = \sigma_l m$$

if m is the mass of the substance in grams and σ_l is expressed in cal/g. Conversely, the heat required for the reverse transformation is

$$Q_{b \to a} = -\sigma_l m$$

(That is, heat is released when steam condenses or water freezes.)

EXAMPLE 5.4
Latent Heat

In an insulated container, 1 kg of water at 28° C is added to an unknown mass of ice at 0° C. The ice melts, and the water comes to a uniform temperature of 10° C. What was the mass of the ice?

Solution

The heat content of the original 1 kg of water, in cooling by 18° C, changed by $Q_1 = -1.8 \times 10^4$ cal. Let x denote the mass of the ice in grams. To melt it required $80x$ cal. To heat the resulting water from 0° C to 10° C, required an additional $10x$ cal. Thus, the heat content of the ice (and the water it became) changed by $Q_2 = 90x$ cal. Heat is conserved, so $Q_1 + Q_2 = 0$. Thus, $90x - 1.8 \times 10^4 = 0$, and $x = 200$ g. ∎

The Rise of the Caloric Hypothesis

Black's investigations of heat posed a challenge. Why should the quantity of heat needed to increase a fixed mass by a fixed temperature increment vary from material to material? Black observed that, just as different substances dissolve in water to a greater or lesser extent, different materials also have varying capacities for absorbing heat: Water's heat capacity is large, mercury's is small. It seemed natural and elegant to regard heat as a material substance—a subtle fluid called *caloric*—to which different materials had different affinities. It was the tenor of the times: Light was also thought to be a form of matter, as were electricity and magnetism.

The concepts of specific heat and latent heat are of great practical and pedagogical importance, but Black wrongly interpreted his results as evidence for caloric theory. He believed that ice melted after it had absorbed enough caloric to become water. A further addition of caloric raised its temperature. Once water reached its boiling point, it continued to absorb caloric (with no change in temperature) until it could become steam.

The density of liquid mercury is 13.6 times that of water, and its specific heat is only a thirtieth that of water. According to Black, mercury is so dense because its atoms are closely packed, leaving little room for caloric. Because iron is less dense, its atoms are further apart and its specific heat is larger. Water is less dense still, with lots of room for caloric, which, Black thought, explains its large specific heat. However, and as you can see from Table 5.1, there is no definite correlation between the densities and specific heats of different substances. For example, paraffin floats on water, but its specific heat is less than 1 cal/g °C.

Black and Lavoisier showed that the total amount of heat (like that of any chemical element) remained unchanged in a thermally isolated system. Since the marvelously simple notion of conservation, which applied so well to each of the chemical elements, seemed to apply to heat as well, Lavoisier included caloric, along with light, in his table of chemical elements. The good and the bad, oxidation and caloric, were integral parts of his new chemistry. The false doctrine of phlogiston was abandoned, but so was the notion (promulgated by Daniel Bernoulli and discussed in the next section) that heat is a form of molecular motion. Lavoisier had taken one giant step forward and a very large step back. Nonetheless, Black and Lavoisier are judged by their many successes, and not by their one colossal blunder.

The Fall of the Caloric Hypothesis

Ancient mariners knew the danger of a rope burn and American Indians (as well as Boy Scouts) once built fires by rubbing sticks together. If heat were a substance, how could it be produced by friction? If rubbing causes the caloric substance to be expelled, and if caloric is conserved, shouldn't the substance that was rubbed be changed in some manner? This question preoccupied one of the most curious characters in the history of science, an American expatriate named Benjamin Thompson (1753–1814), who became known as Count Rumford.

At the end of the eighteenth century, Rumford became the director of the Bavarian arsenal, where he supervised the boring of cannon. It was during this time that Rumford became convinced that the caloric hypothesis was false. Because an enormous amount of heat is generated by the process of boring a cannon, water was used to cool the cannon and the drill. As the water boiled away, it had to be continually replaced. According to caloric theory, caloric was released into the water by the drill, or by the metal being drilled, thereby heating the water. Rumford argued that if heat were a fluid resident in the metal, its caloric should be consumed at some point. Thus, later stages of boring should produce less heat and boil away less water. To test this, Rumford submerged a cannon in a tank of water, began boring the cannon, and measured the time it took the water to boil. Of this demonstration, he wrote: "It would be difficult to describe the surprise and astonishment expressed in the countenances of the by-standers, on seeing so large a quantity of cold water heated, and actually made to boil, without any fire."

Hour after hour, the water boiled furiously as the cannon-boring operation generated heat. It seemed that the water would boil for as long as the treadmill would turn. Where was this heat coming from? According to Rumford, the amount of heat that could be produced by friction "appeared evidently to be inexhaustible." Could the excess caloric somehow have been extracted from the iron filings produced by boring? Rumford compared the metal chips with the material of the cannon itself and found them to be identical. He was compelled to reject caloric theory:

> Anything which any insulated body, or system of bodies, can continue to furnish without limitation, cannot possibly be a material substance; and it appears to me to be extremely difficult, if not quite impossible, to form any distinct idea of any thing, capable of being excited and communicated in the manner the Heat was excited and communicated in these experiments, except it be Motion.

This work, reported in 1798, became an immediate classic, and Rumford became the leader of the anticaloric camp. Two incontrovertible arguments were to prove that heat is not a chemical element or a material substance: (1) Heat could be created without limit by the action of friction, and (2) the addition of heat to a body does not change its mass. Rumford placed the last nails in the coffin of caloric. When Rumford married the widow of Lavoisier, he wrote to his mistress, "I think I shall live to drive caloric off the stage as the late Monsieur Lavoisier (the author of caloric) drove away phlogiston. What a singular destiny for the wife of two Philosophers!"

The Mechanical Equivalent of Heat

If heat were not matter, what could it be? Rumford showed that friction can generate unlimited quantities of heat, but the modern theory of heat did not arise for another half century. The next character in our brief history of heat is another self-taught scholar, James Prescott Joule (1818–1889). Joule became convinced that whenever mechanical work is expended, an exact equivalent of heat is produced. Thus, energy that appears to be consumed when hailstones fall to the ground or waves crest and crash is not lost. Instead, it is converted into an equivalent quantity of heat. Joule demonstrated the conversion of various forms of energy into heat and determined what he called the *mechanical value of heat*—the numerical factor that relates mechanical energy units (such as joules) to heat units (such as calories).

Joule devised many experiments to prove his conjecture. He measured heat with a calorimeter consisting of a known quantity of water in an insulated container. The temperature of the water was monitored by a thermometer. Measured quantities of energy were expended within the calorimeter by several entirely different means. As the energy became heat, the water temperature rose, and Joule measured the heat equivalent to the energy. The energy within the calorimeter came from various devices that were placed within the container and driven by external energy sources. One such apparatus, shown in Figure 5.1 (on page 206), consisted of a paddle wheel, powered by a descending weight, that violently agitated the water. In another, an electric current passed through a resistor in the water. In yet another, air was compressed into a tank within the water. In each case, Joule measured the energy output of the source and the heat deposited in the calorimeter, and in each case he got

FIGURE 5.1
One of Joule's methods to determine the mechanical equivalent of heat. The temperature of a known volume of water in an insulated container is monitored. An eight-bladed paddle is driven by the descent of a mass M. When the mass falls a distance h, its gravitational potential energy decreases by Mgh. An equivalent amount of heat is deposited in the water, whose temperature rises.

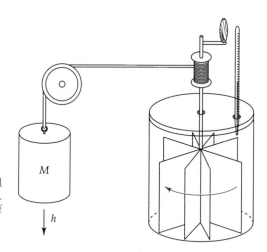

approximately the same result. He found that the mechanical energy needed to raise 1 pound of water by 1 degree Fahrenheit was equivalent to that needed to raise a 1-pound weight by about 800 feet.

Joule concluded: (1) that energy appears in many guises, some of which are kinetic energy, rotational energy, electrical energy, chemical energy, and gravitational potential energy; (2) that heat is another of energy's many forms; (3) that energy may be transformed from one form to another, but the total amount of energy contained within a closed system cannot change. Joule's experiments were reproduced throughout Europe with the same result. Scientists soon agreed that heat is neither a fluid nor is it always conserved—although the sum of all forms of energy within a closed system stays constant. It is a form of energy. Rumford's inexhaustible source of frictional heat reflected the conversion of mechanical energy into heat energy.

It would not be fair to say that Joule discovered the law of conservation of energy.* He was a major player among a dozen European scientists who studied the nature of heat in the decade 1837–1847. Conservation of energy was simply a law whose time had come: from studies of processes linking electricity and magnetism to heat, from the behavior of gases, from the calorimetric analysis of chemical reactions, and from the very practical analyses of the operation of steam engines.

Besides proving that energy is conserved, Joule determined the conversion factor between units of heat and mechanical energy. In honor of his contributions to physics, his name is used for today's SI unit of energy: the *joule* is defined to be 1 newton-meter. The calorie, which was defined as a measure of heat, is in fact a unit of energy:

$$1 \text{ calorie} \equiv 4.184 \text{ joules}$$

---

*In fact, Julius Robert Meyer, a German physician, first proposed in 1842 that heat is a form of energy and that energy is conserved. His work was largely ignored.

EXAMPLE 5.5
An Imaginary Duel

Suppose Dr. Black and Count Rumford set out to settle their scientific differences with pistols at 20 paces. The duel ended in a draw when neither bullet found its mark. The two 10-g projectiles, each fired at 200 m/s, collided in midair to form a single lump of lead, which fell to the ground. Overcome by scientific curiosity, the duelers rushed toward the lump to measure its temperature, thereby knocking one another out. If the temperature of the bullets in flight was 80° C, what was the temperature of the 20-g lump of lead just after the impact?

Solution

The kinetic energy of each bullet was $\frac{1}{2}mv^2 = 200$ J. When the bullets struck one another, they stuck together and came to rest. Their 400 J of kinetic energy was turned into 96 cal of heat. Equation 1, with $m = 20$ g and $\sigma = 0.031$ cal/g °C, may be used to obtain $\Delta T = 154°$ C. The final temperature of the leaden lump was 234° C. ∎

Table 5.2 shows the equivalent of a million joules of energy in various forms. If the total amount of energy in the universe is fixed, how do the people of the world get energy to heat homes, grow and cook food, and make goods? Mostly, they burn the world's age-old, but nonrenewable, store of fossil fuels. The six billion of us consume ~10^{21} J of energy from fossil fuels each year. What shall we do in a few centuries when all of the available natural gas, coal, and oil runs out? The answer could be solar energy. The power falling as sunlight on the state of Florida equals the power produced by all of the man-made fires on Earth. We have the potential to supply all our energy needs from its rays. At the moment, however, the potential is far from actuality.

TABLE 5.2   The Meaning of a Megajoule

Mechanical energy	Car speeding at 60 mph
Potential energy	Car suspended 100 feet off the ground
Chemical energy	One ounce of gasoline
Electrical energy	One-third of a kilowatt-hour
Nutritional energy	Four slices of dry toast
Heat energy	A point of water, boiled away

Exercises

6. In olden times, a bed would be warmed with a hot brick or a hot-water bottle. The brick starts off at the same temperature as the bottle, but it is twice as heavy. Its specific heat is a quarter that of water. Which device warms the bed longer? Justify your answer.

7. A 1-kg brick at 60° C with specific heat 0.25 calories per gram degree Celsius is placed in an insulated container containing 1 kg of water at 20° C. To what temperature does the system come?

8. How many calories would heat a kilogram of mercury by 30° C?

9. A liter of water at 20° C is placed in a well-insulated 1000-watt electric kettle.
   a. How long will it be before the water boils?
   b. How much longer does it take for all the water to boil away?

10. Four hundred grams of ice at 0° C are placed in an insulated container.
    a. What is the least mass of water at 20° C that must be added to melt all the ice? (*Hint*: Let x be the answer in grams. We end up with $400 + x$ grams of water at 0° C. The heat given up by the

---

A BIT MORE ABOUT
Count Rumford

Benjamin Thompson was born in Woburn, Massachussetts, and was indentured to a Salem merchant at age 13. During the next decade, Benjamin learned science on his own, became a schoolmaster, published several scientific papers, led expeditions throughout the White Mountains, became a major in the New Hampshire militia, and, at the same time, an effective secret agent for the British. In 1775, his treachery suspected, he absconded to London.

In England, he contributed to the design of military ordnance and munitions. His research earned him an exalted Fellowship of the Royal Society. His observation that a gun barrel becomes hotter after firing a real bullet than a blank suggested to him that heat may be a form of motion. Returning to America as a British officer, he fought several engagements and recruited a regiment of loyalists. While wintering in Long Island, his troops devastated orchards, razed churches, and used gravestones for dinner tables. A later historian wrote bitterly of Thompson's depredations:

> His acts in this place have given him an immortality which all his military exploits, his philosophical disquisitions and scientific discoveries, will never secure to him among the descendants of this outraged community.

Thompson returned to England as a hero and was knighted by King George III. Soon after sitting for a portrait by Thomas Gainsborough, in his new rank as a British colonel, Thompson set off for the greener pastures of Munich. He enchanted the elector of Bavaria and was promoted to general and ordered to reform and reorganize the army. With this charge, he set up schools for soldiers and their families to receive

Colonel Benjamin Thompson at age 30, painted by William Gainsborough. *Source:* Historical Pictures/Stock Montage, Inc.

water as it cools is $20x$ calories. The heat needed to melt the ice is 80×400 calories. Set these equal and solve for x.)

b. How much boiling water would be needed to accomplish the same task?

11. One kilogram of water at $80°$ C is added to a thermos bottle containing an unknown mass of ice at $0°$ C. The ice melts and all the water in the thermos comes to a uniform temperature of $20°$ C. What was the mass of the ice?

---

a basic education, and increased the salaries of enlisted men so they no were no longer effectively indentured to their officers. The beggars of Munich were rounded up; assured food, warmth, and shelter; and put to work in the manufacture of military equipment. Here is Sir Benjamin as pragmatic social philosopher— and, perhaps, as the inventor of social welfare:

> To make vicious and abandoned people happy, it has generally been supposed necessary first to make them virtuous. But, why not reverse this order? Why not make them first happy, and then virtuous? If happiness and virtue be inseparable, the end will be as certainly be obtained by the one means as by the other.

Thompson also designed clothing to keep soldiers warm. After extensive research, he discovered that air trapped in fur, felt, or feathers is responsible for their insulating power. For the discovery of thermal underwear, Thompson won the Copley Medal of the Royal Society. Other discoveries followed: the drip coffeepot, the built-in kitchen (with cabinets, stove, sink, and so on assembled together rather than free-standing), and the pencil eraser. On May 9, 1792, Thompson was dubbed Count of the Holy Roman Empire, and it is as Count Rumford that Sir Benjamin is known today. His many good works in Bavaria made him as much a hero as he was devil to Long Islanders. In the English Gardens of Munich (which Thompson designed), his bust still stands with the inscription:

To him who rooted out the most disgraceful public evils—Idleness and Mendacity: who gave to the Poor, relief, occupation and good morals, and to the Youth of the Fatherland so many schools of Instruction. Go, saunterer! and strive to equal him in Spirit and Deed, and us in Gratitude.

The Franklin stove and the Rumford fireplace were competing home-heating improvements invented by born-in-America Benjamins. This cartoon was drawn by Isaac Cruikshank and entitled "The Comforts of a Rump ford." *Source:* The Bettmann Archive.

12. Sunlight keeps Earth warm and habitable. Explain why the climate doesn't get hotter and hotter as the Sun pours energy into Earth day in and day out?

13. An incandescent 100-watt light bulb converts electrical energy into radiant light and heat. If it is placed within an opaque and insulated flask containing 1 L of water, all the electric energy consumed is converted into heat and the temperature of the water increases at a fixed rate. (Eventually, it will boil, but let us not concern ourselves with that.) What is the rate of temperature increase in Celsius degrees per minute?

14. In a mixture of gases, the average kinetic energy of the molecules of each species is the same. What is the ratio of the average speed of oxygen and hydrogen molecules in a mixture of these gases? (An oxygen molecule is 16 times heavier than a hydrogen molecule.)

15. The metric unit of power (energy per unit time) is joule per second, or watt. A lit 100-watt light bulb consumes 100 joules of electric energy per second. Electric energy is sold by the kilowatt-hour, which is 1000 watt-hours or 3.6 million watt-seconds, or 3.6 megajoules. In what way is a person on a 2065 Calorie per day diet like a 100-watt light bulb?

5.3 The Kinetic Theory of Gases

We haven't answered the question Black posed: Why does the quantity of heat needed to increase a fixed mass of a substance by a fixed temperature increment vary from material to material? One way to study the nature of heat is to speculate about the invisible internal structure of matter, especially of gases.

Primitive atomistic models of a gas were fashionable in the seventeenth century. Boyle tried to explain his law in terms of particles of air that behaved like coiled-up springs. Descartes suggested rapidly whirling air molecules that repelled neighbors coming too close. Newton conjectured (wrongly!) that a long-range repulsive force acted between molecules to drive them apart, thus producing the effects of air pressure and explaining Boyle's law. The modern and correct kinetic theory of gases depends on six speculative hypotheses about the invisible substructure of gases:

1. Gases consist of molecules
2. Molecules occupy a negligible volume of the gas
3. The number of molecules in a gas is very large
4. Molecules are in random motion

5. Forces between molecules are negligible

6. Collisions between molecules are elastic

All of these statements are true, or true enough, to explain the many empirical successes of the kinetic theory of gases. However, their truth could not be directly established until the early twentieth century. Kinetic theory owes its predictive power to its reliance on a molecular model of gases that is very nearly correct. Its weakness lies in its inability to describe other states of matter. When matter is condensed into a liquid or a solid, most of its volume is occupied by molecules. Furthermore, the molecules touch and exert forces on one another. Whole branches of modern physics are devoted to elucidating the properties of condensed forms of matter. Water, for example, consists of H_2O molecules whose structure is understood in terms of electrons and atomic nuclei. And yet, it is a formidable task to compute the specific heat of water from first principles. Nor can it be done for rock, paper, or scissors. The properties of gases are simpler to understand. We shall partially answer Black's question by deducing, later in this chapter, the specific heats of gases. But first, I describe how the successes of kinetic theory dragged us, kicking and screaming, toward the atomic hypothesis and the notion of heat as chaotic molecular motion.

In the early eighteenth century, Daniel Bernoulli developed a model of gases that is amazingly close to the modern kinetic theory of gases. His theory was proposed 150 years before its time. It was not accepted until late in the nineteenth century, when advances in chemistry reinforced its underlying assumptions. Bernoulli is now regarded as one of the founders of the sciences of mathematical physics and statistical mechanics. In 1738, he published his most famous work, the *Hydrodynamics*, presenting the first correct analysis of the properties of gases in terms of their constituent molecules. It was a subject that would become—much later—one of the proudest achievements of physical science.

The Bernoulli family (Figure 5.2, on page 212) was driven from Antwerp by religious persecution in the late sixteenth century to resettle in Switzerland. According to Eric Temple Bell, a historian of science:

> No fewer than 120 of the descendents of the mathematical Bernoullis have been traced genealogically, and of this considerable posterity the majority achieved distinction—sometimes amounting to eminence—in the law, scholarship, science, literature, the learned professions, administration and the arts.

Daniel Bernoulli (1700–1782) began his career as a professor of physics in Basel in 1733, where he remained (but for a few years as a professor in St. Petersburg, Russia) for the rest of his life. He was an immensely famous and creative physicist and mathematician, accumulating ten prizes from the prestigious French Academy of Sciences. He enjoyed recalling that as a young man introducing himself to a stranger with "I am Daniel Bernoulli," the incredulous response was "...and I am Isaac Newton!"

Bernoulli assumed that air consists of innumerable molecules in rapid and chaotic motion. He believed that the effects of air pressure are due to the many collisions of air molecules with the walls of the container. His proposed model of a gas within a cylinder is shown in the illustration.

Consider a cylindrical vessel ACDB set vertically, and a moveable piston EF in it, on which is placed a vessel of weight W; let the cavity ECDF contain very minute corpuscles, which are driven hither and thither with a very rapid motion; so that those corpuscles, when they strike against the piston EF and sustain it by their rapid impacts, form an elastic fluid that will expand of itself if the weight W is removed or diminished, which will be condensed if the weight is increased. Such, therefore, is the fluid we shall substitute for air. [Here is a very early instance of the description of a real physical substance with a simplified mathematical model—the very essence of modern science.] Its properties agree with those which have already been assumed for elastic fluids, and by them we shall explain other properties which have been found for air and shall point out others which have not yet been sufficiently considered.

At this point, Bernoulli presented a mathematical deduction of Boyle's law from the principles of mechanics. He showed that the volume of a gas is inversely proportional to its pressure, or in terms of his own construction, that if the weight on top of the piston (including the effect of

FIGURE 5.2
Genealogy of the remarkable Bernoulli family. Those in boxes were mathematical physicists distinguished enough to earn entries in the *Dictionary of Scientific Biographies*.

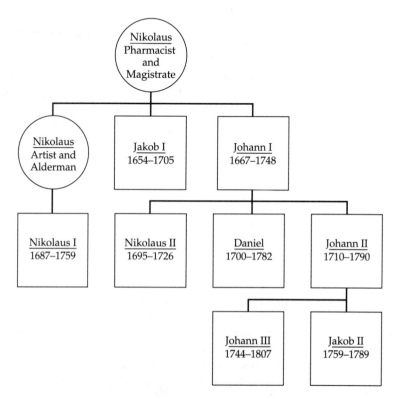

Daniel Bernoulli. *Source:*
The Bettmann Archive.

external air pressure) is doubled, then the volume of the trapped gas is halved. As the volume of gas decreases, its density increases, and the number of collisions of the air molecules with the piston (hence, the pressure) increases. His cogent arguments are easy to reproduce.

*Bernoulli's
Calculation of
Gas Pressure*

Let there be N molecules of gas in a closed cylinder of volume V that zip every which way with a wide range of speeds. To simplify the analysis, suppose there are only six directions in which molecules may move (east, west, north, south, down, and up) and that at any time, equal numbers of molecules move at the same speed v in each of the six directions. This idealization is close enough to the truth to give the right answer. The upward pressure exerted on the piston by the gas within results from the numerous collisions of the upward-moving molecules. We trace the remainder of Bernoulli's argument step by step:

1. Let A be the area of the piston and h be its vertical displacement. Their product is the volume of the gas in the cylinder: $V = Ah$.

2. At any time, $N/6$ of the molecules move upward. In the time interval h/v, all of these molecules will have collided with the piston. The number of collisions per unit time is $N/6$ divided by h/v, or $Nv/6h$. In each collision, an air molecule moving up bounces down, and transfers momentum $2mv$ to the piston. Force is the rate of change of momentum. Therefore, the upward force F of the trapped gas on the piston is the collision rate times the momentum transfer per collision, $F = (Nv/6h)(2mv) = Nmv^2/3h$. The piston does not move because this force is balanced by the weight of the piston and the downward force due to air pressure outside.

3. Pressure is force per unit area, $P = F/A$. The pressure P on the piston due to the gas in the cylinder is $P = Nmv^2/3hA$, or $Nmv^2/3V$. We may rewrite this result as follows:

$$PV = \tfrac{2}{3}N\left[\frac{mv^2}{2}\right] \tag{3}$$

The term in square brackets is the kinetic energy of one gas molecule. In a real gas, some molecules have more energy and some less. Bernoulli's result is correct if the term is replaced by the mean kinetic energy $\overline{KE}$ of a molecule of the gas:

$$PV = \tfrac{2}{3}N\overline{KE}$$

Daniel Bernoulli drew this diagram to illustrate that the pressure on the piston (due to atmospheric pressure and its weight) is balanced by the innumerable impacts of rapidly moving air molecules. The area of the piston is A and its distance above the base of the cylinder is h. The volume of trapped gas is $V = Ah$. *Source:* Courtesy Meyers Photo-Art.

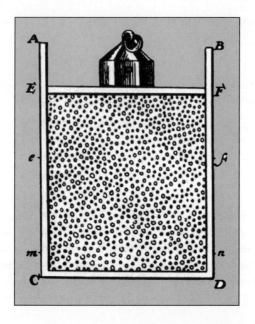

Equation 3 is Boyle's law, and more! Bernoulli deduced that the constant on the right-hand side of Boyle's law is 2/3 times the number of gas molecules times their mean kinetic energy. Thus, he showed that the product of the pressure and volume of a sample of gas is 2/3 of the total translational kinetic energy of all its constituent molecules. (I have to add the word *translational* because some molecules may spin or vibrate, and these motions do not contribute to the pressure.) His logic led him to another brilliant realization:

> The elasticity of the air is not only increased by condensation but by heat supplied to it, and since it is admitted that heat may be considered as an increasing internal motion of the particles, it follows that if the elasticity [i.e., pressure] of air of which the volume does not change is increased, this indicates a more intense motion in the particles of air, which accords with our hypothesis; for it is plain that so much the greater weight W is needed

---

DO YOU KNOW WHY
Nobody Believed Bernoulli?

How could it have been possible that Bernoulli's fabulous prevision of the correct theory of the behavior of gases was rejected for well over a century? Stephen G. Brush, a contemporary historian of science, offers six reasons:

1. Newton's false explanation of Boyle's law in terms of hypothetical repulsive forces between molecules was accepted because of its simplicity and Newton's reputation for infallibility. Although Bernoulli went beyond Newton to link the pressure of a gas to the square of the molecular velocity, he was not able to connect this fact with an accepted temperature scale.

2. Bernoulli's assumption that gas molecules move freely through space contradicted the prevailing notion that space is filled with ether or vortices or with atoms in physical contact.

3. Heat could not be just the energy of molecular motion, since it can be transmitted across empty space (as from Sun to Earth) without any accompanying molecular motions. Infrared radiation does indeed convey heat and not light, but the concept of a wide spectrum of electromagnetic energy-bearing radiations was still over a century ahead. The troubling thing about energy (as with syphilis) is that it may take myriad forms.

4. Bernoulli's theory was one of many theories based on molecular motions. At the time, there was no good reason to prefer one to the other. Only in hindsight do we see that Bernoulli was right and the others misguided.

5. The notion of billiard ball impacts of atoms was losing favor to the Newtonian view of stationary pointlike atoms that act upon one another at a distance.

6. Scientists were skeptical of any and all versions of the atomic hypothesis because there was no direct evidence for a particulate structure of matter. As recently as the 1890s, some eminent physicists regarded atoms as at best a convenient fiction, and at worst an illegitimate ad hoc hypothesis.

to keep the air in the condition ECDF, as the aerial particles are agitated by the greater velocity. It is not difficult to see that the weight W should [vary with the square] of this velocity because, when the velocity increases, not only the number of impacts but also the intensity of each of them increases equally.

Bernoulli understood how and why the gas pressure in a sealed container increases with its temperature. Lacking a systematic method for measuring temperatures, he was unable to complete his analysis and deduce the general law for the behavior of ideal gases. Nonetheless, he made two enormous intellectual strides.

Bernoulli showed that Boyle's law follows from the picture of a gas as a system of tiny and randomly moving particles. Its pressure results from the collisions of molecules against a surface. Suppose that a gas is compressed into a smaller volume but its molecules are not speeded up. The temperature of the gas remains the same, but the collisions of molecules with the walls of the container become more frequent. According to equation 3, the product of pressure and volume of the gas remains constant.

Secondly, Bernoulli identified the equivalence between the temperature of a body and the internal motion of its molecules: The hotter it is, the faster they go. Once the temperature of a gas was recognized to be proportional to the mean kinetic energy of its constituent molecules, the transformation of mechanical energy into heat was demystified. It is the conversion of ordered motion into the disordered motion of molecules.

The Ideal Gas Law

The successes of nineteenth-century chemistry (to be described in Chapter 6), and the realization that heat is molecular motion, eventually confirmed Bernoulli's vision of a gas as an ensemble of randomly moving molecules. How can we reconcile his theoretical result, equation 3, for the product PV in terms of molecular properties, with the combined gas law discussed in Chapter 4?

$$PV = \begin{cases} \frac{2}{3}N\overline{KE} & \text{from Bernoulli} \\ aT & \text{from Boyle and Gay-Lussac} \end{cases} \qquad (4)$$

Recall that $\overline{KE}$ is the average kinetic energy of a gas molecule. Since a does not depend on temperature or pressure, the only way that sense may be made of these two equations is if the Kelvin temperature of a gas is proportional to $\overline{KE}$:

$$kT = \tfrac{2}{3}\overline{KE} \qquad (5)$$

The constant of proportionality k, now known as Boltzmann's constant, is a universal constant. That is, it does not depend on the pressure, temperature, volume, or chemical identity of the particular gas under consideration.

Temperature, which originally was vaguely defined as the intensity of heat, is now identified as a precise measure of the mean kinetic energy of gas molecules. This result is at the root of the modern theory of gases. Boltzmann's fundamental constant is now known to be

$$k = 1.38066 \times 10^{-23} \text{J/K} \tag{6}$$

Thus, the two gas laws in equation 4, with the identification $a = Nk$, become a single powerful equation known as the *ideal gas law*:

$$\text{Ideal gas law (i):} \quad PV = NkT \tag{7}$$

This form of the ideal gas law depends on N and k. A more convenient expression is discussed later on. The word *ideal* needs further comment. An ideal gas is one whose molecules do not interact with one another. No real gas is ideal, and none satisfies equation 7 exactly. However, every gas, when it is at a sufficiently high temperature and low pressure, behaves very nearly as if it were an ideal gas. For almost all purposes, the departures of the behavior of real gases from those expected of ideal gases may be neglected.

EXAMPLE 5.6
Using the
Ideal Gas Law

An unknown gas is initially confined within a 100 L (or 0.1 m³) tank at temperature $T_1 = 300$ K. It is transferred to a smaller, empty tank ($V_2 = 50$ L) and heated to $T_2 = 360$ K. If the pressure P_1 of the gas in the

---

DO YOU KNOW WHY

There Is No Hydrogen in Earth's Atmosphere?

Hydrogen is about 100 times as abundant as oxygen in the universe, and helium is over 200 times as abundant as nitrogen. Yet there is hardly any helium or hydrogen in Earth's atmosphere. Equation 5 explains why it couldn't have been otherwise.

The temperature of a gas, multiplied by Boltzmann's constant, equals 2/3 of the mean kinetic energy of one molecule. Furthermore, in a mixture of different gases, the mean kinetic energy of each molecular species is the same. The mass of a hydrogen molecule is 1/16 that of an oxygen molecule, so that hydrogen molecules in

the atmosphere would move four times faster than oxygen molecules. For example, at 300° C, the average velocity of hydrogen molecules is about 2 km/s, while that of oxygen molecules is about 500 m/s. Some molecules have more than their average velocity, some less.

Recall from Chapter 3 that escape velocity from Earth is about 11 km/s. Molecules in the upper atmosphere with this velocity simply wander off into outer space. This happens much more often for hydrogen and helium molecules than for heavier molecules such as nitrogen and oxygen. Any hydrogen that was once in the atmosphere would have been lost in a few thousand years. There may have been hydrogen and helium in the atmosphere when Earth was young, but they have long ago flown the coop.

larger tank was 5×10^5 N/m^2, how many molecules are in the gas and what does the pressure P_2 in the smaller tank become?

Solution   The number of molecules that we started with can be found from equation 7:

$$N_1 = \frac{P_1 V_1}{k T_1} \simeq \frac{5 \times 10^5 \text{ N/m}^2 \times 0.1 \text{m}^3}{1.38 \times 10^{-23} \text{ J/K}^1 \times 300 \text{K}} \simeq 1.2 \times 10^{25} \text{ molecules}$$

Because all the gas ends up in the smaller tank, we know that $N_1 = N_2$. Equation 7 tells us that $P_1 V_1 / T_1 = P_2 V_2 / T_2$. Substitution into this formula yields $P_2 = (360/300)(100/50) P_1$, and finally, $P_2 = 1.2 \times 10^6$ N/m^2. ∎

The Mole and the Molecule

In Example 5.6, we determined the number of molecules in a sample of an unknown gas solely from its pressure, its volume, its temperature, and the value of Boltzmann's constant. Equation 7 implies the astonishing fact that equal volumes of different gases at the same pressure and temperature contain the same number of molecules. The next chapter tells why this statement was put forward as an ad hoc hypothesis by Avogadro to solve a problem in chemistry several decades before it could be deduced from kinetic theory.

Ordinary objects are made up of enormous numbers of molecules. Until the value of Boltzmann's constant was determined, the number of molecules in an object, or a sample of gas, was unknown. A related means of specifying the quantity of a substance was devised. The *atomic weight* of an element denotes the relative mass of one of its atoms according to an arbitrary scale in which the carbon atom is defined to have atomic weight 12.* In the same way, the molecular weight of a compound denotes the relative mass of one of its molecules in the same standard scale. A quantity of a substance equal in grams to its molecular weight is defined to be a *mole* of that substance. It follows that a mole of any substance contains a definite, fixed, and very large number of molecules. That number is called *Avogadro's number* and is denoted by N_A. The precise determination of N_A was once an important scientific challenge. Today, it has been measured to an accuracy of better than one part per million:

$$N_A = 6.022045 \times 10^{23} \tag{8}$$

Another Form of the Gas Law

Equation 7 is awkward to use because it involves the product of two numbers, N and k, one of which is enormous and the other tiny. Neither quantity was known in the mid-nineteenth century. However, the number of moles n of a sample of gas could be determined from the

---

*The common usage of "atomic weight" rather than the more precise "atomic mass" is unfortunate.

mass of the gas and its molecular weight. From the definition of a mole and of Avogadro's number, we know that

$$n = \frac{\text{Mass of gas in grams}}{\text{Its molecular weight}} \quad \text{and} \quad N = nN_A$$

The ideal gas law may be written in a more convenient form in terms of n (a number of everyday size) rather than the enormous number N:

$$\text{Ideal gas law (ii):} \quad PV = nRT \tag{9}$$

where R is the product of Boltzmann's constant and Avogadro's number,

$$R = kN_A \tag{10}$$

and is known as the molar gas constant. Scientists could determine R from the pressure, volume, and temperature of a sample of gas of known mass and molecular weight. Its value is

$$R \simeq 8.3145 \text{ J/K} \tag{11}$$

or about 2 cal per Kelvin degree.

Chemists often express their results at standard values of pressure and temperature which are designated by the acronym STP:

$$\text{Standard pressure:} \quad P = 1.01325 \times 10^5 \text{ N/m}^2 = 1 \text{ atm}$$

$$\text{Standard temperature:} \quad T = 0° \text{ C} = 273.15 \text{ K}$$

The volume of one mole of any gas at STP is determined from equation 9 and the experimentally determined value of R by the substitution of $n = 1$:

$$\text{Molar volume at STP} = 2.2414 \times 10^{-2} \text{ m}^3 = 22.414 \text{ L}$$

The molecular weight of a gas is proportional to its density. More precisely, the mass in grams of one molar volume of gas at STP equals its molecular weight. This fact made it possible for chemists to measure the molecular weight of any newly discovered gas, thereby helping to identify its chemical composition.

EXAMPLE 5.7
Using the Molar
Gas Constant

A 30-L tank has a mass of 10 kg when it is empty. It is then filled with oxygen gas at a temperature of 20° C and a pressure of $P = 5 \times 10^6$ N/m^2 (about 50 atm). The molecular weight of oxygen is about 32. What is the mass of the tank when it is full?

Solution

We use equation 9 to determine the number n of moles of gas in the tank. Remembering that the Kelvin temperature of the gas is 293 K, we find:

$$5 \times 10^6 \text{ N/m}^2 \times 0.03 \text{ m}^3 = 8.3145 \text{ J/K} \times 293 \text{ K} \times n$$

which yields $n = 61.6$ moles. The mass of the oxygen is n times its molecular weight, or about 2 kg. The filled tank has a mass of 12 kg.

The First Law of Thermodynamics

The total internal energy E (or heat content) of a material object is the sum of the energies of its individual molecules.* In general, E depends on the mass of the object, its intrinsic nature, and the ambient pressure and temperature. In short, it depends on the state of the object. However, E does not depend on how the object was brought to that state. The value of E may be changed in two different ways. The object may be heated. That is, an amount of heat may be added to it. Alternatively, external forces may do work on the object. (Water may be agitated with a stirrer, iron may be pounded with a hammer, and air may be compressed with a pump.) Because energy is conserved, the change in total energy ΔE is the sum of the heat added Q to the object and the work done on it W. This result is called the first law of thermodynamics:

$$\Delta E = Q + W \tag{12}$$

Equation 12 says that doing work on a substance has the same effect as adding to it an equivalent amount of heat. For example, we may bring a vat of water to a boil by building a fire under it (and adding heat Q), by boring a cannon within it (and adding work W), or by doing some of each. The total amount of energy that must be added to bring the water to a boil is the same in all cases.

The Heat Content of Gases

The first law of thermodynamics tells us how the heat content of a substance may be changed. Kinetic theory lets us evaluate the overall heat content E of a gas. A gas consists of molecules in motion. Occasionally they collide with one another, but most of the time the molecules are freely moving and far apart. For this reason, E may be expressed in terms of the temperature and quantity of a gas: It does not depend on the pressure of the gas. The total translational energy of the molecules in a gas is the number of its molecules N times their mean kinetic energy $\overline{KE}$. Equation 4 says that this quantity is equal to $3PV/2$. Using equation 9, we find that the total translational energy per mole of the gas is

$$\text{Total translational energy of a gas:} \quad N_A \overline{KE} = \tfrac{3}{2}RT$$

However, the total translational energy of the gas molecules is not necessarily equal to its heat content because gas molecules may spin around as well as flit about.

The molecules of some gases, such as helium and neon, consist of a single atom. These gases are called *monatomic*. Their molecules do not ordinarily spin. Consequently, the heat content of monatomic gases does equal their total translational energy. The molecules of other gases, such as oxygen and nitrogen, are tiny dumbbells consisting of two molecules

---

*At this point, we are commingling ideas from thermodynamics and kinetic theory. Thermodynamic purists of the nineteenth century would never refer to molecular properties.

bound together. These molecules are called *diatomic*, and they do spin around. The heat content of diatomic gases includes a contribution from the rotational energy of the molecules of these gases. It turns out (and, today, may be proven with quantum-mechanical arguments) that the heat content of a diatomic gas is equal to the total translational energy of its molecules multiplied by the numerical factor of 5/3. Still other gases, such as carbon dioxide, have molecules consisting of more than two atoms. These gases are called *polyatomic*, and their thermal properties are more complicated than those of monatomic or diatomic gases. For the simpler cases, we obtain for the heat content per mole of a gas:

$$\text{Heat content of a gas:} \qquad E = \begin{cases} \frac{3}{2}RT & \text{for monatomic gases} \\ \frac{5}{2}RT & \text{for diatomic gases} \end{cases} \qquad (13)$$

This result can be used to determine the specific heats (or heat capacities) of gases.

The Heat Capacities of Gases

Suppose that n moles of a gas are confined to a volume V at a temperature T. A quantity of heat Q is added to the gas while its volume is kept fixed (Figure 5.3). As a result, the gas temperature increases by ΔT. No work is done in this process, and equation 12 says that the total energy of the gas is increased by $\Delta E = Q$. Equation 13 lets us express ΔE in terms of ΔT:

$$\Delta E = \begin{cases} \frac{3}{2}nR\Delta T & \text{for monatomic gases} \\ \frac{5}{2}nR\Delta T & \text{for diatomic gases} \end{cases} \qquad (14)$$

FIGURE 5.3
One mole of a gas at temperature T is kept in a cylinder whose piston is fixed in place. As heat is added to the gas, T rises but the gas volume V is unchanged. The amount of heat Q that must be added to the gas to raise its temperature, $\Delta T = 1°$ C, is its molar heat capacity at fixed volume, C_V.

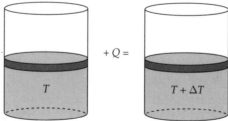

The heat that must be supplied to 1 mole of a gas to raise its temperature $1°$ C in this fashion is called the *molar heat capacity at constant volume* of the gas, or C_V. We find that

$$C_V = \begin{cases} \frac{3}{2}R & \text{for monatomic gases} \\ \frac{5}{2}R & \text{for diatomic gases} \end{cases} \qquad (15)$$

What if n moles of a gas are heated, but the pressure, rather than the volume, is kept constant, as shown in Figure 5.4 (on page 222)? As the gas expands in the cylinder, it does work on the piston. The gas exerts the force PA on a piston of area A. If the piston moves a distance D, the work done by this force is PAD and the change of volume of the gas is $\Delta V =$

FIGURE 5.4
One mole of a gas at temperature T is kept in a cylinder whose piston is free to move. As heat is added to the gas, T rises but the gas pressure P is unchanged. The amount of heat Q that must be added to the gas to raise its temperature, $\Delta T = 1°$ C, is its molar heat capacity at fixed pressure, C_P. The area of the piston is A and the distance it rises is D. Consequently, the change in the gas volume is $\Delta V = AD$. The work done by the pressure of the gas on the piston is $P\Delta V = R\Delta T$.

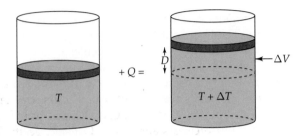

AD. Thus, the work done by the expanding gas is equal to $P\Delta V$. Since the pressure of the gas remains constant, we may use equation 9 to obtain

$$P\Delta V = nR\Delta T$$

The work done by the expanding gas is extracted from its heat content. Equation 12, with $W = -nR\Delta T$, becomes

$$Q = \Delta E + nR\Delta T$$

The heat Q that must be supplied to 1 mole of a gas to raise its temperature $1°$ C at constant pressure involves a sum of two terms: the change in the total energy of the gas, as given by equation 14, and the work done by the gas as it expands. The sum of these terms is called the *molar heat capacity at constant pressure* of the gas, or C_P. We find in general that

$$C_P = C_V + R \qquad (16)$$

and, in particular, that

$$C_P = \begin{cases} \frac{5}{2}R & \text{for monatomic gases} \\ \frac{7}{2}R & \text{for diatomic gases} \end{cases} \qquad (17)$$

The molar heat capacities of gases, as expressed by equations 17 and 15, turn out to be very simple expressions in terms of R. The results depend on the number of atoms in a gas molecule, but are otherwise independent of the chemical nature of the gas. This simplicity of nature would have been obscured had we examined the specific heats per unit mass of the gas rather than the specific heats per mole, their molar heat capacities. Measured values of the molar heat capacities of various gases are shown in Table 5.3. Notice the excellent agreement between these experimentally obtained values and the theoretically deduced values given

TABLE 5.3   Heat Capacities of Various Gases

Type	Gas	C_P/R	C_V/R	$(C_P - C_V)/R$
Monatomic	He	2.50	1.50	1.00
	A	2.50	1.50	1.00
Diatomic	H_2	3.46	2.45	1.01
	O_2	3.53	2.54	0.99

by equations 15 and 17. The kinetic theory of gases provides a precise and predictive description of the thermal properties of gases.

The Specific Heats of the Elements

Kinetic theory does not make quite such explicit and experimentally confirmed statements about solids and liquids. By the molar heat capacity of a substance is meant the amount of heat required to raise the temperature of 1 mole of the substance by 1° C. (The specific heat of a substance, as defined in Section 5.2, is its molar heat capacity divided by its molecular weight.) In 1818, Dulong and Petit discovered a simple and suggestive regularity among the molar heat capacities of solid and liquid chemical elements: They are all roughly the same and about equal to three times the molar gas constant. Table 5.4 lists some examples, with the molar specific heats exhibited as multiples of the gas constant R.

TABLE 5.4   Molar Heat Capacities of Some Elements

Element	Molar Heat Capacity	Element	Molar Heat Capacity
Bromine	$3.38R$	Phosphorus	$3.11R$
Gold	$3.18R$	Gold	$3.18R$
Iron	$3.14R$	Mercury	$3.13R$

One mole of any solid or liquid element contains Avogadro's number of atoms. The law of Dulong and Petit tells us that the heat required *per atom* to raise the temperature of such a substance by 1° C is approximately the same in all cases. This result was an early hint that the temperature of any form of matter, not only a gas, is determined by the mean energy of each of its molecules.

The successes of nineteenth-century kinetic theory were dramatic and impressive, but they were limited. The thermal behavior of gases was quantitatively described, but results for liquids and solids were only crude and qualitative. Not until classical mechanics yielded to quantum mechanics in the twentieth century would scientists begin to understand in detail the properties of bulk matter. Physicists now believe they know all the basic underlying rules governing the behavior of systems containing many atoms. By no means, however, have all of nature's tricks been explained.

EXAMPLE 5.8
Hot Air Ballooning

On the day of the flight, the atmospheric pressure is $P_0 = 10^5$ N/m^2 and the outdoor air temperature is $T_0 = 279$ K. The ground crew has filled a balloon with 1000 m^3 of preheated air at atmospheric pressure and a temperature of 300 K. The mass of the empty balloon and its payload is 240 kg. Under these circumstances, the balloon rises above its burden, but the airship remains grounded. To begin the ascent, the air within the balloon is brought to a higher temperature. It expands but remains at the pressure P_0. No air enters or leaves the balloon. How much heat must be added to the air in the balloon to initiate the flight? The solution is given in four easy steps:

Solution

1. According to the discussion of buoyancy in Chapter 4, the balloon lifts off when its total mass divided by its total volume (its mean density) equals the density of the surrounding air, ρ_0. We begin by calculating ρ_0 at flight time. From equation 9, we find for the volume of 1 mole of the atmosphere:

$$\frac{R \times 279 \text{ K}}{10^5 \text{ N/m}^2} = 2.32 \times 10^{-2} \text{ m}^3$$

The mean molecular weight of air is 29, so that 1 mole of air has a mass of 2.9×10^{-2} kg. The mass of 1 mole of the atmosphere divided by its volume is its density, $\rho_0 = 1.25 \text{ kg/m}^3$.

2. Before liftoff, the balloon contained 1000 m³ of air at a temperature of 300 K and at pressure P_0. We again appeal to equation 9 to determine the number of moles of air n in the balloon:

$$n = \frac{10^5 \text{ N/m}^2 \times 1000 \text{ m}^3}{R \times 300 \text{ K}} = 40,090$$

Thus, the mass of the air in the balloon is $40,090 \times 29$ g, or 1160 kg. The total mass of the balloon and its payload is 1400 kg.

3. As the air in the balloon is heated, it expands from its initial volume of 1000 m³. To achieve liftoff, the mean density of the balloon and its payload must equal ρ_0. The balloon must expand to a volume V such that:

$$\frac{1400 \text{ kg}}{V} = 1.25 \text{ kg/m}^3$$

or $V = 1120$ m³. The confined air remains at pressure P_0. Because its volume was increased by the factor 1.12, so must its temperature T be increased. Liftoff occurs when $T = 1.12 \times 300$ K or 336 K.

4. To fly, the temperature of 40 090 moles of air must be increased from 300 K to 336 K at constant pressure. Because air is a mixture of diatomic gases, its molar heat capacity C_P is $7R/2$ according to equation 17. The heat that must be added to the air in the balloon to initiate the flight is

$$Q = \tfrac{7}{2}R \times 40\,090 \times 36 = 4.2 \times 10^7 \text{ J}$$

We see from Table 5.2 that about $\tfrac{1}{3}$ gal of gasoline is enough to do the job.   ∎

Exercises

Use the following data: atmospheric pressure $= 10^5$ N/m^2; gas constant $R = 8.3$ J/K; absolute zero $= -273°$ C; 1 calorie $= 4.2$ J; atomic weights: $H = 1$, $C = 12$, $O = 16$, $A = 40$

16. Calculate the heat content in joules of the air within a typical lecture room, with a volume of 130 m^3.

17. A sealed tank contains 1 mole of air and 1 mole of liquid water. The pressure in the tank is 1 atm and its temperature is 27° C. The tank is heated to 177° C. The water boils to form steam. What has the pressure in the tank become? (Answer in atmospheres and neglect the small initial volume of the water.)

18. A liter of gas at room temperature and a normal air pressure of $\sim 10^5$ N/m^2 contains $\sim 3 \times 10^{22}$ molecules. Compute the mean kinetic energy of each molecule.

19. Ten million bees are flying randomly about in a box of volume 1 m^3. Each bee weighs 0.3 g and flies at a speed of 1 m/s. What is the pressure (in N/m^2) on the walls of the box due to elastic bee–wall collisions?

20. A certain sample of gas contains 1 mole of CO and 1 mole of H$_2$ at atmospheric pressure and at a temperature of 27° C.
    a. What is the mass of the sample? What is its volume in liters?
    b. How many moles of O$_2$ are needed for the complete combustion of the gas to CO$_2$ and H$_2$O? What is the mass of this quantity of oxygen? What is the total mass of the combustion products?
    c. If the combustion products are returned to atmospheric pressure and $T = 27°$ C, what volume do they occupy?

21. The amount of heat released by the combustion of acetylene gas C$_2$H$_2$ into CO$_2$ and H$_2$O is 50 kilojoules per gram. How much energy is released by the combustion of 1 mole of acetylene? By one molecule of acetylene?

22. One liter of an unknown gas is found to have a mass of 1.78 grams at STP. What is its molecular weight? What is its density in kg/m^3?

23. How much heat in joules must be added to the tank described in Example 5.7 to increase its temperature (and that of the gas within) to 50° C? Assume that the tank is made of iron.

24. What are the specific heats at constant volume of oxygen gas and hydrogen gas in units of cal/g °C? What is the molar heat capacity of water, expressed as a multiple of the molar gas constant R?

5.4 Thermodynamics and the Arrow of Time

Thermodynamics (from two Latin words meaning heat and force) arose in the early nineteenth century to describe the study of the motive power of heat: the capability of a hot body to produce mechanical work. It has since become a wide-ranging discipline concerned with the conditions that mechanical systems may assume, and the changes in these conditions that occur spontaneously, or as the result of interactions between systems, including the transfer of heat or mechanical energy. Thermodynamics approaches the properties of matter differently from kinetic theory. Rather than analyzing matter in terms of what were once highly speculative notions of constituent molecules and their motions, thermodynamics deals in general principles that are independent of any proposed model of the microscopic structure of matter. (Indeed, many of the early contributions to thermodynamic theory were carried out by scientists who did not believe in atoms!) Therein lies the great strength of thermodynamics, for it applies just as well to solids, liquids, and gases. But its lack of specificity is also its weakness: Neither the ideal gas law nor the heat capacities of gases may be deduced from thermodynamic principles alone.

The first law of thermodynamics, introduced in Section 5.3, states that any change in the internal energy of a body results from the heat imparted to it and the work done on it. The first law is an expression of conservation of energy and the equivalence between heat and mechanical energy. In practical terms, it says that there is no such thing as a free lunch: Any machine that does useful work must obtain its energy from an external source. From the standpoint of mechanics (supplemented with the first law of thermodynamics so it may deal with heat), energy may be freely transformed from one form to another as long as its total quantity is conserved. From the standpoint of thermodynamics, however, there are strong constraints on the convertibility of heat to mechanical energy.

The air molecules about us are engaged in rapid chaotic motions. The total heat energy of the air in a large room is about 1 kilowatt-hour, and there is lots more air outside. What's to stop us from extracting this energy and putting it to use? Without an external energy source, we could thereby cool the room in summer, and use the energy to power a television set. What stops us is the second law of thermodynamics.

The Second Law of Thermodynamics

Two equivalent formulations of the second law of thermodynamics were proposed in the nineteenth century. Lord Kelvin postulated:

> A transformation is impossible whose only final result is to convert into work heat extracted from a source at constant temperature,

while the German physicist Rudolf Clausius held:

> A transformation is impossible whose only final result is to transfer heat from a cooler body to a warmer body.

Scientists of the nineteenth century formulated this law as a precise and elegant mathematical statement. Clausius coined the word *entropy* (from the Greek root meaning turning into) to describe the extent to which the energy content of an object has become degraded, and, therefore, less available for conversion into mechanical energy. From another point of view, the entropy of a system is a quantitative measure of its degree of disorder. The second law of thermodynamics, applied to an isolated system, says that its entropy may remain constant for a so-called *reversible* process, or it may increase for an *irreversible* process. Under no circumstances can the entropy of an isolated system decrease. Thus, the second law says something about the sequence, or order, of events that can take place: It defines the direction of the arrow of time. Two everyday examples should clarify the notion of entropy as disorder:

1. A candle is a highly ordered system containing a considerable amount of chemical energy. When the candle burns, its internal energy is not lost, but dissipated as heat throughout the room. The entropy of the room has increased. Without the expenditure of a lot of time and trouble (and far more energy that was in the candle to begin with), the candle cannot be reconstituted from its reaction products.
2. Place a cube of sugar in a glass of water. It soon dissolves and makes the water uniformly sweet. It simply doesn't happen the other way around. Of course, the water will eventually evaporate leaving the sugar at the bottom. But now the water is all gone, and no matter how long you wait, it won't come back.

The Conflict between Mechanics and Thermodynamics

Imagine a video showing the stately dance of the planets and their moons about the Sun. Played in reverse time sequence, it would seem just as realistic as the original version. The planets would turn the wrong way and the Sun would seem to set in the east, but no matter—the reversed motions are in complete accord with Newton's laws. His theory of motion—and most other physical laws as well—do not, in themselves, determine the direction of time's arrow.

The solar system involves only a few dozen heavenly bodies. It's quite another story when a much more complicated process is considered. Another movie, showing the preparation of an omelet from an egg, makes no physical sense at all if it is played backward. How can a theory of physics, whose laws work just as well forward or backward in time, describe a world in which time marches inexorably and unambiguously onward, and Humpty Dumpty may never be put together again?

The answer to this profound question has to do with the complexity of things. An egg is made up of $\sim 10^{23}$ molecules, which are neatly organized into a shell, membranes, a yolk, and many more intricate structures. When the egg is broken, stirred, and fried, its molecules recombine in any one of innumerable patterns, each of which results in a delectable omelet. There are far more ways to turn egg to omelet than there are ways to turn one particular omelet back to an egg. The difference is so great that the path leading back to the egg—although it is entirely compatible with the microscopic laws of physics—simply cannot be retraced.

The second law of thermodynamics is the law of decay. It says that isolated systems tend to become more disordered as time passes: that iced and hot tea soon become indistinguishable, that all living creatures must sicken and die, and that all machines must someday wear out.

Let's study a problem more in keeping with this chapter than astronomy or gastronomy. A room, shown in Figure 5.5, is divided into two sections, A and B, by an impermeable partition in which there is a pinhole. The air on both sides is at the same pressure and temperature. Randomly moving air molecules, some fast and some slow, pass through the pinhole in either direction. It may be unlikely, but surely it is possible, that most of the molecules passing from A to B happen to be fast, while most of those passing from B to A happen to be slow. If the process continues long enough, region B becomes unbearably hot while region A becomes frigid. The second law of thermodynamics is flouted! However, the probability of this occurrence is unimaginably tiny. It is as unlikely as flipping a fair coin a billion billion times and coming up heads every time. It is so unlikely as to be, for any practical purpose or otherwise, impossible.

Enter James Clerk Maxwell, the nineteenth-century physicist whom we shall meet more formally in Chapter 8. He proposed a scheme to evade the second law that does not rely on chance. "Now conceive a finite being," he wrote, "who knows the paths of all the molecules, but who can do no work but to open and close the hole in the diaphragm." This hypothetical being, known as *Maxwell's demon*, allows fast molecules to pass from A to B and slow molecules to pass from B to A (Figure 5.6). All others find the door closed. In this case, a temperature difference between the two regions will certainly develop. The second law is undone "by the intelligence of a very observant and neat-fingered being."

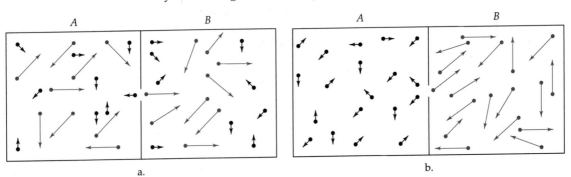

a.   b.

FIGURE 5.5 *a.* The air in each of two rooms is at the same temperature. That is, the air molecules in room A have the same mean kinetic energy as those in room B. *b.* By a freak chance, rapidly moving molecules pass through a hole in the partition from room A to room B, while slower molecules pass from room B to room A. If this perfectly possible but wildly improbable process continues, room B will become noticeably hotter than room A.

FIGURE 5.6
Maxwell's demon tries
to outfox the second law
of thermodynamics by
allowing fast molecules
to pass from *A* to *B*,
but not back. He also
allows slow molecules a
one-way passage from *B*
to *A*.

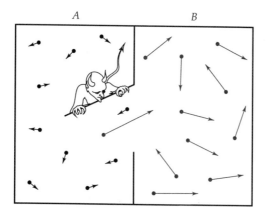

There is a loophole to this argument; you can't fool Mother Nature.
The system consisting of room, diaphragm, door, and demon is not isolated. The metabolism of the imaginary being must be taken into account.
Whether it be microbe, machine, or genetically engineered monster, it
must be connected to the outside world to take its sustenance and eject
its waste products. When these are taken into account, the order created
in the room is necessarily compensated, or more than compensated, by
the disorder produced outside its walls.

Entropy

From the preceding discussion and examples, we see that the second law
of thermodynamics is not implicit in the microscopic laws of physics.
Rather, it is a probabilistic result that emerges when these laws are applied to realistic systems consisting of large numbers of interacting particles. In the end, there is no conflict between the laws of mechanics and
the laws of thermodynamics. But if we are to explore the consequences
of the second law, we must introduce a quantitative meaning to the abstract notion of entropy. Suppose that a small quantity of heat, which
we denote by Q, is added to a body at Kelvin temperature T. It is small
enough that the temperature of the body is hardly changed. I shall pull
a formula out of a hat and say that the entropy of the body, which is
denoted by S, is thereby increased by the amount

$$\Delta S = \frac{Q}{T} \tag{18}$$

If the same small quantity of heat is later extracted from the body, its
entropy is decreased by the same amount.

The entropy of a system of bodies is the sum of the entropies of the
individual bodies. Although we have not defined the total entropy of a
body, equation 18 describes how the entropy of a system changes as heat
is added to, removed from, or transferred among, its constituent bodies.

To see that equation 18 says what we want it to say, consider a
system consisting of two bodies at different temperatures. Body *A* is

FIGURE 5.7
Two bodies, at temperatures T_A and T_B, are thermally isolated from the outside world, but are connected together by a heat-conducting rod. Heat Q flows from the hotter body to the cooler one, but not the other way around.

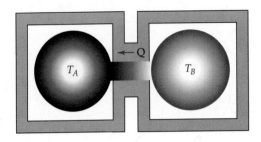

at Kelvin temperature T_A and body B is at the higher temperature T_B. They are insulated from the outside world but connected to one another by a heat-conducting rod, as indicated in Figure 5.7. Let's prove that the requirement that S cannot decrease reproduces what we know from common sense: that heat flows from the hot body to the cold one, and not the other way around.

If B loses a small amount of heat Q, the first law of thermodynamics tells us that A gains the same amount of heat. Energy is conserved. When heat is exchanged between the bodies, the entropy of each body also changes. Our master formula, equation 18, determines these changes to be

$$\Delta S_A = \frac{Q}{T_A} \quad \text{and} \quad \Delta S_B = -\frac{Q}{T_B}$$

Body A gains heat, and its entropy increases. Body B loses heat, and its entropy decreases. The total change in entropy of the system is the sum of the changes of its constituents:

$$\Delta S = \frac{Q}{T_A} - \frac{Q}{T_B} \tag{19}$$

The total entropy increases because $T_B > T_A$. The second law of thermodynamics allows heat to flow from B to A. A flow of heat from A to B would decrease the total entropy, which is impossible. This argument shows that equation 18 is a plausible definition of entropy change. We may now apply the second law of thermodynamics to more complex systems, with powerful and surprising results.

Heat Machines

Figure 5.8 depicts an isolated system consisting of two unspecified bodies, A and B, at temperatures T_A and T_B. They are connected together by an unspecified machine M, which transfers heat between the bodies in accordance with the two laws of thermodynamics. M is also connected to C, which is either a source or a sink of mechanical energy. In the first instance, it might be an electric company that supplies energy; in the second, it might be the drive mechanism of an automobile, which consumes energy.

In each second, the machine M extracts a small amount of heat q from A and deposits a small (but different) amount of heat Q into B. The first law of thermodynamics says that energy accounting must balance. If Q

FIGURE 5.8
The machine M extracts a quantity of heat q from A (at temperature T_A) and deposits a quantity of heat Q in B (at temperature T_B). If $Q > q$, the energy source C supplies the difference $Q - q$ as work done on the system consisting of A and B. If $q > Q$, the energy sink C receives the difference $q - Q$ as work done by the system consisting of A and B.

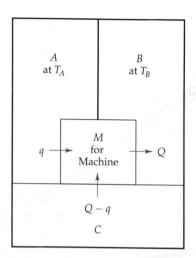

is larger than q, C must supply the energy difference $Q - q$. However, if q is larger than Q, then C receives the energy difference $q - Q$. It's like balancing a checkbook.

Now we turn to the second law. The entropy of each body changes according to equation 18:

$$\Delta \mathcal{S}_A = -\frac{q}{T_A} \qquad \Delta \mathcal{S}_B = +\frac{Q}{T_B}$$

Unlike energy, entropy accounting need not balance. $\mathcal{S}$ may increase or stay the same. Like the national debt, it never decreases. The total change of entropy is the sum of these quantities, and it cannot be less than zero. The second law of thermodynamics says that $\Delta \mathcal{S}_A + \Delta \mathcal{S}_B \geq 0$, or

$$Q \geq \frac{T_B}{T_A} q \tag{20}$$

We need not take C into account in our entropy balance. Useful energy is highly ordered and carries hardly any entropy along with it. Electricity or natural gas is as wonderfully rich in energy as it is deficient in entropy. Similarly, if C is a system on which mechanical work is done, it accepts a negligible amount of entropy.

Equation 20 is all we need to study the thermal behavior of various machines. Let us apply our result to the operation of three down- to-earth devices: a refrigerator, a heat pump, and an automobile, with M, A, and B identified as in Table 5.5.

TABLE 5.5   The Machines We Study

The Machine M	Exhausts q from A	and Deposits Q in B.
M = A refrigerator	A = Its interior	B = The kitchen
M = A heat pump	A = The outdoors	B = The indoors
M = An automobile	A = The burning fuel	B = The exhaust gas

1. *The thermal efficiency of a refrigerator*   Let M be the system of pumps, pipes, and motors that draws heat from A (the interior of a refrigerator, at temperature T_A) to B (the room in which the refrigerator sits, at the warmer temperature T_B). C is a source of power but not of entropy. The system includes the refrigerator, the room, and the power source. Some refrigerators of the same size use more power, and some less. They differ both in their intrinsic thermal efficiency (which we now discuss) and in the quality of their thermal insulation (which we do not).

FIGURE 5.9
Energy balance of the fridge. The inside of a refrigerator is kept at a temperature T_A that is lower than the kitchen temperature T_B. At (i), some heat q leaks from B into A. To keep the food fresh, the mechanism M extracts q from A (ii) and deposits a larger quantity of heat Q in B (iii). The excess energy Q-q is purchased as electricity (iv) from C and is eventually lost from the kitchen. T_A and T_B are maintained at their original values.

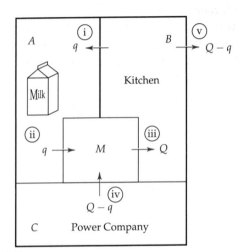

In each small time interval, the refrigerator mechanism must make up for heat q that leaks through its walls or is let in when its door is opened (see item (i) of Figure 5.9). The mechanism of the refrigerator M extracts q from inside the refrigerator [item (ii)] and dumps Q into the room [item (iii)]. We know from energy balance that the power source supplies the difference in energy, $W = Q - q$ [item (iv)]. Subtracting q from both sides of equation 20, we obtain

$$W = Q - q \geq \left(\frac{T_B}{T_A} - 1\right)q = \frac{T_B - T_A}{T_A}q \qquad (21)$$

The thermal efficiency $\mathcal{E}_r$ of a refrigerator is the ratio of the heat extracted from its interior to the energy that must be supplied by the source:

$$\mathcal{E}_f = \frac{q}{W} \qquad (22)$$

The larger is $\mathcal{E}_f$, the lower is your electricity bill. From equation 21, we find that $\mathcal{E}_f$ has a maximum possible value:

$$\mathcal{E}_f \leq \frac{T_A}{T_B - T_A} \qquad (23)$$

To be explicit, choose $T_A = -13°$ C and $T_B = 27°$ C. These values come out nicely in the Kelvin scale, which is the one we must use: $T_A = 260$ K and $T_B = 300$ K. From equation 23, we find $\mathcal{E}_f = 6.5$ for the theoretical maximum of the thermal efficiency of an ideal refrigerator operating between the specified temperatures. It would pump 6.5 J of heat from the refrigerator, while consuming only 1 J of electric energy. Your home refrigerator is far from the thermodynamic ideal, but its $\mathcal{E}_f$ may be as high as 3. It may consume 100 watts of electric power, and pump out heat from the inside at a rate of 300 watts.

2. *The thermal efficiency of a heat pump*   A heat pump is like a refrigerator turned inside out. It warms the house by drawing heat from the cold outside A to keep the inside B warm. In each second, a quantity of heat Q trickles out of the walls and windows [see item (i) of Figure 5.10]. The heat pump extracts q from outdoors [item (ii)] and deposits Q indoors [item (iii)]. Its thermal efficiency $\mathcal{E}_p$ is the heat Q supplied to the house per unit of energy $W = Q - q$ [item (iv)] acquired from the power company:

$$\mathcal{E}_p = \frac{Q}{W} = 1 + \frac{q}{W}$$

If this result is compared with equations 22 and 23, we find:

$$\mathcal{E}_p \leq \frac{T_B}{T_B - T_A}$$

FIGURE 5.10
Energy balance of the heat pump. The temperature outdoors T_A is lower than the temperature inside T_B. At (i), some heat Q leaks from B to A. To keep the house warm, the mechanism M extracts q from A (ii) and deposits the larger required heat Q in B (iii). The excess energy $Q - q$, purchased as electricity from C (iv), ends up in the great outdoors.

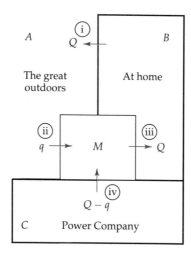

If we choose the same temperatures as before (27° C indoors and −13° C outdoors), we find a value of 7.5 for the thermal efficiency of an ideal heat pump. A real heat pump, which needs a fan to circulate air, may have $\mathcal{E}_p = 3$. It supplies three times more heat than a quartz heater at the same power cost. A great deal of electric energy would be saved if builders installed heat pumps rather than radiant heating in new homes and offices.

3. *The thermal efficiency of an automobile engine*   In Figure 5.11, the heat q is extracted from the combustion of gasoline at a high temperature T_A [item (i)], while a lesser quantity of heat Q is lost in the exhaust gas at the lower temperature T_B [item (ii)]. The difference between these quantities, $q - Q$, is the portion of the fuel energy gainfully used to power the car [item (iii)]. The thermal efficiency of the engine is the ratio of mechanical energy obtained to fuel energy consumed:

$$\mathcal{E}_e = \frac{q - Q}{q}$$

Equation 20 implies $\frac{-Q}{q} \leq \frac{-T_B}{T_A}$ and hence:

$$\mathcal{E}_e \leq \frac{T_A - T_B}{T_A}$$

FIGURE 5.11
Energy balance of the automobile. Heat energy q at the high temperature T_A is extracted from gases in the cylinders (i) by the mechanism M. Part of the heat Q remains in the exhaust gases (ii) at the lower temperature T_B. The rest, $q - Q$, contributes to the mechanical energy of the car (iii) and makes up for energy lost to friction and air resistance (iv). Gasoline is consumed (v) to replenish the heat energy within the cylinders.

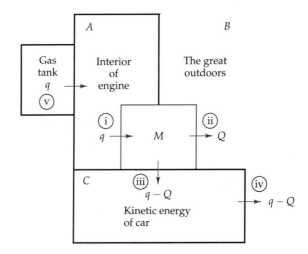

As in the preceding two examples, the second law of thermodynamics, in the form of equation 20, allows us to deduce the maximum possible value of $\mathcal{E}_e$, which would be the efficiency of an ideal automobile engine. If the gasoline is burned at 800 K and the exhaust temperature is 300 K, the thermodynamic maximum is only $\mathcal{E}_e = 62.5$ percent. Over a third of the fuel energy is thrown away. Gas turbines (in which the burn is much hotter) have higher thermodynamic efficiencies. Unfortunately, these higher temperatures produce more pollutants in the form of oxides of nitrogen. Maximum energy efficiency and minimum pollution are admirable goals, but they often conflict with one another.

Thermodynamics vs. Kinetic Theory

Kinetic theory was proposed by Bernoulli in the eighteenth century and developed principally by the Austrian physicist Ludwig Boltzmann (1844–1906). This theory represents an entirely different approach to physics than thermodynamics, because it was founded on unproven hypotheses regarding the invisible substructure of matter. Despite the empirical successes of kinetic theory, scientists of the late nineteenth century were loathe to accept the atomic hypothesis as fact. Boltzmann himself, in the introduction to his monumental treatise on kinetic theory, wrote:

> Away with all dogmatic statements, whether for or against atomistic ideas. When we present the theory of gases as a complex of mechanical analogies, we are indicating how far we are from admitting, in a certain fashion and as a reality, that bodies are constituted of very small particles.

Thermodynamics treats the relationships between heat energy and all other forms, including mechanical energy, chemical energy, and electric energy. Unlike kinetic theory, it deals only with directly measurable quantities, such as temperatures, pressures, and quantities of mass or heat, and involves no speculation regarding the ultimate structure of matter. Thermodynamics is concerned with what can be said about objects consisting of innumerable molecules, whose individual motions we may not be interested in, and are certainly unable to calculate. However, the principal focus of this book is on the particles of matter themselves, and only incidentally on the properties of bulk matter. Thus, we take our leave of thermodynamics at this point.

Many nineteenth-century scientists were reluctant to accept unproven conjectures. They held the view that things that cannot be seen cannot be said to exist. Ernst Mach, an accomplished physicist and renowned philosopher, wrote of heat in 1872:

> If we are astonished at the discovery that heat is motion, then we are astonished at something that has never been discovered. It is quite irrelevant for scientific purposes whether we think of heat as a substance or not.

and of atoms in 1883:

> Atoms cannot be perceived by the senses; like all substances, they are things of thought. Furthermore, the atoms are invested with properties that absolutely contradict the attributes hitherto observed in bodies. However well fitted atomic theories may be to reproduce certain groups of facts, the physical enquirer who has laid to heart Newton's rules will only admit those theories as provisional helps, and will strive to attain, in a more natural way, a satisfactory substitute."

Mach's opposition to atomic theory, and thus to the notion of heat as atomic motion, continued until 1903. At that time, a device was invented that put out a visible light signal upon the arrival of a single subatomic particle emitted by a radioactive material. Visiting the laboratory and seeing the flashes for himself, he is reputed to have said: "Now I believe in the existence of atoms!"

No one any longer doubts that atoms exist. Individual atoms may be seen and manipulated. Thermodynamics and kinetic theory are now seen to be complementary rather than contradictory disciplines. They have joined forces to allow physicists to approach the really hard problems associated with the behavior of condensed forms of matter.

Exercises

25. Why is leaving the refrigerator door open a poor strategy for cooling the kitchen?

26. A house loses heat through its exterior walls and roof at a rate of 5000 J/s. The interior temperature is 27° C and the outside temperature is −13° C.
    a. How much power is required to maintain the interior temperature if heat is obtained by resistance heaters, which convert electricity directly into heat?
    b. Suppose that the resistive heaters are thrown out and replaced by heat pumps with a thermal efficiency of $\mathcal{E}_p = 3$. How much power is now needed to heat the house?

27. American air conditioners display an energy efficiency rating (EER), which expresses the rate at which heat is pumped out of a room per unit of electric power required. The heat flow is measured in BTUs per hour and the electric power in watts. A premium brand of room air conditioner boasts an EER of 10. How many joules of heat are pumped from the room per joule of electric energy used? (1 BTU = 1056. J.)

28. Calculate the entropy change when 10 g of water at 100° C is converted into steam at the same temperature.

29. A gasoline engine has an operating temperature of 900 K and an exhaust temperature of 300 K. Its efficiency is 60 percent of its maximum possible thermodynamic efficiency. If it generates 1 kilowatt of power, how many liters of gasoline does it consume per hour? (The heat content of gasoline is 5×10^4 J/g. Its density is 0.74 kg/L.)

30. What is the thermodynamic efficiency of an ideal heat pump if the indoor temperature is 27° C and the outdoor temperature is 3° C? A heat pump can just as well pump heat out of a room, thus acting as an air conditioner. How would you define the efficiency of such a device? If the indoor temperature were 27° C and the outdoor temperature were 42° C, what would be the device's ideal thermodynamic efficiency?

31. The second law of thermodynamics says that the entropy, or disorder, of an isolated system cannot decrease. However, many familiar systems become more ordered as time passes: Civilizations flourish,

children become slavishly organized adults, and great oaks from tiny acorns grow. Explain how these phenomena may be reconciled with the second law.

Where We Are and Where We Are Going

Lavoisier's understanding of the chemical elements and their atoms was clouded by his conviction that heat is a substance. We have shown how caloric was finally exorcised from physics and chemistry, and how heat was shown to be a form of energy subject to the powerful laws of thermodynamics. Advocates of kinetic theory, by assuming the existence of chaotically moving molecules as a working hypothesis, showed that the temperature, pressure, and specific heats of gases could be deduced from the motions of their supposed molecular constituents.

The next chapter turns from the physical properties of matter to the ways in which different substances combine or react with one another. By means of the quantitative study of chemical reactions, chemists were driven to the atomic hypothesis in the first decade of the nineteenth century, long before their physicist colleagues. Chemical compounds were shown to consist of molecules with specific atomic composition. Chemists could not prove that atoms exist, nor could they determine the mass or size of a single atom. Nonetheless, they were able to determine the *relative* masses (or atomic weights) of atoms of different elements. When the elements were listed in order of ascending atomic weights, a remarkable systematic pattern was discovered—the periodic table of the elements. The appearance of order in the apparent chaos of the chemical bestiary was a prelude to the discovery of the inner structure of the atom itself.

6   Atoms and Elements

Natural causes are at work, which tend to modify, if they do not at length destroy, all the arrangements and dimensions of the earth and the solar system. But though in the course of ages catastrophes have occurred and may occur again in the heavens, though ancient systems may be dissolved and new systems evolved out of their ruins, the atoms out of which the systems are built—the foundation stones of the material universe—remain unbroken and unworn. None of the processes of Nature, since time began, have produced the slightest difference in the properties of any atom. They continue this day as they were created.

James Clerk Maxwell, English physicist (1831–1879)

The Discovery of Phosphorus, as imagined by Joseph Wright a century afterward. It is the first chemical element whose discovery is documented. Phosphorus was extracted from urine by Hennig Brand of Hamburg, Germany, in 1669. It burns spontaneously in air, giving off an intense but eerie glow. *Source:* Courtesy Meyers Photo-Art.

Is matter infinitely divisible or is there a point at which it can be divided no further? To the ancient Greeks, it was a question to be debated rather than tested by experiment. Philosophers scored points by the elegance of their rhetoric or the number of disciples attracted. They imagined the existence of fundamental and unchangeable particles of matter, which they called *atoms,* meaning "uncuttable." Each of the four Greek *elements* were supposed to be made up of a particular variety of atoms. In this chapter, we examine the modern meanings of atoms and chemical elements.

The Greek view of the atomic hypothesis was summarized by the Roman poet Lucretius:

> Atoms must be made of imperishable stuff into which everything can be resolved in the end, so that there may be a stock of matter for building the world anew. The atoms, therefore, are absolutely solid and unalloyed. In no other way could they have survived throughout infinite time to keep the world in being.

Material things neither appear nor disappear: Their forms and textures change while the building blocks of matter survive unscathed. Centuries later, Newton expressed a similar argument for atoms from a religious perspective: "God in the Beginning form'd Matter in solid massy, hard, impenetrable, moveable Particles. That Nature may be lasting, the changes of corporeal Things are to be placed only in the various Separations and new Associations and Motions of these permanent Particles."

Lucretius realized that the persistence of matter was not a sufficient argument for atoms. He looked for direct evidence in everyday phenomena. Of the motions of dust particles in a beam of sunlight, he wrote:

> There you will see many particles under the impact of invisible beams changing their course and driven back upon their tracks, this way and that, in all directions. You must understand that they all derive their restlessness from the atoms. It originates with the atoms which move of themselves.

Lucretius was wrong about dust motes. They dance to turbulent air currents and not as a result of atomic collisions. But he was right about the chaotic motions of air molecules, as we learned in Chapter 5. This chapter tells how it was that chemists were compelled to believe in atoms long before physicists could prove that they really do exist.

Section 6.1 describes the discovery of the laws of chemical combination. These laws convinced chemists of the truth of the atomic hypothesis and let them measure the relative masses of different atoms. Atoms come in many varieties. They comprise about 100 different chemical elements, displaying a bewildering variety of properties. However, the properties of the elements were found to depend in a systematic fashion on their atomic weights. The tale of the discovery of the periodic table of the elements is told in Section 6.2. Section 6.3 is an informal introduction to the names and properties of the elements. The final section summarizes what was known about atoms and elements in the late nineteenth century and peeks at the challenges that would be faced in the twentieth century.

6.1   How Chemists Came to Believe in Atoms

Throughout the ages, the wondrous symmetry of crystals suggested to scientists that there must exist fundamental particles of matter. The French philosopher René Descartes (1596–1650) argued that snowflakes, which are crystals of ice, consisted of spherical particles that

> Are obliged to arrange themselves so that that each has six others surrounding it; one cannot conceive of any reason that would prevent their doing this, because all round and equal bodies that are moved in the same plane by the same kind of force naturally arrange themselves in this manner, as one can see by an experiment, in throwing a row or two of completely round unstrung pearls confusedly on a plate, and shaking them so that they approach one another.

Other seventeenth-century scientists tried to explain the symmetry of crystals in terms of basic particles with simple shapes: spheres, ellipsoids, or regular polyhedra. The Dutch scientist Christian Huygens

Descartes attributed the beautiful symmetry of snowflakes to the existence of spherical water molecules. They arranged themselves into six–sided crystals, much as seven pennies make a regular hexagon. *Source:* © Carl Zeiss, Inc./Photo Researchers (*left*); © Richard B. Hoit/Photo Researchers (*center* and *right*).

(1629–1695), a leading advocate of this vision, knew that his work was purely descriptive and had no dynamical or chemical basis. He wrote:

> I will not undertake to say anything touching the way in which so many corpuscles all equal and similar are generated, nor how they are set in such beautiful order, whether they are formed first and then assembled, or whether they arrange themselves thus in coming into being and as fast as they are produced....To develop truths so recondite, there would be needed a knowledge of nature much greater than that which we now have.

We shall see that the first solid evidence for atoms came from the studies of neither gases nor crystals, but from the newborn exact discipline of chemistry. Quantitative measurements of chemical reactions yielded hints, and later, undeniable (if indirect) evidence for the physical reality of unseen atoms. This is the thread of the tapestry of science we now follow.

The Law of Definite Proportions

Chemists knew that elements combined to form *compounds.* However, it was not clear whether the proportions of the elements in a given compound were fixed or variable. Hollandaise sauce (which is not a compound) is made by beating together egg yolk and olive oil. The proportions are chosen by the chef to produce an appropriate consistency. The French chemist Claude Berthollet (1748–1842) thought chemistry was like cooking. He argued that the ratio of the constituents of a chemical compound is not fixed but depends on the circumstances of its synthesis. Berthollet's countryman and antagonist, Joseph Proust (1754–1826), took the contrary view and enunciated what he believed to be a fundamental law of chemical combination:

> We cannot create compounds as we please. When you believe that you can combine bodies in arbitrary proportions you, short-sighted wretches, are only making mixtures of which you are incapable of distinguishing the

parts; what you are making are monsters. A compound is a substance to which nature assigns fixed ratios; it is a being which nature never creates otherwise than with balance in hand.

Proust's careful experiments found the flaws in Berthollet's work. By 1808, he had won the debate. His *law of definite proportions* has been with us ever since: Every chemical compound, no matter how it is made, contains exactly the same fractions by mass of its various constituent elements.

Proust's law has a natural explanation in terms of atoms. Table salt, for example, is a definite chemical compound. We now know that each salt molecule is made of one atom of sodium and one of chlorine, so that every sample of salt contains equal numbers of atoms of each sort. Salt, whether mined, extracted from seawater, or synthesized from its elements, always contains the same fractions of sodium (39.34 percent by mass) and chlorine (60.66 percent). However, it was (and is) fallacious to argue from this that

> Atoms explain Proust's law.
> Proust's law is true.
> Therefore atoms exist.

Maybe other arguments could explain Proust's law.* The next, and decisive, step toward a conclusive proof of the atomic hypothesis was taken by the English chemist John Dalton.

Dalton's Atoms and Molecules

John Dalton (1766–1844) began his scientific career as a physicist concerned with the properties of gases, and in particular, with the question of how much of various gases could dissolve in water. He wrote in 1803,

> Why does not water admit its bulk of every kind of gas alike? I am nearly persuaded that the circumstance depends on the weight and number of the ultimate particles of the several gases.

Why did different gases dissolve differently? Perhaps the amount of a gas that could enter into solution depended on the size or mass of its atoms. To reinforce this suspicion, Dalton turned his attention to the quantitative study of chemical reactions, in particular to the masses of the reacting substances and their products. By 1810, he published the results of his studies as a *New System of Chemical Philosophy*. We may extract from it Dalton's five chemical principles, which became the foundation of modern chemistry:

1. *Matter is made up of individual atoms.* Dalton noted, "There must be some point beyond which we cannot go in the division of matter. The existence of ultimate particles of matter can scarcely be doubted,

---

*In a similar vein, the theory of evolution predicts that there should be fossils of extinct animals. The discovery of these fossils did not prove the theory, but it was a step in the right direction.

though they are probably much too small ever to be exhibited. I have chosen the word atom to signify these ultimate particles. "

2. *Each chemical element is made up of identical atoms of a particular kind.* The atoms of an element "are perfectly alike in weight and figure, etc. Every atom of hydrogen is like every other atom of hydrogen." Dalton is pressing far beyond the views of his predecessors. Earlier atomists likened atoms to pebbles at the beach, some smaller, larger, shinier, or rounder, but certainly not all the same.

3. *Atoms are unchangeable.* Those of different elements "never can be metamorphosed, one into another, by any power we can control." The failure of centuries of efforts by alchemists to turn base metal into gold was put forward as a principle of chemistry.

4. *Chemical elements may combine to form compounds.* The smallest portion of a compound is made up of a definite number of atoms of each element. He called such groupings of atoms "compound atoms." We call them molecules.

5. *Chemical reactions rearrange atoms into different compounds, but do not change the numbers of atoms of each element.* According to Dalton, "Chemical analysis and synthesis go no farther than the separation of atoms from one another and to their reunion. We might as well attempt to introduce a new planet into the solar system, or to annihilate one already in existence, as to create or destroy an atom of hydrogen."

Chemical Notation

Centuries before Dalton, alchemists used symbols to designate different substances. However, if chemical processes are to be discussed in terms of atoms, symbols are needed to denote individual atoms of different elements. The illustration shows the ideographs Dalton used to indicate different atoms and compounds. The first six symbols stood for single atoms of hydrogen, nitrogen, carbon, oxygen, phosphorus, and sulfur, respectively. Today scientists no longer use Dalton's symbols; instead, they use the letters H, N, C, O, P, and S to stand for these atoms.

Dalton drew a molecule of a particular compound as a grouping of atoms. Item 21 of the illustration stood for a molecule made of one atom of hydrogen and one of oxygen, which Dalton mistakenly identified as the structure of a water molecule. Items 25 and 28 correspond to molecules of the two oxides of carbon, while items 23, 26, 27, 30 and, 34 correspond to five distinct oxides of nitrogen. Subscripts on modern chemical symbols tell how many atoms of a particular kind are in a molecule. The water molecule actually contains 2 H atoms and 1 O atom; its chemical formula is H_2O. The chemical formulas of the oxides of carbon are CO (poisonous carbon monoxide) and CO_2 (carbon dioxide).

Once atomic symbols were decided upon, equations could be used to denote chemical reactions. Thus,

$$C + 2S \longrightarrow CS_2 \tag{1}$$

says that two atoms of sulfur and one of carbon (the reactants) combine to form one molecule of carbon disulfide (the product). Because every compound has a definite atomic composition, and because all the atoms of a given element have the same mass, the law of definite proportions is built in to the notation. In any equation representing a chemical reaction, the number of atoms of every element must be the same on both sides. Thus, Lavoisier's law of mass conservation is built in as well. Many of the laws of chemistry are embodied in Dalton's notation and, as well, in its modern successor.

The concepts of atomic weight and molecular weight were defined in Chapter 5. A *mole* of a substance is a quantity whose mass in grams equals its molecular weight. Thus, 1 mole of every substance contains the same number of molecules: Avogadro's number of them. Equation 1 may otherwise be interpreted to say that 1 mole of carbon combines with 2 moles of sulfur to form 1 mole of carbon disulfide.

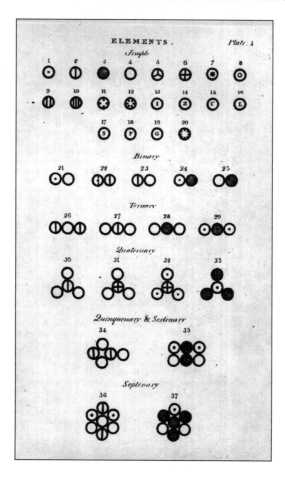

Dalton's symbols for elements and compounds. The modern names for some of his elements are: 1, hydrogen; 2, nitrogen; 3, carbon; 4, oxygen; 5, phosphorus. The compounds numbered 25 and 28 are CO and CO_2. Those numbered 23, 26, 27, 30, and 34 are the various oxides of nitrogen. Notice that Dalton mistakenly regards water, which he numbers 21, as the binary compound HO. *Source:* The Granger Collection.

*The Law
of Multiple
Proportions*

Dalton expressed his new chemical philosophy in terms of unseen atoms and molecules. His hypotheses reproduced the known laws of chemistry, but this would not have been enough to establish the truth of his vision. The capstone of his endeavor was his discovery of a brand new law of chemical combination, the *law of multiple proportions*. This law is an immediate consequence of his chemical hypotheses, and Dalton proved by experiment that it is satisfied in nature.

A BIT MORE ABOUT
John Dalton

Dalton was one of six children of a poor English weaver. A precocious student, he requested and received permission to open a Quaker school at age 12, so that he could supplement his family's meager earnings. Three years later, he replaced his elder brother as an assistant at an established boarding school. Early in his career, he studied his own inability to distinguish colors, a form of color-blindness now known as *Daltonism.* He spent every working day in his laboratory, emerging at precise moments to consult his outdoor thermometers and barometers, except for Thursday afternoons, when he bowled and bet with abandon, and Sundays, when he attended church services twice. Dalton was a man of fixed habits.

In 1793, he became a teacher of mathematics and chemistry in Manchester, where he was regarded as an atrocious lecturer and a sloppy experimenter. His contemporaries said of him, "His voice was thick and gruff, and his articulation thick, indistinct and mumbling," and "His aspect and manner were repulsive, his voice harsh and brawling; his gait stiff and awkward." When he became president of the Manchester Literary and Philosophical Society, he boasted that "he could carry his library on his back and had yet not read half the books," and he said of the Society's literary readings that they "contributed no positive facts to the stock of knowledge and proved nothing."

John Dalton (1766–1844), who is often called the founder of modern atomic theory. *Source:* The Bettmann Archive.

Dalton was a careless experimenter, but his brilliant insights led him to a basically correct theory of chemical phenomena. Nonetheless, as a world-famous scientist, he remained an old codger who was neither literate nor lovable. Partially paralyzed by a stroke toward the end of his life, he continued his conscientious meteorological and religious devotions until the day of his death.

The law of multiple proportions relies on the fact that two elements often can combine in different proportions by mass to form different compounds. When they do so, Dalton discovered, these different proportions are always in simple ratios to one another. The most convincing of his experiments involved the many ways that nitrogen and oxygen can combine. Table 6.1 shows the masses of nitrogen that unite with 100 g of oxygen to form six different oxides of nitrogen.*

TABLE 6.1   The Oxides of Nitrogen

Chemical Formula	Modern Name of the Oxide	Grams of Oxygen per 100 g Nitrogen
N_2O	Nitrous oxide	57
NO	Nitric oxide	$2 \times 57 = 114$
N_2O_3	Nitrous anhydride	$3 \times 57 = 171$
NO_2	Nitrogen dioxide	$4 \times 57 = 228$
N_2O_5	Nitric anhydride	$5 \times 57 = 285$
NO_3	Nitrogen peroxide	$6 \times 57 = 342$

Dalton correctly assumed that the nitric oxide molecule (which he called nitrous gas) is made of one atom of nitrogen and one of oxygen. In modern parlance it is NO. Because nitrogen dioxide contains twice as much oxygen per gram of nitrogen as nitric oxide, Dalton argued that its molecule must be NO_2. Because nitrous oxide contains half as much oxygen per gram of nitrogen as nitric oxide, Dalton argued that its molecule must be N_2O. In each of the cases listed in Table 6.1, the mass of oxygen combining with 100 g of nitrogen is an integer multiple of a certain fixed quantity. The ratio of oxygen to nitrogen in NO is precisely twice that in N_2O; in NO_3 it is six times, and so on. Dalton concluded:

> These facts clearly point out the theory of the process: the elements of oxygen may combine with a certain portion of nitrous gas [nitrogen], or with twice that portion but with no intermediate quantity.

Dalton showed that his new law applied to every chemical reaction he studied. Thus, one of the oxides of carbon has an oxygen-to-carbon ratio precisely twice that of the other. He concluded that the former compound has the chemical formula CO_2 (carbon dioxide); the latter, CO (carbon monoxide). Here are some of the conclusions that Dalton drew from his experiments:

> I call an ultimate particle of carbonic acid [CO_2] a compound atom. Now, though this atom may be divided, yet it ceases to be carbonic acid, being

*One of these gases, nitrous oxide, was discovered by Priestley in 1772. It was one of the earliest recreational drugs—laughing gas. Its bizarre psychological effects were observed by Dalton's colleague, the British chemist Humphry Davy, who offered it to his literary friends, including Samuel Taylor Coleridge, whose poetic imaginings it may have inspired. I doubt if Dalton ever tried it.

resolved by such division into charcoal and oxygen. If two bodies A and B are disposed to combine together, these are the simplest combinations which can take place starting from the simplest:

1 atom of A	+	1 atom of B	=	1 atom of C,	binary
1 atom of A	+	2 atoms of B	=	1 atom of D,	ternary
2 atoms of A	+	1 atom of B	=	1 atom of E,	ternary
1 atom of A	+	3 atoms of B	=	1 atom of F,	quaternary.

Dalton's simple precepts explained the conservation laws for individual chemical elements and Proust's law of definite proportions. Moreover, they explained why elements combine in accordance with his law of multiple proportions! Dalton could think of no other explanation except by means of his atomic theory of chemistry. Neither could anyone else. What could be more elegant, obvious, and yet revolutionary? In all chemical reactions, indestructible and unchangeable atoms simply rearrange themselves into new combinations, producing new chemical compounds.

A half century later, Maxwell—the founder of electromagnetic theory, whom we shall encounter in Chapter 8—contrasted Dalton's atomic theory with Darwin's then-new theory of the origin of the species in a contribution to the *Encyclopædia Brittanica*:

> It has been possible to frame a theory of the distribution of organisms by means of generation, variation, and discriminative destruction. But a theory of evolution of this kind cannot be applied to the case of atoms, for the individual atoms neither are born nor die, they have neither parents nor offspring, and so far from being modified by their environment, we find that two atoms of the same kind, say of hydrogen, have the same properties, though one has been compounded with carbon and buried in the earth as coal for untold ages, while the other has been occluded in the iron of a meteorite, and after unknown wanderings in the heavens has at last fallen into the hands of some terrestrial chemist.

Maxwell compared atoms to manufactured articles, whose uniformity "may be traced to very different motives on the part of the manufacturer: cheapness, serviceability, and quantitative accuracy." Atoms are like "shoes made in large numbers," and because "a number of exactly similar things cannot be each of them eternal and self existent, they must therefore have been made." By whom or by what, Maxwell does not say, but his point is well taken. Dalton's most astonishing realization, and the one that early chemists accepted only reluctantly, is the absolute identity of different atoms of the same element. It's a property we simply never encounter in the everyday world.

What Dalton Didn't Know

What is the mass of one atom? Dalton showed how his atoms and molecules could explain the empirical rules of chemical combination, but none of his experiments told him the size or mass of an individual atom. However, his atomic theory did provide a means of "ascertaining the relative weights of the ultimate particles, both of simple and compound bodies," or what we now call *atomic weights*. (They should, more appropriately, be called atomic masses.) The binary compound CO consists of about 3 parts carbon and 4 parts oxygen by mass. Thus, the O atom must have approximately 4/3 the mass of the C atom. Dalton decided to measure all atomic weight relative to the hydrogen atom, to which he assigned atomic weight 1. For nitrogen he obtained the atomic weight 5, for oxygen 7, for sulfur 13, and so on. His assignments were only tentative, for "after all, it must be allowed to be possible that water is a ternary compound." His measurements, however, were flawed: The atomic weights of nitrogen and oxygen are about 14 and 16, respectively.

One explanation for Dalton's errors was the crudity of his experimental technique. Another explanation was that Dalton did not know that the chemical formula for water is H_2O. He clung to HO as the simplest, and to him, the most reasonable choice.

Dalton made another simple but false assumption. He firmly believed that all chemical elements occur in nature as individual atoms, or as what we now call *monatomic* molecules. This is an innocuous hypothesis for most of the elements. However, the atoms of many gases, including hydrogen, oxygen, and nitrogen, are strongly attracted to one another. They usually form *diatomic* molecules consisting of two atoms: H_2, O_2, and N_2. Dalton could not believe that an oxygen molecule is made of two oxygen atoms. He regarded such a hypothesis as a perversion of nature's elegant simplicity and therefore he never hit on a consistent scheme of atomic weights.*

Because of the ambiguities of his theory, Dalton's further researches became hopelessly ensnarled in rapidly accumulating but seemingly contradictory chemical data. The mess would not be straightened out for decades. Despite this confusion, though, the significance of Dalton's work was recognized, and he was deluged with honorary degrees, medals, pensions, and memberships in learned societies.

How to Balance a Chemical Equation

The skill of balancing a chemical equation may not seem relevant to a course in physics. However, there is not much difference between equations describing chemical reactions and those describing nuclear reactions or reactions among elementary particles. Once the notation is clearly understood, the game is almost trivial. All you've got to do is be sure that the numbers of particles of each kind (atoms, in the case of

---

*The molecules of the so-called inert gases, such as helium and neon, do not combine easily with one another or with other atoms. Their molecules are ordinarily monatomic. However, neon has been observed to form diatomic molecules at very low pressure and temperature. In 1993, chemists succeeded in producing diatomic helium molecules. They are exceedingly fragile and about 100 times larger than air molecules.

chemical equations) are the same on either side of the equation. Let's look at an example. Because oxygen and hydrogen are diatomic gases, the correct chemical equation describing the combustion of hydrogen to form water is

$$2H_2 + O_2 \longrightarrow 2H_2O \qquad (2)$$

The arrow points from the initial reactants to the final products. The subscript numerals are the numbers of atomic constituents per molecule. The prefixed numerals show the minimum number of molecules of each element or compound that may undergo the reaction. In either case, the numeral is omitted when it is 1. Equation 2 says that 2 hydrogen molecules unite with 1 oxygen molecule to form 2 water molecules.

Equation 2 is said to be balanced because the number of oxygen atoms (2) and hydrogen atoms (4) is the same on the left and on the right. In a properly balanced chemical equation, only whole numbers may appear because fractions of atoms or molecules make no sense. Furthermore, the prefixed integers are chosen to be as small as possible: $4H_2 + 2O_2 \rightarrow 4H_2O$ is balanced, but not properly so.

Gay-Lussac and the Chemistry of Gases

The scene shifts from England to the Continent, to the laboratories of Gay-Lussac. His first endeavor (and our first encounter with him in Chapter 5) concerned the effect on gases of changes in temperature.* He then turned from his studies of the physical properties of gases to their chemical properties. Because he worked with gases, Gay-Lussac measured the volumes of the reactants rather than their masses, which led him to a new law, dazzling in its simplicity, that quantitatively described gas chemistry. He wrote:

> I have shown in this Memoir that the compounds of gaseous substances with each are always formed in very simple ratios, so that representing one of the terms of unity, the other is 1, 2, or at most 3. These ratios by volume are not observed with solid or liquid substances, nor when we consider weights.

He found that 2 liters of CO combine with 1 liter of oxygen to form, all together, exactly 2 liters of CO_2. Thus, $2 + 1$ can make 2. He found that 4 liters of NO combine with 1 liter of O_2 to form 2 liters of N_2O_3.

---

*For a time, Gay-Lussac studied the chemical and physical properties of the atmosphere at high altitudes. He ascended to the remarkable height of 23,000 feet in a hydrogen-filled balloon, where he remained to carry out his measurements. This extraordinary feat was accomplished only two decades after the first balloon ascent—Gay-Lussac was the first space scientist.

More generally, suppose that two gases react chemically with one another in such a manner that all of the reacting gases are consumed by the process. The law of combining volumes says that the ratio of the volumes of the reactants, measured at the same pressure and temperature, is always a simple fraction. Moreover, the volumes of the gaseous products of the reaction are simple fractions or multiples of the volumes of the reactants under the same conditions.

Gay-Lussac's experiments puzzled Dalton. Could it be possible that equal volumes of gas contain equal numbers of molecules, as the law of combining volumes suggested? The notion was anathema to him. He thought that the atoms of a gas touched one another, and he assumed that atoms of different gases had different sizes. Dalton's own experiments did not seem to confirm Gay-Lussac's law, and Dalton never accepted its validity.

The results of Gay-Lussac and of Dalton seemed irreconcilable. Until a crucial missing link was provided by the Italian physicist Amadeo Avogadro (1776–1856). His principal observation, known as *Avogadro's hypothesis*, says that equal volumes of all gases, whether elements, compounds, or mixtures of gases, when they are at the same pressure and temperature, contain the same number of molecules.

Avogadro did not prove his assertion. This would be done much later in the nineteenth century by Boltzmann, as a consequence of the kinetic theory of gases. Avogadro simply pointed out that his ad hoc hypothesis, together with one additional unprovable assertion, would reconcile Gay-Lussac's discovery with Dalton's chemical principles. His second, and equally important hypothesis, is this:

> That the constituent molecules of any simple gas whatever are not formed of a solitary elementary atom, but are made up of a certain number of these atoms united by attraction to form a single one.

Molecules of the then-known elemental gases—hydrogen, oxygen, and nitrogen—contain two atoms apiece. They are diatomic gases. Molecules of compound gases, such as CO or CO_2, consist of two or more atoms. Monatomic gases, such as helium and neon, had not yet been discovered.* Equal volumes of any two gases contain the same number of molecules, but not the same number of atoms.

To understand what Avogadro had accomplished, let us return to a chemical reaction that was studied both by Dalton and Gay-Lussac: the combination of carbon monoxide with oxygen to form carbon dioxide. Dalton, thinking that oxygen molecules are monatomic, described the

---

*Oxygen atoms sometimes form the *triatomic* molecule O_3. It is the rare but environmentally important gas *ozone*.

process in terms of the combination of individual atoms and molecules, as shown in Figure 6.1a, which in modern notation would be:

$$CO + O \longrightarrow CO_2$$

Gay-Lussac showed that 2 liters of carbon monoxide combine with 1 liter of oxygen to form 2 liters of carbon dioxide, as shown in Figure 6.1b. This result became explicable when it was realized that oxygen is diatomic, so that Dalton's equation should have been as shown in Figure 6.1c, or:

$$2CO + O_2 \longrightarrow 2CO_2$$

Using this result, we obtain the picture shown in Figure 6.1d, which incorporates both Gay-Lussac's law of combining volumes and the fact that equal volumes contain equal numbers of molecules. (The numbers of molecules shown are smaller than the actual numbers.) Each O_2 molecule services two CO molecules. It seems simple, but it took half a century after the work of Dalton, Gay-Lussac, and Avogadro, for chemists to put it all together. The synthesis needed an accumulation of accurate experimental data, and the genius of one man—the Italian chemist Stanislao Cannizzaro (1829–1910), whose guiding principle was that "the different quantities of the same element contained in different

FIGURE 6.1
a. Dalton's chemical equation for the combination of carbon monoxide with oxygen to form carbon dioxide. *b.* Gay-Lussac finds that 2 L of carbon monoxide reacts with 1 L of oxygen to form 2 L of carbon dioxide. *c.* Dalton's equation, as modified by the fact that oxygen is diatomic: 2 molecules of CO react with 1 molecule of O_2 to form 2 molecules of CO_2. *d.* Because equal volumes of gas contain equal numbers of molecules, Gay-Lussac's result is understood.

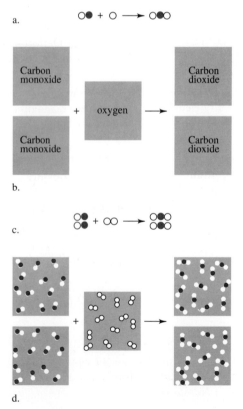

molecules are all whole multiples of one and the same quantity, which, always being entire, has the right to be called an atom."

Once it was recognized that many gaseous elements are diatomic, the conflict between Dalton's findings and those of Gay-Lussac evaporated, and correct atomic weights could be determined. Suppose that hydrogen is assigned atomic weight 1, and a certain volume of H_2 gas (whose molecular weight is 2) weighs 2 g. At the same pressure and temperature, the same volume of O_2 gas weighs 32 g. Because both samples contain the same number of molecules, it follows that the mass of an oxygen molecule is 16 times the mass of a hydrogen molecule and its atomic weight is 16. In another experiment, the same volume of CO_2 is found to have a mass of 44 g, which is therefore its molecular weight. Because we already determined the atomic weight of O to be 16, that of C must be 12. Unambiguous measurements of atomic weights accumulated, setting the stage for the discovery of a correlation between atomic weights and chemical properties. As we shall see in the next section, it took the form of the periodic table of the elements.

What about Energy?

The early history of chemistry was actually much more complicated than I've described here. We learned in Chapter 5 that heat was regarded as a material substance until well into the nineteenth century. Dalton, for example, was a believer in the caloric hypothesis. Today we know that heat is not a substance and that energy is conserved. Chemical equations, such as those we have considered, do not explicitly take energy into account. Some chemical reactions, such as the extraction of metals from ores, require an external source of energy. The heat needed to decompose iron oxide becomes chemical energy stored in metallic iron. Other chemical reactions, such as combustion, release energy in the form of heat.

EXAMPLE 6.1
The Burning Bush

This example brings together what we have learned in this chapter and the preceding chapter. Well-dried wood consists mostly of cellullose, or $C_6H_{10}O_5$. The equation describing its combustion is

$$C_6H_{10}O_5 + 6O_2 \longrightarrow 6CO_2 + 5H_2O \tag{3}$$

However, equation 3 is not the whole story. We must know how much heat energy is stored as chemical energy in the fuel: its heat content per unit mass. The heat content of cellulose (or almost any dry wood) is about 1.7×10^7 J/kg. Using the chemical principles we have learned, we may calculate the energy released by the combustion of one molecule of cellulose.

When a fuel burns, it produces useful energy and environmentally threatening CO_2. The goodness index GI of a fuel can be defined as the ratio of the good to the bad: its energy release in megajoules divided by its CO_2 production in kilograms. The higher the GI, the better is the fuel. What is the GI of cellulose?

Solution   The molecular weight of cellulose is the sum of the atomic weights of its constituent atoms: $6 \times 12 + 10 \times 1 + 5 \times 16 = 162$. As we learned in Chapter 5, 1 mole of any compound contains Avogadro's number of molecules, or about 6×10^{23} molecules. The heat content of 1 mole (or 162 g) of cellulose is about 2.75 megajoules. Thus, the energy released by the combustion of one molecule of cellulose is $2.75 \times 10^6/6 \times 10^{23} = 4.6 \times 10^{-18}$ J. The burning of 1 mole of cellulose, according to equation 3, produces 6 moles of CO_2, or (because the molecular weight of CO_2 is 44) 0.264 kg of CO_2. Thus, the GI of cellulose is 2.75/0.264, or 10.4 megajoule per kilogram. Exercises 8–10 examine the quality of other fuels.   ∎

Exercises

1. One kilogram of sulfur is burned to form SO_2. What is the mass of the SO_2? (The atomic weight of sulfur is 32.)

2. What is the molecular weight of hydrogen peroxide, H_2O_2? What is the mass of oxygen produced by the decomposition of 1 mole of H_2O_2 by the reaction $2H_2O_2 \rightarrow 2H_2O+O_2$?

3. Consider the following reactions among gases. In each case, the products consist of a gas or gases with a total volume of 12 L at standard conditions. What is the initial volume of each of the reacting gases under the same conditions, assuming that all the reactants are consumed in the reaction? [*Hint*: Consider the rection $3O_2 \rightarrow 2O_3$. Three moles (or volumes, according to Avogadro) of O_2 react to form 2 moles (or volumes) of ozone. Thus, 18 L of molecular oxygen do the trick.]
   a. $2CO + O_2 \longrightarrow 2CO_2$
   b. $CH_4 + 2O_2 \longrightarrow CO_2 + 2H_2O$
   c. $CS_2 + 3O_2 \longrightarrow CO_2 + 2SO_2$

4. Measurements are done at STP. If 1 L of CO burns to form CO_2:
   a. how many liters of oxygen are consumed?
   b. what mass of oxygen is consumed?

5. The equation $CH_4 + 2O_2 \rightarrow CO_2 + 2H_2O$ describes the combustion of methane. Answer the same questions as in exercise 4 if 1 L of CH_4 is burned.

6. The equation $CS_2 + 3O_2 \rightarrow CO_2 + 2SO_2$ describes the combustion of carbon disulfide. Answer the same questions as in exercise 4 if 1 L of CS_2 is burned.

7. What volume of oxygen combines with 1 mole of CO at standard conditions? What volume of oxygen combines with 1 mole of CH_4 at standard conditions? (Recall from Chapter 5 that 1 mole of a gas at standard conditions occupies 22.4 L.)

8. In the following table, the heats of combustion and the chemical formulae of four fuels are specified: carbon (from coal), methane (from natural gas), decane (from gasoline), and grain alcohol (ethanol). What is the molecular weight of each fuel? How many moles of CO_2 are produced when 1 mole of each fuel is burned to form CO_2 and H_2O?

Fuel	Formula	Heat Content Megajoule/kg
Carbon	C	34
Methane	CH_4	55
Decane	$C_{10}H_{22}$	47
Ethanol	C_2H_5OH	30

9. Compute the GI index for each of the fuels in the preceeding table; that is, compute the megajoules of energy released per kilogram of CO_2 produced. You will find that the cleanest fuel, from the point of view of the greenhouse effect, produces about half as much CO_2 as coal or wood for the same energy yield.

10. Hydrogen, when burned, releases a considerable amount of heat but generates no CO_2 at all. Explain why its use as a fuel is easier said than done.

6.2 The Periodic Table: A Pattern among the Elements

Between 1790 and 1844, 31 new elements were discovered. Lavoisier's original list was doubled in size. This population explosion reflected the development of powerful new experimental techniques. Electrochemistry evolved from the invention of the electric battery by Alessandro Volta and its rapid exploitation by Humphry Davy. (We discuss these developments in Chapter 7.) By 1807, Davy produced metallic sodium and potassium electrolytically. This achievement was followed by his isolation of magnesium, calcium, strontium, and barium. As the number of known chemical elements increased, chemists searched for a pattern among them. The search climaxed with the development of the periodic table of the elements, whose history and significance is now sketched.

As elements were discovered, they fell into families with distinctive chemical properties. One family includes the elements sodium, potassium, and lithium. They are called *alkali metals*, because they react violently with water to form active chemicals known as alkalis. Another family includes the nonmetals chlorine (a gas), bromine (a liquid), and iodine (a solid). These elements have similar chemical properties and are known as *halogens*, from the Greek for salt producer. One atom of any halogen combines with one atom of any alkali metal to form one molecule of their compound, such as NaCl (table salt), KI (potassium iodide), or LiBr (lithium bromide). All of these salts are colorless, very stable, and very soluble in water. In addition to these families, there are several others. Like halogens and alkali metals, their members have widely different atomic weights, but they share common chemical properties.

As early as 1829, the German chemist Joseph Döbereiner (1780–1849) noticed a correlation between the properties of elements within a family and their measured atomic weights. Among the halogens, for example, he noticed that the atomic weight of bromine (80) is nearly the mean of the atomic weights of chlorine (35.4) and iodine (127). He found another triad among the alkali metals: The mean of the atomic weights of lithium (7) and potassium (39) is 23, which is the atomic weight of sodium. So not only were the chemical properties of the elements in families similar, but there were mysterious numerical relations among their atomic weights as well. These were unlikely to be coincidences and thereby posed a puzzle.

The English chemist John Newlands came closer to the discovery of a regular pattern among the elements. He based his system on an analogy with music. Putting the elements in order of ascending atomic weight, he observed that

> The numbers of analogous elements generally either differ by seven or by some multiple of seven; in other words, members of the same group stand to each other in the same relation as the extremities of one or more octaves in music.

In many cases, Newlands's theory of octaves works quite well. In Figure 6.2, a series of elements is inscribed, in order of increasing atomic weight, on the white keys of a piano. Those identified with the same note of the musical scale are chemically similar to one another. However, Newlands' system failed to describe the heavier atoms, and it did not leave spaces for elements yet to be discovered. His theory—in which there was, in fact, an element of truth—was met with scorn. When he read his paper to the London Chemical Society in 1864, a distinguished colleague asked "whether Newlands had ever tried classifying the elements in the order of the initial letters of their names," which indeed would have been a silly thing to do.

FIGURE 6.2
Newlands's musical analogy. The elements, in order of ascending atomic weights, are written on the white keys of the piano. Elements that are placed one or more octaves apart are chemically similar—for example, lithium, sodium, and potassium (all of them C). The inert gases, such as neon (which belongs between fluorine and sodium), were unknown to Newlands.

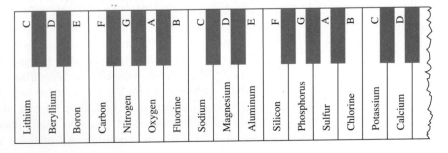

Mendeleev's Periodic Table

Dmitri Ivanovich Mendeleev (1834–1907) was born in Siberia, the last of 17 children. He studied in France and Germany during the great controversy over Avogadro's law, when Cannizzaro presented his methods for precise determinations of atomic weights. Soon after Mendeleev returned to Russia in 1869, he formulated and published the first of many versions of his famous table. He often had to revise his table due to discoveries of new elements and revisions of measured atomic weights.

You can find a contemporary version of the periodic table, which is quite different from any that Mendeleev proposed, in the appendix. Figure 6.3 shows another modern variant, illustrating how successive electron shells are filled. In any periodic table, the elements belonging to a family (such as the halogens or the alkali metals) are found in the same vertical column. The elements along a horizontal row display chemical and physical properties that vary systematically from one to the next.

Mendeleev tabulated the elements in order of ascending atomic weights. (We shall see that this was not exactly the right thing to do.)

FIGURE 6.3
The modern periodic table takes many different forms. This modern variant was devised by E. G. Mazurs in 1967. The groups correspond to electron shells. Each group begins with an alkali metal and terminates with an inert gas. A more conventional periodic table is presented in the appendix.

Dmitri Ivanovich Mendeleev (1834–1907), whose periodic system predicted the existence of the elements scandium, gallium, and germanium, all of which were later found in nature. He also predicted the existence of an element lighter than hydrogen as the substance of the ether. This time he was wrong. *Source:* The Bettmann Archive.

Rather than discussing one or another modern versions of the periodic table, I offer here a brief and annotated description—in Mendeleev's words—of the eight fundamental principles that led him to his discovery.

1. The elements, if arranged according to their atomic weights, exhibit an evident periodicity of properties.

At this point, Mendeleev graciously acknowledged his British colleague: "Newlands had made an approach to the periodic law and had discovered its germ." Newlands's octaves were just a stab in the dark and did not offer a complete systematization of all the chemical elements. Mendeleev's table, however, included all known elements and displayed all the regularities of their properties.

2. Elements which are similar as regards their chemical properties have atomic weights which are nearly of the same value (e.g., platinum, iridium, osmium), or which increase regularly (e.g., potassium, rubidium, cesium).

This is the essence of the periodic table. Mendeleev placed the members of each family of chemically similar elements with very different atomic weights (such as the halogens or the alkali metals) in the same vertical column. The pattern was not quite so simple, however, because Mendeleev noted several instances in which elements with nearly the same atomic weights also had similar properties. (Iron, cobalt, and nickel, for example, which do not lie in the same column, abut one another in the periodic table and are chemically alike.) Significantly, two of the elements Mendeleev mentions (cesium and rubidium) had just been discovered as he was creating his table. Chemistry's growth in the mid-nineteenth century—the development of sophisticated analytical techniques allowing the discovery of new elements and increasingly precise determinations of atomic weights—made the discovery of the periodic table inevitable.

> 3. The arrangement of the elements, or of groups of elements, in the order of their atomic weights, corresponds to their so-called valences as well as, to some extent, to their distinctive chemical properties.

The *valence* of an element is the number of hydrogen atoms that can combine with one atom of that element, or, equivalently, the number of oxygen atoms that chemically combine with two atoms of that element. (Valence has a more profound meaning that was not understood until the structure of the atom was explained in the twentieth century. For our purposes at the moment, however, valence is simply a number from 1 to 7 that characterizes the properties of a chemical element.) Table 6.2 lists the elements between silver and iodine in order of increasing atomic

TABLE 6.2

Element	Atomic Weight	Density	Formula of Oxide	Valence
Silver	108	10.5	Ag_2O	1
Cadmium	112	8.6	Cd_2O_2	2
Indium	115	7.4	In_2O_3	3
Tin	119	7.2	Sn_2O_4	4
Antimony	122	6.7	Sb_2O_5	5
Tellurium	128	6.4	Te_2O_6	6
Iodine	127	4.9	I_2O_7	7

weight. When they are placed in this sequence, other patterns appear: The valences of the elements (a chemical property) increase systematically, and their densities (a physical property) decrease systematically. The next heavier atom after iodine (not known in 1869) is xenon, which does not react with oxygen at all and has valence 0. The next one yet is cesium with valence 1, and it begins a new row. Mendeleev concluded: "Not only are there no intermediate elements between silver and cadmium but, according to the periodic law, there can be none."

> 4. The elements which are the most widely diffused have small atomic weights.

Mendeleev knew that light elements are the most abundant and that most of the elements beyond iron are rare. Today we know that the matter of Earth was made in the stars from primordial hydrogen and helium. Any star can manufacture the elements up to iron; a few make those up to lead in the periodic table; but only the rare supernova synthesizes the heaviest elements.

5. The magnitude of the atomic weight determines the character of the element.

Mendeleev is interpreting his first principle, according to which he tabulated the elements according to their atomic weights. Because the table revealed a pattern among the elements, he concluded that atomic weight somehow determines atomic properties. This is false! The chemical properties of an atom and its correct position in the periodic table are determined by its atomic number Z (the number of protons in its nucleus) and not by its atomic weight. Fortunately for Mendeleev, a list of the chemical elements in order of increasing mass is almost the same as a list in order of increasing atomic numbers. There are, however, two exceptions, one of which may be found in Table 6.2.

6. We must expect the discovery of many yet *unknown* elements—for example, elements analogous to aluminium and silicon, whose atomic weights would be between 65 and 75.

Mendeleev predicted three new elements that would later be found. Figure 6.4 shows a fragment of the periodic table with gaps corresponding to undiscovered elements. Mendeleev predicted what their chemical and physical properties must be: their atomic weights, their densities, the formulæ of their oxides and the properties of their compounds.

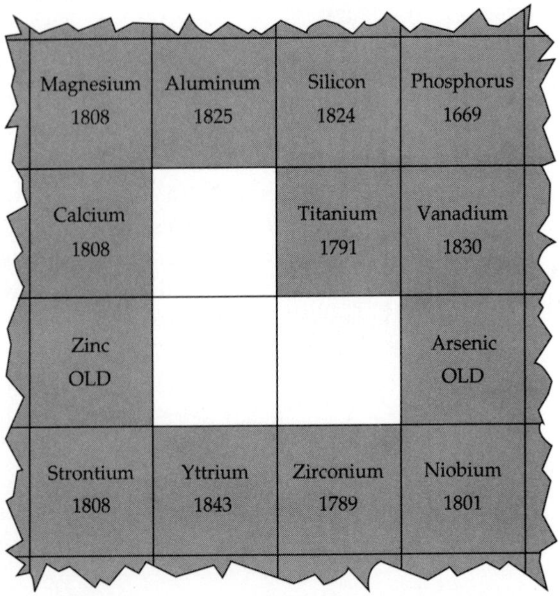

FIGURE 6.4
In 1869, Mendeleev published the first of many versions of what would become the periodic table. This fragment shows a portion of the periodic table with gaps for the three elements whose existence Mendeleev predicted. The dates indicate when elements were first isolated.

Magnesium 1808	Aluminum 1825	Silicon 1824	Phosphorus 1669
Calcium 1808		Titanium 1791	Vanadium 1830
Zinc OLD			Arsenic OLD
Strontium 1808	Yttrium 1843	Zirconium 1789	Niobium 1801

7. The atomic weight of an element may sometimes be amended by a knowledge of those of the contiguous elements. Thus the atomic weight of tellurium must be between 123 and 126, and cannot be 128.

8. Certain characteristic properties of the elements can be foretold from their atomic weights.

Mendeleev is wrong because atomic number, not atomic mass, determines the chemistry of an element. Iodine (atomic number 53) has a smaller atomic weight than tellurium (atomic number 52), whose atomic weight is about 128. To Mendeleev, the measured atomic weights of the elements were only fallible clues to their correct placement in his table. He courageously ignored data that seemed to contradict his vision.

Those were the principles on which Mendeleev based his work. But when his periodic table was first published, it was not taken seriously. Chemists did not set out to search for his predicted new elements. Instead, they happened upon these elements by accident.

DO YOU KNOW
That Most Laws of Physics Have Exceptions?

Mendeleev's fifth principle is wrong! The periodic table, as he presented it, is slightly flawed. Nonetheless, laws that are almost correct have played essential roles throughout the history of science. In explaining the validity of an almost-true law, scientists are often led to a more profound understanding of nature. This is an essential point to keep in mind if one is to comprehend any so-called exact science. Here are a few examples:

• Galileo said that all falling bodies accelerate at the same rate. But the effect of air resistance can never be neglected. Leaves and raindrops fall more slowly than bricks. Nonetheless, Galileo's approximate law was a necessary first step toward Newton's synthesis.

• The ideal gas law was discussed in Chapter 5. All gases obey it approximately, but none obey it exactly. Just because there is no such thing as a perfectly ideal gas does not lessen the importance of the law, because it is the foundation on which today's understanding of real gases is based.

• Newton said that the natural state of motion is uniform motion. Yet, moving bodies on Earth come to rest while heavenly bodies move in ellipses. Newton dealt with the unrealizable abstraction of force-free motion. Furthermore, his theory of gravity, the keystone of physics for more than two centuries, was supplanted by Einstein's theory in the twentieth century. For practical purposes, though, his theory remains as useful as ever.

• A last example concerns Dalton's law of multiple proportions, according to which all compounds consist of an integer number of atoms of various elements. In recent decades, chemical compounds have been made for which the ratio of the number of atoms of various species is not a simple fraction but an irrational number. Fortunately, these exceptions were not known to Dalton, else he may never have persevered.

In 1875, six years after Mendeleev published his first table, the first of his predicted elements was discovered by the French chemist Lecoq Bois-baudan, who called it gallium. Its properties are just what Mendeleev expected. The Swedish physicist Nilson filled the second gap with his discovery of scandium. Finally and triumphantly, Winkler discovered what he called germanium in 1886. All three of Mendeleev's predicted elements were found to exist in nature, and they behaved as Mendeleev foresaw. Figure 6.5 shows the completed fragment of the periodic table with Mendeleev's gaps filled in.

In science, it is not enough to discover something new, but one must recognize and interpret its significance. Mendeleev saw that the order his table brought to the chaos of chemistry signified a deeper level of structure:

> Having thus indicated a new mystery of Nature, which does not yet yield to rational conception, the periodic law, together with the revelations of spectrum analysis, have contributed again to revive an old but remarkably long-lived hope—that of discovering, if not by experiment, at least by mental effort, the primary matter—which had its genesis in the minds of the Grecian philosophers, and has been transmitted, together with many other ideas of the classical period, to the heirs of their civilization.

Mendeleev shared the Davy medal of the Royal Society with Lothar Meyer* in 1882, and Mendeleev won its prestigious Copley prize in 1905. Newlands, whose earlier efforts blazed the way for Mendeleev, copped

FIGURE 6.5
This is what became of the fragment of the periodic table shown in Figure 6.4. It took 17 years, but all of Mendeleev's predicted elements were found in nature.

Magnesium 1808	Aluminum 1825	Silicon 1824	Phosphorus 1669
Calcium 1808	Scandium 1879	Titanium 1791	Vanadium 1830
Zinc OLD	Gallium 1875	Germanium 1886	Arsenic OLD
Strontium 1808	Yttrium 1843	Zirconium 1789	Niobium 1801

*Meyer had published a table similar to Mendeleev's at about the same time. However, Meyer did not leave room in his table for elements that had not yet been discovered. Thus, he missed out on the prediction of new elements.

the Davy prize in 1888—his genius was finally acknowledged by his countrymen.

The Inert Gases Mendeleev was in for a surprise: the discovery of several chemical elements that do not form chemical compounds and that had no place in his table. It is a tale of serendipity. John William Strutt (who became Lord Rayleigh) was fascinated by Prout's hypothesis (which is discussed in Section 6.4) that all atomic weights are exact integers. Rayleigh set out to measure the ratio of the density of oxygen gas to that of hydrogen gas, which was thought at that time to be exactly 16. By 1892, he had established the ratio to be 15.882, and not an integer. Having honed his experimental skills, he set out to measure the density of nitrogen gas, even though its atomic weight was already known not to be an integer. As a careful experimenter, he obtained samples of N_2 in two ways. To his astonishment, they had different densities! Either there was a heavy contaminant in the N_2 he extracted from air or there was a light contaminant in the N_2 he obtained by decomposing ammonia.

Rayleigh enlisted the help of his chemist colleague, William Ramsay. Together, they found a new element comprising almost 1 percent of air. Its atomic weight of about 40 put it between chlorine and potassium, where Mendeleev had left no vacant space. They called it argon, meaning lazy one, since it seemed unwilling or unable to form chemical compounds.

Subsequently, Ramsay and his young assistant Travers found three more inert gases in air: neon (the new one), krypton (the secret one) and xenon (the alien). Helium had been known to exist in the Sun since spectroscopic observations in 1868. It was thought to be a metal absent from Earth, but Ramsay and Travers found helium in terrestrial minerals and discovered it to be a fifth inert gas. These elements (and one more, radon) are chemically inactive. Ordinarily, they do not even stick to one another: Their molecules are monatomic.

Ramsay had to add a new column to the periodic table to make room for the inert gases. Not surprisingly, Mendeleev was outraged. How could there possibly be nonreactive elements with valence of 0, elements for which there were no spaces in his table? For years, he insisted that argon was not an element but a compound. Eventually, however, he was forced to accede to Ramsey's alteration of his table.

In 1904, Rayleigh won the Nobel Prize in physics for "his investigations of the densities of the most important gases and for his discovery of argon in connection with these studies." In the same year, Ramsay won the chemistry prize "in recognition of his services in the discovery of the inert gaseous elements in air, and his determination of their place in the periodic system."

Today, you encounter inert gases in neon signs, argon lamps, and helium balloons. The correct and complete periodic table is shown in the appendix. It is a useful tool as an organizer of chemical lore, but the existence of periodic regularities among the elements posed a puzzle to physicists and chemists alike: The table worked, but why did it work? Its validity would eventually be explained by fundamental physical laws— but not until the substructure of the atom was understood and a new theoretical discipline, quantum mechanics, was invented.

Exercises

11. The valence of aluminum is three. What are the chemical formulas for its oxide and chloride?

12. Show that the atomic weights of silicon, germanium, and tin form a Döbereiner's triad.

13. The seven elements from Li to F and those from Na to Cl form "octaves," as Newlands proposed. Do the seven elements starting at K form such an octave? Explain your answer.

14. Nine chemical elements are mentioned or alluded to in the Bible. Shown below are their atomic weights and their densities (in g/cm^3).

Element	Atomic Weight	Density
Carbon	12.0	2.10
Sulfur	32.1	2.07
Iron	55.8	7.87
Copper	63.5	8.96
Silver	107.9	10.50
Tin	118.7	7.31
Gold	197.0	19.32
Mercury	200.6	13.5
Lead	207.2	11.35

In this table, atomic weights are given in atomic mass units of 1.66×10^{-24} g. Calculate the mass m of each of these atoms in grams. Assume that the atoms are shaped like tiny cubes that are stacked together in the element. Let d be the length of the side of such a cube for a particular atom. The number of atoms per unit volume is $1/d^3$ and the density of the element is m/d^3. Calculate d for each biblical element from the mass m and density m/d^3 of each. In each case, these atomic sizes lie between 2 and 3×10^{-10} m. All atoms, not just those known in antiquity, are of roughly the same size, though they range in mass by a factor of more than 200!

6.3   Getting to Know the Chemical Elements (Optional)

Chemists synthesize substances that never before existed—dyes, plastics, medicines, fabrics, glues, and so on. These materials make the practices of plumbing, dentistry, pole vaulting, and practically everything else, a far cry from what they were a generation ago. I can hardly wait to see the next generation of synthetics. Chemistry is remarkable for its processes as well as its products: strange and wonderful things can be done with a chemistry set or at a laboratory or lecture-demonstration. This section is concerned neither with the science of chemistry nor its applications, but with chemical trivia. It is a restful respite before the tough chapters to follow, rather like the light-hearted kyogen between the acts of solemn Japanese Noh drama, the sherbet served between courses of a French dinner, or the half-time happenings at an American football game.

*Elemental
Etymology*

Chemical elements have curious names deriving from many sources (Figure 6.6). The names of many elements recall geographical locations: towns, cities, states, countries, two continents, an island, and a river.

FIGURE 6.6
The 109 elements derive their names in many different ways.

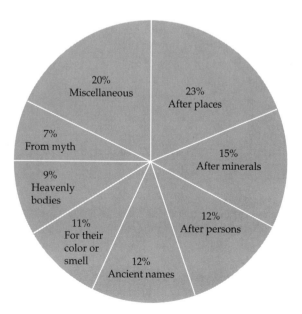

The word *copper* comes from Cyprus, where it was mined in ancient times. Rhenium, honoring the Rhine River, was the last element to be found in nature rather than created in the laboratory. It was discovered by three German chemists in 1925, two of whom subsequently (consequently?) married one another. One of the synthetic elements, number 106, does not as yet have an agreed-upon name.

The Finnish chemist Gadolin found some odd rocks near the Swedish town of Ytterby. In 1793, he extracted a new "earth" from them. About a dozen new elements were eventually isolated from Gadolin's mineral, most of them members of the rare earth family of elements. Four are named after the town: ytterbium, yttrium, terbium, and erbium.

One more element was extracted from Gadolin's rocks. Gadolinium became the first element to be named in honor of a scientist. Gallium, which we mentioned earlier, was the second. Ten more elements are named for scientists, but none of them occur naturally on Earth: They are artificial *transuranic* elements, meaning that they lie beyond Uranium in the periodic table. A complete list of eponymous elements is given in Table 6.3, along with the names and countries of birth of the ten men and two women who have been honored. Eleven different nations are represented—the pursuit of knowledge is a shared endeavor.

Nine elements are named for their colors. Chlorine (from the Greek for green) is a greenish gas. Other names signify the colors of their compounds (e.g., chromium salts display many bright hues) or the colors they impart to a flame (e.g., thallium, from the Greek for bud green). Two elements are named for their smells: bromine from the Greek for stink and osmium from the Greek for odor.

The names of 19 elements suggest other properties. Phosphorus, which burns spontaneously, means light-bearer in Latin. It is the first chemical element with a known discoverer: Hennig Brand, in 1669, isolated phosphorus by distilling a mixture of solid and liquid excrement.

TABLE 6.3   The Eponymous Elements

Element Name	Atomic Number	Honored Scientist
Gallium	31	Lecoq Boisbaudan (France)
Gadolinium	64	Johann Gadolin (Finland)
Curium	96	Marie Curie (Poland)
Einsteinium	99	Albert Einstein (Germany)
Fermium	100	Enrico Fermi (Italy)
Mendelevium	101	Dmitri Mendeleev (Russia)
Nobelium	102	Alfred Nobel (Sweden)
Lawrencium	103	Ernest Lawrence (United States)
Rutherfordium	104	Ernest Rutherford (New Zealand)
Hahnium	105	Otto Hahn (Germany)
Meitnerium	107	Lisa Meitner (Austria)
Nielsbohrium	109	Niels Bohr (Denmark)

The last great alchemist was trying to turn silver into gold, but instead he discovered a pearly-white waxy stuff that glowed in the dark and could light a pipe. Ironically, Brand's home town of Hamburg was demolished by phosphorus bombs during World War II.

Platinum (along with chocolate, tobacco, and maize) is a gift from the New World. Native Americans developed platinum metallurgy in pre-Columbian times. Conquistadors called it platina, or little silver. Six chemically similar metals form a rectangle in the periodic table: ruthenium (for Russia), rhodium (for rose), palladium (for Pallas), osmium, iridium (for colors), and platinum. Curiously, all six members of the "platinum group" are laboriously extracted and separated from the very same ores, only to be carefully recombined with one another to form special alloys for things like dental fillings, pen points, and precision instruments.

The names of two elements reflect the difficulty of their extraction: dysprosium from "hard to get at" and lanthanum from "to lie in hiding." Radium and radon recall the *rad*iation they emit.

Twelve elements, including gold, zinc, and iron, bear ancient names of obscure or unknown etymology. Nine are named for heavenly bodies. (Uranium, cerium, and palladium were found by chemists soon after astronomers discovered Uranus and the asteroids Ceres and Pallas.) An additional seven elements are named for various mythological personages: cobalt for goblins that were said to harass German miners, thorium for the Norse god Thor, titanium for Titans in general, and prometheum for a Titan who was tortured because he gave fire to mortals.

DO YOU KNOW
How Gallium Got Its Name?

The three elements whose existence had been predicted by Mendeleev are scandium, germanium, and gallium. Nilson and Winkler named their new elements after their native lands: scandium for Scandinavia and germanium for Germany. Boisbaudan named his element not for his Gallic homeland, but for himself! Gallium comes from the Latin *gallus* meaning cock, as in its discoverer's name, *Lecoq* Boisbaudan. Gallium is essential for today's electronics industry.

Liquid crystal displays in digital watches and calculators are based on gallium technology. Indeed, gallium devices may supplant "old-fashioned" silicon-based technology. Because of its unique properties, tens of tons of gallium are the active ingredient of two large experiments designed to study neutrinos coming from the Sun, a subject we discuss in Chapter 14. These experiments will reveal the innermost secrets of the Sun and tell us about the tiniest elementary particles. Mendeleev and Boisbaudan would have been amazed and delighted to learn how important to pure and applied science this curious element has become.

Elemental Geography

Consider the lands in which the elements were discovered. Although science is a worldwide collaborative endeavor, there is always an element of national pride. Figure 6.7 shows the nationalities of the discoverers of the 109 chemical elements.* The 13 miscellaneous elements were found by scientists of 9 different nationalities. American elements are mostly synthetic—because of our discovery of the cyclotron and the development of nuclear weapons. Why does Sweden have the largest per capita rate of discovery of chemical elements? Why were most of the elements found by Western Europeans?

FIGURE 6.7
The 109 elements were discovered by scientists of many nationalities.

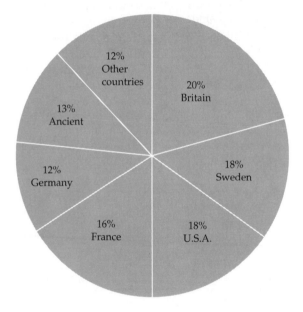

Stable and Unstable Elements

None of the 26 elements beyond bismuth in the periodic table (those with $Z > 83$, whose nuclei contain more than 83 protons) have stable isotopes. All of them are radioactive; that is, sooner or later their atoms decay to form atoms with smaller values of Z. Figure 6.8 shows the *half-lives* of the most stable isotopes of these elements.*

All the elements beyond uranium ($Z > 92$) have half-lives that are much smaller than the age of Earth, which is about 4.5×10^9 y. Although

---

*The concepts of "discoverer" and "nationality" are ill-defined. *A* discovers a compound that he claims contains a new element. *B* isolates the pure element. *C* finds that it is a mixture of two new elements. Who gets what credit? Furthermore, *C* was born in one country, educated in another, finds his element in a third, and settles in a fourth. Which country gets the credit?

*The half-life of a given nuclear species is the interval of time in which about half the nuclei in any sample decay into other nuclear species. We discuss this concept at greater length in Chapter 11.

these elements may have been present when our planet was young, they have long ago transmuted themselves (for that is what radioactive decay is!) into elements further down in the periodic table. Plutonium is the transuranic element with the longest-lived isotope. It may have been as common in the newborn Earth as uranium is today, with an abundance of 3×10^{-6}; that is, 3 parts per million of the mass of Earth may once have consisted of plutonium. Since that time, however, about 55 of its half-lives have passed, and if it was as abundant as we have hypothesized, only $(\frac{1}{2})^{55} \simeq 3 \times 10^{-17}$ of its original atoms will have survived. Its abundance now would be 10^{-22}—far too small to detect. All the plutonium on Earth today is synthetic. It is manufactured in nuclear reactors and put in nuclear bombs or used as nuclear fuel. Uranium is the heaviest of the elements to have survived since Earth's creation.

Seven elements below uranium are also very short-lived: polonium, astatine, radon, francium, radium, actinium, and protoactinium. They are present on Earth, but their atoms have not been here since the beginning: They are continually decaying and being replenished. They are the *daughters*, or by-products, of radioactive decays of uranium and thorium. They are produced after one or more steps in the decay chain of U or Th, and they are eventually transmuted into lead. The alchemists' dream is

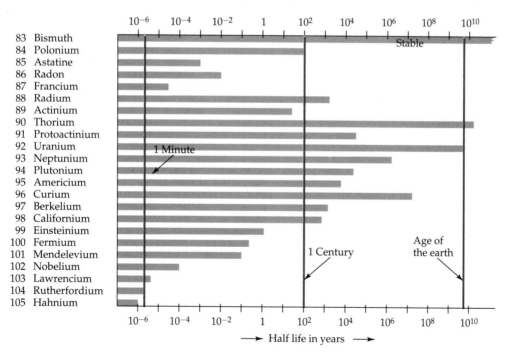

FIGURE 6.8 None of the elements past bismuth in the periodic table have stable isotopes: They are all radioactive. This plot shows the half–lives of the longest–lived isotopes of elements from $Z = 83$ to $Z = 105$. No isotope of elements with $Z > 105$ lives for more than one second. Uranium and thorium, which have isotopes with half-lives comparable to the age of Earth, are the radioactive parents of many short-lived elements that may be found in their ores.

realized in nature, but in reverse: expensive radium becomes worthless lead. Astatine is the rarest of the radioactive daughters of uranium. It was first synthesized in 1940. Much later, and with great difficulty, it was found to exist in uranium ores. Francium is also exceedingly rare: the total amount of it on Earth at any time is less than an ounce!

Two relatively light elements have no long-lived isotopes: technetium ($Z = 43$) and promethium ($Z = 61$). To be studied, they must be synthesized in the laboratory because neither element occurs naturally on Earth.

Elemental Abundances

If we exclude elements that are either absent from Earth or extremely short-lived, we are left with 83 chemical elements with which to build living creatures and their artifacts. Which are the most abundant of these? The answer depends on where you look. Four possibilities are: Earth's crust, its oceans, its atmosphere, and the Sun. Table 6.4 lists the Top Ten elements (showing their abundances by mass) in each of these places. These data are reproduced as pie graphs in Figure 6.9. Not as much information can be presented in this way because the slices corresponding to rare elements are too thin to show. However, the striking differences in chemical composition among Earth, air, water, and Sun are made apparent.

TABLE 6.4 The Top Ten Elements

In Earth's Crust			In Earth's Oceans	
O	46.4%	(1)	O	85.7%
Si	28.2%	(2)	H	10.7%
Al	8.3%	(3)	Cl	1.9%
Fe	5.6%	(4)	Na	1.1%
Ca	4.2%	(5)	Mg	1×10^{-3}
Na	2.4%	(6)	S	9×10^{-4}
Mg	2.3%	(7)	Ca	4×10^{-4}
K	2.1%	(8)	K	4×10^{-4}
Ti	6×10^{-3}	(9)	Br	7×10^{-5}
H	1×10^{-3}	(10)	C	3×10^{-5}

In Earth's Atmosphere			In the Sun	
N	75.5%	(1)	H	75%
O	23.1%	(2)	He	23%
A	1.3%	(3)	O	9×10^{-3}
C	1.3×10^{-4}	(4)	C	4×10^{-3}
Ne	1.3×10^{-5}	(5)	Fe	1.4×10^{-3}
Kr	3×10^{-6}	(6)	Si	1×10^{-3}
He	7×10^{-7}	(7)	N	9×10^{-4}
Xe	4×10^{-7}	(8)	Mg	8×10^{-4}
H	4×10^{-8}	(9)	Ne	6×10^{-4}
S	1×10^{-9}	(10)	S	4×10^{-4}

FIGURE 6.9
a. The composition of Earth's crust. *b.* The composition of Earth's oceans. *c.* The composition of Earth's atmosphere. *d.* The composition of the Sun.

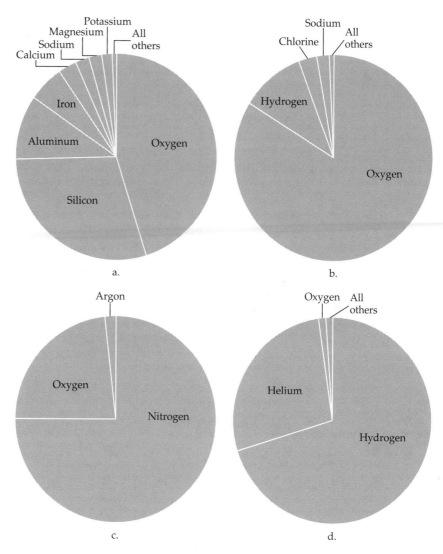

Why are these four bodies so wildly different in composition? The chemical constitution of the Sun is roughly the same as that of the universe as a whole, which emerged from the big bang as ~80 percent hydrogen and ~20 percent helium, plus a smidgin of larger atoms. Stars have been fusing hydrogen and helium nuclei together to form larger atoms ever since, but only very slowly. The inner planets of the solar system, including Earth, are made of what was left after the Sun formed—heavier elements from the ashes of long-dead stars.

Why is the chemical composition of the planets and their atmospheres what it is? This is a challenging puzzle involving astrophysics, geophysics, geochemistry, and biology. At the moment, we have only partial answers. For example, scientists believe that the foul primordial atmosphere of Earth was drastically changed by weird organisms that lived billions of years ago. They ate up the noxious CO_2 and CH_4 and released oxygen for air-breathing latecomers like birds, fish, and bees. We also know that as we pave our forests and burn our fossil fuels, we are systematically undoing this process, as we put CO_2 back into the atmosphere.

Human life requires many elements, but not all of them: just the 21 shown in Table 6.5. The list is a bit surprising: two of the commonest elements on Earth, silicon and aluminum, are not needed by any of its life forms. Iodine is very rare in the crust. In the sea, where life began, its abundance is about 1 percent that of bromine. Why do we need iodine, but not bromine? Selenium is 38th in order of abundance on land and 68th in the sea. Nonetheless, it is an essential ingredient of at least one human enzyme.* Who knows how selenium got into the 21-club?

Ingenious physicists, chemists, and biologists put almost every element to work—like germanium for electronics, niobium for superconductors, and bismuth for bellyaches. Every one of the 109 known chemical elements has its own exotic tale to tell: of its discovery, its unique properties, its role in life or medicine, or its applications to science and technology.

TABLE 6.5   The 21 Elements of Human Life

Hydrogen	Oxygen	Nitrogen
Carbon	Sulfur	Phosphorus
Calcium	Sodium	Potassium
Copper	Chromium	Cobalt
Chlorine	Iodine	Fluorine
Manganese	Magnesium	Molybdenum
Iron	Selenium	Zinc

Exercises

15. List the nine chemical elements named for heavenly bodies.

16. What seven elements have mythic names not associated with heavenly bodies? Sketch one myth and explain its relevance to its eponym.

17. What elements among the Top Ten do Earth's crust, its oceans, its atmosphere, and the Sun have in common?

18. Research your favorite element. How was it discovered and by whom? Where is it found and in what form? What are its chemical and physical properties?

19. The recommended daily dose of selenium is 10^{-7} g. How many selenium atoms is this? (Selenium has atomic weight 79.)

---

*You can buy selenium supplements at a corner drugstore because it is believed by some people to protect against cancer. Beware! Too much selenium gives *alkali disease* to unfortunate cattle in South Dakota and can be fatal to you.

20. A tumbler containing 140 g of holy water is thoroughly mixed with all 1.4×10^{21} kg of the waters of Earth. How many of its molecules are to be found in any tumbler of water afterwards?

21. The early universe consisted of 75 percent hydrogen and 25 percent helium by mass. What fraction of its atoms were helium atoms?

6.4 What's in an Atom?

Chemists and physicists—each in their own way—were forced to accept the existence of unseen atoms. Atoms explain the striking regularities of chemical combination: the laws of definite proportions, multiple proportions, and combining volumes. Atoms of different sorts form elements, and those of any element are identical in shape, form, and structure. The properties of elements depend in a systematic and periodic manner on their atomic numbers. However, individual atoms could not be seen until very recently. Many nineteenth-century scientists regarded atoms as little more than a useful artifice, a convenient theoretical framework having no necessary relevance to the real structure of the microworld.

The power of the atomic hypothesis extends far beyond its ability to describe chemical reactions. Most of the properties of matter are now understood, at least in principle, in terms of atoms and the forces of electricity and magnetism. Consider the three states of matter: solid, liquid, and gas. In gases, molecules are far apart and move freely through the intervening vacuum. A quantity of gas can expand to fill a larger container or be compressed into a smaller volume. In liquids, the molecules are free to move about, but they are unable to escape the mutual attraction of their neighbors. Liquids adjust their shape to fill any container, but the total volume of a given quantity of liquid is not easily changed. Solids are bodies in which the molecules are fixed in position to form a rigid structure. The molecules of a solid often prefer to adopt a symmetrical arrangement, like tiles on a bathroom floor: They form crystals. Although molecules in solids occupy definite positions, they are nonetheless capable of rotations and vibrations whose intensity is a measure of temperature.

The atomic hypothesis brings logical unity to the material discussed in this chapter and in Chapters 4 and 5, which deal with gases and heat. Atoms did much more than explain the rules of chemistry, however. They became the framework on which a true science of the properties of matter was built.

Avogadro's Number and the Size of Atoms

If the atomic hypothesis were to be more than a useful tool and a guess about the nature of matter, someone had to determine the size and mass of a single atom. Atomic weights are only relative weights. To say that the atomic weight of oxygen is 16 means only that the O atom is 16 times heavier than the H atom. But how heavy is an H atom? Chemists knew that X grams of an element with atomic weight X (1 mole of it) contained a certain fixed but unknown number of atoms. They knew that the number was the same for every element, and that it had to be very large because atoms are small. It had a name, Avogadro's number, but what was its value? These questions illustrate the decisive difference between philosophical speculation and exact science, as expressed forcefully by the British physicist Lord Kelvin (1824–1907):

> When you can measure what you are speaking about and express it in numbers, you know something about it, and when you cannot measure it, when you cannot express it in numbers, your knowledge is of a meagre and unsatisfactory kind. It may be the beginning of knowledge, but you have scarcely in your thought advanced to a stage of a science.

Let's put ourselves in the place of these scientists. What do we already know and how can we proceed logically from this information to a determination of atomic size and Avogadro's number? Newton knew that the volume of steam is about 1000 times greater than the volume of the water that is boiled. Thus, 1 cm^3 of water yields roughly 1 L of steam. Because water molecules touch one another, it follows that steam molecules are far apart. The same is true for all gases. Under ordinary conditions, the molecules of any gas occupy only about one-thousandth part of the volume of the gas. Consequently, the average distance between neighboring gas molecules is about ten times the diameter of one molecule. (If air were greatly magnified, it would resemble a bunch of randomly moving Ping-Pong balls that are, on the average, about 1 ft apart.) This is a helpful bit of data, but we need to know something more in order to pin down the size of the molecule.

We find another clue to atomic sizes in the process of diffusion. Imagine an open bottle of perfume in a tranquil room. The velocity of the perfume molecules is hundreds of meters per second, yet it takes hours before the odor permeates the room. Why is this? Perfume molecules do not proceed directly to the nose of the victim. They make many collisions with intervening air molecules, thus performing a circuitous random walk about the room. The rate of diffusion depends both on the molecular velocity and the *mean free path*, the average distance a perfume molecule moves between successive collisions. From observed diffusion rates, scientists of the nineteenth century determined the mean free path of air molecules to be approximately 10^{-7} m, which is small by human standards but much larger than the size of a single atom.

As a result of this accumulating knowledge, we are able to make very rough estimates of both the size of a molecule and the value of Avogadro's constant. (We do the same in Examples 6.2 and 6.3.) Today, these quantities are much better known. Avogadro's number, N_A, has been determined to a precision better than 1 part per million:

$$N_A = 6.022\,136 \pm 0.000\,004$$

EXAMPLE 6.2
How Big Is
a Molecule?

Suppose that the diameter of an air molecule is $\sim R$. Thus, its area is $\sim R^2$ and its volume is $\sim R^3$. From the preceding discussion, we know that in air there is, on the average, one molecule in each volume of $\sim 1000R^3$. We also know that the mean free path d of an air molecule is $\sim 10^{-7}$ m. The volume swept out by a molecule in traversing one mean free path is its area times d or $R^2 d$, as shown in Figure 6.10. Thus, it must be true that $R^2 d = 1000R^3$, from which it follows that $d = 1000R$ or $R = 10^{-10}$ m. This crude calculation gives about the right answer for the size of an air molecule.   ∎

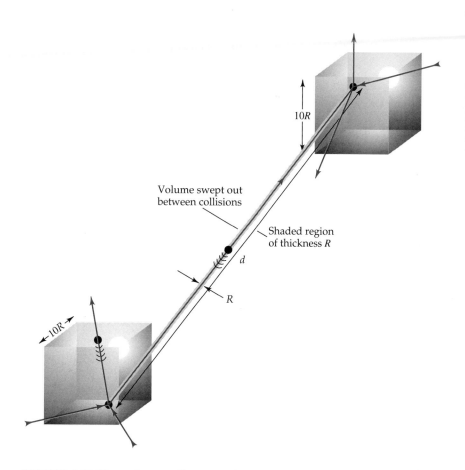

FIGURE 6.10 The unknown diameter of an air molecule is R, its area is about R^2, and its volume is about R^3. The mean path length between collisions is d. There is, on the average, one molecule per box of side $10R$ and volume $1000R^3$. The volume swept out by a molecule between collisions is typically equal to $R^2 d$. Setting this quantity equal to $1000R^3$, we find $R = d/1000$ as a rough estimate for the diameter of an air molecule.

EXAMPLE 6.3
How Can
We Estimate
Avogadro's Number?

Avogadro's number, N_A, is the number of molecules in 1 mole of any substance. The molecular weight of water H_2O is $2 + 16 = 18$, so that the mass M of 1 mole of water is 18 g. The density of water is 1 g/cm^3. Therefore, the volume V of 1 mole of water is 18 cm^3, or $V = 1.8 \times 10^{-5}$ m^3. The diameter of a water molecule can be estimated from arguments like those in Example 6.2 to be $R \simeq 3 \times 10^{-10}$ m. The volume occupied by one water molecule is $\sim R^3$, so the volume of a mole of water is $V \simeq N_A R^3$. Setting the two estimates for V equal to one another, we find $2.7 \times 10^{-29} N_A \simeq 1.8 \times 10^{-5}$. Solving this equation, we find $N_A \simeq V/R^3 \simeq 7 \times 10^{23}$. Keep in mind, however, that although our result is approximately correct, it is an off-the-cuff "guesstimate." Water molecules are not like bricks that are neatly stacked together: They are irregularly shaped objects with spaces left between them. ∎

Brownian Motion

At the dawn of the twentieth century, atomic theory began to satisfy Kelvin's injunction. It had "advanced to the stage of a science." Faith in atoms, along with lots of hard work, allowed scientists to find answers to many perplexing questions concerning the behavior of gases, the formation of crystals, and the rules of chemistry. They could even tell why the sky is blue and sunsets are red. Still, there were doubters. The successes of physics and chemistry, however impressive they were, did not prove that atoms truly existed. Could individual molecules, or at least their effects, be detected in a more direct and satisfying fashion?

The pressure that a fluid exerts on a piston or any other surface seems to be a continuous elastic force. In reality, it is due to the impacts of many individual molecules. Imagine something so tiny that the discrete nature of pressure becomes apparent, say a tiny pollen grain. In 1827, the English botanist John Brown watched pollen grains floating on water through his microscope. He was amazed to see them engage in complex motions, randomly moving hither and thither while erratically rotating (Figure 6.11). Was this a sign of the latent life force within pollen? Brown answered his own question by studying other suspended materials, including the dust of igneous rocks, which were created in the bowels of the earth and are as inorganic as could be. Brown saw that any finely divided material suspended on water experiences mysterious jerks and twists. This peculiar motion of particles on a fluid surface became known as *Brownian motion*.

Brownian motion had to have a rational explanation in terms of the impacts of individual water molecules on dancing dust motes. But it was not until 1905 that Albert Einstein succeeded in giving a mathematical analysis of the effect. Einstein's calculations agreed with quantitative observations. In as direct a fashion as one might hope, the effects of discrete atomic collisions had been observed and measured. The cynics were silenced. Everyone agreed that atoms are real things and not mere mathematical abstractions.

FIGURE 6.11
Brownian motion. The positions of three pollen grains floating on water, as viewed through a microscope at 30-second intervals.

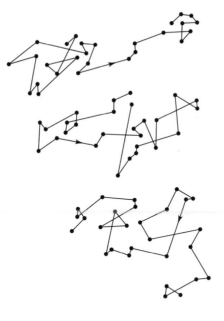

How Do Atoms Differ from Each Other?

The evidence for atoms was now overwhelming. Yet questions remained. Why are there so many different varieties of atoms? How are they related to one another?

One of the first scientists to ask these questions was William Prout (1785–1850), a physician and amateur chemist, who was also known for his discovery that gastric juices contain hydrochloric acid. In 1815, Prout published a paper pointing out that the densities of several gases, and thus their atomic weights, seemed to be exact integer multiples of the density of hydrogen. In a sequel, Prout proposed that every atom is formed from a number of hydrogen atoms that are somehow bound together. Consequently, all atomic weights had to be integers. Prout's hypothesis remained an active subject of inquiry throughout the century. It was soon found that the hypothesis is not exactly correct—atomic weights are not integers. However, most atomic weights are very nearly whole numbers—far too many for the phenomenon to be simply the result of chance.

The element of truth in Prout's conjecture could not be understood until the structure of the atom was discovered. Let us depart from this historical narrative and look ahead. In the twentieth century, scientists learned that atoms are composite structures consisting of a tiny *atomic nucleus* surrounded by *electrons*. Most of an atom's mass resides in its nucleus, which in turn is made up of particles called *protons* and *neutrons*.

The chemical properties of an element are determined by the number of electrons in an atom or, equivalently, the number of protons in the nucleus of the atom. This integer, usually denoted by Z, is called the *atomic number* of an element and tells its place in the periodic table. The atomic weight of an atom is determined by the total number of protons and neutrons found in its nucleus. However, this integer, denoted by A, is only approximately equal to the actual atomic weight of the atom.

The nucleus hydrogen is called a proton: It has $A = 1$ and $Z = 1$. All other atomic nuclei contain both protons and neutrons. Neutrons and protons have about the same mass and, as a result, the mass of every atom is roughly an integer multiple of the mass of the hydrogen atom. Thus, Prout's hypothesis is approximately true for all atoms. For example, nitrogen's nucleus contains 7 protons and 7 neutrons. Its atomic weight is about 14. Similarly, the carbon atom contains 6 protons and 6 neutrons and has an atomic weight of about 12.

However, the atomic weight of chlorine is measured to be 37.5. It lies almost halfway between two whole numbers. How can this be? The answer is that the atoms of chlorine are not all the same. The atoms of almost all chemical elements occur in several varieties with different masses; these are known as *isotopes*. The nuclei of different isotopes of the same element contain the same number of protons but different numbers of neutrons. The atomic weight of any pure isotope is approximately an integer.

Naturally occurring (or *native*) chlorine is a mixture of two isotopes, both with $Z = 17$. About 75 percent of its atoms have $A = 35$ and atomic weight 34.97. The remaining 25 percent have $A = 37$ and atomic weight 36.96. (Notice that both isotopic masses are nearly integers.) Isotopes of a given element are indicated with a superscript to the left of the chemical symbol giving the value of A. Thus, the two isotopes of chlorine are ^{35}Cl and ^{37}Cl. The mean atomic weight of chlorine is $0.75 \times 34.97 + 0.25 \times 36.96$, which comes out to be approximately 37.5.

Most native elements are mixtures of isotopes. Hydrogen is mostly ^{1}H, but it contains 1.5×10^{-4} of *heavy hydrogen* or ^{2}H. Oxygen is mostly ^{16}O, but it contains tiny admixtures of isotopes with $A = 17$ and $A = 18$. A rare isotope of uranium, ^{235}U, is laboriously extracted from native uranium (which is mostly ^{238}U) for use in nuclear reactors. Tin, with $Z = 50$, occurs in nature as a mixture of ten different isotopes, with A ranging from 112 to 124.

Today's internationally accepted convention for atomic weights was established in 1961. All atomic weights are now measured in what are called *atomic mass units*, or amu. The amu is defined to be 1/12 the mass of a single atom of a specific isotope of carbon, namely ^{12}C. The value of the atomic mass unit is known to an accuracy better than one part per million:

$$1 \text{ amu} = 1.660540 \times 10^{-27} \text{ kg}$$

The atomic weight of ^{12}C, in this convention, is exactly 12. However, the atomic weight of ^{1}H is not 1 but 1.007825. In terms of amu, the atomic weight of any pure isotope is approximately equal to the integer A.

EXAMPLE 6.4
Isotopic Abundance

Native carbon contains a small admixture of ^{13}C, whose atomic weight is 13.003. The mean atomic weight of native carbon is 12.011. What fraction of carbon atoms is ^{13}C?

Solution

Let the unknown fraction be p. Thus, 1 mole of native carbon contains pN_A ^{13}C atoms and $(1-p)N_A$ ^{12}C atoms. Its total mass is $\{13.003p + 12.000(1-p)\}N_A$ amu. The average mass of an atom of native carbon is this result divided by N_A, or $12 + 1.003p$ amu. When this expression is set equal to 12.011 amu, we find that $p = 0.011$. That is, 1.1 percent of native carbon atoms consist of the rare isotope ^{13}C. ∎

Elementary Particles?

The age-old quest for the ultimate constituents of matter converged on atoms. Each element was believed to be made up of identical atoms that were indivisible, eternal, and immutable. All these beliefs turned out to be wrong. The atoms of an element are not all the same: Most elements have several different isotopes with different masses. Atoms are not indivisible because they have parts: electrons (discovered in 1897) and nuclei (discovered in 1912). Many atoms are not eternal: They spontaneously explode or decay in the process of radioactivity, thereby transmuting themselves into atoms of other elements. No atom is immutable: Its

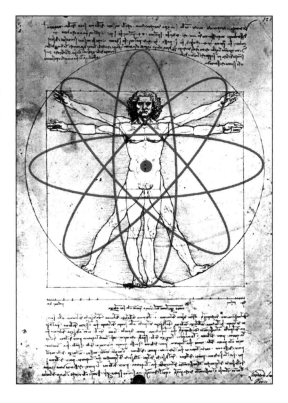

An atomic logo is superposed on Leonardo da Vinci's drawing of ideal human proportions. In reality, a person is ten powers of ten larger than an atom. The atomic nucleus is tinier yet. It is to its atom as a single cell is to the human body. *Source:* The Granger Collection.

electrons may be stripped away and its nucleus may be torn asunder by the process of nuclear fission or assembled from scratch by the process of nuclear fusion.

In the 1930s, scientists learned that protons and neutrons somehow stick together to form atomic nuclei. In the 1970s, they realized that not even protons and neutrons are elementary. They are made of *quarks*, which are curious entities that do not exist except as parts of other particles. Who knows whether quarks and electrons are elementary? Before asking such questions, we had better specify what properties a truly elementary particle should have. Here is my personal list of qualifications:

1. *Number* The underlying rules of nature should be simple, and there shouldn't be too many different kinds of elementary particles. Atoms do not satisfy this criterion. Mendeleev knew of some 60 different chemical elements, and today we know of 109. There are too many atomic species for them to be elementary. Today's particle physicists have a list of 17 particles that seem to be truly elementary, but once again, there are too many for comfort.

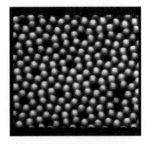

Gert Binnig and Heinrich Rohrer, working at the IBM Zurich Research Laboratory, shared the 1986 Nobel Prize in physics for their design of the scanning-tunneling microscope. The individual atoms on the surface of a silicon crystal are clearly visible in this photograph. *Source:* Dr. Phaedon Avouris, IBM Corporation.

2. *Size* Anything with a measurable size is likely to be made up of smaller things. Truly elementary particles should be pointlike. Atoms do not satisfy this criterion. Every atom, give or take a factor of 2, has a radius of about 10^{-10} m. Atoms are big enough to be seen individually with modern scanning-tunneling microscopes, as shown in the photo. Electrons and quarks, however, do seem to be pointlike particles.

3. *Pattern* The existence of any systematic pattern among a collection of elementary particle candidates suggests that they are not really elementary and that there is another layer of structure remaining to be discovered. Mendeleev believed that the success of his periodic table demanded that there be a more primal form of matter than the atom. Indeed, there are several more layers of the onion. Today's elementary particle candidates are called quarks and leptons. They also fill out a new kind of periodic table. Does this mean they are composite systems, too? We do not know.

4. *Vibrations* How can you tell if a gift box is empty or not? Wiggle it, listen to see whether the contents jiggle. If they do, the sound may tell you what's inside. (Cigars may be easily told from perfume.) An elementary particle mustn't jiggle when wiggled, but an atom does! When struck, an atom's parts vibrate and light of characteristic frequency emerges. Nuclei, when struck by an energetic particle, also produce characteristic radiation. They are composites of neutrons and protons. But even these tiny particles produce characteristic vibrations when they are struck. Protons and neutrons are made of quarks.

As the twentieth century began, the atom scored zero for four on my scoreboard. Its structure was about to be revealed. According to *The Encyclopædia Britannica*, 11th edition, published in 1910:

Modern discoveries in radioactivity are in favor of the existence of the atom, although they lead to the belief that the atom is not so eternal and unchangeable as Dalton and his predecessors imagined.

Clues suggesting the existence of atomic substructure emerged from the studies of electricity, magnetism, and light. For that reason, the next two chapters are devoted to those disciplines. We will learn:

- That a quantity called electric charge may neither be created nor destroyed
- That electric charge is carried by tiny particles such as protons and electrons
- That the motions of electrons within atoms produce light
- Which is a phenomenon intimately linked to electricity
- Which is the force holding atoms together

These notions came together, and they clashed! Classical physics does not apply to the physics of the microworld. It was replaced by *quantum mechanics* (and later by *relativistic* quantum mechanics). By the early twentieth century, the atom was known to consist of a tiny massive nucleus surrounded by a cloud of electrons described by a quantum mechanical *wave function*. Atomic nuclei were found to consist of protons and neutrons. However, neither the parts of atoms nor the parts of its parts satisfy my specifications, which remain, even now, the proper guidelines for the continuing search for the ultimate constituents of matter.

Exercises

22. A tiny drop of oil falls on a still pond. It spreads out, forming a thin film of area 1 m^2. The volume of the drop is 2 mm^3 (2×10^{-9} m^3), and the thickness of the film is 3 molecular diameters. From this data, calculate the diameter of an oil molecule.

23. A certain protein molecule has a molecular weight of 2 million. What is the mass of a mole of it? What is the mass of one molecule of it? (Express your answers in kilograms.)

24. The conversion factor from amu to grams is the reciprocal of Avogadro's constant, N_A. Explain why this is.

25. How many protons and neutrons are found in the nuclei of ^{238}U?

26. Native gallium consists of 60.4 percent ^{69}Ga with atomic weight 68.93, and 39.6 percent ^{71}Ga with atomic weight 70.93. What is the mean atomic weight of native gallium?

27. About 1.5×10^{-4} of native hydrogen atoms are ^{2}H. What fraction of the mass of native hydrogen does ^{2}H form?

28. Rhenium is a mixture of two isotopes with atomic masses 185.0 and 187.0. Its mean atomic weight is 186.2. What fraction of native rhenium atoms is ^{185}Re?

29. Native magnesium is a mixture of three isotopes. It consists of 79 percent ^{24}Mg with atomic weight 23.985, 10 percent ^{25}Mg with atomic weight 24.986, and 11 percent ^{26}Mg with atomic weight 25.983.
    a. What is the mean atomic weight of native magnesium?
    b. What is the ratio of the weight of a ^{24}Mg atom to that of a ^{1}H atom? Notice that it is not close to a whole number.

Where We Are and Where We Are Going

In the late nineteenth century, the sizes and masses of atoms had been measured, and their properties were seen to vary systematically with atomic weight. There was little doubt that atoms were real objects and not mere mathematical conveniences. Their identities were preserved by all known physical and chemical processes and they could not be taken apart. For good reason, then, most scientists believed that atoms were the ultimate building blocks of all matter.

Why are there so many different kinds of atoms and what can explain their curious properties? The first solid clues that the atom is a composite system made of simpler parts emerged from the study of electrical and magnetic phenomena, to which we turn in the next chapter. At first, electricity and magnetism seemed to have little to do with one another and even less with the nature of the atom. Only very gradually did it become apparent that electromagnetism is the key to atomic structure and the root cause of almost all of the everyday properties of earthly matter.

7 Tell Me What Electricity Is, and I'll Tell You Everything Else*

I hope that the insight which you have here gained into some of the laws by which the universe is governed may be the occasion of some among you turning your attention to these subjects; for what study is there more fitted to the mind of man than that of the physical sciences? And what is there more capable of giving him an insight into the action of those laws, a knowledge of which gives interest to the most trifling phenomenon of nature, and makes the observing student find

Tongues in trees, books in the running brooks,

Sermons in stones, and good in every thing?

Michael Faraday (1791–1867), British physicist

*The title of this chapter is a challenge to physicists offered in 1862 by the British physicist William Thomson (1824–1907), who is also known as Lord Kelvin.

Lightning is nature's way of demonstrating the power of electricity. *Source:* © Dr. Vic Bradbury/Science Photo Library/Photo Researchers.

The wonders of electricity are displayed in countless devices that make modern life what it is. Electricity, however, is more than a useful technology. It provides the force that holds atoms and molecules together and gives form and substance to matter. Electric forces underlie almost everything we sense or do. Electricity is inextricably linked to its kissing cousin magnetism—not only historically, but at a far deeper level. Today we know that electricity and magnetism are different manifestations of the same underlying force.

Section 7.1 describes early observations of electrical and magnetic phenomena that led to the concepts of magnetic poles, electric charge, and electric current. Magnets and electrified bodies exert forces on one another even though they are physically separated. Electric and magnetic forces were found to satisfy laws analogous to Newton's law of gravitation. Section 7.2 examines the behavior of electric charges at rest. The notions of electric field and electric potential (voltage) are introduced. Section 7.3 examines the behavior of electric charges in motion, or electric currents. The discovery of the electric battery made steady currents available to pure and applied science. Electric currents passing through thin wires produces heat (as in a toaster) or light (as in a lamp bulb). Section 7.4 examines the passage of electric currents through chemical solutions. The study of this process not only explained how electric batteries work, but gave the first hints of the electrical substructure of the atom and the existence of a smallest unit of electric charge.

7.1 Magnetism and Electricity

Our language reveals that the story of magnetism and electricity began in ancient Greece. *Electrum* is Greek for amber, a precious and decorative fossil resin. In English and Greek, it came to mean the color amber or a name for a woman with that color hair, as in *Forever Amber* and *Mourning Becomes Electra*. The Greeks observed that a piece of amber, when rubbed with fur or flannel, attracted bits of wood or fabric. William Gilbert (1540–1603), the English scientist who first studied electric and magnetic phenomena systematically, adapted the word *electric* from electrum in 1600 to describe the mysterious property displayed by amber. The word *electricity* arose in 1646.

Magnesia, a part of ancient Greece now in Turkey, was a rich source of minerals. Lumps of iron ore from Magnesia were found to attract bits of iron. We call these curious bodies lodestones, but the Greek playwright Euripides coined the word *magnets* for them. (The province of Magnesia also lent its name to two chemical elements: magnesium and, through a process of linguistic inversion, manganese.)

Magnetism and Magnetic Poles

Many Greeks were fascinated by the mysterious power of magnets and strove to understand their properties. Here are some of the things they found:

- Magnetism is associated with iron and things that contain iron. Other substances are unaffected by magnets. (This is not quite true; most materials display magnetic properties, but they are more intense and easier to detect for iron and certain other elements.)
- Magnetism may be communicated from a lodestone to pieces of unmagnetized iron. Greek scientists found that an iron object will hang suspended from a magnet, and from that object another, and so on.
- The magnetic attraction can pass through nonmagnetic materials. A magnet attracts a piece of iron even though a piece of parchment or wood is interposed.

The ancient Greeks were not scientists in the modern sense because they did not always test their hypotheses about nature. Aside from their correct findings, they claimed that magnets cause melancholy and cure disease, that a magnet could be used to test the chastity of a woman, and that the power of a magnet is destroyed if it is rubbed with garlic.

The next development in magnetism was more technological than scientific. Possibly as early as the fourth century A.D., the Chinese found that a magnet, when suspended by a string, orients itself in a certain direction. They realized that a magnetized needle afloat in water can act as a navigational instrument. The magnetic compass slowly wended its way West—helped by Marco Polo and Arab traders—to guide Columbus to America a millennium later.

Gilbert has been called "the Galileo of magnetism." He was the personal physician to Queen Elizabeth I and one of Britain's earliest converts to the Copernican cosmology. Aside from his own investigations of magnetism and electricity, Gilbert examined the accumulated knowledge of the past and separated the wheat from the chaff. The following are among his correct conclusions about magnetism:

- Every magnet has two *poles*, a north pole and a south pole. If a magnet is suspended so that it is free to rotate, its north pole points approximately toward the north geographical pole of Earth, and its south pole points toward the south.
- The force that magnets exert on one another may be attractive or repulsive. The north pole of one magnet is attracted to the south pole of another. However, two north poles push one another apart, as do two south poles. Thus, similar magnetic poles repel, while opposite magnetic poles attract.
- The behavior of a magnetic compass results from the fact that Earth either is a huge magnet or that it contains one, as seen in Gilbert's illustration. The force of Earth's magnetism on a compass needle causes it to point north.

Gilbert published this figure in 1600. *A* and *B* are Earth's poles, whose magnetic field is imagined to arise from a great lodestone within. *Source:* Chapin Library, Courtesy Jay M. Pasachoff.

Magnetism and electricity showed differences as well as similarities. Magnets were known to maintain their magnetic properties indefinitely, while electricity had to be aroused by friction and its effects seemed to be transient. Although magnetism is limited to iron-bearing materials, Gilbert discovered that electricity is less discriminating. He developed a delicate instrument to detect electric forces and added 23 materials to the list of substances that were electrified when rubbed with another material. Among these "electrics," as he called them, were sulfur, glass, and sealing wax. Gilbert's treatise on magnetism, published in 1600, so impressed Galileo that he described its author as "great to a degree that is enviable." Let's savor a sample of Gilbert's flowery Elizabethan prose:

> A few words must be said to point out the vast difference between electric and magnetic actions; for men still continue in ignorance, and deem that inclination of bodies to amber to be an attraction, and comparable to the magnetic coition. In all bodies everywhere are two causes: matter and form. Electrical movements come from matter, but magnetic from the prime form; and these two differ widely from one another—the one ennobled by many virtues, and prepotent; the other lowly, of less potency, and confined in certain prisons, as it were; whereby its force has to be awakened by friction. A loadstone attracts only magnetic bodies; electrical attract everything. A loadstone attracts great weights. Electrics attract only light weights. Now,

a loadstone does repel another loadstone; for the pole of one is repelled by the pole of another that does not agree naturally with it, but all electrics attract objects of every kind; they never repel. [Yes they do!] The matter of the earth's globe is brought together by itself electrically. The earth's globe is directed and revolves magnetically.

How eerily close on the mark is that last bit! Material bodies, such as bricks and boulders, are held together by electric forces, although gravitational forces are more important for objects the size of Earth. Gilbert knew that Earth's axis of rotation is aligned with its magnetic axis. However, rotation is the cause and magnetism the effect, rather than the other way about. Electricity and magnetism seemed to have much to do with one another, as indeed they do. In both cases, physically separated objects could exert forces on one another. How and why they do this would remain a mystery until the nineteenth century.

Hardly anything more was discovered about magnetism for another two centuries. Magnetism posed an intractable puzzle to serious scientists, but it was a source of endless speculation for amateurs and charlatans. (See the feature about good and bad science). During the seventeenth and eighteenth centuries, ingenious devices were devised to generate and store static electricity. Scientific dilettantes amazed and bemused their audiences with huge sparks and painful shocks:

> The Carthusian monks at the convent in Paris were formed into a line 900 feet long, by means of iron wires between every two persons, and the whole company gave a sudden spring at the same instant.

My shop teacher at Public School 52 in Manhattan pulled the same sadistic stunt on my 7th grade class!

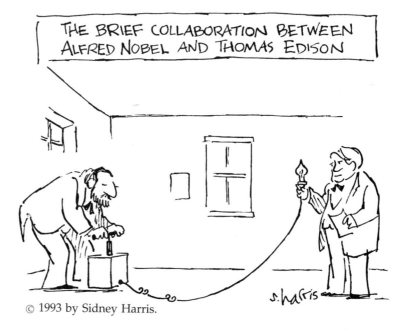

THE BRIEF COLLABORATION BETWEEN ALFRED NOBEL AND THOMAS EDISON

© 1993 by Sidney Harris.

DO YOU KNOW

*The Difference between Good
Science and Bad Science?**

Literate Europeans were bewildered by the revelations of the late eighteenth century. Newly found gases, vital or fixed, noxious or inflammable, swirled about us and even allowed man to fly. "The most subtle spirit," that Newton had described, "which pervades and lies hid in all gross bodies," seemed to be not one but a multitude. Besides electrical fluids, there were the four humors of conventional medicine (blood, choler, phlegm, and bile), various ethers to transmit the Sun's light and gravity, a phlogiston that Lavoisier denied and a caloric he advocated. And there was magnetism.

"Never have so many systems, so many new theories of the universe, appeared as during the last few years," remarked the *Journal de Physique* in 1781, adding that these systems and theories were mutually inconsistent. How was one to tell fact from fancy? Priestley, speaking of electricity, offered a warning:

> Here the imagination may have full play, in conceiving of the manner in which an invisible agent produces an almost infinite variety of visible effects. As the agent is invisible, every philosopher is at liberty to make it whatever he pleases.

Lavoisier issued another: "It is with things that one can neither see nor feel that it is important to guard against flights of imagination." But it was too late! The threat materialized in the person of Franz Anton Mesmer, who came to Paris in 1778 with his "theory" of animal magnetism and his "discovery" of another invisible fluid. Sickness, said Mesmer, resulted from an obstacle to the flow of this fluid through the body, which was like a magnet. Parisian women with illnesses real or imagined paid huge sums to visit Mesmer's clinics and sit in vats with hands linked together to form a "magnetic circuit," to be poked on their poles by mesmerized iron wands. Miraculous cures and notorious scandals abounded. Mesmer prospered, but like the sorcerer's apprentice, he had tapped forces beyond his control. Mesmer's animal magnetism was, Darnton writes:

> debated in the academies, salons and cafés. It was investigated by the police, patronized by the queen, ridiculed on the stage, burlesqued in popular songs, doggerels and cartoons, practiced in a network of secret societies, and publicized by a flood of pamphlets and books.

What began as a hoax became a challenge to the establishment and a medium of communication among radicals. In 1784, a concerned French government enlisted Lavoisier and Franklin to investigate the threatening new fad. The renowned scientists concluded that Mesmer's fluid did not exist. Its effects were due to the overheated imaginations of the mesmerists.

In 1785, Thomas Jefferson, representing the United States in France, remarked that mesmerism was "an imputation of so grave a nature as would bear an action at law in America. Mesmerism is dead, ridiculed." More alive than Jefferson realized, mesmerism became a force to impel the French Revolution and its bloody aftermath. "Mesmerism had taken such a grip on France that its place in history cannot be limited to the 1780's; it continued to mold popular attitudes and interests well into the 19th century." But it would never again be confused with science.

*This material deals with pseudoscience, and is abstracted from *Mesmerism and the End of the Enlightenment in France* by Robert Darnton (Cambridge, MA: Harvard University Press).

Electricity and
Electric Charge

A few years after Gilbert's death, Nicolo Cabeo, an Italian Jesuit and outspoken opponent of the Copernican heresy, discovered another property shared by electricity and magnetism. He saw that an electrified rod at first attracted bits of paper to it, but flung them away as soon as they touched the rod. Thus, the electric force between two electrified bodies—like the magnetic force between magnets—could be either an attraction or a repulsion.

Magnetism could be communicated from a magnet to a piece of unmagnetized iron by contact, but it seemed that the only way an object could be electrified was by friction. The English physicist Stephen Gray (1666–1736) found otherwise—thereby revealing yet one more similarity between electricity and magnetism. When he rubbed a glass tube with wood, the tube became electrified and could transfer its state of electrification to other bodies that it touched; for example, a cork in the end of the tube became electrified even though it had not been rubbed. Subsequently, Gray managed to transport electricity from a rubbed object, along a wire for a distance of hundreds of feet to another object, which became electrified without being rubbed.

Where Gilbert spoke of an "electric effluvium," Gray referred to the property that could attract or repel other bodies and pass through a wire as the "electrick virtue." Much later, in 1767, Priestley spoke of an electrified body as one possessing a charge of electricity or an *electric charge*. Gray's experiments showed that electric charge can pass through some materials, but not others. Substances such as copper and other metals, through which electric charge may freely pass, are today called *electric conductors*. Other materials, such as amber and glass, through which electric charge cannot pass, are called *insulators*.

Gilbert determined whether objects were electrifiable by holding them in his hand and rubbing them. This procedure worked for insulators. It failed for conductors because their charge escaped through the good doctor's electrically conducting body. By suspending electric conductors with insulating silk strings, Gray showed that they could, in fact, be given an electric charge. He applied his procedure to various metallic objects, and even to a chicken and an unfortunate young orphan, who was more startled than shocked when his body was made to attract bits of feathers and lint (and the attention of the assembled gray-bearded scientists).

Gray's work attracted the attention of the French scientist Charles Dufay (1698–1739), who proved that all bodies, if they are insulated, may be electrified (that is, charged). Furthermore, he showed that there are two kinds of electric charge. When two glass rods are rubbed with silk, they repel one another. When two amber rods are rubbed with fur, they also repel one another. However, an electrified amber rod attracts an electrified glass rod. Dufay discovered a correct and important principle of electrical science:

> There are two distinct kinds of electricities, very different from one another: one of these I call vitreous electricity; the other resinous electricity. The first

is that of rubbed glass, rock crystal, precious stones, hairs of animals, wood, and many other bodies. The second is that of rubbed amber, copal, gum lac, silk, thread, paper, [and so on]. The characteristics of these two electricities is that a body of, say, the vitreous electricity repels all such as are of the same electricity; and on the contrary, attracts all those of the resinous electricity.

Dufay regarded the state of electrification of an object as an intrinsic property of matter elicited by friction. By the middle of the eighteenth century, Dufay's observations were reinterpreted in terms of two material fluids: the "resinous fluid" and the "vitreous fluid." Ordinarily, these fluids were in balance, but when amber was rubbed with fur, the amber acquired an excess of resinous fluid, while the fur acquired an excess of the vitreous fluid.

Benjamin Franklin (1706–1790) had not heard about the "two–fluid theory" of electricity. He found another explanation of Dufay's results that is both simpler and more correct. Franklin believed that the so-called electric fluid consisted of electric particles that repelled one another but were attracted to ordinary matter. The fluid was supposed to be present in all material bodies. When one body is rubbed against another, some electric fluid is displaced. Franklin designated the effect produced by rubbing glass as "positive electricity" and that produced by rubbing amber as "negative electricity." When a glass rod is rubbed with a silk handkerchief, the rod acquires an excess of fluid (or positive charge) and the handkerchief is left with a deficit of fluid (or negative charge). Similarly, when amber is rubbed with fur, the amber becomes positively charged and the fur (which has lost some electric fluid) becomes negatively charged.* One mysterious fluid, by its excess or absence, did the work of two. Friction does not create electric charge, but only moves it from one body to another.

The flow of electric charge from one place to another is called an *electric current*. Electric currents take many forms: a flow of atomic electrons through a wire or other conducting medium, a beam of electrons passing through the vacuum in a TV tube, the migration of positive and negative ions toward the electrodes within a battery, a discharge of lightning from the heavens to Earth, and so on.

Frankin was convinced that lightning is big-time electricity and thunder the snap of an enormous spark. The illustration shows Franklin and his son flying a silk kite during a violent thunderstorm in the summer of 1752. "Electrical fire" from the clouds streaked down the wet silk string to a dangling key that sparked at Franklin's knuckles. By such daring and dangerous deeds, Franklin collected electric charge from the sky and showed that both positive (vitreous) and negative (resinous) charges are

*Franklin's choice of which kind of charge to call positive was and is inconvenient. The electron's charge is negative, so an electric current going one way is a flow of electrons in the opposite direction. In another way, he was lucky: G. W. Richmann tried the kite trick a year later in St. Petersburg, Russia, but he was struck dead by a lightning bolt. "It is not given to every electrician to die in so glorious a manner," wrote Priestley in an obituary.

Benjamin Franklin and his son are shown collecting some of the electric charge from a thunderbolt. *Source:* The Bettmann Archive.

found in storm clouds. The charge that Franklin collected from storm clouds passed to the kite, then through the wet string as an electric current, and to whatever object Franklin held in his hands, which thereby became electrically charged.

Having a practical bent, Franklin invented the lightning rod to protect ships, buildings, and their inhabitants from the destructive effects of lightning. Lightning is an electric discharge by which a large and potentially destructive electric charge may travel from a thundercloud to the ground. A lightning rod consists of a metallic device extending from the ground to the highest point of a building. It provides an alternative and innocuous route for the electric discharge to follow. Today, the role of the lightning rod is played by the internal plumbing system of a building.

Franklin's theory of electricity was close to today's view. Matter is made of atoms, which consist of heavy, positively charged nuclei and light, mobile electrons. When two different substances are rubbed together, friction displaces some of the electrons from one body (which is left with a net positive charge) to the other (which acquires a negative charge). It follows from this view that electric charge—like matter itself—can neither be created nor destroyed, but merely moved about.

Electric currents transport charge from one part of an isolated system to another, but the total quantity of charge contained in the system cannot change or be changed. This law, called the *conservation of electric charge,* is believed to be an absolute, inviolate and sacrosanct law of nature.

Each constituent particle of a body carries a well-defined electric charge, and the charge of the body is the sum of the charges of all its parts. Ordinary matter contains as many positive as negative charges and is *electrically neutral*. Macroscopic bodies become electrically charged when their atoms lose or gain some of their constituent electrons. Because electrons carry negative charge, a flow of electrons from A to B is an electric current passing from B to A.

The Quantitative Study of Electricity

In the eighteenth century, scientists discovered an astonishing indication of the unity of physical science. The electric force between two charges depends on distance in the same way as the force of gravity depends on distance. Both forces satisfy an inverse-square force law. The history of this revelation involves three scientists we have met—Franklin, Priestley, and Cavendish—and one more, the French physicist Charles Augustin Coulomb (1726–1806), for whom the electric force law is named.

Franklin was elected a Fellow of the (British) Royal Society in 1756 for his studies of electricity. By then, however, his interests had turned from science to politics. He spent most of the next 15 years in England representing colonial Pennsylvania, although he remained an active participant in the affairs of the Society. Priestley was elected a decade later. At one of its meetings, Franklin told Priestley of a "curious observation" he had made years before. He noticed that an electrically charged metal cup strongly attracts a small cork ball on a silk string when the ball is brought near the exterior of the cup but not if the ball is brought near its inside wall. By 1767, Priestley not only confirmed Franklin's result but pointed out its deep meaning:

> May we not infer that the attraction of electricity is subject to the same laws with that of gravitation? It is easily demonstrated that were the earth in the form of a shell, a body in the inside of it would not be attracted to one side more than another.

Cavendish, who was also a member of London's learned Society, heard of Priestley's argument and agreed that the electric force between two charges must satisfy an inverse-square law. He proved this result in the 1770s by means of a sensitive experiment based on Franklin's chance observation and Priestley's explanation of it.

As indicated in Figure 7.1, Cavendish charged an insulated metal sphere and surrounded it by a larger spherical metallic shell. The two bodies were connected by a wire that let charge flow freely between sphere and shell. Subsequently, he removed the shell and the wire and found that all of the charge of the sphere had leaked to the shell.

Suppose the electric field of a spherical charge varies inversely with the Nth power of the distance, where $N = 2$ corresponds to an inverse-square law. Using Newton's calculus, Cavendish deduced that if N had any value other than 2, then some charge would have to remain on the sphere. Because Cavendish found no detectable remnant of charge, he

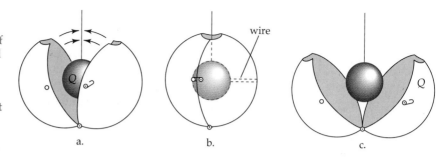

FIGURE 7.1
Cavendish establishes the inverse-square law of electric force. *a.* A metal sphere bearing charge Q is placed within an uncharged conducting shell. *b.* Electrical contact is established between the charged sphere and the conducting shell via the fiber. *c.* When the shell is opened, all of the charge that was on the sphere is now found to be on the shell. This result implies that the electric force satisfies an inverse-square law.

concluded that N differs from 2 by less than 0.03. He never published this work. Cavendish was an unsociable fellow who was indifferent to public recognition. The discovery remained hidden away among his papers for a century, until it was rescued from oblivion by Maxwell.

Coulomb, who was not aware of Cavendish's earlier work, directly measured the force exerted by one charge on another in 1785. To accomplish this, he used a device known as a torsion balance.* Coulomb placed a spherical pith ball on one end of an insulating rod, which was suspended horizontally by a long thin fiber. A second insulated pith ball was brought nearby and held fixed, as shown schematically in Figure 7.2. When the pith balls were given electric charges of the same sign,

FIGURE 7.2
The apparatus with which Coulomb measured the electric force between two charges. One charged ball is insulated and suspended from above by a fine fiber. Another charged ball is brought nearby and held fixed. The electric force displaces the mobile ball until the torsion of the fiber balances the electric repulsion. The angle by which the suspended ball turns measures the electric force between the two charges.

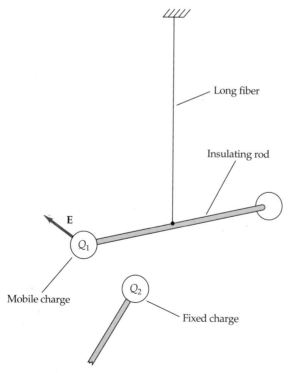

*Cavendish used just such a device to detect and measure the gravitational attraction between two large metal balls, thereby measuring Newton's constant G. His measurement was a marvel of experimental technique because the force of gravity is terribly weak and is easily masked by other effects. Coulomb's task was easier.

the suspended pith ball was pushed away from the fixed ball, causing the fiber to be twisted. It came to rest when the electrical repulsion was compensated by the restoring force of the twisted fiber. By rotating the support of the suspended ball, Coulomb was able to measure the force as a function of the distance between the balls. In a similar way, he measured the attractive force between balls that had been given opposite charges.

When the charges of the two bodies are fixed, Coulomb discovered, the magnitude of the electric force F_e varies with the inverse square of the distance R between the centers of the balls:

$$F_e \propto \frac{1}{R^2} \qquad \text{Charges kept constant} \qquad (1)$$

In another series of experiments, Coulomb found how the electric force depends on what he called "the degree of electrification" of each body, or what we now call their electric charges. When the distance between two charges is fixed, Coulomb found, the magnitude of the electric force F_e varies with the product of the charge Q of one ball with the charge q of the other:

$$F_e \propto q \, Q \qquad \text{Separation kept constant} \qquad (2)$$

Equations 1 and 2 are combined as a general expression for the force between two charges:

$$F_e = \frac{k_c \, q \, Q}{R^2} \qquad (3)$$

where q and Q are the values of the two charges and R is the distance between them. The numerical value of k_c, the constant of proportionality in this result, depends on the choice of units.

Equation 3 is known as *Coulomb's law*. The factor of R^2 in the denominator is the mathematical statement of the inverse-square law. The direction in which the force acts is along the line connecting them (Figure 7.3). The force is repulsive for charges of the same sign and attractive for charges of opposite sign. Coulomb's law is a correct expression for the force between two pointlike or spherically symmetric charged objects. It is approximately valid if the sizes of the charged objects are small compared to the distance between them.

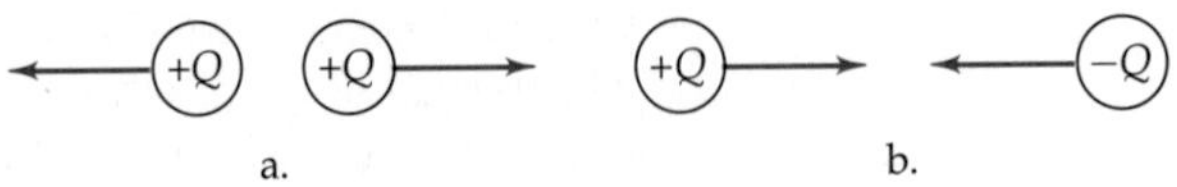

FIGURE 7.3
a. If the charges of two bodies have the same sign, each body exerts a repulsive force on the other. *b.* If the charges of two bodies have opposite signs, each body exerts an attractive force on the other.

a.

b.

To use equation 3, we must define a unit of electric charge with which to express q and Q and evaluate the constant k_c. Today's SI unit of charge is called the *coulomb* and denoted by C. Its precise definition is given in Chapter 8. If charge is measured in coulombs, force in newtons, and distance in meters, k_c takes the value

$$k_c = 9 \times 10^9 \text{ N m}^2/\text{C}^2 \qquad (4)$$

Expressed in SI units, Coulomb's law becomes:

$$F_e = 9 \times 10^9 \, \frac{q \, Q}{R^2} \tag{5}$$

One coulomb is a huge charge. Imagine two equal and opposite charges with $Q = \pm 1$ C placed 1 km apart. According to equation 5, the electric force attracting them to one another is $k_c \, (1\,\text{C})^2/(1000\,\text{m})^2$, or 9000 N, about the weight of a ton of bricks! Yet, 10 C of charge flows through a typical toaster or hair dryer per second. No enormous electric forces come into play because the innumerable negatively charged electrons flowing through the device are not spatially segregated from the neutralizing background of positive atomic charges. As electric charge flows through the wires of the toaster, they become hot, but they remain electrically neutral. The brute strength of the electric force makes it difficult to rend apart the positive and negative constituents of matter.

EXAMPLE 7.1
The Force Due to
Several Charges

The total electric force on a charged body is the vector sum of the forces due to all other nearby charged bodies. Suppose that three small charged bodies are fixed on the x-axis at $x = 1$, $x = 4$, and $x = 7$, in meters. The middle body has charge $Q_M = 2 \times 10^{-4}$ C, that on the left has $Q_L = -1 \times 10^{-4}$ C, and that on the right has $Q_R = 3 \times 10^{-4}$ C, as shown in Figure 7.4. What is the electric force acting on Q_R?

Solution

The problem is solved by using equation 5 twice. The repulsive force on Q_R due to Q_M is $k_c Q_M Q_R / (3\,\text{m})^2$, or $+60$ N. The attractive force on Q_R due to Q_L is $k_c Q_L Q_R / (6\,\text{m})^2$, or -7.5 N. Thus, the total electric force acting on Q_R is $+52.5$ N.

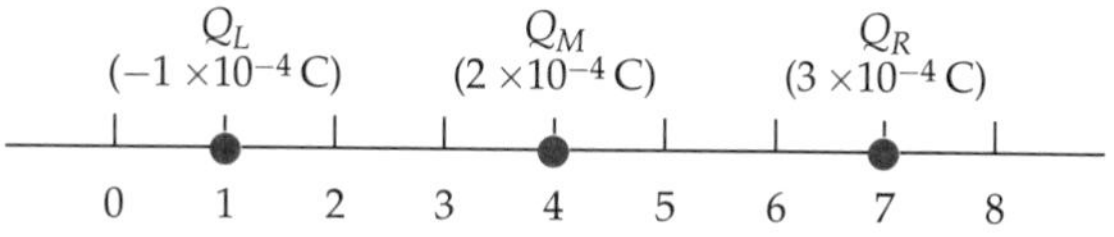

FIGURE 7.4
Three charged bodies, with charges Q_L, Q_M, and Q_R, are placed as shown. The electric force on each body is the sum of the forces due to the other two bodies.

Exercises

1. Gilbert correctly pointed out that Earth itself acts like a gigantic magnet aligned with its axis of rotation. Is the geographical north pole of Earth a north magnetic pole or a south magnetic pole? Explain your answer.

2. Two equal and opposite charges Q attract one another with a force of 1 N when they are 10 cm apart. What is Q in coulombs?

3. The force acting between two charges of 1 μC (microcoulomb) is 90 N. How far apart are the charges?

4. In Example 7.1, calculate the force on the charges Q_L and Q_M. Verify that the sum of the electric forces acting on the three charges vanishes.

5. One charged body exerts an electric force of 1000 N on a second charged body. If both charges were tripled and their spatial separation were doubled, what would the force become?

A BIT MORE ABOUT
Electric and Magnetic Charge

Electric charges of the same sign repel each other, and opposite charges attract. The same statements pertain to magnetic poles. Similar poles repel one another, and opposite poles attract. Coulomb showed that the force between magnetic poles varies with the product of their strengths and depends on the inverse square of the distance between them—just like the electric force between charges!

The analogy between electricity and magnetism is imperfect. A piece of amber rubbed with fur acquires a positive charge, and the fur acquires a negative charge. Positive and negative charges can be separated from one another. Not so for magnetic poles! Every magnet has a north pole and a south pole of the same strength. If you saw off one pole of a magnet, two new magnetic poles are created. Each fragment becomes a magnet with its own equal and opposite poles. Continue this process many times, and you find that each atom of the magnet is itself a tiny magnet with two poles. A magnet with one pole would be akin to a piece of string with one end.

Physicists refer to an isolated magnetic pole as a *magnetic monopole*. It is the magnetic analog to an electric charge. Just because nobody has seen a magnetic monopole or knows how to make one doesn't mean there is no such thing.

Perhaps we haven't tried hard enough. Ordinary matter is made of atoms consisting of positively charged nuclei and negatively charged electrons. A body with an excess of one or the other displays a net charge. Because electrically charged particles exist, should not magnetic monopoles hide somewhere among nature's wonders as well?

Physicists try to make monopoles at large accelerators or find them in ancient ores, clam shells, or among cosmic rays, but nobody has seen one for sure. The most recent possible sighting took place at Stanford University on February 14, 1982. Blas Cabrera prepared a superconducting loop of wire connected to a sensitive "superconducting quantum interference device." A monopole coming from outer space and passing through the loop would generate a recognizable signal. After 151 days of running the experiment, Cabrera detected an event that might have been due to a monopole. I sent him a Valentine's day message:

> *Roses are red,*
> *Violets are blue,*
> *The time has come*
> *For monopole two.*

Monopole two has not shown up, and monopole one was almost certainly caused by a clumsy graduate student or a glitch in the apparatus. Using much larger detectors, neither Cabrera nor anyone else has reported another event. Nonetheless, the great monopole hunt continues with gusto.

6. The hydrogen atom is made of a proton and an electron separated by a distance of about 10^{-10} m. The charges of these particles are equal and opposite, with magnitude 1.6×10^{-19} coulombs. How big is the electric force holding them together?

7. Avogadro's number, the number of molecules in a mole, is about 6×10^{23}. What is the charge in coulombs of a mole of electrons?

8. A pint of milk contains about 10^{26} electrons and an equal number of protons. Imagine that all its electrons are put into one box and all its protons into another. What is the force between the boxes if they are placed at diametrically opposite points on Earth?

7.2 The Electric Field and the Electric Potential

Earth exerts a force on the Moon, even though the Moon is far from Earth. A magnet exerts a force on a piece of iron it does not touch. Electric forces act between physically separated charges. How are these forces communicated through the intervening empty space?

One point of view is called *action at a distance*. Each body is imagined to act directly on another, even though the two bodies are at a distance from one another. Direct action poses conceptual problems. By what mysterious mechanism does one body affect a remote body? If one body moves and the distance between them changes, does the force on the other change in synchrony? This would imply an instantaneous transmission of information from one body to the other by an unknown mechanism. This is implausible in classical physics. It is intolerable today, because the special theory of relativity tells us that no signal can travel faster than the speed of light. In the seventeenth century, Newton addressed this question in the context of gravity:

> It is inconceiveable that inanimate brute matter should, without the mediation of something else, which is not material, operate on and affect other matter without mutual contact. That gravity should be innate, inherent and essential to matter, so that one body can act upon another at a distance, through a vacuum, without the mediation of anything else by and through which their action and force may be conveyed from one point to another, is to me so great an absurdity, that I believe no man who has in philosophical matters a competent faculty of thinking can ever fall into it.

Two hundred years later, Maxwell posed the same question:

> I must ask you to go over very old ground, and to turn your attention to a question which has been raised again and again since men began to think. The question is that of the transmission of force. We see that two bodies at a distance from each other exert a mutual influence on each other's motion. Does this mutual action depend on the existence of some third thing, some medium of communication occupying the space between the bodies, or do the bodies act upon each other immediately, without the intervention of anything else?

If action at a distance is an unreasonable hypothesis, then what does explain the ability of one body to exert a force on a remote body? Michael Faraday (1791–1867), whose investigations laid the groundwork for modern electromagnetic theory, detested the notion of action at a distance as exemplified in Newton's law of gravity or Coulomb's law of electrical attraction. He was convinced that the space separating two objects was not inert but played an active part in electrical, magnetic, and gravitational phenomena.

Fields

Faraday argued that the region about an electric charge, a magnet, or a massive body is permeated by something that "weaves a web through the sky." That something is what we call today an electric, magnetic, or gravitational *field*. The "lines of force" of the field produced by one body fill all space and can exert a force on a second body, wherever it may be. From the field point of view, object *A* exerts a force on object *B* through a two-step process:

1. Object *A* generates a field everywhere.
2. The field of object *A* exerts a force on object *B*.

Faraday demonstrated the reality of magnetic lines of force by sprinkling iron filings on a sheet of paper lying on a bar magnet so the lines could be seen, as in the illustration. The field has become one of the most important constructs of modern physical science.

Let's focus on electric phenomena and Coulomb's law. Consider the electric force $\mathbf{F}_e$ acting on a test charge q. The electric field $\mathbf{E}$ at the

Iron filings are scattered on a glass plate atop a bar magnet. They align themselves with the magnetic field, showing its lines of force. A magnet has two opposite magnetic poles, and the pattern of iron filings resembles that of the electric field of two equal and opposite charges. *Source:* © Richard Megna, 1986/Fundamental Photographs.

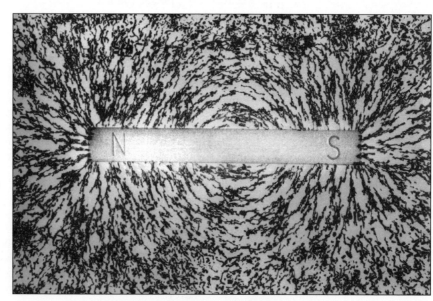

location of the charge is defined to be the ratio of the electric force acting on the charge to the value of the charge. Conversely, the electric force on a charged body is the product of the body's charge with the electric field due to other charged bodies:

$$\mathbf{E} \equiv \mathbf{F}_e/q \qquad \mathbf{F}_e \equiv q\,\mathbf{E} \qquad (6)$$

The concept of an electric field is once removed from directly observed phenomena: The value of the electric field at any point determines what the force would be if a body were placed at that point. The SI unit of electric field strength is newtons per coulomb, or N/C.

Force is a vector, so the electric field is a vector as well. We can determine the electric field $\mathbf{E}$ at a point a distance R from a charge Q by Coulomb's law. Comparing equations 6 and 3, we find that the magnitude of the electric field at a distance R from a charge Q is

$$E = \frac{k_c Q}{R^2} = 9 \times 10^9 \frac{Q}{R^2} \qquad (7)$$

The vector $\mathbf{E}$ produced by a charge points directly away from a positive charge and directly toward a negative charge. The electric field produced by a charge is defined at every point in space, whether or not a test charge is there.

Let the directions of the electric field of a charge at different points be denoted by tiny imaginary arrows. If these arrows are connected together, they form continuous directed lines of force that emerge from positive charges and enter negative charges. This construction is shown in Figure 7.5. More properly, the figure should be three-dimensional and look like a spherical porcupine. At any point, the lines of force tell the direction of the electric field.

FIGURE 7.5
a. The lines of force of a positive charge. *b.* The lines of force of a negative charge.

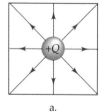 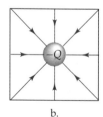

a.   b.

EXAMPLE 7.2
The Electric Field of a Point Charge

What is the magnitude of the electric field at a distance of 3 m from an electric charge 2 μC (microcoulomb)? What is the field at a distance of 1 m?

Solution

The electric field of a point charge is given by equation 7. If $R = 3$ m, we find $E = k_c\,(2 \times 10^{-6}\ \mathrm{C})/(3\,\mathrm{m})^2$, or 2000 N/C. At $R = 1$ m, the field is nine times larger, or 18000 N/C. ∎

The Electric Field of Several Charges

The electric field is a vector. Thus, the electric field produced by several charges is the vector sum of the fields due to each charge. Figure 7.6 illustrates the calculation of the electric field due two electric charges. If Q_1 is at A and Q_2 is at B, the electric field at a third point C is given by

$$\mathbf{E} = \mathbf{n}_1 \frac{k_c \, Q_1}{R_1^2} + \mathbf{n}_2 \frac{k_c \, Q_2}{R_2^2} \tag{8}$$

R_1 denotes the distance from A to C, while $\mathbf{n}_1$ (a vector of length 1) denotes the direction from A to C. Similarly, R_2 and $\mathbf{n}_2$ denote the distance and direction from B to C. The electric field at C is obtained by adding together the fields of the two charges.

FIGURE 7.6
A negative charge Q_1 is located at A and a positive charge Q_2 is located at B. The electric field $\mathbf{E}$ at point C is the vector sum of $\mathbf{E}_1$ (due to Q_1) and $\mathbf{E}_2$ (due to Q_2). The unit vectors $\mathbf{n}_1$ and $\mathbf{n}_2$ point from the charges to point C.

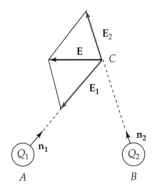

Figure 7.7 shows the lines of electric force produced by two charges that are either equal or equal and opposite. In the second case, every line of force begins at the positive charge and returns to the negative charge. None of the lines of force of an electrically neutral system of charges can wander off to infinity.

FIGURE 7.7
a. The lines of force of two equal and oppo-site charges. *b.* The line of force of two equal charges.

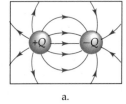

a.

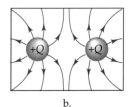

b.

EXAMPLE 7.3
The Electric Field Produced by Two Charges

Two electric charges are fixed on the x-axis. A charge of 1×10^{-4} C is put at point A, with $x = 0$. A charge of 3×10^{-4} C is placed at point B, with $x = 4$ m (Figure 7.8).

a. What is the electric field at point C, located on the x-axis at $x = 1$ m?

b. What is the electric field at point D, located on the y-axis at $y = 3$ m?

FIGURE 7.8
Charges are placed at A and B as shown on the x-axis. We are asked to determine the electric field at points C and D. Note that the electric field at D due to the charge at B points along the hypotenuse of a 3-4-5 right triangle.

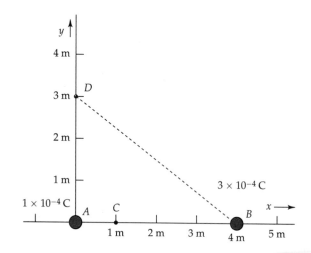

Solution

a. To solve these problems, we use equations 8 and 4. The electric field at C due to the charge at A points in the positive x direction. Its magnitude is $k_c(10^{-4}\ \text{C})/(1\ \text{m})^2$, or $+9 \times 10^5$ N/C. The field due to the charge at B points in the negative x direction and is $k_c(3 \times 10^{-4}\ \text{C})/(3\ \text{m})^2$ or -3×10^5 N/C. The components of the electric field at C are $E_x = 6 \times 10^5$ N/C and $E_y = 0$.

b. This problem is more difficult because the points A, B, and D do not lie along a line. However, I've chosen them to form a 3–4–5 triangle so we may avoid trigonometry. The electric field at D due to the charge at A points in the positive y direction. Its magnitude is $k_c(10^{-4}\ \text{C})/(3\ \text{m})^2$, or $+1 \times 10^5$ N/C. The distance from B to D is 5 m, so the field due to B at D has magnitude 1.08×10^5 N/C. Its x component is $-\frac{4}{5}$ of its magnitude (-8.64×10^4 N/C) and its y component is $+\frac{3}{5}$ of its magnitude ($+6.48 \times 10^4$ N/C). The components of the electric field at D are $E_x = -8.64 \times 10^4$ N/C and $E_y = 1.65 \times 10^5$ N/C. ∎

Gravity at Ground Level

The gravitational field $\mathbf{g}$ is analogous to the electric field. As in equation 6, $\mathbf{g}$ is defined as the ratio of the gravitational force acting on an object to its mass. For objects near the surface of Earth, $\mathbf{g}$ points down and its magnitude is Galileo's universal acceleration g. At or near Earth's surface, the gravitational field $\mathbf{g}$ is approximately constant in magnitude and is directed toward the center of Earth.

In Chapter 3, the quantity $-m\mathbf{g} \cdot \mathbf{r}$ was interpreted as the gravitational potential energy of a body of mass m relative to the origin of the coordinate system. Suppose that the body moves from $\mathbf{r}_1$ to $\mathbf{r}_2$. The work done by the gravitational force $m\mathbf{g}$ is

$$\text{Work done by } \mathbf{g} \ = \ m\mathbf{g} \cdot (\mathbf{r}_2 - \mathbf{r}_1) \ = \ m\mathbf{g} \cdot \mathbf{r}_2 - m\mathbf{g} \cdot \mathbf{r}_1$$

The total energy of the object is conserved, so the work done by the force is equal and opposite to the change in its potential energy. Thus, the gravitational potential energy of an object is the product of its mass m (an attribute of the object) and $-\mathbf{g} \cdot \mathbf{r}$ (a function defined at all points in space). The latter quantity is defined to be the *gravitational potential* V_g, relative to the origin:

$$V_g = -\mathbf{g} \cdot \mathbf{r}$$

The gravitational potential energy of an object is the product of its mass and the value of the gravitational potential at its position. For example, the gravitational potential at an altitude of 1000 m is approximately 10^4 J/kg relative to its value at ground level. The gravitational potential energy of an 80-kg skydiver at that altitude is 8×10^5 J.

The Electric Potential of a Constant Field

What we said about gravity also applies to the behavior of a charge q in a constant electric field $\mathbf{E}$. The work done by the electric force $q\,\mathbf{E}$ as the charge moves from $\mathbf{r}_1$ to $\mathbf{r}_2$ is

$$\text{Work done by } \mathbf{E} \; = \; q\mathbf{E} \cdot (\mathbf{r}_2 - \mathbf{r}_1)$$

and is equal to the difference between the initial electric potential energy and the final potential energy of the charge. The quantity $-q\,\mathbf{E} \cdot \mathbf{r}$ is the *electric potential energy* of the charge relative to the origin.

The electric potential energy $-q\,\mathbf{E} \cdot \mathbf{r}$ is the product of q (an attribute of the object) and $-\mathbf{E} \cdot \mathbf{r}$ (a function defined at all points in space). The latter quantity is defined to be the *electric potential V_e* relative to the origin:

$$V_e = -\mathbf{E} \cdot \mathbf{r} \tag{9}$$

In general, the electric potential energy of an object is the product of its charge and the value of the electric potential at its position. The SI unit of electric potential is the *volt*. The volt is defined to be 1 joule per coulomb. The dimensions of electric field are newtons per coulomb. From the definition of the volt, we see that

$$1 \text{ newton per coulomb} \equiv 1 \text{ volt per meter}$$

Electric fields are usually stated in volts per meter; electric potential is measured in volts and often called *voltage*. The concept of electric potential applies to many circumstances other than motion in a constant field. For example, the potential difference between the terminals of a battery is the voltage of the battery. When 1 C of electric charge flows from the + terminal to the − terminal of a 9-volt battery, the electric potential energy of the battery decreases as it provides 9 J of useful work.

EXAMPLE 7.4
Constant Electric Fields

A constant electric field $E = 2 \times 10^5$ volts per meter points in the positive x direction.

a. If the electric potential is 0 at $x = 0$, what is its value at $x = 10$ cm?

b. A charged body with mass $m = 10$ g and charge $q = 4$ μC starts at rest at $x = 0$. What is its velocity at $x = 10$ cm?

Solution

a. Equation 9 says that the electric potential at a point on the x-axis is $-Ex$. Thus, the potential at $x = 0.1$ m is $V_e = -2 \times 10^4$ volts.

b. The electric potential energy of the charge is q times the electric potential. It decreases from 0 at $x = 0$ to -0.08 J at $x = 0.1$ m. Its kinetic energy increases by the same amount. Thus, $\frac{1}{2}mv^2 = 0.08$ J, from which we find $v = 4$ m/s.   ■

Fields and Potentials of Point Charges

Coulomb's law for the electric force $\mathbf{F_e}$ between two point charges, q and Q, is similar in form to Newton's law for the gravitational force $\mathbf{F_g}$ between two point masses, m and M. If the separation of the bodies is R, the magnitudes of the forces are

$$F_e = \frac{k_c q Q}{R^2} \qquad F_g = \frac{G m M}{R^2} \tag{10}$$

Suppose that the charge Q (or the mass M) is fixed at the origin, while the charge q (or the mass m) is located at an arbitrary point P at a distance R from the origin. The electric force exerted on q is the product of the charge q with the electric field $\mathbf{E}$ at P produced by Q. In the same way, the gravitational force exerted on a mass m is the product of its mass with the gravitational field $\mathbf{g}$ at P produced by M. The formulae describing the two forces are analogous:

$$\mathbf{F_e} = q\mathbf{E} \qquad \mathbf{F_g} = m\mathbf{g} \tag{11}$$

If we compare equations 10 and 11 and recall the rules for the signs of electric and gravitational forces, we obtain expressions for the fields $\mathbf{E}$ and $\mathbf{g}$ at any point P in space:

$$\mathbf{E} = +\mathbf{n}\frac{k_c Q}{R^2} \qquad \mathbf{g} = -\mathbf{n}\frac{G M}{R^2} \tag{12}$$

where $\mathbf{n}$ is a unit vector directed from the object to the point P, a distance R away.

Notice the sign difference in the expressions for $\mathbf{E}$ and $\mathbf{g}$. Because m and M are positive numbers, the gravitational force due to a massive body is always attractive. However, charges may be positive or negative. Equation 12 tells us that charges of the same sign repel and charges of opposite sign attract. This is not the only difference between gravity and electromagnetism. Particle for particle, the electromagnetic force is enormously more powerful than the gravitational force. For example, the electrical attraction between the electron and proton in a hydrogen atom is about 40 powers of 10 larger than their gravitational attraction.

We learned in Chapter 3 that a mass m at a distance R from another mass M has gravitational potential energy:

$$\text{Gravitational potential energy} = -\frac{GmM}{R} \qquad (13)$$

A charge q at a distance R from another charge Q has the electric potential energy:

$$\text{Electric potential energy} = +\frac{k_c q Q}{R} \qquad (14)$$

The difference in sign between equations 13 and 14 reflects the sign difference in equations 12. Because the force between charges of the same sign is repulsive, their potential energy is positive. If two nearby charges of the same sign are released, they fly apart as potential energy becomes kinetic energy. Conversely, if two charges of opposite sign are near one another, energy must be supplied to separate them.

EXAMPLE 7.5
Electric
Potential Energy

Two charges are located on the x-axis: $Q_1 = +1 \times 10^{-4}$ C is put at $x = 0$, and $Q_2 = 3 \times 10^{-4}$ C is put at $x = 4$ m, as in Example 7.3 and Figure 7.8.

a. What is the electric potential energy of the system of two charges?
b. What does the electric potential energy become if the charge at the origin is moved to the point $y = 3$ on the y-axis? How much energy must be supplied to the system to do this?

Solution

a. The relevant formula is equation 14. The electric potential energy is $k_c Q_1 Q_2 / R$. In the initial configuration, $R = 4$ m, and the electric potential energy is 67.5 J.
b. When Q_1 is moved 3 m in the y direction, it is brought to a point 5 m from Q_2. The electric potential energy becomes 54 J. Because the potential energy decreases, no energy is needed to move the charge. On the contrary, the motion of the charge releases 13.5 J of energy.

■

Equations 13 and 14 determine the gravitational potential produced by a point mass M, and the electric potential produced by a point charge Q:

$$V_g = -\frac{GM}{R} \qquad V_e = +\frac{k_c Q}{R} \qquad (15)$$

If a mass m is placed a distance R from M, its gravitational potential energy is $m V_g$. If a charge q is placed a distance R from Q, its electric potential energy is $q V_e$. Notice that the gravitational and electric potentials are scalars, not vectors. Thus, the potential at a point P due to a system of many bodies is the arithmetic sum of the potentials due to each body.

EXAMPLE 7.6
The Electric
Potential Produced
by Two Charges

Once again, consider the configuration of two charges shown in Figure 7.8.

a. What is the electric potential at point C?
b. What is the electric potential at point D?

Solution

a. This exercise illustrates how equation 15 is applied to a system of two charges. The electric potential at C due to the 1×10^{-4} C charge 1 m away is 9×10^5 volts. The electric potential at C due to the 3×10^{-4} C charge 3 m away is also 9×10^5 volts. The electric potential at C is the sum of these results, or 1.8×10^6 volts.

b. The electric potential at D due to the 1×10^{-4} C charge 3 m away is 3×10^5 volts. The electric potential at D due to the 3×10^{-4} C charge 5 m away is 5.4×10^5 volts. The electric potential at D is the sum of these results, or 8.4×10^5 volts. If you compare this example with Example 7.3, you see how much simpler it is to calculate electric potentials than electric fields.   ■

The Potential of a Conductor

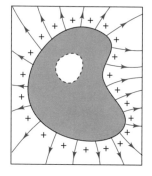

FIGURE 7.9
This irregularly shaped conductor has a cavity within it. When any conductor is electrically charged, the charge distributes itself on the outer surface. The electric field near the conductor is perpendicular to its outer surface, as shown by the lines of force. The electric field vanishes within the conductor and within its cavity.

A conductor (such as a piece of scrap metal found at a junkyard) is something through which electric charge may move. Figure 7.9 shows an irregular conducting object with a hollow space within it. Suppose that this conductor is given the electric charge Q. How does the charge distribute itself? What is the electric field produced by the charged conductor? These are difficult questions in *electrostatics*, the study of electric charges at rest. However, some of the properties of charged conductors are wonderfully simple and easy to explain.

1. Suppose that two points on the surface of a conductor are at different electric potentials. Charge will flow from regions of higher potential to regions of lower potential. Charge continues to flow until all points on the conductor are at the same potential. Thus, the charge on a conductor distributes itself in such a way that every point at the surface of the conductor is at the same electric potential V.

2. The electric field within the conductor must vanish, for otherwise it will cause atomic charges to move about. In particular, the electric field must vanish within any hollow spaces within the conductor. It follows that all of the charge on the conductor must reside on its outer surface. Furthermore, the electric field at a point just outside the surface of a conductor points in a direction perpendicular to the surface at that point.

3. The electric potential V of a conductor is proportional to the total charge Q it carries:

$$Q = cV$$

The constant c, so defined, is called the *capacitance* of the conductor. The SI unit of capacitance is the coulomb per volt and is known as the *farad*. In general, it is difficult to compute the capacitance of a conductor from its size and shape.

The electric charge of an insulated conductor cannot escape because air is a very poor conductor of electricity. The charge spreads out over the surface of the conductor so that individual elements of charge are as far apart as possible. The electric potential energy of the charge on the conductor becomes as small as possible, but it cannot vanish. The amount of electric energy $\mathcal{E}$ that is stored in a charged conductor is

$$\mathcal{E} = \frac{Q^2}{2c} = \frac{cV^2}{2} = \frac{QV}{2} \tag{16}$$

where Q is the charge on the conductor, c is its capacitance, and V is its electric potential.

The concept of capacitance is illustrated by a spherical conductor of radius R_0 (see Figure 7.10). In this case, the charge Q distributes itself uniformly over the surface of the sphere. The electric potential of such a charge distribution is given by:

$$V_e(R) = \begin{cases} k_c Q/R & \text{for } R \geq R_0 \\ k_c Q/R_0 & \text{for } R < R_0 \end{cases}$$

Outside the conductor, the electric potential and field are the same as those of a point charge Q placed at the center of the sphere. Inside the conductor, the electric field vanishes. The potential of the conductor is $k_c Q/R_0$. Thus, the capacitance of a conducting sphere is

$$c = R_0/k_c$$

The conducting sphere is one of a very few instances of a conductor whose capacitance is easily determined.

Every electronic device contains small circuit elements called *capacitors* because they are designed to have large capacitance. Many of them consist of two sheets of tightly wrapped metal foil with a sheet of insulator between. Wires connected to each of the metal foils lead out of the capacitor. A capacitor may be used to store electricity, or to act like a spring to influence the oscillations of an electric current passing through.

FIGURE 7.10
A spherical conductor of radius R_0 and charge Q. The charge is uniformly distributed on the surface, and the electric field outside is identical to that of a charge Q placed at its center. Thus, the electric potential of the conductor is $k_c Q/R_0$.

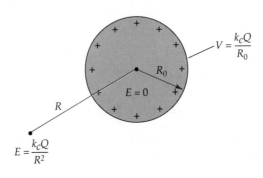

EXAMPLE 7.7
A Spherical
Conductor

A conducting sphere with a radius of 3 cm is given an electric charge of 5×10^{-7} C. What is the electric potential at its surface? What is the magnitude of the electric field at its surface? How much electric energy is stored by the conductor?

Solution

From the outside, a charged spherical conductor behaves like a point charge. The potential at its surface is $k_c Q/R$, which, in this case, is 1.5×10^5 volts. The field at its surface is $k_c Q/R^2$, which is 5×10^6 volts per meter. The energy stored in the conductor is given by the last equality in equation 16: $\mathcal{E} = 3.75 \times 10^{-2}$ J. ∎

Exercises

9. What is the value of the electric field at a distance of 30 cm from a charge of 7×10^{-6} C?

10. Equal charges of 10^{-6} C are placed on the x-axis at $x = \pm 1$ m.
    a. What is the electric field at the origin?
    b. In what direction is the electric field at a point on the positive y-axis?
    c. What is the magnitude of the electric field at the point $y = 1$ on the y-axis?
    d. Describe in words how the magnitude of the electric field changes as a function of y along the positive y-axis.

11. Equal and opposite charges are placed on the x-axis: a charge of 10^{-6} C at $x = -1$ m and a charge of -10^{-6} C at $x = +1$ m.
    a. What is the electric field at the origin?
    b. In what direction is the electric field at a point on the positive y-axis?
    c. What is the magnitude of the electric field at the point $y = 1$ on the y-axis?
    d. Describe in words how the magnitude of the electric field changes as a function of y along the positive y-axis.

12. A charge of 10^{-5} C is 6 m from a charge of -4×10^{-5} C.
    a. What is the electric potential energy of the system of two charges?
    b. What is the electric field at the point midway between the two charges?
    c. What is the electric potential at the point midway between the two charges?
    d. At what point on the line connecting the two charges does the electric potential vanish?
    e. At what point on the extension of the line connecting the two charges does the electric field vanish?

13. The gravitational force between two masses and the electric force between two charges both vary with the reciprocal square of the

distance R between them—that is, with R^{-2}. Two 1-kg bodies carry identical electric charges. Calculate what this charge is if their electrical repulsion precisely cancels their gravitational attraction no matter what their separation.

14. A charge $4Q$ is fixed at a distance L from a second charge Q. At exactly one point on the line between them, the electric field of the one body just cancels the field of the other. How far is this point from the smaller charge?

15. What is the electric potential at a point P that is 15 m from a charge of 5 μC? Suppose that a second charge of 2 μC is located at P. What is its electric potential energy? Suppose that the first body is held fixed and the second body, whose mass is 0.2 g, is released. What does its speed become when it is far away from the first body?

16. What is the electric potential in volts at a distance of 10^{-10} m from a proton whose charge is 1.6×10^{-19} C?

17. An isolated conducting sphere with a 10-cm radius is brought to a potential of 9×10^5 volts.
    a. What is the electric field at its surface?
    b. What is the charge on the sphere?
    c. How much electric energy is stored by the sphere?

18. A spherical drop of mercury with a charge $8Q$ splits into eight identical spherical droplets, each with the same charge and radius. The droplets fly far apart and do not affect each other.
    a. Determine the following ratios:
        i. the radius of the drop to the radius of a droplet
        ii. the potential of the drop to the potential of a droplet
        iii. the electric energy of the drop to the electric energy of a droplet
    b. What percentage of the electric energy of the original drop has been converted to other forms of energy by its fission?

7.3 From Frogs to Batteries

Electrical machines of the eighteenth century generated electric charges that could be stored in devices such as charged conductors, whose discharges produced brief spurts of electric current. But no continuous source of steady electric current was available to research or technology until the first electric battery, the *voltaic pile*, arose out of a fascinating controversy between two Italian scientists: Luigi Galvani (1737–1798) and Alessandro Volta (1745–1827). The properties of steady electric currents, as opposed to transient discharges, could be examined once batteries became available.

Alessandro Volta and Luigi Galvani. The discovery of the electric battery resulted from the curiosity of Galvani, the skepticism of Volta, and the extraordinary observations of both. *Source:* Historical Pictures/Stock Montage, Inc.

Animal Electricity vs. Metallic Electricity

Galvani, who trained in anatomy and medicine, suspected that there was a relation between electricity and the activity of nerves and muscles. His moment of truth happened by accident in 1786. Here is a lightly edited version of the first paragraph of his classic paper:

> I had dissected a frog, and laid it on a table on which there was an electrical machine, while I set about doing other things. The frog was entirely separated from the machine, at no small distance away from it. While one of my assistants, by chance, touched the point of his scalpel to the crural nerve of the frog, the muscles of the limb were suddenly and violently convulsed. Another [assistant] noticed this happening only at the instant a spark came from the electrical machine. He was struck by the novelty of the action. I was mentally preoccupied at the time, but when he drew my attention to it I immediately repeated the experiment. I touched the other end of the crural nerve with the point of my scalpel, while my assistant drew sparks from the electrical machine. At the very moment when the sparks appeared the animal was seized as it were by tetanus.

Aware of Franklin's proof of lightning's electrical nature, Galvani tied small lightning rods to the nerves of frog's legs and suspended them outdoors, with the feet on the ground. Dead frogs dancing to thunder and lightning confirmed the relation between muscular activity and electricity. A further chance observation confused the matter. Galvani's frogs were held upright by brass hooks, with their feet on an iron trellis. When the hooks were pressed to the trellis, muscle contractions occurred even on sunny days! Indoors, a dead frog with a brass hook in its marrow jerked and twisted when the hook touched the iron plate on which the frog had been placed. To isolate the phenomenon, Galvani joined two wires of different metals to form a bimetallic arc. When both ends were touched to the frog, its muscles convulsed. Did the metal arc somehow release electricity inherent in the frog? Or, did the effect result from the production of electricity by the junction of dissimilar metals? Galvani chose what we now know to be the wrong answer:

Galvani's laboratory, as drawn in 1791. Several of his experiments with frogs are illustrated. *Source:* Courtesy, Meyers Photo-Art.

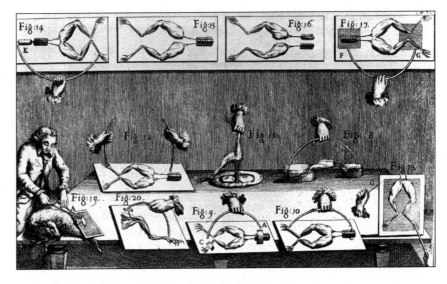

> In the animal itself there was an indwelling electricity. We were strengthened in such a supposition by the assumption of a very fine nervous fluid that during the phenomena flowed into the muscle from the nerve, similar to the electrical current.

Galvani announced his results on "animal electricity" in 1791,* the very year that Volta was elected to the Royal Society for his invention of a sensitive detector of tiny electrical effects, a device now called, oddly enough, a *galvanometer.* Volta confirmed Galvani's results, but came to the opposite conclusion. He believed that electricity is produced not by animals, but by contact between different metals. He proposed a rival theory of "metallic electricity." Meanwhile, assisted by his nephew Giovanni Aldini, Galvani succeeded in eliciting muscle contractions solely by the manipulation of nerves and muscles without the use of metals. Aldini, on a visit to England, showed that an electric discharge seemed to briefly reanimate a decapitated felon. For this remarkable feat, he was awarded the Copley Medal of the Royal Society.

The two theories seemed irreconcilable and European scientists split into camps of bitter rivals—some supporting Volta of Padua, others Galvani of Bologna.† Each scientist saw part of the truth. Galvani's pioneering studies began the science of electrophysiology. Nerves do produce

---

*Galvani's epochal discovery came at a most unpropitious time. Just a few years earlier, the French Academy of Sciences had denounced Mesmer's "animal magnetism" as the fraud it was.

†Meanwhile, Napoleon's imperial ambitions began. By 1797, both cities were forced to join the Cisalpine Republic, a puppet government controlled by France. Volta favored Napoleon while Galvani remained loyal to the old régime Their differences became political as well as scientific.

The world's first electric batteries, as drawn by Volta in 1800. The device on top consists of several brine-filled cells, Volta's *crown of cups*, each with a copper and zinc electrode. "In regard to the columnar apparatus," Volta wrote, "I endeavored to discover the means of lengthening it a great deal by multiplying the metallic plates in such a manner as not to tumble down." *Source:* Courtesy, Meyers Photo-Art.

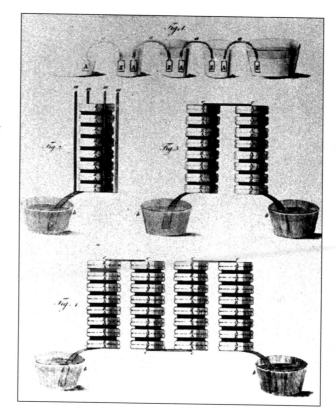

and transmit electrochemical impulses. Today, the electrical activity of the brain and the heart are routinely measured—with electroencephalograms and electrocardiograms—to diagnose human illnesses. However, the juxtaposition of two dissimilar metals generates electricity as well. Volta's endeavors led him to invent the electric battery, with immediate and tremendous consequences for chemistry, physics, and industry. In 1800, Volta's work with metallic electricity finally paid off. Here is an edited extract from his letter to the Royal Society:

[My] apparatus will, no doubt, astonish you. It is only the assemblage of a number of good conductors of different kinds arranged in a certain manner: 30, 40, 50 or more pieces of copper applied each to a piece of zinc and as many strata of cardboard soaked in salt water or lye, such strata being interposed between every pair of two different metals in an alternate series, and always in the same order are all that is necessary for constituting my new instrument. It has the great advantage of needing no outside charge, and instead of a momentary discharge provides a continuous flow of electricity.

This endless circulation of the electrical fluid (this perpetual motion) may seem paradoxical or even inexplicable, but it is no less true and real, and you may feel it, as I might say, with your hands. I found myself obliged to combat the pretended animal electricity of Galvani, and declare it external electricity moved by the mutual contact of metals of different kinds.

English scientists exploited Volta's discovery. Rumford, the destroyer of Lavoisier's notion of caloric, returned to England in 1798. His many practical inventions had made him wealthy. He founded the Royal Institution in London and appointed as its first director the 23-year-old Humphry Davy. This ambitious chemist built an immense electric battery in the cellar of the Institution patterned after Volta's device, but consisting of a bank of 2000 double plates of zinc and copper. Davy set out to explore the virgin discipline of electrochemistry. He produced a brilliant arc between two carbon electrodes—the first electric light. A modified arc in which lime is heated by the electric discharge—a device that gave rise to the word *limelight*—is still used in some theaters. In 1807, Davy used the great battery to decompose molten alkalis, producing sodium and potassium in elemental form for the first time. Using these as chemical reagents, he isolated many elements for the first time, including barium, calcium, strontium, and magnesium. The histories of physics and chemistry are inextricably tied together.

In 1801, Volta was invited to Paris where he presented three lecture–demonstrations of his electrical researches. Napoleon himself attended all of them.* After Volta's first talk, according to the written record of the occasion:

> Citizen Bonaparte proposed that the Class should demonstrate its desire to gather together the wisdom of all those who cultivate the sciences, and should present a gold medal to citizen Volta. He also proposed that a commission should be given the task of carrying out on a large scale all experiments which might throw a new light on the important branch of physics, which citizen Volta has just expounded.

Years later, when news of Davy's discoveries reached France, Napoleon (now his Imperial Majesty) ordered that there should be constructed at the Ecole Polytechnique voltaic piles of different sizes, including one that should be bigger than any used hitherto. The age of big science had begun.

Electric Currents

At the beginning of the nineteenth century, batteries provided scientists and tinkerers with a steady source of electricity that could be put to work to do many amusing, enlightening, and useful things. Batteries supply electric current, which is much like a current of air or water.

When water falls, gravitational potential energy becomes other forms of energy—heat, sound, and splash in the wild; mechanical energy at a water wheel; or electric energy at a hydroelectric plant. The energy of the Colorado River, which flows from the Rocky Mountains to the Gulf

---

*The emperor-to-be was keen on science. He claimed that had he not become a general with the destiny of leading France, he would have been a scientist. In his own words, "I would have thrown myself into the study of the exact sciences...following the route of Galileo and Newton...I know of no better way for a man to spend his life than to know nature...I would have left behind me the memory of great discoveries." Maybe.

of California, carved out the Grand Canyon. In the same way, charge flows as an electric current from a higher electric potential to a lower electric potential. Its energy accomplishes more mundane tasks: It becomes light (in lamps), heat (in toasters), sound (in stereos), or motion (in fans). The flow rate of a river is the mass of water that passes per second. The flow rate of charge through a circuit is the size of an electric current. The SI unit of electric current is 1 coulomb per second and is called the *ampere*.

An *electric circuit* involves a source of electric energy (such as a battery), a device that consumes electric energy (such as a lamp), and wires to bring the current from the source to the device and back to the source. An ideal wire is a long, thin piece of perfectly conducting material—that is, material that allows an electric current to pass through it freely. When electric charge moves along an ideal wire, it encounters no resistance. The electric potential all along the length of a continuous wire is the same. The same may not be said for the electrical device, which is designed to extract energy from the flow of current through it. Thus, the two wires shown in Figure 7.11 are at different electric potentials. There is a voltage difference between them.

FIGURE 7.11
An electric circuit. Electric current *I* flows from the power source, at a potential V_2, to a lamp or other electrical device. The same electric current *I* returns to the power source at a lower potential V_1.

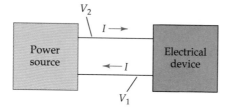

Complex chemical processes taking place within a battery maintain its terminals at different electric potentials. The difference between them is called the *voltage* of the battery. When the terminals are attached to a light bulb, for example, an electric current flows from the + terminal of the battery, through a connecting wire to the light bulb, and back through another wire to the − terminal. The electric potential energy of the charges making up the current becomes heat and light in the filament. Darkness is conquered, but at the same time the store of electric energy in the battery is depleted.

Electric Power

At what rate does falling water release its gravitational potential energy? Imagine a 10 m high waterfall at which 10^3 kg of water fall per second. The gravitational potential energy of a mass *m* at a height *h* is *mgh*, where $g = 10$ m/s². Thus, the gravitational potential energy of the water that falls in 1 s is 10^5 J. Power is the rate of change of energy and is measured in J/s, or *watts* (W). The power released by the falling water in this example is 100 kilowatts. It is available to be harnessed to a waterwheel or an electric generator.

At what rate is the stored energy of a battery consumed? Let V be the voltage of the battery and I be the current passing through an electric circuit to which the battery is connected. The positive terminal of the battery is at an electric potential V greater than its negative terminal. I coulombs of charge per second pass through the circuit across the potential difference V, thereby converting IV joules of electric potential energy per second into other forms of energy. The battery provides the power IV to the circuit.

This is a general result. An electric current I flowing through a device from a higher potential V_2 to a lower potential V_1, provides the electric power P in watts:

$$P = I\,(V_2 - V_1) \qquad (17)$$

where I is in amperes and potentials are in volts.

You don't need a course in physics to know that some lamp bulbs cast more light than others. A 60-watt bulb is designed to consume 60 watts of power when it is connected to a 120-volt power supply. According to equation 17, an electric current of 0.5 ampere flows through its filament when it is turned on. A 150-watt bulb consumes 150 watts of power when it is connected to a 120-volt power supply. According to equation 17, a current of 1.25 amperes flows through its filament when it is turned on. What physical property distinguishes a 60-watt bulb from a 100-watt bulb? The answer to this question involves the concept of electrical resistance.

Electrical Resistance

Let's pursue the analogy between the flow of electric charge through a circuit and the flow of water through a hose. Imagine watering the garden. Water is supplied to the tap at a fixed pressure. (Its pressure in analogous to the fixed voltage supplied by the electric company.) If the hose is stepped on, the flow of water decreases. The crimp in the hose offers resistance to the free flow of water. The more it is crimped, the less space there is for water to pass, and the less water passes. A perfectly straight hose offers resistance as well: the longer or narrower the hose, the less water flows from the nozzle.

Electricity behaves the same way. Of course, we don't get all our electricity from batteries. Electric power companies produce and distribute electricity. In the United States, the two terminals of most electrical outlets are maintained at a potential difference of 120 volts.*

A lamp lights when an electric circuit is established between the power source and the lamp filament, which is a long skinny coil of tungsten wire offering resistance to the free flow of charge through the circuit.

*Commercial electricity is AC or *alternating current*. This means that the voltage difference between the two electric terminals swings back and forth from + to − 60 times per second. Never mind this nonessential complication.

The ends of the filament are maintained at a fixed potential difference of 120 volts. (That's what the electric company does.) According to equation 17, the power consumed by the bulb is equal to the product of the fixed voltage and the current through the filament. However, the current (like the rate at which water passes through a crimped hose) depends on the degree of resistance offered by the filament. The greater the filament's resistance, the smaller the current, the less power consumed, and the less light produced.

In short, the power consumed by a bulb depends on the electrical resistance of its filament. At a qualitative level, we've answered the question we started with. The filament of a 60-watt bulb is longer, or thinner, or both, than the filament of a 150-watt bulb. It offers greater resistance to the passage of an electric current than does the filament of a 150-watt bulb.

Ohm's Law

The German physicist Georg Simon Ohm (1787–1854) found the relationship between the current that passes through a particular object and the voltage across it. From the results of careful experiments, he concluded that "the force of the current is as the sum of all the tensions, and inversely as the entire length of the circuit." In modern language, Ohm's law connects the potential difference V across a device to the current I passing through it. The relationship could not be simpler: V is proportional to I:

$$V = IR \qquad (18)$$

The constant of proportionality is a characteristic of the device through which current is passing, called its *electrical resistance R*. The unit of resistance, 1 volt per ampere, is the *ohm*. It is abbreviated by the symbol Ω (capital omega, recalling the name of the law's discoverer).

To understand the meaning of Ohm's law, we return to our favorite light bulbs. A 60-watt bulb consumes 60 watts of power when the potential difference across its filament is $V = 120$ volts. We saw from equation 17 that a current of $I = 0.5$ ampere passes through it. According to equation 18, the resistance of its filament is $R = 240\ \Omega$. A current of $I = 1.25$ amperes flows through the filament of a 150-W bulb. Because the potential difference across its filament is $V = 120$ volts, its resistance is $R = 96\ \Omega$.

EXAMPLE 7.8
Ohm's Law

A current of 3 amperes passes through a device whose resistance is 40 Ω. What is the potential difference across the resistor? How much electrical power does it consume?

Solution

Equation 18 expresses the potential difference across the resistor as $V = (3\text{ amperes}) \cdot (40\ \Omega)$, or 120 volts. Knowing V and I, we may use equation 17 to determine the power P dissipated by the resistor: $P = (120\text{ volts}) \cdot (3\text{ amperes})$, or 360 watts. ∎

Resistivity

In our hose analogy, the rate of flow of the water varies with the pressure at the tap with a constant of proportionality that depends on the properties of the hose. In the electrical case, the resistance of a wire is expressed in terms of its length, its cross-sectional area, and a parameter intrinsic to the material of the wire, its *resistivity*:

$$R \text{ (wire)} = \frac{\text{Length} \times \text{Resistivity}}{\text{Area}} \tag{19}$$

The resistivity of copper is much smaller than that of iron. That's why most electrical wires are made of copper. Some materials, at very low temperatures, have no resistivity at all. These are called superconductors.

Ohm's result was not immediately accepted. Decades after his work, Maxwell tested Ohm's results and found that equation 18 is roughly satisfied even with currents so powerful as to almost fuse the wire. Ohm's law became an indisputably useful and important description of the properties of bulk matter. However, it is an empirical equality that is only approximately satisfied by most materials. It is not a fundamental law, but the kind of relationship whose validity physicists try to understand in terms of deeper principles and the microscopic substructure of matter.

EXAMPLE 7.9
Resistivity

The resistivity of copper is about 1.7×10^{-8} ohm–meters. Use equation 19 to answer the following questions:

a. What is the resistance of a copper wire with a radius of 0.5 cm and a length of 10 m?

b. The wire in (a) is evenly stretched to become a much thinner wire 1 km long. What is its resistance?

Solution

a. The resistance of the wire is its resistivity (1.7×10^{-8} ohm–meters) times its length (10 m) divided by its cross-sectional area a. The cross-sectional area of the original wire is πr^2, where $r = 5 \times 10^{-3}$ m. Thus, $a = 7.9 \times 10^{-5}$ m^2. We obtain for the resistance of the wire, $R \simeq 2 \times 10^{-3}$ Ω.

b. The wire's length was increased by a factor of 100. Because its volume does not change, its cross-sectional area is decreased to 10^{-2} of its original value. The resistance of the extruded wire is about 20 Ω.

■

*Simple Circuits
(Optional)*

The circuits we discuss in this section use a battery, one or more resistors, and some wire with which to connect them. The battery is defined to be a device with two terminals between which there is an invariable potential difference V. A wire is a conductor with no resistance. A resistor is a device with a fixed electrical resistance. Figure 7.12 illustrates three simple circuits: (a) consists of one resistor connected to the battery;

FIGURE 7.12
Three simple circuits. The commonly used symbol for a battery recalls the shape of Volta's pile. The symbol for a resistor is a zigzag line. The solid lines represent perfectly conducting wires. *a.* A resistor with resistance R is connected to a battery of voltage V. *b.* Two resistors with resistances R_1 and R_2 are connected to the battery in series. *c.* Two resistors with resistances R_1 and R_2 are connected to the battery in parallel.

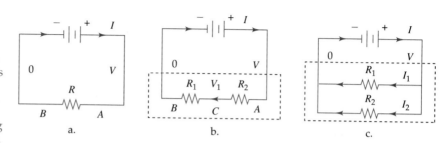

(b) consists of two resistors connected to each other *in series* and to the battery; and (c) consists of two resistors connected to each other *in parallel* and to the battery.

Circuit (a) reviews what we have already learned. A potential difference between points A and B is provided by the battery. The current flowing through the circuit is given by equation 18 to be V/R. If a 100-Ω resistor is hooked to a 6-volt battery, the current is 60 milliamperes.

In circuit (b), the resistors are wired together in series. This means that the current I flowing through R_1 is also the current flowing through R_2. Point C is at a potential intermediate between points A and B. Let the voltage difference between points A and C be V_2 and that between C and B be V_1. Their sum is the voltage difference between points A and B, namely V. Equation 18 is applied to each of the two resistors to yield:

$$V_1 = I\,R_1 \qquad \text{and} \qquad V_2 = I\,R_2$$

If we add these two equations together, we obtain a relation between the current supplied by the battery and its voltage:

$$V = V_1 + V_2 = I\,(R_1 + R_2)$$

The result is Ohm's law for the "black box," surrounded by dotted lines. We have proven a simple but important theorem: The combined resistance of two resistors R_1 and R_2 wired in series is the sum of the two resistances:

$$R = R_1 + R_2 \qquad \text{for two resistors in series} \qquad (20)$$

In fact, the result is more general. The combined resistance of any number of resistors wired together in series is the sum of the individual resistances.

In circuit (c), the resistors are wired together in parallel. This means that the potential difference across each of them is equal to V. However, the currents passing through them are different. Equation 18 is applied to each of the resistors to yield:

$$I_1 = V/R_1 \qquad \text{and} \qquad I_2 = V/R_2$$

Because the current supplied by the battery is the sum of the currents passing through each resistor, we add these two equations to discover that $I = V(1/R_1 + 1/R_2)$. If this result is solved for V and algebraically simplified, we find:

$$V = I \left(\frac{R_1 R_2}{R_1 + R_2} \right)$$

The result is Ohm's law for the "black box," surrounded by dotted lines. We have proven another important theorem. The combined resistance of two resistors R_1 and R_2 wired in parallel is given by the formula:

$$R = \frac{R_1 R_2}{R_1 + R_2}$$

or by the equivalent formula:

$$\frac{1}{R} = \frac{1}{R_1} + \frac{1}{R_2} \qquad \text{for two resistors in parallel} \qquad (21)$$

This equation, too, can be generalized. The reciprocal of the combined resistance of any number of resistors wired together in parallel is the sum of the reciprocals of the individual resistances. For example, if ten 7-Ω resistors are wired in parallel, their combined resistance is 0.7 Ω.

EXAMPLE 7.9
Resistors in Series
and Parallel

We are given four resistors with restances 1, 2, 3, and 6 Ω. If they are connected together in series to a 12-volt battery, what is the total resistance and how much power is consumed? If they are connected together in parallel to a 12-volt battery, what is the total resistance and how much power is consumed?

Solution

Connected in series, the combined resistance is $1 + 2 + 3 + 6 = 12 \ \Omega$. We find, from equation 18, that the battery supplies a current of 1 ampere and, from equation 17, a power of 12 W. Connected in parallel, the combined resistance is the reciprocal of $\frac{1}{1} + \frac{1}{2} + \frac{1}{3} + \frac{1}{6}$, or 0.5 Ω. The battery supplies a current of 24 amperes and a power of 288 W. ∎

Let's get tricky. Suppose that two 60-watt bulbs are connected together in series, as shown in Figure 7.13, with a terminal of one bulb linked to a terminal of the other. What happens when the remaining two terminals are connected to an electric wire and plugged in? The electric current must wend its way through the filaments of both bulbs.

FIGURE 7.13
Two 60-watt bulbs are
wired to one another in
series.

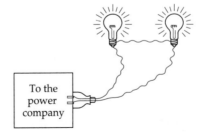

To the
power
company

Two filaments are twice as long as one. Equation 19 says that the total resistance of the conjoined bulbs is twice that of either one, or 480 Ω. When the linked bulbs are connected to a 120-volt line, a current of 0.25 ampere passes through them, according to Ohm's law. The bulbs consume 30 watts of power according to Equation 17. Each of the two bulbs glows dimly.

Exercises

19. What is the resistance of the filament of an American 100-W bulb, where the line voltage is 120 volts? What is the resistance of the filament of a French 100-W bulb, where the line voltage is 240 volts?

20. How much current passes through a 1500-W hair dryer, if the line voltage is 120 volts?

21. Resistors of 2 Ω and 3 Ω are connected together in series. If 10 amperes of current flow through the smaller resistor, what is the voltage across it? What is the current passing through the larger resistor? How much power is dissipated by the system of two resistors?

DO YOU KNOW

All the Electrical Units?

Probably not! There are more electrical units than you can count with fingers and toes. Six of them have been defined so far. Some of those we have not and shall not define are:

Gauss, Gilberts, Henries, and Oersteds;
Maxwells, Siemens, Teslas, and Webers.

Let's review the electrical units we have introduced. The coulomb was defined by equations 3 and 4. Other units are defined in terms of this fixed quantity of electric charge.

- The *ampere* measures the rate of flow of charge—that is, the size of an electric current. It equals 1 coulomb per second.

- The *volt* measures the electric potential. It equals 1 joule per coulomb. A charge of 1 C at a potential of 1 volt has an electric potential energy of 1 J.

- The *ohm* is a unit of electrical resistance. It equals 1 volt per ampere. If a 1-Ω resistor is connected to the terminals of a 1-volt battery, 1 ampere of current flows through it.

- The *watt* is not exclusively an electrical unit. It is a unit of power equal to 1 joule per second. A current of 1 ampere passing through a voltage difference of 1 volt consumes (that is, changes to other forms) 1 watt of electric power.

- The *farad* is a unit of capacitance. It equals 1 coulomb per volt. A conductor with a capacitance of 1 farad at a potential of 1 volt carries a charge of 1 C.

22. Resistors of $2\,\Omega$ and $3\,\Omega$ are connected together in parallel. If 10 amperes flow through the smaller resistor, what is the voltage across it? What is the current passing through the larger resistor? How much power is dissipated by the system of two resistors?

23. A toaster is rated at 1200 W at a line voltage of 120 volts. It is plugged into a receptacle in the kitchen of a house whose old wiring presents a resistance of $3\,\Omega$ to the circuit. This means that the power company is supplying 120 volts to a series circuit consisting of the toaster and the house wiring.
    a. What is the resistance of the toaster?
    b. How much power does the toaster dissipate?
    c. At what rate is heat dissipated in the wiring?

22. I spent a sabbatical year in Marseilles, where the line voltage is 240, twice the American voltage. My apartment was provided with a fixture consisting of a translucent globe containing two American 100-W bulbs wired in series. Explain why the lamp worked perfectly well until one of the bulbs blew out. What happened then and why?

25. Prove the following assertion: The resistance of two resistors in series is always larger than that of either resistor, and the resistance of two resistors in parallel is always smaller than that of either resistor.

7.4 Electrochemistry

Michael Faraday played the most important role in the story of electricity and magnetism. In this section, I describe his discoveries in *electrochemistry*, the study of the chemical effects of an electric current. But first, let us get acquainted with the young Faraday.

As an apprentice bookbinder with an amateur's interest in science, Faraday attended lectures at the Royal Institution in London. At one lecture, Davy explained how he decomposed muriatic acid into hydrogen and a green gas he named chlorine. Davy could decompose chlorine no further. He argued that it was a new element, and that the acid was a compound of hydrogen and chlorine. This was scientific heresy because most chemists accepted Lavoisier's claim that oxygen is the essential ingredient of all acids.

Faraday became an ardent fan of Davy and begged to be taken on as his assistant. On Christmas day, 1812, he was offered the post at a salary of one guinea a week, with fuel, candles, and two rooms in the garret of the Royal Institution. He would be given laboratory aprons and free access to the laboratories. First, however, Faraday was to accompany Dr. and Mrs. Davy on a two-year-long scientific tour of Europe.

Michael Faraday (1791–1867) as a young man. *Source:* AIP/Emilio Segré Visual Archives.

The excitement began in Paris where Ampère and his colleagues presented Davy with a curious purple chemical smelling a bit like chlorine, a substance the French chemists had just isolated. Davy was never without his portable chemical laboratory and set to work. On that very day, he concluded that the new substance did not contain chlorine. A month later, being unable to decompose it further with a borrowed battery, Davy suggested the name iodine and insisted "from all the facts that have been stated, there is every reason to consider this new substance as an undecomposable body." Faraday, who had been a vicarious observer of the discovery of chlorine, was now an active participant at the birth of another new chemical element.

In Genoa, Davy turned to the question of animal electricity. At its famous aquarium, he examined certain fish to see whether they discharged a form of electricity and could therefore decompose water. Davy found no effect, but Faraday, decades later, returned to this problem and established that all of the so-called electricities then known—static, voltaic, animal, thermoelectric, and magnetoelectric—were manifestations of the same underlying physical principles.

Davy borrowed a superb lens from the Grand Duke of Florence. With focused sunrays, he burned diamonds and determined their combustion product to be carbon dioxide. He showed that diamonds are pure carbon. We know today that diamond, graphite, charcoal, and soot are all made of carbon atoms, just differently arranged. Both the power and frailty of scientific reductionism are revealed. Soot and shining stone are atomically identical. How their myriad atoms are put together is the thing: Pattern alone tells the worthless from the priceless. Not all knowledge resides at the level of the ultimate structure of matter.

When war clouds gathered over Europe in 1814, Davy ended the tour. Faraday returned at last to his beloved London and escaped from the most insufferable Mrs. Davy. He would never again forsake his laboratory for the glitter of the salon. Faraday began his career as the protégé of a chemist and remained a promising but unexceptional chemist until Davy's death in 1829. As a lowly caterpillar becomes a butterfly, so he emerged as one of the greatest physicists of all time. We learned earlier how he was led to propose the concept of the electric field. Chapter 8 describes how he discovered the phenomenon of electromagnetic induction, by which magnetism produces electricity, and by which we now convert mechanical energy to electrical energy with the dynamo and vice versa with the electric motor. He also deduced two marvellously simple and suggestive laws governing electrolysis.

Electrolysis

In Section 7.3, we studied the passage of electric currents through conductors. Metals resist the flow of current and convert electric energy into heat. The thin filament of a lamp bulb becomes hot enough to radiate light. When a current passes through a gas, atoms are excited and light can be produced more directly, as in a fluorescent fixture. It is natural to ask what happens when a current passes through a liquid, such as a molten salt or a salt solution. In Volta's electric battery, chemical reactions take place in the salt solution between dissimilar metal disks, thereby generating an electric current—chemical energy becomes electrical energy. *Electrolysis* is the inverse process, by which an electric current sent through a fluid causes chemical transformations to take place.

Two English scientists, Anthony Carlisle and William Nicholson, heard of Volta's discovery and hurriedly built their own battery out of British coins and wet cardboard. They wired its terminals to platinum plates immersed in a dilute salt solution, as in Figure 7.14. Platinum was used because it is chemically inert. The scientists found that hydrogen gas bubbled up at the negative plate (the *cathode*), and oxygen gas at the positive plate (the *anode*). They showed that an electric current flowing through a salt solution decomposes water into its elements. Here is how it happens, in five easy steps, as viewed from a twentieth-century perspective:

1. Electrons, denoted by e^-, are pumped into the cathode by the electric battery.

2. They combine with water molecules according to the cathode reaction:

$$2\,e^- + 2H_2O \longrightarrow H_2 \uparrow + 2OH^-$$

FIGURE 7.14
Water is decomposed by the passage of an electric current through it. Hydrogen gas collects at the platinum cathode and oxygen gas collects at the platinum anode.

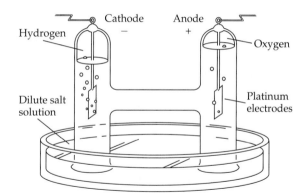

The upward arrow indicates that a molecule of hydrogen gas is liberated. At the same time, two OH^- ions are produced in the solution. Each consists of an oxygen atom, a hydrogen atom, and one extra electron.

3. An excess of negative charge develops at the cathode and an electric current flows through the liquid. Here's where the salt comes in. Pure water contains only a few ions and conducts electricity poorly. Salt comes apart into charged ions when it is dissolved. These facilitate the passage of the current.

4. Electrons jump from water molecules to the anode according to the anode reaction:

$$2\,H_2O \longrightarrow O_2 \uparrow + 4\,H^+ + 4\,e^-$$

A molecule of oxygen gas is liberated, while four H^+ ions (hydrogen atoms that have lost an electron) are produced in the solution.

5. Electrons migrate from the anode toward the battery, thus completing the electric circuit. Leftover ions in the solution find one another and recombine according to

$$H^+ + OH^- \longrightarrow H_2O$$

The net result of the cathode reaction, the anode reaction and the recombination process, is to decompose water:

$$2\,H_2O \longrightarrow 2\,H_2 + O_2$$

The energy required to do this is furnished by the battery.

The electrochemical decomposition of common salt proceeds similarly. Molten salt is made up of Na^+ ions and Cl^- ions. When an electric current is passed through it, as in the industrial device shown in Figure 7.15, the following electro-chemical reactions take place:

$$e^- + Na^+ \longrightarrow Na \uparrow \qquad 2Cl^- \longrightarrow Cl_2 \uparrow + 2e^-$$

where electrons are supplied at the cathode and extracted at the anode.

Davy, Carlisle, and Nicholson did not understand the microphysical processes underlying their discoveries. Neither did Faraday, but his precise studies of electrochemical processes in 1832 led him almost all the way to the modern view.

FIGURE 7.15
An electric current passes through molten salt in this commercial reaction vessel. As the salt is electrolyzed, chlorine gas is collected at the anode. At the same time, molten metallic sodium rises up at the cathode.

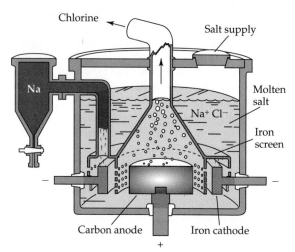

The Laws of Electrolysis

Faraday searched for quantitative laws governing electrolysis. Rather than a dilute salt solution or a molten salt, he electrolyzed concentrated solutions of salts such as copper sulphate ($CuSO_4$) or silver sulphate (Ag_2SO_4). In these cases, the ensuing electrochemical reactions produce a deposition of metal at the cathode and evolve oxygen at the anode:

$$Cu^{++} + 2e^- \longrightarrow Cu \uparrow \qquad 2H_2O \longrightarrow 4H^+ + O_2 \uparrow + 4e^-$$

He measured the current passing through the electrolyte and the time interval over which the process was continued. The product of the current and the time interval was the electric charge that had traversed the cell. By weighing the cathode both before and after, Faraday found the mass of metal that had come out of solution. From the volume of oxygen gas that was collected, he determined its mass. After many experiments of this kind, Faraday formulated two precise laws that are of astonishing simplicity:

• The quantity of material (say, copper) liberated at each electrode depends solely on the quantity of charge that has passed through the cell. It does not depend on the area of the electrodes, the rate of the electrolytic process, the particular copper salt in the solution or its concentration, or on anything else at all.

- The charge that must pass through the electrolyte to liberate 1 mole of any material is always a small integer multiple of a certain fixed charge. It is approximately 96500 coulombs and is now referred to as a *faraday* of charge.

Let's apply these laws to the electrochemical processes we have already discussed. Suppose that 9.65 amperes of current are sent through molten salt for 10000 seconds (about 3 hours). A charge of 1 faraday passes through the electrolyte. One mole of sodium (23 g) is produced at the cathode and 1 mole of elemental chlorine (35.5 g) is collected at the anode. A sodium ion, we know today, is a sodium atom that has become charged by losing an electron. It is positively charged be-

---

DO YOU KNOW

The Words Faraday Coined for His Discoveries?

The new technology of electrochemistry needed a new language as well. To present his extraordinary findings, Faraday had to invent many new words. In this philological enterprise, he sought counsel from a professional wordsmith, William Whewell of Trinity College, Cambridge. All of these words remain with us today:

Anion (From the Greek for upward way). A negatively charged ion. Anions migrate to the anode of an electrolytic cell.

Anode The positive electrode of an electrolytic cell.

Cation (From the Greek for downward way). A positively charged ion. Cations migrate to the cathode of an electrolytic cell.

Cathode The negative electrode of an electrolytic cell.

Electrode One of the two bodies that is connected to a terminal of a battery or other source of electricity and immersed in the electrolyte.

Electrolysis The process of electrochemical decomposition.

Electrolyte A material (usually a salt solution or a molten salt) that conducts electricity and is thereby decomposed.

Electrolytic cell A device consisting of an electrolyte in which the anode and cathode are immersed, in which electrolysis takes place.

Electrolyze To be electrochemically decomposed.

Ion An atom, such as may be found in an electrolyte, that has become electrically charged.

In addition to the words that Faraday introduced, two electrical units are named for him. The *farad* measures the capacity of a body to carry charge. The *faraday* is a quantity of electric charge equal to that of Avogadro's number of electrons.

cause the charge of the electron is negative. Because 1 faraday of charge produces 1 mole of sodium, the electric charge of Na^+ must be 1 faraday divided by the number of atoms in 1 mole, to wit:

$$Q(Na^+) = \frac{96,500 \text{ C}}{N_A} \simeq 1.6 \times 10^{-19} \text{ C}$$

At the same time, 1 mole of chlorine atoms (half a mole of molecular Cl_2 gas) evolves at the anode. It follows that:

$$Q(Cl^-) = -\frac{96,500 \text{ C}}{N_A} \simeq -1.6 \times 10^{-19} \text{ C}$$

which is also the electric charge of the electron. One faraday of electric charge corresponds to a mole of electrons.

If 2 faradays of electric charge pass through a dilute salt solution, 1 mole of H_2 gas is produced at the cathode, but only half a mole of O_2 appears at the anode. This is because it takes two electrons to liberate an oxygen atom. Similarly, it takes 2 faradays to extract a mole of copper or zinc from a salt solution, because the ions of these metals are doubly charged.

Faraday's experiments shed light on the operation of an electric battery because it is also an electrolytic cell. He showed that the electric battery is not the perpetual motion device Volta imagined it to be, and that every child with an electric toy knows it is not. Electrochemical reactions taking place within convert chemical energy into electric energy. When its store of energy is exhausted, a battery must be recharged or recycled.

EXAMPLE 7.11
Electrolysis

Two metal plates, one of zinc and the other of the nonreactive metal platinum, are placed in a slightly acidic solution (Figure 7.16). The two

FIGURE 7.16
A platinum bar and a zinc bar are wired together and placed in a dilute acid. Zinc atoms of one bar become Zn^{++} ions in the electrolyte. Hydrogen ions in the electrolyte migrate to the platinum bar, where they are neutralized to form hydrogen gas. A current passes through the wire from the Pt bar to the Zn bar.

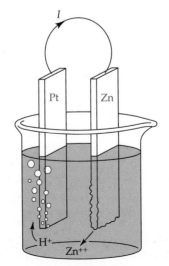

plates are connected by a wire. The device is now a primitive battery, and a current I flows through the wire. Meanwhile, hydrogen gas is produced at the platinum plate while the zinc plate is gradually eaten away as Zn^{++} ions enter the solution. Suppose that the current is measured to be 0.1 ampere and that the process is carried out for 24 hours.

a. How much charge flows through the circuit?

b. What mass of hydrogen is produced?

c. What mass of zinc enters into solution?

Solution

a. The charge that flows through the wire from the Pt plate to the Zn plate is the product of current and time: 0.1 ampere times 86,400 s (24 hours) yields 8640 coulombs or $\sim 9 \times 10^{-2}$ faraday.

b. One faraday of charge liberates 1 mole of elemental hydrogen from the electrolyte. Thus, 0.09 mole of hydrogen was released. Because the atomic weight of hydrogen is 1, the hydrogen gas that bubbled up at the Pt plate had a mass of 0.09 g.

c. Two faradays of charge will convert 1 mole of zinc atoms of the electrode into Zn^{++} ions in the electrolyte. Thus, 0.09 faraday dissolve 4.5×10^{-2} moles of zinc. Because the atomic weight of zinc is 65, the mass of the zinc electrode is reduced by 2.9 g.   ∎

Faraday neither knew about electrons nor believed that atoms have parts. He thought of electricity as a fluid that is entangled with matter. He approached (but did not quite make it to) the notion of a discrete atomic or subatomic unit of charge in his statement:

> The equivalent weights of bodies are simply those quantities of them which contain equal quantities of electricity, or have naturally equal electric powers; it being ELECTRICITY which determines the equivalent number because it determines the combining force. Or, if we adopt the atomic theory or phraseology, then the atoms of bodies that are equivalents to each other in their ordinary chemical action, have equal quantities of electricity naturally associated with them. But I confess that I am jealous of the term atom; for though it is very easy to talk of atoms, it is very difficult to form a clear idea of their nature.

Half a century passed before physicists saw that Faraday's laws of electrolysis demand that electric charge comes in packages of a fixed and universal size—the amount needed to neutralize a single ion. At a famous Faraday Memorial Lecture in 1881, the German physicist Helmholtz dared to put forward the idea whose time had finally come:

> If we accept the hypothesis that elementary substances are composed of atoms, we cannot well avoid concluding that electricity also is divided into elementary portions which behave like atoms of electricity."

By the close of the nineteenth century, the atom of electricity was observed in the laboratory and given its name: the electron. In the early years of the twentieth century, scientists developed a clear idea of the nature of the atom. Lord Kelvin's challenge of 1862, the title of this chapter, was amazingly prescient.

————◆ · ◆——————

Exercises

26. A current of 965 amperes, sent through a molten salt of aluminum, yields the metal at the rate of 5.4 g/min. The atomic weight of aluminum is 27. From this information, determine how many electrons combine with an aluminum ion in the electrolyte to produce an aluminum atom.

27. A piece of zinc is put into a solution of hydrochloric acid. The zinc is eaten away as 10 L of hydrogen gas at STP is liberated. What mass of zinc has dissolved?

28. The atomic weight of silver is 108 and its ions are singly charged. How much silver is deposited at the cathode if a current of 10 amperes is sent through a solution of a silver salt for 100 hr?

29. The unit of nuclear superiority, the "kiloton," originally meant the energy released by a thousand tons of TNT. Today, the kiloton is redefined to be exactly 10^{12} calories ($\sim 4.2 \times 10^{12}$ J). Nuclear weapons range in power from kilotons to megatons. What is the electric potential energy in megatons of two 1–faraday charges (i.e., moles of electrons) separated by a distance of 1 km?

30. A faraday of charge passing through an electrolytic cell liberates 1 mole of any monovalent element. Explain why this result suggests the existence of a fundamental unit of electric charge.

31. A typical automobile battery stores 4 kilowatt-hours of energy, while the heat content of a fluid ounce of gasoline is about a million joules. The energy available from the battery is the equivalent of how many gallons of gasoline?

Where We Are and Where We Are Going

This chapter dealt with electricity and electrical phenomena. Little was said about magnetism beyond a few parenthetical remarks. Magnetism, it turns out, is produced by electricity. Electric charges at rest are surrounded by electric fields. However, we shall learn in the next chapter that when they are in motion, they produce magnetic fields as well. (This is heady stuff, since Galileo assured us that the laws of physics are the same for all uniformly moving observers. If an electric charge flits by me, I see both an electric and magnetic field. But if I fly with the charge, I see only an electric field. The simple act of changing one's state of motion has the remarkable effect of mixing up magnetism and electricity!)

The magnetism of an electromagnet, such as is used to lift up old cars, results from electric currents circulating in a coil. That of a permanent magnet results from the motions of atomic electrons within. Not only can electricity produce magnetism but, Faraday showed, magnetism can make electricity. His discovery made it possible to generate electric power from fuels and falling water. Even more surprisingly, electricity and magnetism can bootstrap one another to produce electromagnetic waves, among whose many forms is light. And it is through the study of light, on Earth and from the stars, that scientists began to discern the simplicity and beauty of physical laws.

8 The Marriage of Electricity and Magnetism

Faraday discovered the facts of nature; Maxwell, taking the known facts, endowed them with a theoretical setting. Just because this kind of work lies deep down, a long time must often elapse before its utility first becomes apparent. An example is Maxwell's electromagnetic theory, and the waves which he saw were implied in the theory. Nearly a quarter of a century were to pass before these waves were detected in the laboratory, and yet another quarter of a century before they took their place in everyday life. While they exemplify the time lag which must inevitably intervene between seed-time and harvest, between the research in pure science and its utilitarian application, they provide an even more outstanding illustration of how great the value may be—even if often a deferred value—of research in pure science carried out for the sake of pure science alone, and with no motive other than to understand the innermost workings of nature.

Sir James Jeans, British physicist

Many great scientists contributed to the sciences of electricity and magnetism. James Clerk Maxwell (1831–1879), whose unified theory of electromagnetism revealed light to be an electromagnetic wave, was the greatest of them all. *Source:* The Bettmann Archive.

Equals repel, opposites attract, for both magnetic poles and electric charges. Magnetic forces are described by magnetic fields, electric forces by electric fields. And yet, electricity and magnetism seem to be entirely disparate phenomena. We wander through a forest on a stormy, moonless night, where a flash of (electric) lightning gives a glimpse of the terrain and our (magnetic) compass guides us home, but the profound relationship between electricity and magnetism is obscured by the complexity of the everyday world.

This chapter examines the hidden but intimate links between electricity and magnetism that were revealed in the nineteenth century. Section 8.1 describes the startling discovery that an electric current influences compasses and behaves like a magnet. We describe how electric currents generate magnetic fields that exert forces on other electric currents. Section 8.2 describes the discovery of electromagnetic induction: how changing magnetic fields produce electrical effects. We shall see how this discovery led to Maxwell's great synthesis of electricity and magnetism. He deduced (and Hertz demonstrated) that there exist self-sustaining electromagnetic waves, of which light is one example, radio waves another. Section 8.3 examines the relationship between electric currents and pointlike charges in motion.

8.1 Electricity Makes Magnetism!

As electric batteries spread throughout Europe and America, physicists acquired strong and sure sources of electric current. Two new sciences emerged: electrochemistry, of which we have spoken, and electromagnetism, whose story begins in Denmark, the land of two famous Hans Christians: Andersen, who told fairy tales of frog-kissing princesses, and his close friend Oersted, who started us on our way from the frog-inspired researches of Volta and Galvani to the triumphs of classical physics.

Hans Christian Oersted (1777–1851) was born on the charming island of Langeland. He became a pharmacist, then a philosophically inclined professor of physics and sometime poet at the University of Copenhagen. In 1813, influenced by Kant's idea of the unity of natural phenomena, Oersted set his own challenge:

> One has always been tempted to compare the magnetic forces with the electrical forces. The great resemblance between electrical and magnetic attractions and repulsions and the similarity of their laws necessarily would bring about this comparison. An attempt should be made to see if electricity has any action on the magnet as such.

The Current and the Compass Needle

Oersted was teaching a course on electricity and magnetism. The table was littered with various devices with which he could show the sparks and arcs produced by electric currents and the remote effects of magnets

Oersted observes that an electric current, provided by a crude voltaic cell, can deflect a nearby compass needle. *Source:* Courtesy, Meyers Photo-Art.

on compass needles. When he sent a current through a wire that happened to be near a magnetic compass, as in the illustration, he noticed that the current affected the orientation of the compass needle. Never before was electricity known to produce magnetic effects—Oersted had found the first direct link between electricity and magnetism.[*]

Oersted investigated the new phenomenon exhaustively during the rest of the academic year. His triumphant paper entitled *Experiments on the Effect of a Current of Electricity on the Magnetic Needle* was published on July 21, 1820. In it, he asked and answered the following important questions regarding the effect of an electric current on a suspended needle:

- *What should the new phenomenon be called?* "To the effect which takes place in the conductor and in the surrounding space we shall give the name 'conflict of electricity'." Oersted uses the phrase "conflict of electricity" to describe what would later be interpreted as the production of a magnetic field by an electric current. A nearby compass needle responds to the magnetic field generated by the current.
- *Does an electric current affect unmagnetized needles?* Not at all: "A brass needle, suspended like a magnetic needle, is not moved by the effect of the wire. Likewise, needles of glass or lacquer remain unacted upon."
- *Can the effect produce a current that can penetrate through intervening matter?* "The effect passes to the needle through glass, metals, wood, water, resin, stonewear, stones. The transmission of effects through all these materials has never before been observed in electricity and galvanism. The effects, therefore, which take place in the conflict of electricity are very different from the effects of either of the electricities."[†]
- *How does the effect depend on the distance between the current-carrying wire and the compass?* "If the distance is increased, the angle [by which the needle is deflected] diminishes proportionately." That is, the magnetic field produced by a current passing through a long straight wire depends inversely on the distance to the wire.
- *How does the effect depend on the size of the current?* Oersted could not give a precise answer to this question because instruments were not yet available to measure currents. However, he established that the magnitude of the twisting force on a magnet needle "varies with the [voltage] of the battery."
- *Does the effect of a current on a compass needle depend on the material of which the conducting wire is made?* It does so only in that the currents

---

[*]Oersted's discovery may be the only instance of a major scientific breakthrough taking place during a lecture to students.

[†]Oersted alluded to static and current "electricities." Faraday had not yet established their equivalence—he was still fooling with chemistry when Oersted made his epochal discovery.

passing through wires with different resistances are different: "The nature of the metal does not alter the effect, but merely the quantity. Wires of platinum, gold, silver, brass, iron, ribbons of lead or tin, a mass of mercury, were employed with equal effect."

- *In what direction is a compass needle made to point by a nearby electric current?* "From the preceding facts we may likewise collect that this conflict performs circles. The magnetic needle was moved from its position by the [electric current,] but the galvanic circles must be complete." This cryptic comment demands explication.

The magnetic field at a point near the current-carrying wire points neither toward the wire nor away from it. Nor does it point along the wire. The lines of force of the magnetic field produced by the current "perform circles" about the wire. That is, compass needles orient themselves tangentially to circles around the wire. The direction of the magnetic field of a wire is given by the *right-hand rule*. Imagine grasping the wire so that your right thumb points in the direction of the current. Your fingers indicate the direction of the circular magnetic field lines, as shown in Figure 8.1.

Within weeks of its submission, Oersted's short paper was translated from Latin to English, German, French, Italian, and Dutch. His discovery awakened two sleeping giants: Ampère in France and Faraday in England. In the remainder of this section, we discuss how Ampère and his colleagues founded the new science of electromagnetism. The remaining

FIGURE 8.1
Oersted said that the magnetic effect of a current "performs circles," by which he meant that compass needles line up as shown about a current. The arrows on the needles denote their north poles. Today, we say that the magnetic field lines produced by the current form circles. The north pole of a needle indicates the direction of the magnetic field.

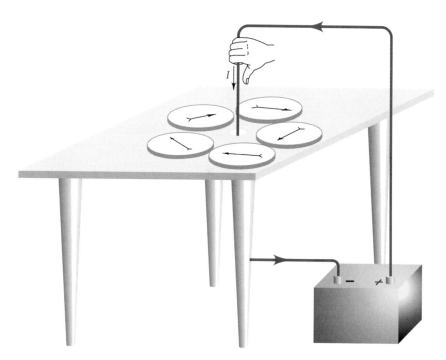

André-Marie Ampère (left) and his friend François Arago. Arago brought Oersted's work to the attention of Ampère, who began the quantitative study of electromagnetism. *Source:* Historical Pictures/Stock Montage, Inc.

pieces of the puzzle were found by Faraday and put in place to form a complete theory by Maxwell. These developments are described in Section 8.2.

Electromagnetism

André-Marie Ampère (1775–1836) was born and reared in the French provinces. His father coached the precocious boy, so that by age 12 he had mastered the scientific treatises of Laplace and Lagrange, and had learned Latin to study those of Euler and Bernoulli. But when he was 18, he witnessed the execution of his father in the Terror of 1793. He recovered from this trauma, only to suffer the loss of his beloved wife a decade later. At this point, Ampère and his mother moved to Paris, where he became a respected and diligent scholar but accomplished nothing noteworthy until the remarkable autumn of 1820. How did a middle-aged scientist suddenly blossom into the founder and master of a new science?

François Arago, a younger scientific colleague of Ampère's, read Oersted's paper and rushed to Geneva, Switzerland, to see Oersted's experiment demonstrated. On September 4, 1820, Arago described the effect to the elite of French science. A week later, he reproduced the original experiments for the astonished savants. Arago showed that an electric current could magnetize steel permanently and soft iron temporarily. His enthusiasm inspired Ampère to make his own gargantuan contributions to the new science.

The theory of electromagnetism was created by Ampère, Arago and their colleagues. The autumn of 1820 was a pinnacle of French science. It represented *"a turning point of history no less significant than the hundred days of Napoleon which brought him to Waterloo."**

Ampère immediately set out to study the magnetic effects produced by electric currents. Where Oersted saw only the effect of a current on a

* Norman Feather, *Electricity and Matter* (Edinburgh: Edinburgh University Press, 1968), 73.

FIGURE 8.2
A current-carrying coil is suspended so that it is free to rotate. It behaves like a compass needle. One side is attracted to the north pole of a bar magnet *a*, the other to the south pole of the magnet *b*.

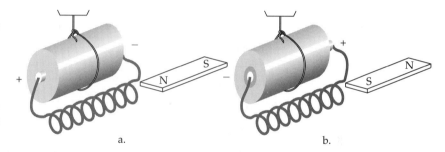

a.   b.

magnet, Ampère spoke of "the mutual action of a current-carrying wire and a magnet." If a current exerted a force on a magnet, Newton's third law required that the magnet must exert an equal and opposite force on the current-carrying wire.

To show that a magnet acted on a current, he arranged a current-carrying coil to be free to rotate on its axis, as shown in Figure 8.2. When he brought a bar magnet nearby, the loop turned to face the magnet (Figure 8.2a). When the current was reversed, the loop did an about-face (Figure 8.2b). When precautions were taken to isolate the current loop from extraneous effects, it oriented itself in Earth's magnetic field as if it were a magnetized compass needle. It was just that, Ampère believed.

Ampère found that a loop or coil of current behaves in all respects like a magnet. He conjectured that, conversely, all magnetic behavior, even that displayed by lodestones, results from electric currents circulating within magnetized bodies. One of today's masters of magnetism writes:*

> [Ampère's arguments] did not logically justify the belief that permanent magnetism is caused by internal electric currents. Nevertheless, this was the hypothesis most economical in concepts which could be put forward, and as it turned out, the most fruitful in stimulating new discoveries and in creating 'insight' into the 'physics' of magnetism. It was Ampère and his followers who acted in the modern style, which we may describe as the harmonious union of theory and experiment.

It took another century before the internal currents within a magnet were identified with the motions of atomic electrons. Ampère's guess was basically correct, although the properties of magnetic materials still mystify practicing physicists. Theoretical physicists have yet to understand how recently discovered high-temperature superconductors do their thing.

The Force between Currents

If an electric current behaves like a magnet, it seemed plausible to Ampère that two electric currents should exert forces on one another. However, he wrote:

*Daniel C. Mattis, *The Theory of Magnetism* (New York: Harper & Row, 1965), 14.

When M. Oersted discovered the action which a current exercises on a magnet, one might certainly have suspected the existence of a mutual action between two circuits carrying currents; but this was not a necessary consequence; for a bar of soft iron also acts on a magnetized needle, although there is no mutual action between two bars of soft iron.

FIGURE 8.3
Two parallel currents, I and J, attract one another with a force F that is proportional to the product of the currents and inversely proportional to the separation D between them.

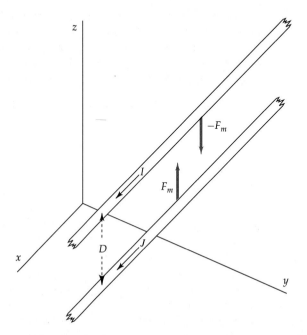

Ampère performed experiments to determine whether currents exert forces on one another. He found the force, measured it, and formulated the theory by which the force may be computed. A simple example of Ampère's general result is shown in Figure 8.3, in which two current-carrying wires are suspended one above the other and parallel to the x-axis. The upper wire carries current I, the lower one current J. If their length L is much greater than the distance between them D, the force between the wires satisfies simple laws:

- The force is an attraction if both currents have the same sign. It is a mutual repulsion if the currents have opposite signs.
- The magnitude of the force is proportional to the product of the two currents. Furthermore, it is proportional to the length of the wires and inversely proportional to the distance between them. If current is measured in amperes, lengths in meter, and force in newtons, the constant of proportionality comes out to be 2×10^{-7}. The force that each current exerts on the other is

$$F_\mathrm{m} = (2 \times 10^{-7})\, IJL/D \qquad (1)$$

In fact, equation 1 provides the modern definition of the unit of electric current. The ampere is the steady current that, when flowing in straight parallel wires of infinite length and negligible thickness, separated by a distance of 1 meter in free space, produces a force between the wires of 2×10^{-7} newtons per meter of length.

EXAMPLE 8.1
The Force between
Two Wires

Two copper wires carrying equal currents I are strung parallel to one another 10 m apart. One wire hangs one cm above the other. The uninsulated wires have circular cross sections one mm in diameter. The density of copper is 9 g/cm^3. For what value of I does the upward magnetic force on the lower wire equal its weight?

A BIT MORE ABOUT
Arago

Scientists cherish superstars like Ampère and Faraday, whose brilliance outshines all others, just as they dote on the rare supernova blazing with the power of a billion suns. But many staid and steady luminaries set the stage on which genius plays. François Arago (1786-1853) was the man behind the scenes for a three-decade run of French science. Graduating at the head of his class from Ecole Polytechnique, Arago was sent by the Bureau des Longitudes to survey the strategically important Balearic Islands. He was arrested as a spy but escaped to Algiers. He tried to return home, but his boat was seized by a Spanish warship. After years of struggle, he returned to Paris in 1809 with the precious survey in hand! At age 23, Arago was welcomed into the prestigious Société d'Arcueil, elected to the Institut de France, appointed professor of mathematics at his *alma mater,* and later made Director of the Paris Observatory.

When Arago learned of Louis Daguerre's new photographic process, he championed the invention and befriended its shy inventor. Through Arago's patronage, Daguerre earned a generous pension and a lion's share of the credit for inventing photography. Arago, even

as his sight was failing, became the first astronomer to use photographic techniques in conjunction with a telescope. In 1850, the well-known physicist Sir David Brewster said of Arago in an address to the British Association:

> Threatened with the loss of that sight which has detected so many brilliant phenomena, and penetrated so deeply the mysteries of the material world—he is now completing, with the aid of other eyes than his own, those splendid researches which will immortalize his own name and add to the scientific glory of his country.

The blind physicist developed experimental techniques (such as rapidly rotating multiple mirrors) that were used triumphantly by Fizeau and Foucault to measure the speed of light (see Chapter 9). As for himself,

> I can only, in the present condition of my sight, accompany with my good wishes, the experimenters who desire to follow my ideas.

Arago was a catalyst to his peers and an inspiration to his younger colleagues. In 1845, he encouraged a promising mathematician, Urbain Leverrier, to study the anomalous motion of Uranus. This endeavor led to the discovery of Neptune, and Arago's protégé succeeded his mentor as Director of the Paris Observatory.

Solution   The weight of the wire is the product of its volume and density. It is a circular cylinder of length $L = 10^3$ cm and radius $r = 5 \times 10^{-2}$ cm. Its volume is $\pi r^2 L$ or 2.5π cm^3. Its mass is ~ 70 g and its weight is ~ 0.7 N. The upward magnetic force due to the current in the upper wire is $2 \times 10^{-4} I^2$ N according to equation 1. We set this force equal to the weight of the wire and solve for I to obtain $I \simeq 60$ amperes. If the current is larger, the lower wire will spring upward.   ∎

Electromagnetism   From 1820 to 1826, Ampère refined the discipline he named electromagnetism. Along with Biot and Savart, he showed how magnetic effects may be described by the magnetic fields produced by currents and the magnetic forces exerted by magnetic fields on currents. Maxwell later wrote of Ampère's work:

> The whole, theory and experiment seems as if it had leaped, full-grown and full-armed, from the brain of the 'Newton of electricity.' It is perfect in form and unassailable in accuracy; and it is summed up in a formula from which all the phenomena may be deduced, and which must always remain the cardinal formula of electrodynamics.

Electric Fields and Electric Forces, Reviewed   The laws governing the forces between currents are very much like those governing the electric forces between charges. We learned in Chapter 7 how to find the electric force one charged body exerts on another charged body. It is a two-step process. (1) We compute the electric field produced by one of the bodies; (2) then we compute the force on the other body exerted by the electric field of the first body. Let the total charge of the first body be Q and that of the second be q. In Figure 8.4, each body is divided into many tiny regions or *charge elements*, which we designate as ΔQ and Δq. Because each charge element is small, we may use Coulomb's law to determine its electric field. The electric field produced by the shaded charge element ΔQ at the charge element Δq is

$$\Delta \mathbf{E} = k_c \frac{\Delta Q \, \mathbf{n}}{R^2} \qquad (2)$$

where $\mathbf{n}$ is a unit vector pointing from ΔQ to Δq, which are separated by the distance R. The electric field $\mathbf{E}$ produced by the body with charge Q at a remote point is the vector sum of the electric fields produced by each of its charge elements. If the charge Q resided on a conducting sphere, its electric field anywhere outside the conductor would be the same as that of a point charge Q placed at the center of the sphere. Calculating the electric field produced by a lopsided distribution of charge is simple in principle but difficult to carry out in practice.

Let There Be Light!

In a sense, Maxwell's equations were smarter than he was. Once he found the correct equations, he saw that they told of the existence of electromagnetic waves. Even in the absence of any charges and currents, in a vacuum (or the ether, according to Maxwell), a changing magnetic field creates an electric field, and the changing electric field recreates the changing magnetic field. This bootstrap process continues in the form of an electromagnetic wave spreading throughout space.

Maxwell showed that electromagnetic waves are broadcast in all directions by oscillating electric currents. Their frequency is equal to that of the electrical oscillations. If he had known of the existence of charged particles such as electrons, he might have said, in a pedagogical spirit:

> Everyone knows that an electron at rest produces an electric field, and that an electron in uniform motion produces a magnetic field as well. My equations have a further consequence. When an electron is accelerated, it produces electromagnetic waves. Thus, if it is made to go faster, or slows down, or vibrates, or moves in a circle, it will lose some of its energy in the form of electromagnetic radiation.

What were these waves? Maxwell found that he could compute the velocity of his "self-sustaining undulations of the ether" in terms of the electric and magnetic properties of currents and stationary charges that had been measured in the laboratory. The calculated velocity of his waves came out to be exactly the same as the directly measured speed of light. Imagine his surprise and delight! His conclusion was revolutionary but inescapable. Light is an electromagnetic phenomenon consisting of rapidly changing electric and magnetic fields oriented transversely to the direction of propagation of the light ray. The unification of electricity and magnetism produced an unexpected bonus: It was also a theory of light!

The Discovery of Radio Waves

Not everyone believed Maxwell's theory at first. Some theorists tried to do without displacement currents. No one had directly shown that electromagnetic waves exist by producing and detecting them with electromagnetic devices. Maxwell did not live to see the ultimate triumph of his theory.

Heinrich Hertz (1857–1894) was comfortable with both book and bench—he contributed both to theoretical and experimental physics. He may have been the first person on the Continent to master Maxwell's theory . His discovery of radio waves made him its staunchest advocate. After internships at Munich, Berlin, and Kiel, Hertz married and set up shop in 1886 as Professor of Physics at the Technical University of Karlsruhe. Within two years, by means of a beautiful series of experiments, he found electromagnetic waves.

Maxwell's displacement current allows changing electric fields to

Transmitter   Receiver

Radio waves

FIGURE 8.16
How Hertz discovered radio waves. Sparks made by a high-voltage source of electricity produced an oscillating current that radiated electromagnetic waves. These waves propagated several meters to an isolated electric circuit, the receiver, where they induced a current. Hertz saw sympathetic sparks being produced at the receiver. Parabolic metal mirrors focused the waves from the transmitter and to the receiver.

generate magnetic fields. Faraday's electromagnetic induction is the converse statement—that changing magnetic fields generate electric fields. Together, these two relations lead to an electromagnetic "bootstrap" effect that makes possible the propagation of waves through empty space.

Hertz set out to generate electromagnetic waves in his laboratory. He knew that a spark, jumping between two conductors, transfers charge from one to the other. The system overshoots, so that the conductors develop opposite charges. There is a second spark, and a third, and so on. If the circuit is carefully designed, the current flows back and forth with any desired frequency. It is said to resonate at that frequency, like a tuning fork or an organ pipe. Hertz's apparatus involved a transmitter (to generate waves) and a receiver (to detect them). Both devices were tuned to the same frequency of about 10^8 cycles per second, or Hz.

His transmitter consisted of a spark gap powered by a high–voltage source placed at the center of a parabolic metal mirror (Figure 8.16). A second spark gap, connected to a simple loop of wire to act as an antenna, was placed about five feet away within its own reflecting mirror. Zap! A spark is generated at the first spark gap, and a smaller spark is produced at the electrically inert receiver. Electromagnetic waves were produced. They sped from the transmitter to the receiver to cause the sympathetic spark. Hertz wrote his classic paper entitled *On Electromagnetic Waves in Air and their Reflection* only after his careful studies proved that the effect could only be attributed to electromagnetic waves. It concluded with a tribute to Maxwell:

> It is clear that the experiments amount to so many reasons in favor of that theory of electromagnetic phenomena which was first developed by Maxwell from Faraday's views. It also appears to me that the hypothesis as to the nature of light which is connected with that theory now forces itself upon the mind with still stronger reason than heretofore.

Hertz discovered radio waves in 1888.

Guglielmo Marconi, sensing a feasible commercial development, strove to improve upon Hertz's merely pedagogical demonstration. In a series of experiments, he sent a radio signal for a distance of 10 m, then 300 m, then 3000 m, then across the English Channel. By 1901, he sent the letter *s* (dit-dit-dit in Morse code) across the Atlantic from Cornwall to Newfoundland, and, by 1918, he succeeded in sending telegraph messages across the world from Wales to Australia. Marconi shared the Nobel Prize in physics in 1909. In November 1920, the first commercial radio station, KDKA, began broadcasting from Pittsburgh, Pennsylvania.

Exercises

12. A bar magnet is suspended from one of its poles and allowed to fall through a copper hoop. Describe qualitatively the electric current that is induced in the hoop.

13. This exercise, and exercises 14 and 15, involve the device shown in Figure 8.14 and described by equations 9 and 11. Suppose that $B = 0.5$ tesla, $L = 0.4$ m, and $R = 6\,\Omega$. The conducting bar moves outward at the constant velocity $v = 12$ m/s.
    a. What is the *emf* of the circuit (the potential difference across the resistor)?
    b. Use Ohm's law to find the current I flowing through the circuit.
    c. What is the power P consumed by the resistor?
    d. What force must be applied to the bar to maintain its uniform velocity?

14. As the conducting bar moves outward at velocity v, the force maintaining its motion stops. The mass of the bar is $m = 0.02$ kg and its contacts with the parallel wires are frictionless. The current through the circuit decreases and the bar slows down. It can be shown that the bar slows to half its initial velocity in the time $t = 0.69\, mR/(BL)^2$. What is this "decay time" for the circuit described in the previous exercise?

15. The conducting bar moves outward at the constant velocity v. How would the current I and the power consumption P change if:
    a. the velocity v is reversed in direction?
    b. the velocity v is doubled?
    c. the magnetic field B is doubled?
    d. the resistance R is doubled?

16. Explain why the dynamo in Figure 8.15 generates a current that alternates in direction as the loop rotates.

17. In a region in which there are neither charges nor currents, Maxwell's equations (in modern notation) take the form:

$$\nabla \cdot \mathbf{E} = 0 \qquad \nabla \times \mathbf{E} + \dot{\mathbf{B}} = 0$$

$$\nabla \cdot \mathbf{B} = 0 \qquad \nabla \times \mathbf{B} - \dot{\mathbf{E}}/c^2 = 0$$

Starting with these equations, replace $\mathbf{B}$ by $\mathbf{E}/c$, and replace $\mathbf{E}$ by $-c\mathbf{B}$. Show that the transformed equations are identical to the original equations. You do not need to understand the mathematical symbols to do this exercise and discover the remarkable symmetry between electricity and magnetism.

8.3 Particles and Fields

Ampère, Faraday, and Maxwell suspected that an electric current consists of electric charges in motion. However, none of them knew it for sure. Faraday came close to the notion when he found that a discharge of static electricity—which he might have regarded as charge in motion—produces electrochemical effects just like an electric current. He also showed that a static discharge passing through a coil produces a momentary deflection of a compass needle. Nevertheless, Faraday, not one to speculate unduly, committed himself to no specific model: "By current," he wrote, "I mean anything progressive, whether it be a fluid of electricity, or two fluids moving in opposite directions, or merely vibrations, or, speaking still more generally, progressive forces."

Maxwell, like Faraday, was uncommitted to any specific physical picture of electric charge. To him, charge was an abstract and hypothetical construct. It could be a fluid, or two contrary fluids, or some unspecified condition of matter. Of the hypothetical molecule of electricity hinted at by Faraday's laws of electrolysis, Maxwell wrote:

> This phrase, gross as it is, is out of harmony with the rest of this treatise. It is extremely improbable that when we come to understand the true nature of electrolysis we shall retain in any form the theory of molecular charge, for then we shall have obtained a secure basis on which to form a true theory of electric currents and so become independent of these provisional hypotheses.

Wrong! Molecular charges do exist. There is neither an "electric fluid" nor any reality to electric charge in and of itself. The concepts of matter and charge are inseparable. Matter is made up of particles, and electric charge is an intrinsic attribute of these particles.

If the current in a wire does consist of electric charges in motion, then a moving charged body must produce a magnetic field. The trouble is that it is hard to get enough charge moving fast enough to produce much of a magnetic field. That challenge was met by the American physicist Henry Rowland (1848–1901), but not until 1878, long after this point in our narrative. Rowland's paper began:

> These experiments were made with a view of determining whether or not an electrified body in motion produces magnetic effects. There seems to be no theoretical ground upon which we can settle the question, seeing that the magnetic action of a conducted electric current may be ascribed to some mutual action between the conductor and the current. Hence an experiment is of value. Professor Maxwell has computed the magnetic action of a moving electrified surface, but that the action exists has not yet been proved experimentally or theoretically.

Rowland found the effect! According to my colleague Edward Purcell,[*]

> Rowland made many ingenious and accurate electrical measurements, but none that taxed his experimental virtuosity as severely as the detection and measurement of the magnetic field of a rotating charged disk. The field to be detected was something like 10^{-5} of the earth's field in magnitude—a formidable experiment, even with today's instruments.

The Fields Produced by Moving Charges

An electric current consists of a flow of charged particles. When a current passes along a conducting wire, mobile, negatively charged electrons flow relative to the fixed, positively charged atomic constituents of the conductor. The beam of electrons flowing through the vacuum in a TV picture tube toward the screen is an electric current. So is a beam of protons whirling about in a particle accelerator, or the migration of ions through the electrolyte of a battery.

In the nineteenth century, electrons and protons had not yet been discovered. Faraday had coined the word *ions,* but these were ill-defined constructs. Maxwell's equations were framed in terms of macroscopic distributions of electric charges and currents. Nonetheless, the notion of particulate charges was in the air. If matter is made of such particles, and if we are concerned with the fundamental properties of nature, we have been barking up the wrong tree. Instead of asking about the electric and magnetic fields produced by distributions of charge and current and the forces they exert on one another, we should study the electromagnetic interactions of individual particles.

The charge of a macroscopic body is the sum of the charges of its constituents. If the sum is 0, the body is electrically neutral. If it contains a surplus of N particles with individual charge q, then its charge is Nq. If some of the charged constituents of a body are moving in the same direction, the flow of charge is an electric current. The size of the current is the rate of the flow. If there are n moving charges per unit length in a wire, each with charge q and velocity v, the current in the wire is $I = nqv$.

In 1895, the Dutch physicist Hendrik Lorentz reformulated electromagnetism in terms of hypothetical, structureless charged particles bearing positive and negative electric charges. The questions he asked are:

- What is the electric field produced by a point particle carrying electric charge q?
- What is the magnetic field produced by a point particle carrying electric charge q?

The answers to both questions are implicit in equations 2 and 5:

$$\mathbf{E} = k_c \frac{q\,\mathbf{n}}{R^2} \quad \text{and} \quad \mathbf{B} = \frac{k_c}{c^2}\frac{q\,\mathbf{v} \times \mathbf{n}}{R^2} \tag{12}$$

The velocity of the charged particle is $\mathbf{v}$, and the unit vector $\mathbf{n}$ is directed

[*]Edward M. Purcell, *Electricity and Magnetism*, 2d ed. (New York: McGraw Hill, 1985), 241.

from the particle to the point at which the fields are evaluated. A charged particle, whether it is at rest or in motion, is surrounded by a radial pattern of electric field lines. However, the magnetic field of a charged particle is proportional to its velocity. An electric charge at rest produces no magnetic field. If it is moving at velocity v, the lines of magnetic field form circles along its path. The strengths of the electric and magnetic fields produced by a moving charge satisfy inverse-square laws.

If the charged particle is moving uniformly, equation 12 is the whole story. If, however, the charged particle is moving in a circle or otherwise being accelerated, the situation becomes enormously more complicated. The electric and magnetic fields produced by the particle at one time can act on the charge at a slightly later time. The net result of this *self-interaction* leads to the production of electromagnetic waves. When charged particles are accelerated, they lose some of their energy by radiation.

The Force Exerted on a Moving Charge

Suppose that a charged particle with charge q and velocity $\mathbf{v}$ is moving through a region in which there are both electric and magnetic fields. What is the force $\mathbf{F}$ that is exerted on the particle by the fields? The answer to this question is implicit in equations 3 and 4:

$$\mathbf{F} = q\,(\mathbf{E} + \mathbf{v} \times \mathbf{B}) \tag{13}$$

where $\mathbf{E}$ and $\mathbf{B}$ are the electric and magnetic fields at the position of the particle. Equation 13 lets us determine the trajectory of a charged particle moving through electric and magnetic fields.

DO YOU KNOW ABOUT
Synchrotron Radiation?

Charged particles moving through a region of constant magnetic field describe circular orbits. At the same time, they radiate electromagnetic energy. The process is called *synchrotron radiation* because of its relevance to the design and construction of synchrotrons. *Synchrotrons* are particle accelerators in which charged particles are made to whirl about in circular orbits with higher and higher energies. The faster they travel, the more energy they lose by synchrotron radiation. Accelerating electrons is a bit like pumping up a leaky tire: The more air (or energy) you put in, the faster you lose it.

One way to beat the problem is to make the synchrotron bigger and hence less curved. The world's most powerful electron synchrotron is 17 miles long. It lies half in France and half in Switzerland.

Synchrotron radiation is important to astronomy as well as particle physics. Astronomers first learned that Jupiter's magnetic field is ten times larger than Earth's by detecting and measuring the synchrotron radiation produced by electrons trapped by its magnetism. The Crab Nebula is a remnant of a supernova whose light first reached Earth in A.D. 1054. Although it is almost 1000 years old, synchrotron radiation produced by high-energy electrons trapped by strong magnetic fields are still reaching us. It was a titanic explosion!

The electric force on a particle points along the electric field and has magnitude qE. The magnetic force is more difficult to visualize, since it involves a cross product. The direction of the magnetic force is perpendicular to both $\mathbf{v}$ and $\mathbf{n}$, and is determined by the right-hand rule if q is positive. If a positively charged particle is moving in the x direction and the magnetic field points in the y direction, the magnetic force acts in the positive z direction.

If the velocity of the charge is not perpendicular to the magnetic field, the force is reduced by the sine of the angle between them. Here are some examples. In each case, the magnetic field points north:

- If a positive charge moves northeast or southeast, the force acts up, with magnitude $QvB/\sqrt{2}$. (The sine of 45° or 135° is $1/\sqrt{2}$.)

- If a positive charge moves southwest or northwest, the force acts down, with magnitude $QvB/\sqrt{2}$. (The sine of 225° or 315° is $-1/\sqrt{2}$.)

- If a positive charge moves north or south, it is not subject to a magnetic force. (The sine of 0° or 180° vanishes.)

- In each of these cases, the force acts in the opposite direction if the charge of the particle is negative.

EXAMPLE 8.3
Ions in Crossed Electric and Magnetic Fields

In Figure 8.17, a beam of positively charged ions emerges from an ion source. They travel in the y direction and pass through a region of electric field E in the vertical z direction and magnetic field B in the x direction. Describe what happens to the beam.

Solution

The force experienced by an ion of positive charge q and speed v points in the z direction. Its value is $F = q(E - vB)$. All the ions in the beam are deflected upward or downward except those whose speed is $v = E/B$. Only those ions with this speed find their way through the aperture in the screen. By properly choosing the strengths of the fields, physicists can prepare a beam of ions moving at any desired speed. ■

FIGURE 8.17
A beam of positive ions is directed through crossed electric and magnetic fields. **E** exerts an upward force on the ions. **B** exerts a downward force on the ions whose magnitude is proportional to the ion velocity. Ions whose velocities are just right pass through this device without being deflected.

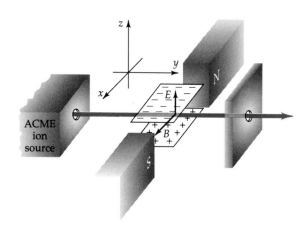

Exercises

18. A narrow beam of (negatively charged) electrons races from the cathode to the anode in an evacuated glass tube. Describe what happens when the north pole of a bar magnet is placed near the beam at the side of the tube. What happens if the magnet is turned around to present its south pole to the beam? (The magnetic field lines of a magnet run from its north pole to its south pole.)

19. A beam of ions is sent through the velocity selector described in Example 8.3. The electric field is 10^4 volts per meter. What must the magnetic field be if the ions that pass through are to have speeds of $10^{-3} c$?

20. You are presented with a beam of ions of various electric charges but all with the same mass and velocity. Describe how you would measure their charges.

21. A singly charged uranium ion ($q = +1.6 \times 10^{-19}$ C) moves east through the upper atmosphere. Earth's magnetic field of 10^{-5} tesla at the ion points north. What must the speed of the ion be for the upward magnetic force on it to equal its weight? (The mass of the ion is $\sim 4 \times 10^{-25}$ kg.)

22. Each ion in a beam has the same positive charge q, mass m, and velocity v. The ions move along the horizontal x-axis. They encounter a region in which there is a constant magnetic field B pointing up along the z-axis. Their speeds are unchanged, but the ions are deflected so as to describe circular arcs in the horizontal plane.
    a. Do the ions move clockwise or counterclockwise?
    b. The magnetic force on an ion divided by its mass is its centripetal acceleration. Express the radius R of its circular trajectory in terms of B, q, m, and v.
    c. The ion is singly charged ($q = 1.6 \times 10^{-19}$ C) with mass 3.2×10^{-26} kg and speed 10^6 m/s. What is R if $B = 0.1$ tesla?

Where We Are and Where We Are Going

Ampère learned how electric currents produce magnetic fields and how magnetic fields exert forces on currents. Faraday found that changing magnetic fields produce electric fields that can generate electric currents. Maxwell completed the picture by formulating the theory of electrodynamics, which made possible the generation, transmission, and utilization of electric power. Electricity rapidly grew from a laboratory curiosity to a vital utility to be used by everyone, not only for lighting and heating but to power the engines of industry.

The useful constructs of physics are the electric and magnetic fields generated by the motions of electrically charged particles. There are no electric and magnetic fluids. Electric charge is an attribute borne by the particles of which matter is made. Electrons are negatively charged and atomic nuclei are positively charged. Maxwell predicted, and Hertz established, that oscillating currents (or accelerated charges) emit electromagnetic waves. Radio waves, which also began as a laboratory curiosity, soon evolved into the technology of electronics: as wireless communication, commercial radio and television, radar, computers, and so on.

Light is also an electromagnetic wave, as are X rays, gamma rays, and so on. Most of the properties of matter are electromagnetic phenomena. The explosive development of physics in the early twentieth century began with the problems posed by electromagnetic waves. Quantum mechanics (which we discuss in Chapter 11) was born in 1900 from the study of the electromagnetic waves emitted by hot objects. The special theory of relativity (which we discuss in Chapter 12) was born in 1905 from the study of the propagation of electromagnetic waves through a vacuum. Wave phenomena are the key to the understanding of modern physics, and it is to the properties of waves we turn in the next chapter.

9   Waves

Waves are everywhere. Everything waves. There are familiar, everyday sorts of waves in water, ropes and springs. There are less visible but equally pervasive sound waves and electromagnetic waves. Even more important is the wave phenomenon of quantum mechanics, built into the fabric of our space and time. How can it make sense to use the same word—"wave"—for all these disparate phenomena? What is it that they all have in common?

Howard Georgi, contemporary American physicist

What are the wild waves saying?

J. E. Carpenter, British playwright

All sorts of metaphorical waves beset us: waves of guilt, joy, and relief; remorse, enthusiasm, and grief. But the waves we are interested in are more like the one pictured here. *Source:* The Bettmann Archive.

Count the ways in which the creatures of the world sense their habitats! Canine brains are mostly olfactory lobes: Smells make the dog's world. Bats use sonar, cats and cockroaches probe the dark with whiskers, sharks and skates detect electric fields, and pigeons consult their built-in compasses. Humans depend almost exclusively on their eyes and ears. Sight and hearing bring word of matter to mind, but the messengers themselves are as interesting as the messages. Sound and light consist of waves. Waves are among the most common of nature's wonders. They are found on the ocean's surface, from tiny ripples to the wakes of ships to giant tsunami or tidal waves. Earthquakes make waves in the solid ground. The sound of music results from waves produced on drumheads, columns of air, or taut strings by various acts of beating, blowing, bowing, or plucking.

Not only sound and light, but matter itself is a form of wave. Quantum theory describes atoms and their parts in terms of matter waves. The study of waves is central to all of physical science. Students of physics spend much of their time mastering wave behavior. Nonetheless, the fascinating features of waves can be described in words, with only a little help from algebra and trigonometry. Section 9.1 focuses on mechanical waves in one dimension, such as those produced by a piano string or a taut rope. In Section 9.2, we concentrate on some of the properties shared by sound and light waves. The phenomena of reflection, refraction, and diffraction are discussed in the context of optics—the science of light—in Section 9.3. Section 9.4 is a brief introduction to the science of spectroscopy, the study of the characteristic electromagnetic radiation produced by different chemical elements and compounds.

9.1 What Are Waves?

There are many ways to transport energy or information. A carrier pigeon brings a handwritten letter from one place to another, but a beeper makes do with radio waves. Waves are the means by which energy (but not matter!) is transmitted from one point to another. Waves are generated when two conditions are met:

- Energy must be injected into a previously quiescent medium: by a pebble falling into a pond, a finger strumming a harp string, a ship plowing through the sea, an explosion setting up seismic waves in Earth, an oscillating electric current, and so on. The added energy creates a disturbance in the medium.
- The medium into which energy is deposited must be somewhat elastic, in the sense that it does not quickly gobble up the energy and turn it into heat—it's hard to make waves propagate through a wet sponge or whipped cream. If the medium can exert a restoring force to undo the effect of the initial disturbance, the message will pass through it in the form of a wave.

Speech, song, and the screech of brakes are heard when transient fluctuations in the density of the air arrive at our eardrums. Sound is produced when something vibrates, producing a disturbance in the air. The spring of the air tries to smooth out the disturbance—where the density is high, the air expands; where it is low, it contracts. As the fluctuations are undone at one point, they are displaced to another point as a wave. The wave travels through the air, which itself remains at rest. Pressure is the restoring force for sound waves that lets them move through a medium. In the case of ripples on a lake, the disturbance is a fluctuation in the shape of the water surface, and the surface tension of the water tends to keep the surface flat. For large ocean waves, gravity is the restoring force. For water waves, the motion of the water making up the wave is up and down, and not in the direction the wave is moving. Electromagnetism is special; the medium is empty space, which does not move at all. The analog to a restoring force is provided by the electric and magnetic fields of the waves themselves.

Although they appear under disparate circumstances, waves have many common features. The behavior of waves is simplest when they move along a line, rather than in two or three dimensions. We begin our study of waves by examining the propagation of mechanical waves along a taut string or rope.

Waves in One Dimension

A long and taut horizontal rope is a medium along which wave disturbances may pass. If one end of the rope is jiggled up and down, a disturbance is created. Because it has a beginning and an end, we refer to such a disturbance as a *wave pulse*. The pulse consists of one or more up and down displacements of the rope. The tension of the rope provides a restoring force that tries to keep the rope straight and undo the disturbance. As one part of the rope straightens, its kinetic energy passes to an adjoining portion. The pulse displaces itself horizontally along the rope without much change in its shape, just like a wave through water or the sound of a cough through a room. The energy of the wave pulse propagates along the rope as a wave disturbance. Three general properties of wave pulses are illustrated in Figure 9.1.

FIGURE 9.1
Wave pulses moving along a string. *a.* A disturbance is set up at one end of the string. It propagates to the right as a wave pulse. Its amplitude is *A* and its velocity is *v*. *b.* A disturbance created at the other end of the string moves to the left at velocity −*v*.

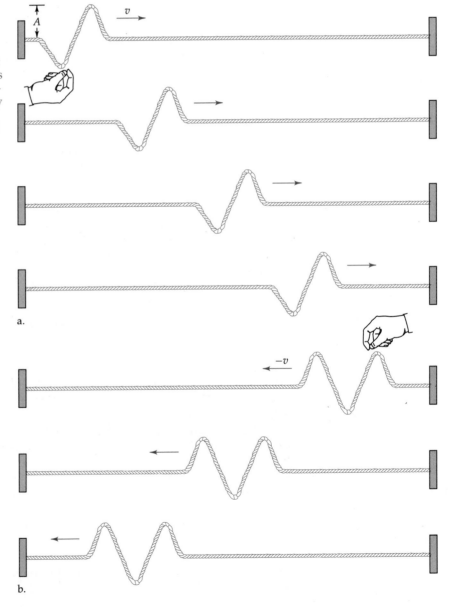

- *Amplitude* The amplitude of the pulse measures the size of the disturbance in the medium. For a pulse on a rope, it may be taken to be A, the maximum vertical displacement of the rope from its equilibrium position. The amplitude of an ocean wave is its height above sea level; that of a sound wave is the magnitude of the pressure fluctuation about the ambient air pressure; that of an electromagnetic wave is the strength of its electric or magnetic field.

- *Direction* Waves travel in definite directions. For waves on a rope, there are only two possible directions: to the right or to the left. For ocean waves, the direction lies on the plane of the ocean. Light and sound waves may propagate in any direction.

- *Speed* Waves travel at definite speeds. The speed of a wave may be determined in two ways: by experiment or by theoretical analysis. Newton's calculation of the speed of sound in air did not agree with observations. Two centuries later, Maxwell computed the speed of light and kinetic theorists computed the speed of sound: Both results agreed flawlessly with experiment.

The speed of a wave on a taut string or rope may be calculated from the principles of mechanics. The result is

$$v = \sqrt{T/\rho} \tag{1}$$

where T is the tension of the string (with dimension of force) and ρ is the mass per unit length of the rope. The greater the tension of the rope, the more rapidly the wave progresses. Similarly, the skinnier or lighter the rope, the faster is the velocity of the wave. For a given string and tension, the velocity of the wave does not depend on the shape or size of the disturbance moving along it.

Suppose that a pulse is propagating along a long rope to the right, in the positive x direction. If we take a snapshot of the rope at $t = 0$, we can measure the instantaneous vertical displacement $A(x)$ of the rope from its equilibrium position at every point x. Assume that the displacement is largest at $x = 0$. What is the displacement of the rope at a later time t? The point at which the rope is most displaced will have moved to $x = vt$, where v is the speed of the wave. Thus, a wave that moves to the right is described by $A(x - vt)$. Its shape does not change—it simply shuffles off to the right. The same reasoning lets us describe a left-moving wave. If the wave form is $B(x)$ at $t = 0$, it will be $B(x + vt)$ at later times. The notions of amplitude, speed, and direction of a one-dimensional wave are neatly expressed in precise mathematical form:

$$\begin{aligned} A(x - vt) &\quad \text{is a wave moving to the right} \\ A(x + vt) &\quad \text{is a wave moving to the left} \end{aligned} \tag{2}$$

The function A describes the shape of the propagating disturbance—its waveform.

Superposing Waves

What happens when two wave pulses moving in opposite directions meet one another, as shown in Figure 9.2. Pulse *A* is moving to the right while pulse *B* is moving to the left. The first frame depicts the shape of the rope at $t = 0$. The resultant disturbance is the sum of the two pulses: $A(x) + B(x)$. At later times, the shape of the rope is given by

$$A(x - vt) + B(x + vt)$$

The two waves simply pass through one another. In passing, the two waves briefly overlap to form a complex and time-dependent pattern. Afterward, the two disturbances separate from one another and continue along as they were, ever onward.

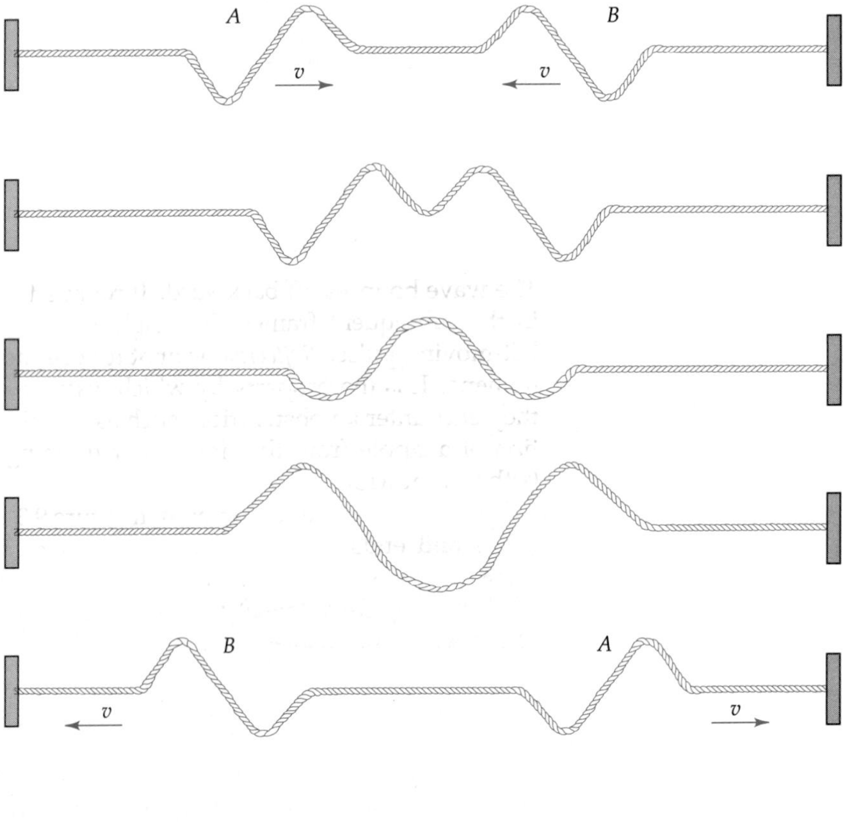

FIGURE 9.2
The frame at the top shows two pulses moving toward one another, the right-moving pulse *A* and the left-moving pulse *B*. The other frames show the development of the wave disturbance at equally spaced subsequent times. The pattern is complex as the pulses overlap. The last frame shows that the pulses have passed through one another unscathed.

A similar phenomenon takes place when a wave in a rope encounters an obstruction. In the top frame of Figure 9.3, a wave pulse moving to the right comes to the end of its rope, which is held firmly fixed.

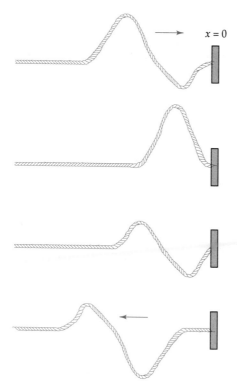

$x = 0$

The wave bounces off backward. It reflects from the endpoint, as shown
in the subsequent frames. The right-moving pulse is converted into a
left-moving pulse. *Reflection* is another common feature of all wave phenomena. It is the property by which waves change their direction when
they encounter an obstruction, such as an echo from a cliff, or the reflection of a ripple from the side of a swimming pool, or your face in the
bathroom mirror.

Suppose that the rope shown in Figure 9.3 extends along the negative
x-axis and ends at $x = 0$. Let the incident wave be described by the
known function $A(x-vt)$ and the reflected wave by the function $B(x+vt)$.
It's easy to figure out what B must be because the end of the rope is fixed.
There can be no displacement of the rope at the point $x = 0$:

$$A(-vt) + B(+vt) = 0 \qquad \text{so that} \qquad B(+vt) = -A(-vt)$$

This constraint is uniquely satisfied by the left-moving wave $B(x+vt) =
-A(-x-vt)$. The shape of the rope at all times and at all positions along
the negative x-axis is given by a single function $S(x,t)$ consisting of the
sum of an incident and reflected wave:

$$S(x,t) = A(x-vt) - A(-x-vt). \tag{3}$$

Sine Waves

If the end of a rope at $x = 0$ is continually jiggled up and down at a constant frequency, it produces a periodic wave form along the positive x-axis resembling Figure 9.4. At $t = 0$, suppose that the displacement of the rope from its equilibrium position is given by a sine function:

$$A(x) = A_0 \sin(2\pi x/\lambda)$$

The maximum displacement of the rope is $\pm A_0$ and is called the amplitude of the sine wave. The factor of 2π is not put in to confuse you. The period of a sine function is 2π, so x must increase by λ to complete one cycle of the disturbance. The distance between consecutive crests is λ, the *wavelength* of the wave.

FIGURE 9.4
A sine wave, viewed at one time as a function of x, the position along the rope. Its wavelength is λ and its amplitude is A.

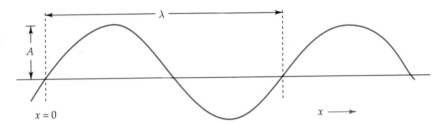

The source of energy is at $x = 0$, and the wave thereby generated propagates to the right. If the velocity of the wave is v, the displacement of the rope must depend on the combination $x - vt$, as in equation 2. Thus, the disturbance, as a function of space and time, must be

$$A(x, t) = A_0 \sin(2\pi(x - vt)/\lambda) \tag{4}$$

There's a lot to this result. Figure 9.5 shows the shape of the waveform at several consecutive times. The entire pattern moves to the right at velocity v.

Let's examine the time dependence of the jiggled end of the string. Putting $x = 0$ in equation 4, we find

$$A(0, t) = A_0 \sin(-2\pi vt/\lambda)$$

The argument of the sine function completes one cycle when it changes by 2π. For this to happen, t must increase by λ/v. Thus, the *period* of the wave is $P = \lambda/v$. The number of waves generated at $x = 0$ per second is the same as the number of crests passing any point per second and is called the *frequency* of the wave; it is denoted by f. The frequency of the wave is the reciprocal of its period: $f = 1/P = v/\lambda$, or

$$v = f\lambda \tag{5}$$

Sine waves are waves of a specified frequency or wavelength, whether they are waves on a string, a sound wave of a definite musical note (a pure tone), or light of a definite color (monochromatic light). The velocity of a sine wave is the product of its wavelength and its frequency.

Sine waves differ from the wave pulses we considered earlier. They

FIGURE 9.5
The sine wave is propagating to the right. The dashed line draws your attention to a particular crest of the wave as it moves along the rope. (In each frame, the crest appears at a different point of the rope, which itself has no horizontal motion.) In the last frame, the crest has advanced by one wavelength and the original pattern is restored. The time between the first and last frame is one period of the wave. The velocity of the crest is one wavelength per period, or $v = \lambda f$.

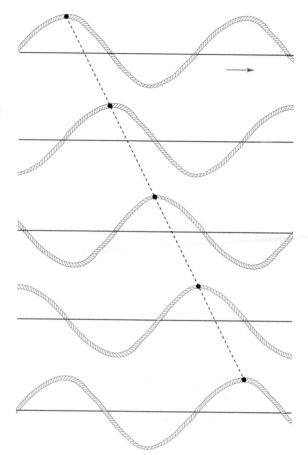

are mathematical idealizations that are both infinite and eternal: The sine wave has no beginning and no end, neither in space nor in time. Why do we consider such mathematical abstractions, rather than simple wave pulses? We do so because sine waves of different frequencies and amplitudes may be added together, or *superposed* to form more intricate wave forms. In fact, any pattern of waves, no matter how complex— from a Beethoven symphony (sound waves) to the glint in a lover's eye (light waves) to a wave pulse on a rope—may be expressed as a superposition of many different sine waves. To study wave phenomena, it is sufficient to examine the behavior of endless trains of monochromatic or sine waves.

Superposing Sine Waves

Waves with different frequencies can be superposed on one another, resulting in complex wave patterns. While flutes produce nearly sinusoidal sound waves, other instruments blend many frequencies with the dominant note to give them their characteristic timbres. Several such notes are sounded simultaneously to produce chords, which follow one another in harmonic progressions. A TV screen is made up of numerous tiny pixels of three colors: red, green, and blue. By activating them in various combinations, hundreds of distinct hues are produced.

If two tones are sounded at the same time but their frequencies are not quite the same, we hear a periodic change in the loudness of the sound. These periodic changes are called *beats*. We can see how beats are created if we add up two sine waves of slightly different frequencies, as shown in Figure 9.6. At times, the two waves add together construc- tively and the sound is louder. At other times, the two waves interfere destructively. They tend to cancel one another, and the sound intensity is small. The frequency of the beats is the difference between the two frequencies. When two musical instruments sound the note A, but one is tuned to 440 Hz and the other to 441 Hz, we hear beats at a rate of one per second.

Suppose that one wave of frequency $f + \epsilon/2$ is superposed with a wave of the slightly smaller frequency $f - \epsilon/2$. The difference between their frequencies ϵ is small compared to their average frequency f. If waves with these frequencies and equal amplitudes are superposed, the result (at a fixed point in space, say $x = 0$) becomes

$$
\begin{aligned}
A(t) &= A_0 \sin\left(2\pi(f + \epsilon/2)t\right) + A_0 \sin\left(2\pi(f - \epsilon/2)t\right) \\
&= A_0 \sin\left(2\pi f t\right) \cos\left(\pi \epsilon t\right)
\end{aligned}
\tag{6}
$$

The second form of this equation follows from the trigonometric equal- ity:*

$$
\sin(A + B) = \sin A \cos B + \cos A \sin B
\tag{7}
$$

The sine factor in equation 6 is a rapidly oscillating wave with fre- quency f. The cosine factor is a slowly oscillating function. The ampli- tude of the wave is greatest when $t = 0, 1/\epsilon, 2/\epsilon, \ldots$ Halfway between these points, the amplitude $A(t)$ vanishes. The time between beats is $1/\epsilon$ and the beat frequency is ϵ.

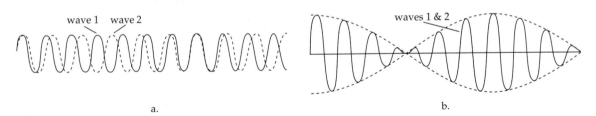

FIGURE 9.6 *a.* Two waves with nearly the same frequency are shown. *b.* The result of superposing these waves (adding them together) produces a signal with a modulated amplitude. The beat frequency is the difference between the frequencies of the two waves.

*Trigonometry emerged from the study of triangles. Its utility and importance is far greater than its humble origin suggests.

FIGURE 9.7
Both ends of the rope are jiggled at the same rate. If the frequency is properly chosen, the right-moving wave combines with the left-moving wave to produce a pattern of standing waves. These propagate neither to the left nor to the right. In part *a*, the wavelength of the standing wave is twice the length of the oscillating rope. In part *b*, the wavelength equals the length of the rope. In part *c*, one of the shakers has left and the endpoint of the rope has been fixed in place. The original and reflected waves combine to form a standing wave whose wavelength is 4/3 the length of the rope. The points along the rope that remain stationary are called nodes. There is one node in part *a* and two in each of *b* and *c*.

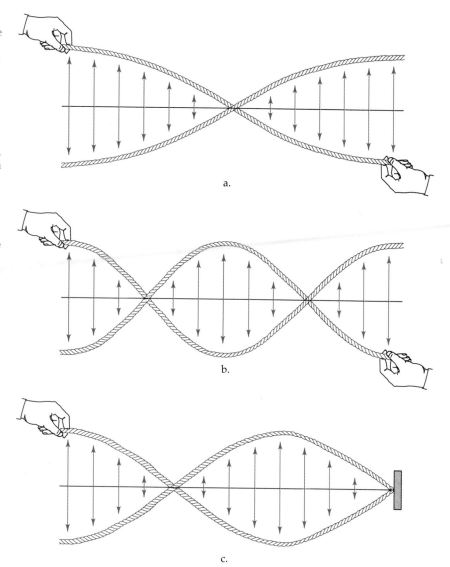

a.

b.

c.

Standing Waves

Suppose that a rope is held by two experimenters. The experimenter to the right could jiggle his end of the rope, thereby producing left-moving waves. The experimenter to the left could do the same, producing right-moving waves. If the two synchronize their actions, they can produce a wave pattern such as that shown in Figure 9.7. In this case, there is no evident wave propagation. A few special points remain fixed, and all other points jiggle up and down, to a greater or lesser extent. Motion of this type is known as a *standing wave* and its fixed points are called *nodes.* A standing wave consists of a superposition of right- and left-moving waves with the same amplitude:

$$A(x, t) = A_0 \sin (2\pi (x - vt)/\lambda) + A_0 \sin (2\pi (x + vt)/\lambda)$$

$$= 2A_0 \sin (2\pi x/\lambda) \cos (2\pi f t)$$

Once again, we have used equation 7 to simplify the result.

EXAMPLE 9.1
The Piano

When a piano is played, the keys activate hammers that strike the piano strings and set up standing waves. The frequencies of the standing wave and the tone it makes are the same. The two ends of a particular string are held fixed 0.5 m apart as it vibrates in a standing wave with no nodes except at the endpoints. The tension of the string is 605 N and its mass per unit length is 3.125 g/m.

a. What is the velocity of waves on the string?
b. What is the frequency of the sound it produces?
c. What is the wavelength of the musical tone it produces?

Solution

a. Equation 1 gives the wave velocity on the string to be $v = \sqrt{605/(3.125 \times 10^{-3})}$, or 440 m/s.
b. The wavelength of the standing wave is twice the length of the string, or 1 m. Equation 5 tells us that the frequency of the standing wave is 440 Hz (which is A above middle C).
c. To answer this question, we need to know the speed of sound in air, which is about 330 m/s. Because it is also equal to the product of the wavelength λ of the sound wave and its frequency (440 Hz), we find $\lambda = 0.75$ m.   ■

Exercises

For some of these problems, you will need to know the speed of sound (330 m/s) or the speed of light (3×10^8 m/s) in air.

1. The tension on a ship's cable is 300 N. If 100 m of the cable have a mass of 300 kg, what is the velocity of a wave on the cable?

2. The speed of mechanical waves on a certain piano wire is equal to the velocity of sound in air. If the wire has a mass of 1 g/m, what is the tension in the wire?

3. Electromagnetic radiation whose wavelength lies in the interval 3.5×10^{-7} m $< \lambda < 6.5 \times 10^{-7}$ m can be seen by the human eye. What is the range of visible frequencies?

4. Young and healthy people can hear sound waves with frequencies in the interval 30 Hz < f < 20,000 Hz. What is the range in wavelength of audible sound?

5. The fixed ends of a guitar string are 54 cm apart.
   a. What are the three largest wavelengths of standing waves that may be set up on it?
   b. Sketch these standing waves.
   c. What are ratios of the frequencies associated with these standing waves?

6. Two musical tones are played at the same time. One has a wavelength of 1 m, the other a wavelength of 1.003 m. What is the frequency of the beats that are heard?

7. Consider how a piano makes sound. The wavelength of a disturbance on a string is determined by the length of the piano string. However, the pitch of the produced tone is the frequency of the string's vibration. According to equation 1, the velocity of the disturbance on the string increases with the square root of its tension. Thus, the pitch of the keys can be adjusted by tightening or loosening the strings, which is what piano tuners do.
   a. Explain why the bass piano strings are thicker than the treble strings.
   b. Suppose that the tension on a piano string that normally produces middle C is doubled. Show that the note is changed to F# above middle C. (Refer to Table 1.9 for the relation between the frequencies of these notes.)

9.2 Sound and Light

The pitch of a tone and the color of a lamp signal their frequencies f, the number of complete cycles of a sine wave reaching us each second. An ideally transparent medium allows waves of all frequencies to pass with no loss of energy. The speed of waves in such a medium cannot depend on frequency.*

The speed of a wave depends on the properties of the medium through which it passes. The speed of sound is about 330 m/s in air at a temperature of 0° C. At room temperature sound travels about 3.5 percent faster. The speed of sound in a gas, like the mean velocity of individual air molecules, varies with the square root of the absolute temperature. Because the molecules of solids and liquids are crowded together, the restoring force is much larger in these media than it is in

*The speed of light, however, often does depend on frequency: That's why diamonds are beautiful and prisms make rainbows. Neither diamond nor glass is perfectly transparent.

gases—the speed of sound is about 1500 m/s in water and about 5000 m/s in aluminum.

The speed of light is also affected by the medium. Light travels at $c \simeq 3 \times 10^8$ m/s in a vacuum, at $0.9997c$ in air, and at about $c/2$ in glass. Not all wavelengths of light or sound are detectable to unaided eyes or ears. Audible sound waves have wavelengths spanning ten octaves, from 1.5 cm to 10 m. The wavelengths of visible light barely fill one octave.

The Speed of Sound

The association of sound with the rapid motions of material bodies has always been known: oscillating vocal cords, strings, membranes, reeds,

---

DID YOU KNOW
That Newton Made Mistakes?

Newton knew the speed of sound, and he set out to explain why it is what it is. Science is more than a collection of measurements. "Why?" and "How?" must complement "What?" and "When?" Theory and observation are inseparable. Newton came close to a correct calculation of the speed of sound, even though his argument was flawed. For example, his atoms were fixed in space, not in chaotic motion. His result was

$$s = \sqrt{P/\rho} \qquad (8)$$

where P is the air pressure and ρ is its density. Putting known values of these quantities into equation 8, Newton obtained for the speed of sound $s = 280$ m/s. However, he knew that s had been measured to be 330 m/s. Newton could not admit that he got the wrong answer, and devised ingenious but absurd corrections to his result.

Water—whose molecules touch one another—when boiled yields about 1000 times its volume as steam. Steam molecules (and those of other gases) occupy only about a thousandth of the gas volume at normal conditions. Therefore, an imaginary straight line passing through

air cuts through air molecules for about 10 percent of its length. Newton asserted that sound waves plunge instantaneously through the molecules in its path, thus speeding its way along. This spurious analysis gave a 10 percent correction, but the result was still too small. Newton next excluded "foreign airs and vapors" from the density of air, which should be that of "good English air" alone. These, he argued, amounted to about 15 percent of ordinary air. After both revisions, Newton boasted of precise agreement with experiment. Stuff and nonsense!

In his initial analysis, Newton erred by using Boyle's law, which applies to a gas at uniform temperature. But sound waves leave no time for heat to flow—temperature oscillates along with pressure and density. A correct calculation required modern kinetic theory. In 1816 —a century and a half later—Laplace corrected Newton to obtain

$$V_s = \sqrt{\gamma \, P/\rho} \qquad (9)$$

where γ is a measurable (and much later, calculable) constant depending on the nature of the gas. For air, $\gamma = 1.4$, and equation 9 gives the observed sound velocity. The only remaining mystery in the science of sound is how to design a great concert hall!

tuning forks, and columns of air make sounds. A bell rings. Its vibrations speed through the room as sound waves—that wiggle your eardrums, that wiggle a series of tiny bones in your skull, that wiggle the minuscule hairs of your inner ear, that generate nerve signals for your brain to decipher as: "A bell is ringing!" Boyle showed that a material medium is needed to convey sound: a ringing bell set on soft cotton in an airtight enclosure cannot be heard when the air within is removed. Sound waves consist of alternate compressions and rarefactions traveling through the air. If there is no medium, there is no message.

Sound travels at a finite speed. You hear thunder seconds after lightning strikes, about five seconds per mile of distance. Should lightning strike a tree at a known distance to the observer, the speed of sound may be estimated—but waiting for a thunderstorm is not an ideal way to do a precise measurement. Furthermore, the speed of sound depends on temperature (a measure of molecular velocity) and on wind speed (because sound, like a swimmer, has a characteristic speed relative to the medium).

How do we measure the speed of sound? In 1708, William Derham timed the delay between firing a cannon and hearing the report at a distance of 12.5 miles. His results varied from 55 to 63 s, depending on the wind. Why? Sound waves travel at a certain speed relative to the air. If the air is moving, sound is carried along with it. When the wind is steady, the speed of sound may be measured by the "reciprocal method" adopted by the French Academy in 1738. Academy members placed two cannon 30 km apart and fired them alternately. They measured the delays for both directions of transit. The effect of the wind is opposite in the two cases, and the average of the two determinations gives an accurate measurement of the speed of sound.

EXAMPLE 9.2
Measuring the
Speed of Sound

Two cannon are alternately fired from sites 30 km apart. The time delays between flash and sound are measured from each site. The result is 92 s in one case, 88 s in the other. What do you deduce for the speed of sound, in meters per second? If the wind were blowing directly from one station toward the other, what would be the wind speed?

Solution

Let the wind speed be v, the sound speed s. The ground speed of sound moving against the wind is $s - v$. If the sound moves with the wind, it travels with a ground speed of $s + v$. But the speed of sound is also the distance it travels divided by the elapsed time:

$$s - v = 30\,000/92 \qquad s + v = 30\,000/88$$

in the units requested. Adding and subtracting these simultaneous linear equations, we find $s = 333$ m/s and $v = 7.4$ m/s. ∎

The Doppler
Effect

Have you ever stood by the tracks as a train passes and noticed that its whistle sounds higher in pitch as the train approaches, lower as it recedes? In 1842, Christian Doppler explained this phenomenon as a wave phenomenon. A train is moving at velocity v toward a stationary listener at position $x = 0$. The train is at position $D - vt$. The train's whistle oscillates at a frequency f (and with a period of $1/f$). Because the train is moving toward the listener, the perceived sound frequency f_a is greater than f. Let's examine the effect quantitatively.

Suppose that the whistle generates a compression of the air at $t = 0$ when the train is at a distance D from the listener. The compression propagates to the listener at the velocity of sound s relative to the air (and, hence, to the listener). It arrives at the listener at time

$$t_1 = D/s$$

The next compression is produced one period later, at time $t = 1/f$, when the train is at position $x = D - v/f$. It begins its trip to the listener at $t = 1/f$ and arrives at time

$$t_2 = \frac{1}{f} + \frac{D - v/f}{s} = \frac{D}{s} + \frac{1 - v/s}{f}$$

The time difference $t_2 - t_1$ is the period of the sound heard by the observer. Its frequency f_a is the reciprocal of this period:

$$f_a = \frac{f}{1 - v/s} \qquad \text{sound source approaching} \qquad (10)$$

The faster the train is moving, the greater is the perceived frequency of the whistle. An identical argument treats the case of a receding train. In this case, the result is

$$f_r = \frac{f}{1 + v/s} \qquad \text{sound source receding} \qquad (11)$$

Figure 9.8 illustrates the Doppler effect in both cases. If the listener stands near the tracks, the pitch of the whistle suddenly decreases as the train passes by.

The Doppler effect also applies to electromagnetic waves. If a light source is approaching an observer with velocity v, its frequency is shifted upward and toward the violet end of the spectrum. If a light source is receding from an observer, its frequency is shifted toward the red. Equations 10 and 11 may be used to describe the Doppler shift of light if the velocity of sound s is replaced by the velocity of light c.* This effect is of practical importance because it lets us measure velocities of moving objects. Policemen bounce radar signals off speeding cars. The amount by which the frequency of the signal is Doppler-shifted determines whether the driver gets a summons.

---

*The exact equations describing the Doppler shift of light differ from equations 10 and 11 (with s replaced by c) because of effects due to the special theory of relativity. The classical equations may be used if v is much smaller than c.

FIGURE 9.8
Two stationary observers and a moving train are shown. Wave crests emanate from the train's whistle at each of the indicated points. The frequency of the whistle f is relative to the train. Each crest spreads out at velocity s as a circle. The train is moving to the left at half the speed of sound. The sound wave moving to the left has frequency $f_a = 2f$, while that moving to the right has frequency $f_r = 3f/2$.

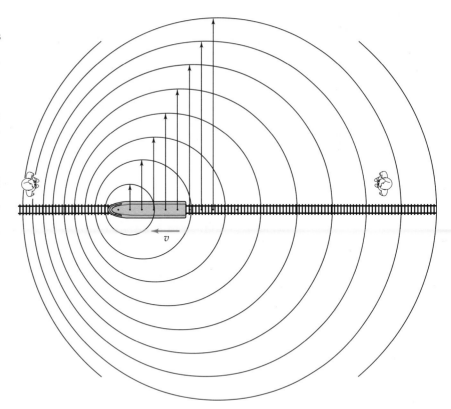

We can't bounce light or radar signals off stars or galaxies. However, because stars are made of the same kind of matter we are, starlight is similar to the light produced by the Sun. As we discuss in Section 9.4, many of the the same characteristic spectral lines are seen in the light of all stars. If a distant galaxy is moving relative to us, its entire spectrum is Doppler-shifted in frequency. Its spectral lines are displaced relative to those of stationary light sources. Thanks to this effect, we know that distant galaxies recede from the solar system at speeds proportional to their distances from us. That's the effect that told us of the expanding universe, and of its birth, long ago, in the Big Bang.

EXAMPLE 9.3
The Doppler Effect

A train blows its whistle while approaching and receding from an observer. As the train passes, the pitch of the whistle descends one full tone of the musical scale. How fast is the train traveling?

Solution

The ratio of the frequency of the whistle as it approches to its frequency as it recedes is given by equations 10 and 11 to be

$$\frac{f_a}{f_r} = \frac{1 + v/s}{1 - v/s} \simeq 1 + 2v/s$$

where s is the speed of sound and v is the train's speed. The final approximate equality is valid for values of v much smaller than s. We learned in Chapter 1 that the musical scale is divided into 12 intervals one-half tone apart. The ratio of consecutive frequencies is $2^{1/12}$. Thus, the ratio of two frequencies one tone apart is $2^{1/6}$ or 1.1225. The speed of the train satisfies the equation $2v = 0.1225s$, so that $v \simeq 20$ m/s.   ■

The Speed of Light

Galileo tried to measure the speed of light by timing the flashes of a distant lantern. He failed because light travels a million times faster than sound. However, Galileo's discovery that Jupiter has moons rotating about it led to an indirect demonstration of light's finite speed.

Io is the innermost of Jupiter's four large satellites. It circles Jupiter every 42.5 hours and is briefly eclipsed by the giant planet once in each turn. In 1675, the Danish astronomer Ole Roemer measured the times between eclipses of Io. He found that the intervals are slightly shorter when Earth moves toward Jupiter than when it moves away. This could only be understood, he argued, if light took time to travel from Io to Earth. From the size of the effect, Roemer concluded, the speed of light, which physicists designate as c, is about 10000 times greater than the speed of Earth in its orbit. He could not determine c because the value of the astronomical unit—the distance from Earth to the Sun—was not yet known.

EXAMPLE 9.4
The Eclipses of Io

By how many seconds is the time between consecutive eclipses of Io shorter when Earth is moving toward (rather than away from) Jupiter (Figure 9.9)? Use 1.5×10^{11} m for the radius of Earth's orbit, $\pi \times 10^7$ s for the length of the year, and $c = 3 \times 10^8$ m/s.

FIGURE 9.9
Earth, Jupiter, and Io. At one time of the year, Earth is at A and is heading directly toward Jupiter. About six months later, Earth is at B heading away from Jupiter. Io is eclipsed by its planet about once every two days. The time between its eclipses, as measured from Earth, is about 30 s longer at B than at A. This was the first empirical proof that light travels at a finite speed.

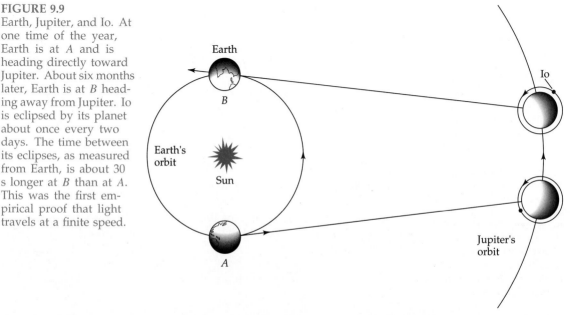

Solution   This problem offers another illustration of the Doppler effect. The circumference of Earth's orbit is 2π times its radius. The result, divided by the length of the year in seconds is the orbital speed of Earth, about $v = 10^{-4}\,c$. The period between consecutive eclipses, as measured by an observer at rest with respect to Jupiter, is $P = 42.5$ h. An observer approaching Jupiter at speed v sees a shorter period P_a between eclipses. A receding observer sees a longer period P_r. These periods are determined from equations 10 and 11:

$$P_r = (1 + v/c)\,P \qquad \text{and} \qquad P_a = (1 - v/c)\,P$$

The difference between these expressions is $P_r - P_a = 2(v/c)P$. As Earth recedes from Jupiter, Io's eclipse period is $2 \times 10^{-4} \times 42.5$ h (or about 30 s) longer than it is six months later when Earth approaches Jupiter.   ■

The Aberration of Starlight

Roemer's work was greeted with skepticism by many other scientists because he was unable to measure the effect of Earth's velocity on the eclipses of Jupiter's other moons. In 1727, the British astronomer James Bradley published "Account of a new discovered motion of the Fix'd Stars". His careful observations revealed a curious shift, or *aberration*, of a star's position in the sky due to Earth's velocity in its orbit.

We can understand the aberration of starlight by examining a down-to-earth situation: A man holds up a long, cylindrical open-ended tube while standing in the rain on a windless day. Raindrops falling straight down (at speed u) enter the tube directly. He begins to run at speed v. If the rain is to pass through the tube, it must be tilted forward by an angle θ relative to the vertical such that $\tan \theta = v/u$ (Figure 9.10).

FIGURE 9.10
Raindrops are falling vertically with speed u. The long test tube is held at an angle as it moves horizontally. *a.* If the tube is tilted just right, a raindrop is caught. The raindrop and the tube are shown at four consecutive times as seen by an observer at rest. *b.* From the point of view of the figure carrying the tube, the raindrop is not falling vertically but is headed directly into the tube. The motion of Earth around the Sun produces a similar shift in the apparent positions of stars in the sky.

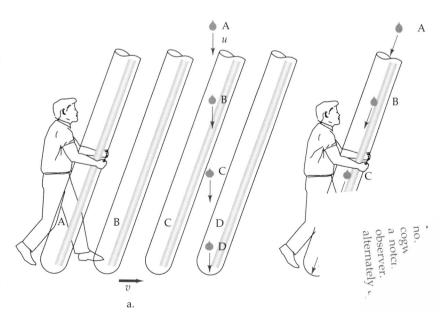

Swap tube for telescope and rain for light from a star directly overhead. Because Earth moves, the telescope must be tilted: A star that is directly overhead will not appear to be so. The amount of its angular displacement is θ (in radians) $\simeq v/c$, because Earth's speed is small compared to light speed. As the year progresses, the direction of v changes and so does the direction of the apparent displacement of stars. If the seasonal displacement of a star's position is measured, the speed of light can be determined.

The discovery of the aberration of starlight was a convincing proof that light is not transmitted instantaneously, but rather travels at a very large, but determinable, velocity. Referring to Roemer's work half a century later, Bradley wrote: "It is as it were a Mean betwixt what at different times been determined from the eclipses of Jupiter's satellites." Bradley deduced that light takes 8 minutes 12 seconds to travel from the Sun to Earth. More accurate measurements of c, the speed of light in a vacuum, awaited the development of nineteenth-century technologies that made direct laboratory measurements feasible.

A series of ingenious experiments performed together and in competition by the French physicists Armand Fizeau and Jean-Bernard Foucault finally pinned down the value of c. Fizeau's apparatus is shown in Figure 9.11. A narrow beam of light is directed at a half-silvered mirror at a 45° angle. The transmitted portion of the beam passes through a

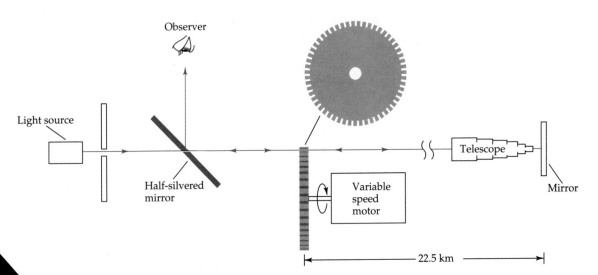

FIGURE 9.11 The Fizeau experiment. A beam of light is directed through a half-silvered mirror to a rotating notched disk. Light passing through a notch in the disk travels along a 45-km round trip and returns to the wheel. A telescope at the remote mirror is used to focus the beam. If the returning light signal meets a notch of the disk, a signal passes between the teeth and is reflected by the half-silvered mirror to the observer. If it strikes a tooth of the disk, no signal is seen. As the speed of the cogwheel is increased, signals appear and disappear.

rapidly rotating notched disk to a distant mirror, from which it is reflected back, passing once again through the rotating disk and the tilted
mirror, and finally to an observer. As the disk begins to rotate, a point
is reached where the returning beam is blocked by a tooth of the disk
and the signal is extinguished. As the disk rotates faster, the beam finds
its way through the next notch and the signal reappears. As the speed
is increased further, the light is periodically extinguished. If the speed
of the disk is monitored, the velocity of light can be determined. Fizeau
obtained a result for c correct to within a few percent.

EXAMPLE 9.5
The Fizeau
Experiment

How did Fizeau measure the speed of light with the apparatus shown
in Figure 9.11?

Solution

Suppose that the round-trip time of the light signal is the same as
the time it takes for the disk to turn by one notch:

$$f2L/c = 1/N$$

where f is the rotation rate of the disk in revolutions per second, L is the
distance between the disk and the remote mirror, and N is the number
of notches in the disk. If this condition is satisfied, the observer sees
the light. In Fizeau's experiment, $L = 22.5$ km and $N = 200$. When the
disk began to rotate, the signal at the observer disppeared. It reappeared
when the angular speed of the disk was $33\frac{1}{3}$ rps. Fizeau deduced that
$c = 33.33 \times 45 \times 200$ or 3×10^5 km/s. As the speed of the disk was
increased further to its maximum value of 900 revolutions per second,
the signal appeared and disappeared 27 times!   ∎

Foucault devised a scheme in which a narrow beam of light is directed to a rapidly rotating mirror (Figure 9.12). The light is periodically

FIGURE 9.12
The Foucault experi-
ment. A light beam is
directed toward a rotat-
ing mirror. Once per
revolution, the beam is
reflected toward a fixed
mirror. When the sig-
nal returns, the mirror
has rotated. The signal
seen by the observer is
deflected by an angle
θ. The speed of light is
determined by the angu-
lar speed of the rotating
mirror, the distance be-
tween the mirrors, and
the value of θ.

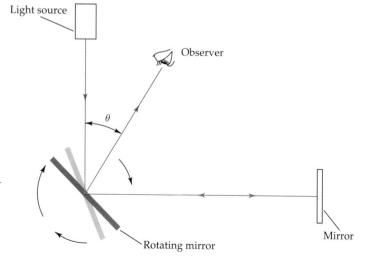

Light source

Observer

θ

Mirror

Rotating mirror

reflected toward a distant fixed mirror. When the beam returns, the rotating mirror has moved so that the beam is reflected in a direction dependent on the rotation speed of the mirror. A careful measurement of the deflection of the beam allowed Foucault to measure c with better than 1 percent accuracy.

In the latter decades of the nineteenth century, more refined experiments determined the speed of light to 1 part in 10^4. At this accuracy, the difference between the speed of light in air and in a vacuum becomes relevant. Light travels 82 km/s more slowly in air than in a vacuum, a difference of about 3×10^{-4}.

Exercises

8. How do we know that the speed of sound in air is independent of frequency? (Think of echoes or the crack of a well-hit baseball.) How do we know that the speed of light in empty space is independent

A BIT MORE ABOUT
Hearing and Vision

We hear sound waves that are produced by vibrating bodies. We can hear beats, experience the Doppler effect, and create standing sound waves in a room. Light also consists of waves although we have not yet discussed intrinsic wave properties of light.* Our eyes and ears are wave detectors that tell us about the world in sound and light. The study of these senses involves not only physics, but chemistry, physiology, ophthalmology, and psychology as well. This book can only scratch the surface of an exciting, rapidly growing, and multidisciplinary endeavor.

We can tell roughly where sounds

*Light displays some of the properties of particles as well. We shall see in Chapter 11 how the schizoid nature of light led to the discovery of quantum mechanics.

come from by using our two ears in tandem and by turning our heads. Blind people can do a bit better in this respect, but the sense of hearing, unlike the sense of sight, cannot give us a precise picture of the world around us in space. Hearing's strong point is its ability to analyze the time sequence and frequency composition of what we hear. The inner ear contains innumerable tiny detectors. Each of them responds to one particular frequency of sound waves.

We hear sound waves over a span of almost ten octaves. But if Beethoven's A Minor Quartet is transposed to B minor, most of us (except for those blessed with perfect pitch) would not hear the difference. The ratios among perceived frequencies, not their exact values, are what counts in hearing. Any two notes an octave or a fifth apart are consonant when played together or in succession. Any two tones whose frequencies are not a simple fraction clash. Musical theory describes which progressions of chords of frequency? (Think of what the appearance of a double star or a nova would be if light speed depended on frequency.)

9. Aberration of starlight: Show that the apparent position of a star lying on the celestial equator is displaced by about 20 arcseconds in a direction depending on the season. Consider only the two times of the year when Earth's velocity is normal to the line-of-sight to the star. (*Hint:* Recall that Earth's orbital speed is about $10^{-4} c$.)

10. A spherical sponge absorbs every drop of rain that hits it. Four identical dry spherical sponges were caught outdoors during a summer shower on a windless day. Three of the sponges were moving at the same speed v: A was moving up, B was moving east, C was moving down. The fourth sponge D did not move at all. Put the sponges in order of ascending wetness and justify your result.

---

seem natural and inevitable and which seem surprising and dramatic. We delight in music because our ears can find the integer-based relationships in notes of chords (harmony), in the timing of their presentation (rhythm), and in their sequence (melody).

Vision shows us the world in space and time. We see things with a spatial resolution incomparably finer than hearing can provide. Furthermore, the movies in our brains are shown in color. The rainbow is said to consist of seven colors: red, orange, yellow, green, blue, indigo, and violet. In the study of physics, the color of light in any portion of the visible spectrum is unambiguously linked to its wavelength or frequency. A beam of light with wavelength 4.8×10^{-7} m is blue, while one of 5.9×10^{-7} m is yellow. The subjective identification of color is subtler.

When it comes to frequency discrimination, the ears have it over the eyes, which respond to a mere octave of frequencies. Color-sensitive cones in the retina contain one of three different dyes. Rather than measuring individual frequencies as ears do, eyes take snapshots through three differently colored filters. However, the frequency distribution of daylight varies dramatically from noon to twilight and even more at night, in the harsh blue glare of a mercury streetlamp or the garish yellow of a sodium vapor lamp. If colors sensed were directly related to frequencies presented, we would have no firm notion of the colors of things. Yet, the McDonald's arch is always golden and the color of money is green. The colors we see are not directly determined by the ratios of the three signals received at the eye and processed by the brain. The perception of color depends on how these ratios relate to those of other images in the visual field. The relation among optical frequencies makes a picture, just as the relation among sound frequencies makes a song.

---

11. A beeper produces brief tones with a pitch of 1260 Hz. It makes 63 beeps per minute. A beeper carried by an approaching motorcyclist is heard by a pedestrian to beep at a rate of 72 beeps per minute.
   a. What is the speed of the approaching motorcycle?
   b. What is the pitch of the tone heard by the pedestrian?
   c. What beep rate does the pedestrian hear when the motorcycle is speeding away?

12. Foucault's mirror rotated at 1000 revolutions per second, and the light beam traversed a total distance of 24 m on its path from the rotating mirror to the fixed mirror and back. (His apparatus, unlike Fizeau's, fitted entirely within his laboratory.) Calculate the angle between the incident and emergent light beams in degrees, minutes, and seconds of arc.

13. You are speeding toward a red traffic light. At what fraction of the speed of light must you be traveling if the frequency of light is shifted upward by 20 percent so that the stoplight appears to be yellow?

14. Show that Newton's formula, equation 8, is dimensionally correct. Using his original result of 280 m/s, and an air pressure of 10^5 N/m, compute the density of the air that he used.

15. From equation 9 and the ideal gas law, show that the speed of sound varies with the square root of the Kelvin temperature. Let V_0 be the speed of sound at $0°$ C. Show that the velocity of sound at a temperature of $20°$ C is $V_{20} \simeq 1.036 V_0$.

9.3 The Science of Optics

Optics is the quantitative study of the propagation of light. Among many other things, optics deals with microscopes, telescopes, eyeglasses, rainbows, and sunsets. It is not only a science, but an art and a technology. In this section, we discuss the reflection, refraction, and diffraction of light, and its dissection into colors. Mirrors reflect and lenses refract; these are familiar phenomena. Diffractive phenomena are less familiar, but they are just as important. Reflection can be understood in terms of waves as particles—after all, billiard balls reflect from the cushions of a table just as light reflects from a mirror. Refraction can also be explained in terms of particles of light. However, diffraction is an essentially wave-like phenomenon.

The Reflection of Light

A light wave consists of rapidly alternating electric and magnetic fields. Figure 9.13 shows a schematic view of a monochromatic light wave moving to the right. The vertical lines with arrows indicate the direction of the electric field of the wave at one instant of time. We refer to up arrows as crests and down arrows as troughs of the wave. The distance between successive crests is the wavelength of the light. The magnetic field of the wave points in a direction perpendicular to the page and is not indicated. The entire pattern of electric and magnetic fields moves to the right with the speed of light.

FIGURE 9.13
A schematic drawing of a light wave moving to the right. The arrows transverse to its direction of propagation denote the instantaneous direction of the electric field of the light wave. The entire pattern moves to the right at the velocity of light of the medium through which the beam passes.

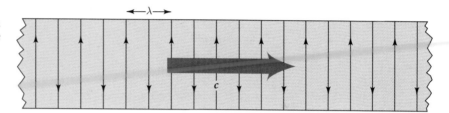

The law of reflection is simple to state, and it was discovered long before light was identified as an electromagnetic wave: The angle of incidence of a light beam on a reflecting surface is equal to the angle by which the beam is reflected. This law follows from the electromagnetic theory of light. We learned in Chapter 8 that the electric field at the surface of a conductor is perpendicular to it. When a light beam encounters a conductor, something must happen to guarantee this result. What happens is reflection. Figure 9.14 shows a light beam striking a polished plane conductor: a mirror. It also shows a reflected beam. Focus on point A on the mirror. The electric field of the incident beam

FIGURE 9.14
Reflection. A light wave falls on a mirror. Because the electric field at the surface of the mirror must be perpendicular to the mirror, a reflected wave is generated whose angle of reflection θ_f equals the angle of incidence θ_i. Notice that the electric field vectors at A, B, and C (and all along the mirror) add up to vectors that are perpendicular to the reflecting surface.

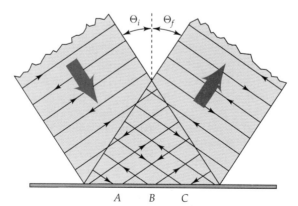

FIGURE 9.15
Two mirrors are joined
at right angles. Every
incident beam strikes
both mirrors in turn, and
returns in the same di-
rection it came from.

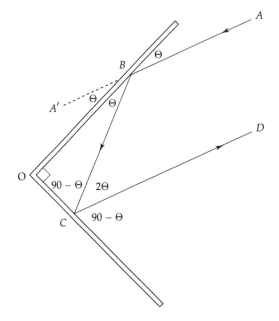

points to the lower left, while that of the reflected beam points symmet-
rically to the lower right. Their vector sum (the electric field at A) is
perpendicular to the surface. The same is true at all times and at every
other point along the mirror, but only if the reflected beam emerges at
the same angle as the incident beam enters,

$$\theta_i = \theta_f \qquad \text{for reflection} \tag{12}$$

When a light beam reflects from a mirror, the angles of incidence
and reflection are equal.

EXAMPLE 9.6   In many bathrooms, you can find two plane mirrors joined together at
The Corner Mirror   right angles, as shown in Figure 9.15. A beam of light lying in the
plane perpendicular to both mirrors reflects from one mirror to the other
mirror. Show that the emergent light beam is parallel to the incident
beam.

Solution   The path of the light beam is $A \to B \to C \to D$. From the law of
reflection, we see that the three angles at B labeled θ are equal to one
another. Thus, $\angle A'BC = 2\theta$. Triangle $\triangle BOC$ is a right triangle, so that
$\angle OCB = 90° - \theta$. The law of reflection, used a second time, tells us
that $\angle BCD = 2\theta$. Thus, the line BC cuts the lines $A'A$ and CD such
that opposite interior angles are equal. Consequently, $A'A$ and CD are
parallel. The direction of the incident beam is reversed.   ■

Figure 9.16 compares the effect of a corner mirror to a plane mirror. The
plane mirror (Figure 9.16a) produces a left–right reversed image, while
the corner mirror (Figure 9.16b) lets you see yourself as others see you.

FIGURE 9.16
a. The plane mirror. To make the analysis simpler, Josie is peeping at herself through her left eye. However, the image she sees is peeping through its right eye. Plane mirrors produce a left-right inverted image. *b.* The corner mirror. Josie is peeping through her right eye, and so is her image! Corner mirrors let you see yourself as you look to others.

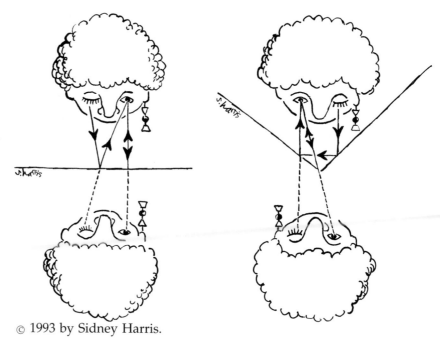

© 1993 by Sidney Harris.

A more elaborate corner mirror consists of three mutually perpendicular plane mirrors. This device reflects any incident light beam back to where it came from. Tiny corner mirrors of this kind are embedded in road signs. They illuminate themselves at night by reflecting light from automobile headlights directly back to the driver's eyes. Corner mirrors placed on the Moon do not have to be carefully positioned to return radar signals to observers on Earth. Scientists used such mirrors to measure the Earth and the Moon to an accuracy of a few meters.

Refraction

The velocity of light is different in different media. Because of this, light changes its direction as it passes from one transparent medium to another. Figure 9.17 (on page 396) shows a light beam penetrating a plane boundary from air into still water or glass. Let v_i be the velocity of light in the upper medium and v_f be its velocity in the lower medium. For the case shown, $v_i > v_f$, because light travels more slowly in water or glass than it does in air.

For light waves, or any other waves, the frequency of the incident beam is equal to the frequency of the refracted beam. Otherwise, the two beams would not match one another at the boundary. It follows from equation 5 that the ratio of the wavelengths of the two beams equals the ratio of their velocities:

$$\lambda_i / \lambda_f = v_i / v_f$$

FIGURE 9.17
Refraction.   A beam
passes from a region in
which the wave velocity
is v_i to one in which it is
v_f. The direction of motion of the wave and its
wavelength change, but
the frequency$_f$ remains
the same.   The angle
of incidence is θ_i and
the angle of refraction is
θ_f. They are related by
Snell's law.

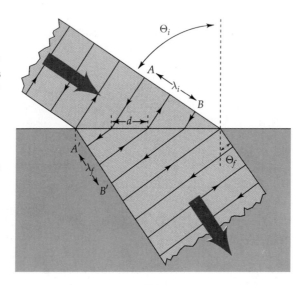

This result lets us relate the angle of incidence θ_i to the angle of
refraction θ_f.

From the geometry of the figure, we see that

$$\lambda_i = d \sin\theta_i \qquad \text{and} \qquad \lambda_f = d \sin\theta_f$$

Putting these results together, we find

$$\frac{\sin\theta_i}{\sin\theta_f} = \frac{v_i}{v_f} \qquad \text{for refraction} \tag{13}$$

This result is known as *Snell's law of refraction*. Equations 12 and 13 are
all of the the basic physical principles you need to know to design your
own telescopes and microscopes.

EXAMPLE 9.7
The Jewel in the
Swimming Pool

Figure 9.18 shows a diamond ring lying at the bottom of a swimming
pool 2.4 m deep at a distance of 1.8 m from its edge.   Because light is
refracted at the water's surface, the ring appears to lie 3.2 m from the

FIGURE 9.18
A diamond ring lies at
the bottom of a swim-
ming pool. The observer
lines up the image of the
ring with the edge of the
pool. The ring appears
to be much further from
the edge than it really is.

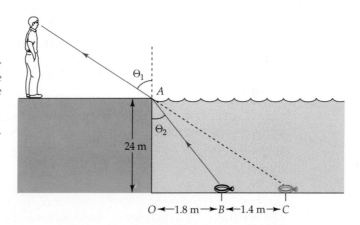

edge of the pool. From this information, deduce the ratio of the speed of light in water to its speed in air.

Solution The triangle $\triangle AOB$ is a right triangle. The length of its hypotenuse is $\sqrt{2.4^2 + 1.8^2}$, or 3 m. Thus, $\sin \theta_2 = 0.6$. The triangle $\triangle AOC$ is a right triangle. The length of its hypotenuse is $\sqrt{2.4^2 + 3.2^2}$, or 4 m. Thus, $\sin \theta_1 = 0.8$. Snell's law of refraction, equation 13, says that $c \sin \theta_2 = v \sin \theta_1$, where v is the speed of light in water. We find that $v/c = 0.75$.

■

Diffraction

Waves propagating through a medium—not just light and sound, but waves on the sea, the vibrations of a drumhead, and the tremors of an earthquake—share many features, some of which we have discussed. Now we turn to the interference effects produced when obstacles intervene or when several sources of waves are present at the same time.

Think of the pattern of waves obtained by jiggling one of the boundaries of a very large swimming pool. A linear pattern of ripples moves to the right, as shown in Figure 9.19 (on page 398). Suppose the oncoming wave encounters an obstacle with a gap through which waves may pass. What happens depends on the size of the gap. If the width of the gap is large compared to the wavelength, as in the middle figure, the waves pass straight through the opening. The obstacle leaves a shadow just as you and I do on a sunny day.

However, if the gap is small compared to the wavelength, the waves spread out in a semicircular pattern to all points on the right side of the pool. There is no shadow at all! This is the phenomenon of diffraction. If the gap is neither very large nor very small, the resulting diffraction pattern is more complex.

Let us turn from water waves to light waves. Light undergoes diffraction whenever it passes through any small aperture. Suppose that a parallel beam of monochromatic light with wavelength λ encounters an opaque plate on which there is a narrow slit of adjustable width d. A screen is placed a distance L from the plate. If d is large, an image of the slit of thickness d appears on the screen. As d is made smaller, a curious thing happens. A pattern of alternating bright and dark lines appears on the screen, such as that shown in the photo. The size of the pattern grows as d is further diminished. This phenomenon results from the constructive or destructive interference of light at the screen coming through different parts of the slit.

The width of the central bright line in a one-slit diffraction experiment is $2\lambda L/d$. Suppose that $d = 1$ mm and $L = 5$ m. If yellow light is used ($\lambda = 6 \times 10^{-7}$ m), the bright line is six times larger than the width of the slit. If the room is darkened, fringes can be seen extending for several centimeters in each direction.

We can more easily understand diffraction if we consider two very narrow slits rather than a single one. Imagine a plate with two narrow slits separated by a distance d. A monochromatic beam enters at the left.

FIGURE 9.19
Waves are produced in a swimming pool by periodically moving its boundary back and forth. *a.* The waves move to the right with no hindrance. The crests of the waves are shown as vertical lines. *b.* Waves passing through a gap that is wide compared to their wavelength. The waves maintain their original direction except for the region very near the edge of the gap. There are no waves in the shadow of the obstacles. *c.* Waves passing through a narrow gap spread out in a circular pattern. There is no shadow; the waves reach all points of the swimming pool.

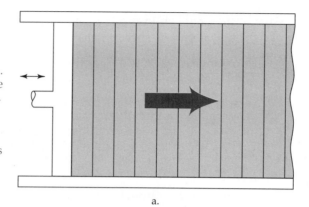

a.

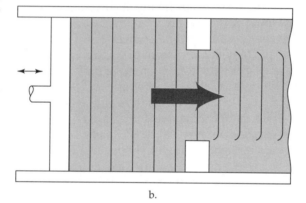

b.

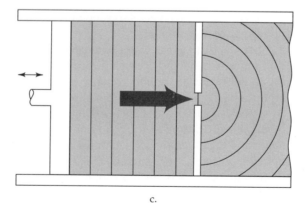

c.

Each narrow slit spreads the beam over a wide area. Light from the two slits recombines at the screen. Where the two beams interfere constructively, a bright stripe appears. Where they interfere destructively, a dark stripe appears. A pattern of alternating bright and dark bands is seen on the screen.

Whether the interference is constructive or destructive depends on the difference in path length between the slits and any point on the

One-slit diffraction. The diffraction pattern seen on a screen by light passing through a single slit. If the width of the slit is d and its distance from the screen is L, the width of the central bright line is $2\lambda L/d$, where λ is the wavelength of the light. *Source:* © Ken Kay, 1987/Fundamental Photographs.

screen, as shown in Figure 9.20. If the length difference is a whole number of wavelengths, the two waves are said to be "in phase." Their electric and magnetic fields complement each other and a bright line is produced. If the path length difference is half a wavelength more than a whole number of wavelengths, the fields cancel one another and no light reaches the screen. A dark line results.

Consider point C on the screen at a distance y above its midpoint. The distances light must travel from A or B to C are the hypotenuses of the right triangles $\triangle ACD$ and $\triangle BCE$. They are given by

$$P_A = \sqrt{L^2 + (y + d/2)^2} \qquad P_B = \sqrt{L^2 + (y - d/2)^2}$$

If d and y are both very small compared to L, the following approximate expressions may be used:

$$P_A \simeq L + \frac{(y + d/2)^2}{2L} \qquad P_B \simeq L + \frac{(y - d/2)^2}{2L}$$

By subtracting these expressions, we obtain an approximate formula for the difference in length of the two paths:

$$\Delta = P_A - P_B = yd/L + d^2/4 \simeq yd/L$$

where, in the last expression, we have dropped the term $d^2/4$ because d is much smaller than y. Whenever Δ is an integer multiple of λ, the

FIGURE 9.20
Two-slit diffraction. A beam of light is sent through two narrow slits. Where the difference between the path lengths AC and BC is a whole number of wavelengths, the two beams interfere constructively. A pattern of alternating bright and dark bands is seen on the screen.

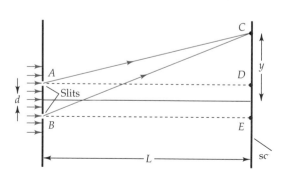

waves coming from A and B are in phase. They add up to make a bright line. This situation arises for $y = n\lambda L/d$, where

$$n = \ldots, \ -2, \ -1, \ 0, \ +1, \ +2, \ \ldots$$

However, whenever Δ lies halfway between two integers, the two waves are opposite in phase. They cancel one another to make a dark line. This occurs for $y = \left(n + \frac{1}{2}\right) n\,\lambda/d$. In the two-slit experiment, the bright lines are equally spaced, with a separation of $\lambda L/d$.

In a similar fashion, the diffraction of light caused by its wavelike nature limits the resolving power of microscopes. No matter how well it is built, things smaller in size than the wavelength of the light that is used cannot be seen. Shorter wavelengths can reveal smaller details. That is why ultraviolet microscopy is more powerful than visual microscopy. Electron microscopes are more powerful yet. They use the wavelike properties of particles and are intrinsically quantum mechanical devices.

The Splendor of Color

Sloths and squids, but not dogs and cats, see the world in color. So do most of us.* The importance of color—in nature, art, design, and daily life—is shown by the multitude of words we use for different hues. Browsing through a dictionary for shades of red, we find

Rose, Cerise, Carnation, Scarlet, Pink, and Crimson, Maroon, Magenta, Fuchsia, Ocher, Rubicund, Ruby, Vermilion.

The wonder and beauty of color fascinates us and challenges painters, photographers, physicians, and cosmeticians. Color is also the key to our understanding of the nature of matter, both here on Earth and in the stars. Stephen Hawking, an incomparably brilliant cosmologist and best-selling author, is presently the Lucasian Professor of Mathematics at the University of Cambridge. Centuries ago, Isaac Barrow addressed the physics of color from that very chair. According to his once definitive textbook on optics:

Red emits a light more clear than usual but interrupted by shady interstices. Blue discharges a rarefied light, as in bodies which consist of white and black particles arranged alternately. Green is nearly allied to blue. Yellow is a mixture of white and a little red. Purple consists of a great deal of blue mixed with a small portion of red. The blue color of the sea arises from the whiteness of the salt it contains, mixed with the blackness of the pure water in which the salt is dissolved; and the blueness of the shadows of bodies, seen at the same time by candle or daylight, arises from the whiteness of the paper mixed with the faint light of blackness of twilight.

Newton succeeded Barrow as Lucasian Professor and was unsatisfied with his predecessor's vague thoughts. When he was 23, he bought a

---

*Some people are color-blind. Their retinas are deficient in one or another of the three visual pigments.

FIGURE 9.21
Blue light is refracted more than red light as it passes from glass to air. A continuous line drawn in two colors appears broken when viewed through a prism placed on the paper.

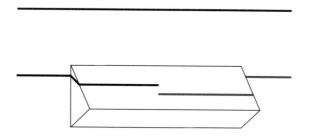

glass prism "to try therewith the phenomena of colors." You can repeat his first experiment by drawing a straight line on a piece of paper, half in red and half in blue. View it through a prism and it appears, as in Figure 9.21, as a broken line. The blue portion is bent, or refracted, more than the red. From Snell's law, it follows that blue light travels more slowly through glass than red light does. The higher the frequency of light, the more it is deflected by a prism.

Newton sent a narrow shaft of sunlight through a prism and then to a screen. He discovered that the white light of the Sun was spread out into a rainbow, as shown in Figure 9.22. He wrote that "the Sunbeams passing through a Glass Prism to the opposite Wall there exhibited a Spectrum of divers colors." Each tinted beamlet, when it was sent through a second prism, was deflected even more but it was divided no further. Newton realized that light rays of different colors travel at different speeds through glass and are thereby refracted at different angles. He wrote that "the Sun's light is a Heterogeneous Mixture of Rays" which are "parted or sorted from one another" by the action of the prism. Newton used the word *spectrum* (then meaning an apparition) to refer to the dissection of a light beam according to the wavelengths of its components—its colors.

Little more was learned about the scientific nature of color until 1800, when Herschel (who had already discovered Uranus, the seventh planet) examined the solar spectrum with the aid of a sensitive thermometer. He found the radiant heat of the Sun to be strongest at a point beyond the red end of the visible spectrum. He had discovered *infrared* radi-

FIGURE 9.22
White light is separated into its component colors by a prism. The higher the frequency of the light, the more it is deflected.

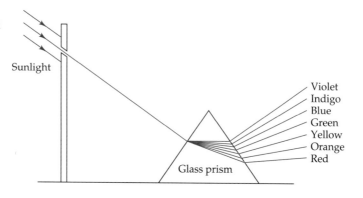

A BIT MORE ABOUT
Color

Physicists are primarily concerned with purely sinusoidal light waves or monochromatic light. The rest of us, especially painters and cinematographers, are much more interested in the effects obtained by the mixing of colors. The fact that our eyes process visual information in only three color channels is both a curse and a blessing. How much richer would our art be if our eyes took four or more color samples? Or, if we shared with squids the ability to sense the polarization of light in addition to its color? However, since the eye has only three color sensors, three colors suffice. When mixed together, these colors produce all the shades and hues we see. Colors may be added together, as in the image of a TV tube. Or they may subtracted away, as in an oil painting viewed in white light.

Chemicals that produce light of a definite color when they are struck with electrons are called *phosphors*. Color television tubes use three different phosphors to produce the glints of red or green or blue that, in combining, entrance us.

Figure 9.23 shows the colors obtained by projecting three colored beams of the same intensity onto a screen in a darkened room. When colors are combined additively in this fashion, we see that

$$\text{red} + \text{green} = \text{yellow},$$

$$\text{green} + \text{blue} = \text{cyan},$$

$$\text{blue} + \text{red} = \text{magenta},$$

$$\text{and} \quad \text{red} + \text{green} + \text{blue} = \text{white}$$

Color film uses three different dyes to achieve verisimilitude. Each dye absorbs light waves far from its characteristic frequency. Light passing through two of the dyes, or reflecting from a blend of two paints, combines together subtractively. Figure 9.24 shows the colors obtained by mixing equal portions of cyan, magenta, and yellow paints:

$$\text{yellow} + \text{cyan} = \text{green},$$

$$\text{cyan} + \text{magenta} = \text{blue},$$

$$\text{magenta} + \text{yellow} = \text{red},$$

$$\text{and} \quad \text{yellow} + \text{cyan} + \text{magenta} = \text{black}$$

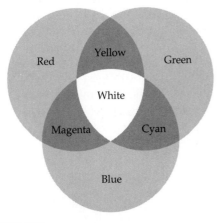

FIGURE 9.23
Color addition. Red, blue, and green spotlights impinge on a white screen. The region in which all three beams overlap appears white. The regions in which two beams complement one another are cyan, magenta, and yellow.

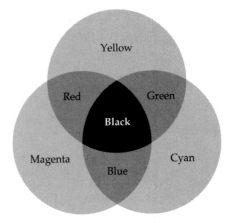

FIGURE 9.24
Color subtraction. Assume that this figure is painted on a canvas with cyan, magenta, and yellow paints. Equal mixtures of the appropriate two paints are used where two circles overlap. They appear red, green, and blue. In the central region, a mixture of all three paints absorbs all the incident light. It appears black.

ation. A year later, a portion of the spectrum beyond the violet was found to blacken silver chloride—*ultraviolet* light had made its appearance. Thomas Young, an early advocate of the wave theory of light rather than Newton's corpuscular theory, measured the wavelengths of the seven colors recognized by Newton. He found that the spectrum of visible light includes wavelengths extending between about 4 and 7×10^{-7} m.

Exercises

16. Blue paint reflects light of a wide range of frequencies centered on spectral blue. Other frequencies are absorbed. Yellow paint absorbs frequencies far removed from yellow.
    a. Explain why a mixture of blue and yellow paints is green.
    b. Explain why the overlap of blue and yellow spotlights appears to be a shade of lilac, but is certainly not green.

17. Explain why all metals are shiny when they are polished.

18. Prisms bend blue light more than they do red. Which color of light travels faster through glass?

19. Why do shadows have fuzzy edges?

20. A parallel beam of blue light ($\lambda = 4 \times 10^{-7}$ m) is directed through a slit 1 cm wide to the Moon, which is about 4×10^{8} m away. How wide is the central blue stripe on the Moon?

21. A beam of light with frequency 5×10^{14} Hz is sent through two parallel slits that are 6×10^{-4} m apart.
    a. What is the wavelength of the light?
    b. How far apart are the bright lines appearing on a screen 8 m from the slits?

22. Your seat at an outdoor concert at Lewisohn Stadium is behind a pillar. Both sound and light consist of waves. Yet, you can hear the orchestra but not see it. Explain why.

23. Although you can hear the music in your cheap seat of exercise 22, the sound is somewhat muffled. Which instrument is affected more, the piccolo or the bassoon? Why?

9.4 Spectral Lines of Stars and Atoms

Newton saw that sunlight could be taken apart into its component colors. However, his crude prism did not reveal an amazing property of the solar spectrum. In 1814, when Joseph Fraunhofer examined the solar spectrum with a large prism, he found to his astonishment that it is interrupted by numerous dark lines, as in the photo. The lines would

Absorption lines in the solar spectrum lying between 3900 and 4600 angstrom units in wavelength. *Source:* The Observatories of the Carnegie Institution of Washington. Courtesy Jay M. Pasachoff.

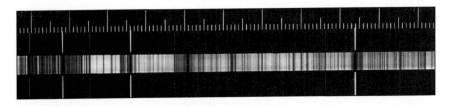

be useful for his trade as lensmaker: They would serve to standardize the colors of light at which the refractive indices of various glasses were measured. He designated eight of the most conspicuous lines with consecutive letters: *A* (deep red), to *D* (bright yellow), to *H* (violet).

Because a prism bends shorter wavelengths more than longer ones, it can be used to compare the wavelengths of different light beams. However, it does not yield an absolute measurement of wavelength. To this end, Fraunhofer invented a new instrument: the *diffraction grating*. The modern diffraction grating consists of a flat piece of glass on which thousands of fine and equally spaced parallel lines are engraved. When white light is sent through this device, some of it is deflected because the lines of the grating mimic the slits of the two-slit experiment. Because there are thousands of lines on the grating, the net result is simply to bend light of wavelength λ by an angle θ given by

$$\sin \theta = \lambda/d \qquad (14)$$

where *d* is the spacing between the slits. A measurement of the angle by which a spectral line is deflected determines its wavelength.

EXAMPLE 9.8
The Diffraction Grating

Suppose that a diffraction grating has 3000 parallel lines per centimeter engraved on it. By what angle θ is yellow light, with $\lambda = 5.8 \times 10^{-7}$ m, deflected by the grating?

Solution

The spacing between the lines of the grating is $d \simeq 3.33 \times 10^{-6}$ m. Equation 14 tells us that $\sin \theta = 0.174$, so $\theta \simeq 10°$. ∎

A diffraction grating or a prism used to measure the wavelengths of spectral lines is called a *spectrometer*. Fraunhofer used his spectrometer to measure and catalog the wavelengths of 586 dark lines of the solar spectrum.* He resolved his *D* line into two nearby dark lines with wavelengths of 5.890 and 5.896 $\times 10^{-7}$ m. When he sent the light produced by a flame through his spectrometer, he observed bright yellow lines with exactly the same wavelengths, but he never found out why.

Today, we know that the yellow doublet of lines is associated with the chemical element sodium. The dark *D* lines of the Sun are produced

---

*Fraunhofer's best diffraction grating had a spacing between lines of $d = 3$ microns. By 1887, American physicist Henry Rowland (1848–1901), built them with spacings twenty times smaller. With such a device, he produced a solar spectrum 40 ft in length in which he identified over 20,000 Fraunhofer lines! From the lines, he identified the presence of 36 chemical elements in the Sun. By 1928, the number was 51 and growing.

by the selective *absorption* of light by sodium atoms in the outer and cooler envelope of the Sun. The bright *D* lines of a flame result from the selective *emission* of light by hot sodium atoms. Sodium's emission lines are the reason the flame turns yellow when the pot boils over, and they explain the unearthly yellow pall inflicted on us by highway engineers. Fraunhofer never learned that each and every chemical atom and molecule has its own characteristic spectrum of electromagnetic wavelengths.

Putting his grating to a telescope, Fraunhofer studied the spectra of heavenly bodies. Those of planets closely resembled that of the Sun, but those of stars were subtly different. Fraunhofer created the science of stellar spectroscopy, which has become one of the most powerful tool of astrophysicists.

What were the strange dark lines in the spectra of starlight? John Herschel (son of the discoverer of Uranus) hadn't the foggiest notion, but upon observing the bright colors produced when chemicals were intensely heated in a flame, he concluded, "The colors thus communicated

---

A BIT MORE ABOUT
Fraunhofer

Joseph Fraunhofer (1787–1826) began his career as an optical wizard at the tender age of 11, when he was apprenticed to a Munich mirror maker. His many important contributions to optical science were accomplished in a tragically short life of 39 years. It sometimes pays to start early.

For the same reason that a prism produces a spectrum, a simple glass lens does not focus all incident light rays to the same point. Different colors are differently deflected. When the red part of the picture is clear, the blue is blurry, and conversely. This effect is known as *chromatic aberration* and Newton declared it to be uncorrectable. He regarded chromatic aberration as an innate obstacle to the construction of better refracting telescopes. He also believed, wrongly, that all sources of light display continuous spectra. Newton's misconceptions, together with his extraordinary prestige, may have retarded the science of optics as much as his successes impelled that of mechanics.

It took a century and a half, but Fraunhofer caught both of Newton's optical errors. Aside from discovering spectral lines and inventing the diffraction grating to map those of solar, stellar, and earthly spectra, he developed the achromatic lens. Fraunhofer showed that two properly configured lenses made of different kinds of glass, when used in tandem, do a better job of focusing light of different colors.* The compound lenses found in today's cameras, camcorders, and binoculars are virtually free of chromatic aberration.

*John Dolland produced such lenses by 1757, but only by trial and error. Fraunhofer made lensmaking into a precise science.

by the different bases to flames afford, in many cases, a ready and neat way of detecting extremely minute amounts of them." He didn't understand the lines, but he could use them as a technique for chemical analysis.

By the middle of the nineteenth century, physicists began to suspect that spectral lines were associated in some way with the internal structure of atoms or molecules. Fraunhofer's D doublet of dark lines in the solar spectrum appears as two bright lines when a sodium-containing material is placed in a flame. This result suggested that sodium is present in the Sun. But why did materials in a flame produce bright lines while the Sun's lines (at the identical frequencies) were dark? The German physicist Gustav Kirchhoff (1824–1887) found the answer in 1859.

When an object is heated, it radiates light. For reasons that were not yet understood, a given atom in the object can radiate only certain definite wavelengths of light (like sodium, whose spectrum includes the yellow D doublet). Conversely, if light passes through a gas, certain specific wavelengths of light are selectively absorbed by the atoms of the gas. Foucault had already shown that when a bright light passes through vaporized sodium, two dark lines appear at just the locations of the solar D doublet. Kirchhoff proved from fundamental thermodynamic principles that a body's emissive and absorptive abilities are related to one another. A substance that selectively emits a certain color when hot absorbs the same color when cool. The emission lines and absorption lines associated with each chemical element must coincide. Since the outermost layer of the Sun is cooler than its interior (and since its inner layers do produce a more-or-less continuous spectrum of light), the dark lines of the Sun reveal the presence of those atoms that produce the same bright lines when heated in a laboratory on Earth.

Kirchhoff deduced and demonstrated that the dark lines of the solar spectrum have the same wavelengths as the bright lines produced by elements such as hydrogen, iron, sodium, potassium, and magnesium. Thus, he proved that these elements are present in the Sun. Not long before, the French philosopher Auguste Comte had written of the quest to know the stars:

> We understand the possibility of determining their shapes, their distances, their sizes and motions, whereas never, by any means, will we be able to study their chemical composition.

Philosophers should never say never.

Kirchhoff collaborated with Robert Bunsen, the man behind the Bunsen burner. They developed a spectrometer to study the light produced by various chemicals when they are intensely heated in a flame. This led to a powerful new method for chemical analysis by which they discovered two new chemical elements. In 1860, Bunsen wrote to a friend:

> I have been very fortunate with my new metal. I have got 50 grams of the nearly chemically pure chloro-platinate obtained from no less that 40 tons

of mineral water. I am calling my new metal "cæsium" from "cæsius" blue, on account of the splendid blue line in its spectrum.

Soon after, they found rubidium (with a brilliant red line), and others found thallium and indium spectroscopically.

Hints of
Subatomic
Structure

Light is an electromagnetic wave. Each chemical element can emit or absorb light of certain definite frequencies or wavelengths—its spectrum. At the end of the nineteenth century, many scientists believed that these frequencies were (or had something to do with) the frequencies at which atoms could vibrate. And, if atoms can vibrate, they must have parts. Surely the study of atomic spectra could shed light on atomic structure. Here is the state of the science of atomic structure according to the ninth edition of the *Encyclopædia Britannica,* published in 1890:

> The spectrum of a body is due to periodic motion within the molecules. It seems probable that there is a numerical relation between the different periods of the same vibrating system. In [organ pipes or stretched strings] the relation is a simple one, these periods being a submultiple of the fundamental period. The harmony of a compound sound depends on the fact that the different times of vibration are in the ratio of small integer numbers. We may with advantage extend the expression "harmonic relation" to the case of light. We shall therefore define an "harmonic relation" between different lines of a spectrum to be a relation such that the wavelengths are in the ratio of [small] integers.

The article tells of many ingenious but unsuccessful attempts to find the harmonies of atoms. In fact, the spectra of the chemical elements do not show such regularities as the resonant frequencies of pipes or strings. However, there was a clear-cut relation among the spectral lines of hydrogen, the simplest of all atoms. It was found by a Swiss schoolteacher, Johann Jakob Balmer, in 1885. The *Britannica* entry continues:

> Even if the wavelengths of two lines are found to be occasionally in the ratio of small integer numbers, it does not follow that the vibrations of molecules are regulated by the same laws as those of an organ pipe or a stretched string. Balmer has indeed lately suggested a law which differs in an important manner from the laws of vibration of the organ pipe and which still leaves the ratios of the periods of vibration integer numbers. According to him, the hydrogen spectrum can be represented by the equation:

$$\lambda = \frac{n^2}{n^2 - 4}\lambda_0, \qquad n = 3, 4, 5, \ldots \tag{15}$$

[where $\lambda_0 = 3645.6 \times 10^{-10}$ m.] The agreement is a very remarkable one, for the whole of the hydrogen spectrum is represented by giving to n successive integer values from 3 up to 16.

Remarkable indeed! Balmer's inspiration was based on careful measurements of four visible hydrogen lines: one red, one green, one blue, and

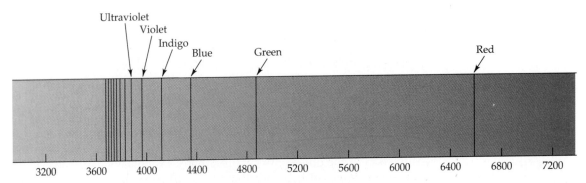

FIGURE 9.25 The Balmer series of spectral lines of hydrogen. Five of the lines lie in the visible portion of the electromagnetic spectrum. All the rest of the Balmer lines are ultraviolet. The series approaches a limiting frequency of 3645.6 angstroms.

one indigo. All but one of the rest lie in the near ultraviolet, more and more closely approaching 3545.6 angstroms.* The violet line was quickly found and many other ultraviolet lines were observed soon thereafter. Every line was just where Balmer's formula said it should be. His formula worked like a charm and hinted at the inner structure of the hydrogen atom (Figure 9.25).

No such simple formula could be found to describe the spectra of other elements. However, another curious regularity showed up in 1908. All the frequencies of any atom's spectrum can be expressed in terms of a series of fundamental frequencies, f_i. These may be put in descending order such that $f_1 > f_2 > f_3 > \cdots$. Every spectral frequency of an atom is a difference between two of its fundamental frequencies, and to every difference $f = f_i - f_j$ (where $i < j$) there corresponds a spectral line of the atom with frequency f. This result is known as the Ritz Combination Principle.

To see the power of the principle, let us rewrite Balmer's formula, equation 15, in terms of spectral frequencies:

$$f = \left(\frac{1}{2^2} - \frac{1}{n^2} \right) f_0 \qquad \text{the Balmer series: } n = 3, 4, 5, \ldots \qquad (16)$$

where

$$f_0 = 4\,c/\lambda_0 \simeq 3.29 \times 10^{15} \text{ Hz} \qquad (17)$$

and n is any integer larger than 2. This formula gives the spectral frequencies as differences in accordance with the Ritz principle, and the fundamental frequencies of the hydrogen atom are read off to be

$$f_n = \frac{f_0}{n^2} \qquad n = 1, 2, 3, \ldots \qquad (18)$$

*Careful measurements of the spectrum of hydrogen were carried out by the Swedish physicist Anders Ångström (1814–1874), for whom the spectroscopists' unit of wavelength is named. The angstrom unit of length, Å, equals 10^{-10} m.

Total solar eclipse. On rare occasions when the Sun, Moon, and Earth are aligned, the Moon's disk fully obscures our view of the Sun and solar prominences may be seen. The spectral analysis of these prominences during the total eclipse of 1868 revealed the emission lines of a then unknown chemical element—helium. *Source:* © François Gohier/Photo Researchers.

According to the Ritz principle, another series of spectral lines must exist with the infrared frequencies:

$$f = \left(\frac{1}{3^2} - \frac{1}{n^2} \right) f_0 \qquad \text{the Paschen series: } n = 4, 5, 6, \ldots \qquad (19)$$

The Paschen series was duly found and named for its discoverer. A few years later, a series of ultraviolet lines was discovered by Theodore Lyman at Harvard. These lines correspond to the frequencies:

$$f = \left(\frac{1}{1^2} - \frac{1}{n^2} \right) f_0 \qquad \text{the Lyman series: } n = 2, 3, 4, \ldots \qquad (20)$$

The spectrum of hydrogen satisfied the Ritz combination principle with the fundamental frequencies determined by equation 18. The long sought "harmonic relations" had finally shown up, but what they meant was another matter. Niels Bohr would amaze the world in 1913 with a new theory that explained both the mystifying simplicity of the hydrogen spectrum and the intricacies of the periodic table. Quantum theory was about to be born.

The Trouble with Helium

During a total eclipse of the Sun, its faint and far-reaching corona, as well as its small pink, pulsating prominences, can be seen, Russian monks described the eclipse of May 1185:

The Sun became like the crescent of the Moon, from the horns of which a glow similar to that of red-hot charcoals was emanated. It was terrifying for men to see this sign of the Lord.

Not until August 18, 1868, did scientists study solar prominences with their newly discovered spectrometers. A team of astronomers packed their instruments and sailed to India where the eclipse was total. The solar prominences showed only bright lines, telling them that they consisted of glowing streamers of intensely hot gas. The brightest lines were the red and green hydrogen lines (Fraunhofers C and F) and a mysterious yellow line that did not quite coincide in frequency with sodium's yellow doublet D_1 and D_2. They called the new line D_3 and attributed it to a new chemical element not yet found on Earth. It was called helium from *helios* for Sun and *-ium*, indicating a metal. Terrestrial helium was discovered by Ramsay in 1895, but it turned out to be an inert gas, not a metal.

In 1896, a curious new series of spectral lines was discovered in the spectrum of an exceptionally hot star. Alternate lines of the series were the familiar Balmer lines, but there were additional lines in between. The so-called Pickering series of frequencies satisfied Balmer's formula,

$$f = \left(\frac{1}{2^2} - \frac{1}{n^2} \right) f_0 \qquad (21)$$

but with n equal to an integer or a half integer: $n = \frac{5}{2}, 3, \frac{7}{2}, 4, \ldots$ Lockyer, who was the codiscoverer of helium in the Sun, attributed these lines to the breaking up of hydrogen atoms into more primitive "proto-atoms." In fact, these lines had nothing to do with hydrogen! Bohr would show that the Pickering series was the analog to the Paschen series for singly ionized helium atoms. Hevesy (a Hungarian friend of Bohr) wrote to Rutherford:

> When I told him [of this result] the big eyes of Einstein looked still bigger, and he told me, "Then it is one of the greatest discoveries." I felt very happy hearing Einstein saying so.

Exercises

24. Show that no spectral line of hydrogen has a frequency exceeding f_0.

25. Suppose that the diffraction grating described in Example 9.8 is placed 10 m from a screen. The D doublet of yellow lines impinges on the device. How far apart do the two lines of the doublet appear on the screen?

26. Compute the wavelengths of the first four lines ($n = 3, 4, 5, 6$) of the Balmer series of hydrogen's spectral lines. (*Hint:* The frequency of the $3 \to 2$ transition is $\left(\frac{1}{4} - \frac{1}{9} \right) f_0$, or 4.57×10^{14} Hz. The wavelength

is obtained by dividing this result into the speed of light: $\lambda = 6.565 \times 10^{-7}$ m, or 6565 angstroms. You do the rest.)

27. The nth series of spectral lines of hydrogen consists of the frequencies

$$\left(\frac{1}{n^2} - \frac{1}{m^2}\right) f_0 \tag{22}$$

where m is any positive integer larger than n. Show that all of the Lyman lines (the series obtained with $n = 1$) have higher frequencies than all of the Balmer lines (the series with $n = 2$). They all lie in the ultraviolet part of the electromagnetic spectrum.

28. Show that all of the Paschen lines ($n = 3$) have lower frequencies than all of the Balmer lines. They all lie in the infrared.

29. Determine if the following statement is true or false: The smallest spectral frequency in the nth spectral series of hydrogen exceeds all of the frequencies in the $(n + 1)$th series.

30. The singly charged helium ion is a system consisting of a single electron bound to a nucleus of charge 2. It is much like a hydrogen atom but with a larger nuclear charge. Bohr showed that its spectral lines have frequencies given by a formula that is very similar to equation 22:

$$\left(\frac{1}{n^2} - \frac{1}{m^2}\right) 4 f_0 \tag{23}$$

with f_0 multiplied by the square of the nuclear charge. In equation 23, m and n are any positive integers with $m > n$. Show that this formula, with n set equal to 4, reproduces the Pickering series given by equation 21.

31. On the basis of the information in exercise 30, write a formula for the frequencies of the spectral lines of doubly charged lithium ions, another system consisting of a single electron electrically bound to a nucleus. (The atomic number of lithium is 3.)

Where We Are and Where We Are Going

Maxwell deduced and Hertz proved that light consists of electromagnetic waves. Atoms of a given element can absorb or emit light waves of certain characteristic colors. Like organ tubes or drumheads, they can vibrate at specific frequencies. Thus, they must be composite structures with moving parts carrying electric charges. But what is their structure? At this point, classical physics threw in the towel. Chapter 10 describes three great experimental surprises that mark the birth of modern physics. Studies of electric discharges in gases allowed scientists to find the electron and X rays. They also inspired the discovery of the spontaneous disintegration of certain atoms: radioactivity. These novelties quickly became the tools with which we learned how atoms are made.

In the beginning of the twentieth century, the science of light was faced with two apparently insuperable difficulties: the refusal of light to be pigeonholed as either a wave or a particle, and the failure to detect Earth's motion through the ether. Both problems were solved. The wave-particle duality of light led to quantum mechanics, which totally changed the logical basis of physics and is the subject of Chapter 11. The invariability of the speed of light through the vacuum led Einstein to his special theory of relativity, which totally changed our understanding of space and time and is the subject of Chapter 12.

10 Inside the Atom

Curiosity drives us on to try to discover the secrets of nature, those secrets which are beyond our understanding, which can avail us nothing and which men should not wish to learn.

Augustine of Hippo, Christian saint

By research in pure physics I mean research made without any idea of application to industrial matters but solely with the view of extending our knowledge of the Laws of Nature. I will give just one example of the "utility" of this kind of research, one that has been brought into great prominence due to the War—I mean the use of X rays in surgery. It was not the result of a [search for] an improved method of locating bullet wounds. This might have led to improved probes, but we cannot imagine it leading to the discovery of X rays. No, this method is due to an investigation in pure science, made with the object of discovering the nature of Electricity.

J. J. Thomson (1916), British physicist

The last ten years of the nineteenth century are known as the Gay Nineties or the Mauve Decade. They marked the finale to Queen Victoria's long reign. It was a time for Monet's lilies and Art Nouveau; for venerating Sarah Bernhardt and villifying Oscar Wilde. Millions of immigrants thronged to the land of opportunity, where a few robber barons became immensely wealthy. America fulfilled its manifest destiny by remembering the Maine, occupying Cuba and the Philippines, and annexing Puerto Rico, Hawaii, and Guam. The world celebrated the seemingly unbounded promise of science and technology—at the 1889 Paris Centenary, with Eiffel's immense tower and its American-made Otis elevators, and at Chicago's 1892 Columbian Exposition, where 21 million people danced all night in an electrically lit square mile, thrilled to the giant Ferris wheel, and tittered at imported Egyptian belly dancers. The marvels of technology ushered in the twentieth century.

Turbine–powered steel ships plied the Atlantic in less than a week, gasoline–powered vehicles took to the streets, and brand-new electric subways slithered beneath Paris, London, and Boston. The first self-propelled airplane flew in 1895, soon followed by manned aircraft at Kitty Hawk. Niagara Falls was harnessed to power New York's factories, movie projectors, and electric chairs. Aspirin was hailed as the pain remedy it is, heroin as the cure for drug addiction it is not. Tun-

The Eiffel Tower. This 300 m high steel structure was built in Paris in 1889 to celebrate the 100th anniversary of the French Revolution and the power of modern technology. It was twice as tall as the Great Pyramid and remained the world's highest edifice until the Chrysler building was erected in 1930 in New York City. *Source:* © Pierre Berger/Photo Researchers.

nels, bridges, trolleys, and air conditioned skyscrapers with automatic elevators transformed city life.

Who would deny that science had conquered nature, that our knowledge of the material world was almost complete? Newton's synthesis of celestial and terrestrial mechanics was joined by Maxwell's unification of electricity, magnetism, and optics. The forces of nature were understood and could be controlled. The effects of heat on matter followed from the chaotic but statistically predictable motions of immutable atoms, and the rules of chemistry were systematized by Mendeleev's marvelous chart. It seemed as if all of the big problems of physical science had been solved. What remained, or so it seemed, were mere footnotes to the grand accomplishments of the past. "The future of physics is in the fifth decimal place," said some saddened scientists of the late nineteenth century, who felt that nothing much was left for them to discover. How wrong they were!

This chapter focuses on the experiments that led to the realization that the atom is neither elementary, nor immutable, nor necessarily eternal. Rather, it is a structured system made up of an atomic nucleus surrounded by electrons. Section 10.1 describes three great surprises of the 1890s: the serendipitous observations of X rays and radioactivity and the discovery of the first elementary particle, the electron. Section 10.2 describes another surprise: the discovery that most of the mass of an atom resides in a tiny nucleus whose positive charge determines the chemical identity of the atom.

10.1 Three Surprising Discoveries

Three astonishing revelations of the 1890s opened our eyes to the world within the atom and began a still-unfolding scientific and technological revolution: X rays were discovered in 1895, radioactivity in 1896, and the electron in 1897. Scientists often fancy that their researches form the driving force of modern technology. Yet, these monumental achievements of three renowned and highly trained scientists depended on the perseverance of the three modest artisans (none with university degrees) who developed photography, spark coils, and vacuum pumps. These inventions were crucial, not only to the world–shaking discoveries of the 1890s, and not only to physics.

Three Inventions That Paved the Way

The first crude and unstable photographic image was made in 1802. The process remained a useless curiosity until 1839, when Louis Daguerre, a painter of no great renown, invented a way to make enduring photographs. A lens focused a scene on a copper plate coated with silver iodide. (Iodine had been discovered only a few decades earlier.) Subsequent chemical treatment of the plate produced a permanent image, a *daguerreotype*. The first news photograph showed the great Hamburg

An early photograph (c. 1843) and a modern stop-time photograph of a bullet passing through a playing card. *Source:* The Bettmann Archive (left) and © Gary S. Settles/Photo Researchers.

fire in 1842. Photographic film was invented in 1850, dry plates 30 years later, and George Eastman's *Kodak*, with its self-contained roll of film hit the market in 1893. Photography was the key to the discovery of radioactivity and allowed X rays to become an instant medical miracle. Throughout the twentieth century, photographic techniques were as essential to astronomy and physics as they were to the arts and the media.

The spark coil was a spin–off from Faraday's electromagnetic induction device. Recall from Chapter 8 that the device consisted of an iron ring about which two coils were wrapped. When the primary coil was connected to or disconnected from a battery, an electric current surged briefly in the secondary coil. If the secondary coil had more turns than the primary, the voltage at the secondary was amplified. By the 1850s, Heinrich Rühmkorff (a German engineer residing in Paris) developed Faraday's primitive induction device into a marketable product that converted the low voltage supplied by a battery into a reliable source of electricity at voltages as high as 100 kilovolts. In the nineteenth century, these Rühmkorff coils became highly prized scientific instruments: They represented the state of the art in high-energy physics. They enabled Hertz to generate radio waves, Marconi to develop wireless telegraphy, Röntgen to discover X rays, and Thomson to find the electron. Similar devices, much improved, are found under the hoods of most cars and trucks today. They provide the high voltage necessary to actuate the spark plugs of gasoline engines. Rühmkorff won a 50,000 franc prize for his work, but he died in poverty. He had given all his money to science and charity.

Electric discharges in Geissler tubes. Electrons, streaming through a near vacuum, collide with and excite atoms in their path. The atoms emit light, producing colorful patterns. The same process takes place in many modern lighting devices, such as neon signs and mercury vapor streetlights. *Source:* The Granger Collection.

The first vacuum pump was built by von Guericke in 1657. No significant improvement was made for another two centuries, when Johann Geissler invented the far more powerful mercury pump. A talented glassblower as well as a skilled mechanic, he used his new pump to manufacture evacuated glass tubes with two electrodes sealed within them. They were the first vacuum tubes, precursors of the cathode ray tubes in today's TV sets and computer displays. Weird patterns of light, such as those in the photo, appeared when Geissler's tubes were connected to an induction coil. Various minerals, if included within the tube, glowed in brilliant colors. His tubes became a popular novelty. Faraday, in his later years, experimented with Rühmkorff's coils and Geissler's gadgetry.

The Discovery of X Rays

William Conrad Röntgen (1845–1923) was fascinated by the curious goings-on within Geissler tubes, but during his investigations he found something so entirely unexpected that he said to his wife, "People will say that Röntgen has probably gone crazy." He discovered a mysterious and invisible form of radiation that streamed from the positive electrode or anode of the tube, through its glass wall, and into his laboratory. Röntgen found that the radiation caused minerals to glow and could penetrate materials that were opaque to light. In particular, they would expose a photographic plate that was wrapped in black paper. He called the new radiation "X rays" (for x, the unknown quantity in mathematics) and submitted his first paper about them in December 1895. It began:

> If the discharge of a fairly large Rühmkorff induction coil is allowed to pass through a [Geissler] vacuum tube, and if one covers the tube with a fairly close-fitting mantle of thin black cardboard, one observes in the completely darkened room that a paper screen painted with barium platinocyanide placed near the apparatus glows brightly, or becomes fluorescent with each discharge, regardless of whether the coated surface or the other side is turned toward the discharge tube. This fluorescence is still visible at a distance of two meters from the apparatus.

FIGURE 10.1
X-ray tube. A potential difference between the electrodes accelerates electrons released by the hot filament, or at the cathode. The electrons strike a target at the anode and produce X rays.

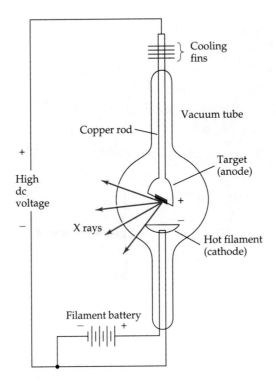

When an electric discharge streaks through a vacuum tube, a beam of electrons flows from the cathode toward the anode.* When the electrons collide with the anode or with the glass wall of the vacuum tube, they excite atoms and electromagnetic radiation is emitted. (Certain minerals, placed within the tube, emit light or fluoresce when they are struck by electrons: these are prototypes of the phosphors used in TV screens.) If the electrons are energetic enough, their collisions with atoms produce X rays (Figure 10.1).

Chance may have played a part in Röntgen's discovery. A fluorescent material may have been lying about at the right time and in the right place—physics labs are not known for neatness. Yet, Röntgen followed up his initial discovery with the care and alacrity one expects of a great scientist. He proved that X rays are different from the cathode rays within the tube. Cathode rays could be easily deflected by a nearby magnet. Röntgen showed that magnets did not affect X rays at all. Cathode rays would later be shown to consist of electrically charged particles; X rays are electrically neutral. Because metal and bone block the passage of X rays, Röntgen was able to produce pictures of a set of weights sealed in a small box and of the bones of his wife's hand. They were the first X-ray photographs. Soon after his discovery, a reporter asked,

---

*These so-called *cathode rays* were the original focus of Röntgen's research. Later in this chapter, we describe how cathode rays were shown to be rapidly moving electrons.

One of Röntgen's first X-ray photographs shows the bones of his wife's hand. *Source:* The Bettmann Archive.

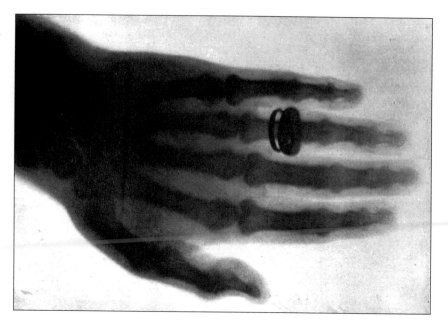

"What did you think?" Röntgen said, "I did not think; I investigated." To "What are your rays?" he confessed, "I do not know!"

The suspicion soon arose that X rays were ultra-ultraviolet light—electromagnetic waves of exceedingly short wavelengths—but this was not proven until 1912. At that time, Max Von Laue (1879–1960) proposed using a crystal to diffract X rays. The atoms of a crystal are arranged in a regular pattern in three dimensions. If X rays actually were electromagnetic waves, the waves reflected by individual atoms should add together constructively in some directions and interfere destructively in others. That is, X rays should diffract from crystals. A beam of X rays could be sent through or reflected from a crystal, and the diffraction pattern could be captured on photographic film. A complex array of spots should be produced, such as that shown in the photo. Laue convinced two of his colleagues to do the experiment. They bombarded a crystal with X rays, observed a diffraction pattern, and thereby proved that X rays are a form of electromagnetic radiation.

Immediately after the observation of X-ray diffraction, W. H. Bragg (1862–1942) showed how the diffraction pattern can be calculated from the spacings between the atoms in a crystal. By observing the angles at which X rays reflect from a crystal of known structure, scientists were able to prove they are waves and measured their wavelength.

Bragg showed how the structure of an unknown crystal or molecule can be determined from its X-ray diffraction pattern. Mathematical analysis of the diffraction pattern can reveal the placement of individual atoms. Using X-ray diffraction techniques, biologists found the double helix structure of DNA, the molecule that carries genetic information. Röntgen's mysterious rays became a powerful new technology for the

The X-ray pattern pro-
duced by a crystal of
titanium dioxide. *Source:*
Omikron/Photo
Researchers.

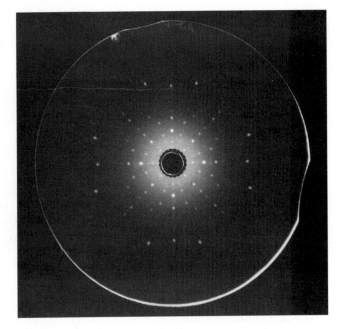

study of the inner structure of matter. It often happens in science that
today's discovery becomes tomorrow's tool. The importance of X-ray
diffraction to science has been recognized by the award of no less than
five Nobel Prizes to its practitioners.

Bragg Reflection   A simple example of X-ray diffraction is the phenomenon of Bragg reflection from crystal surfaces. Figure 10.2 shows several equally spaced
layers of atoms a distance d apart lying near a crystal surface. X rays incident on the crystal at an angle θ relative to its surface reflect from each
layer. The reflected waves add up constructively at certain special angles
of reflection. Waves proceeding along the trajectories AF and $BB'F$ arrive at F at the same phase if the difference of the two path lengths is an

FIGURE 10.2
X-ray diffraction from
a crystal surface. Each
layer of atoms acts like
a mirror. The reflection
is strongest when waves
reflecting from succes-
sive layers add construc-
tively. This occurs at
angles θ such that the
path lengths AF, $BB'F$,
$CC'F$, and so on, differ
by an integer number of
wavelengths.

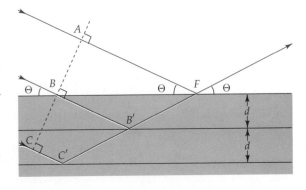

integer multiple of the X-ray wavelength λ. A trigonometric calculation shows that this happens if

$$\sin\theta = \frac{n\lambda}{2d} \qquad \text{for} \quad n = 1, 2, 3, \ldots \qquad (1)$$

If the X-ray wavelength is sufficiently small—that is, if $\lambda < 2d$—equation 1 is satisfied by one or more angles. At each such angle, the waves reflecting from all the layers of atoms arrive at F and continue onward in phase with one another. Reflection is strong. At other angles, the waves cancel one another and reflection is weak. If the value of d is known, Bragg reflection can be used to measure the wavelength of a beam of X rays. A crystal, in the discipline of X-ray spectroscopy, plays the role of a diffraction grating in optical spectroscopy.

EXAMPLE 10.1
X-Ray Diffraction

The atomic layers of a nickel crystal are 4.5 angstroms apart. X rays with a wavelength of 2 angstroms are incident on the crystal at an angle θ relative to the crystal planes. At what angles is there strong reflection?

Solution

From the information we are given, $\lambda/2d = 2/9$. Equation 1 becomes $\sin\theta_n = n/4.5$, where n is an integer. For $n = 1$, we obtain $\theta_1 = \arcsin 2/9 \simeq 12.8°$. In a similar fashion, we obtain

$$\theta_2 = 26.4°, \qquad \theta_3 = 41.8°, \qquad \text{and} \quad \theta_4 = 62.7°$$

However, if n is 5 or greater, we obtain $\sin\theta_n > 1$, which has no solution. Thus, there are a total of four angles at which the X rays reflect from the crystal surface.   ■

The Discovery of Radioactivity

Henri Becquerel devoted himself to the study of luminescence—the production of light by means other than heat—as had his father and grandfather before him, and as would his only son.*

Röntgen observed: "It is certain that the spot on the wall [of the vacuum tube] which fluoresces the strongest is to be considered as the main center from which the X rays radiate in all directions." Could there be some linkage between X rays and fluorescence? Becquerel later reminisced,

> I thought immediately of investigating whether the [X rays] could not be a manifestation of the vibratory movements which gave rise to the [fluorescence] and whether all phosphorescent bodies could not emit similar rays. The very next day I began a series of experiments along this line of thought.

Certain chemicals are phosphorescent. After being placed in sunlight for a few hours and taken to a dark room, they continue to glow. This transient effect is a chemical process having nothing to do with radioactivity. The particular phosphorescent material that Becquerel had at hand was

---

*A *luminescent* body, such as a fire-fly, produces light but not heat. A *fluorescent* body, such as a TV screen, produces light while it is exposed to a source of radiation. A *phosphorescent* body continues to emit light after the source of radiation is removed.

potassium uranyl disulfate, which luckily contained uranium and its radioactive decay products. To find out if phosphorescent materials emit X rays, Becquerel carried out the following experiment:

> One wraps a photographic plate in thick black paper so that the plate does not fog in a day's exposure to sunlight. A plate of phosphorescent material is laid above the paper on the outside and the whole is exposed to the sun for several hours. When the photographic plate is subsequently developed, one observes the silhouette of the phosphorescent substance, appearing in black on the negative. If a coin, or a sheet of metal is placed between the phosphorescent material and the paper, then the image of these objects can be seen to appear on the negative. The phosphorescent material emits radiations which traverse paper opaque to light.

Becquerel thought he had confirmed his hypothesis. He believed that phosphorescent chemicals exposed to sunlight produce X rays in addition to light. A week later, to his amazement, he found that the image on the film had nothing to do either with phosphorescence or sunlight! Here is what happened.

In a modification of his earlier trials, Becquerel placed a copper cross between the uranium salt and the photographic plate (Figure 10.3). The sky was overcast that day, so he put his wrapped plate in a dark desk drawer to await the winter sun that never came. A few days later (who knows why?), he developed the plate. His son Jean, who collaborated with his father, wrote that "Becquerel was stupefied when he found that his silhouette picture was even more intense than the ones he had obtained the week before."

Sunlight was irrelevant! Becquerel's subsequent experiments showed that his rays were not the same as Röntgen's. Any uranium compound, phosphorescent or not, even pure uranium, ceaselessly produced what Becquerel called uranic rays. "All experiments," he wrote, "show the seeming permanence of the phenomenon. One has not been able to

FIGURE 10.3
Becquerel's discovery of radioactivity. A copper cross is placed atop a photographic plate and wrapped in opaque paper. A piece of radioactive ore produces emanations that penetrate the wrapping but not the metal cross. When the plate is developed, an image of the cross appears.

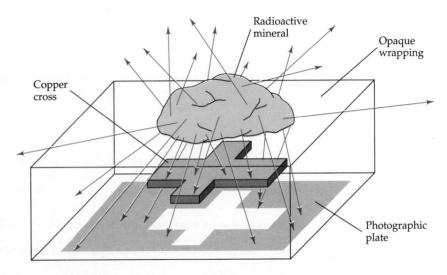

recognize wherefrom uranium derives the energy which it emits with such persistence." Becquerel was the first to see the release of nuclear energy in the form of natural radioactivity. The emanations that passed through the black paper to expose the photographic plate were produced by the intrinsic radioactivity of the uranium salt.

Becquerel believed that radioactivity was a property unique to uranium and its compounds. His young colleagues, the husband and wife team of Pierre and Marie Curie, showed that the element thorium (which had been discovered in 1828) was also radioactive. Furthermore, they discovered that pitchblende (a rich uranium ore) produces more radioactive emanations than uranium itself. From a large quantity of this ore, they isolated two new radioactive elements, which they called polonium and radium. Nothing could retard or affect the spontaneous activity of uranium—neither heat nor cold, acid nor base, sunlight nor X rays, nor even time itself!

Was Becquerel's discovery of radioactivity an act of genius or was it just dumb luck? Again, according to Becquerel's son, "[My father] said that the investigations, which during 60 years had followed one another in this same laboratory, formed a chain, which at the propitious hour, were ineluctably to end up with radioactivity." The Nobel committee leaned toward genius. The first of the Nobel Prizes was awarded to Röntgen for his discovery of X rays. Two years later, the Prize was split between Becquerel and the Curies. But the puzzles posed by radioactivity were far from being solved. The story of radioactivity will be resumed in Chapter 13.

Röntgen set out to study cathode rays but was rewarded, on or about Christmas 1895, with the discovery of X rays. Becquerel searched for a suspected but nonexistent link between Röntgen's rays and phosphorescence. Instead, he found something totally unexpected: radioactivity. In January 1896, Ernest Rutherford, a young scientist who was working at the renowned Cavendish Laboratory in Cambridge, England, wrote to his fiancée:

> The Professor [his advisor, J. J. Thomson] has been very busy lately over the new method of photography discovered by Professor Röntgen. [Thomson] is trying to find out the real cause and nature of the waves, and the great object is to find the theory of matter before anyone else, for nearly every Professor in Europe is now on the warpath.

The professor and his assistant would find the parts of the atom. Thomson was about to discover the electron. Sixteen years later, Rutherford would discover the nucleus of the atom.

The Discovery of the Electron

Joseph John Thomson (1856–1940), called J. J. by his students and colleagues, discovered the first "elementary particle" in 1897. Forty years later, Thomson published his *Recollections and Reflections*. His description of his own work is more lucid than any I could write:

Joseph John Thomson at work in the Cavendish Laboratory. J. J., as he was universally known, was director of the Cavendish Laboratories at Cambridge University from 1884 to 1918. His predecessor was Maxwell, his successor Rutherford. Thomson discovered the electron in 1897. Sixteen years later, he showed that some neon atoms weigh 10 percent more than others, thereby discovering stable isotopes. Any scientist can be lucky enough to make one great discovery. With two to his credit, Thomson joins Faraday and Lavoisier in my short list of the immortals of science. *Source:* The Granger Collection.

The research which led to the discovery of the electron began with an attempt to explain the discrepency between the behavior of cathode rays under magnetic and electric forces. [Hertz had shown that] magnetic forces deflect the rays in just the same ways as they would a negatively electrified particle moving in the direction of the rays.

[Perrin had shown that a metal] cylinder receives a copious negative charge when the cathode rays are deflected by a magnet into the cylinder. This would seem to be conclusive evidence that the rays carried a charge of electricity had not Hertz found that when they were exposed to an electric force they were not deflected at all. From this he came to the conclusion that they were not charged particles. He took the view that they were flexible electric currents flowing through the ether, and that they were acted upon by magnetic forces in accordance with the laws discovered by Ampère for the forces exerted on electric currents.

Such currents would give a charge of negative electricity to bodies against which they struck. They would be deflected by a magnet. They would not be deflected by electric forces. These are just the properties which the cathode rays were for a long time thought to possess.

Cathode rays were identified as electric currents because they were deflected by magnetic fields and they caused objects in their path to become electrically charged. At the time, however, many scientists did not believe that electric currents consist of electric charges in motion. The failure of Hertz to observe the deflection of cathode rays by electric fields seemed to confirm their view. Thomson was convinced otherwise. He set out to see for himself whether cathode rays are affected by electric fields.

FIGURE 10.4
The device with which
J. J. Thomson discovered
the electron. Electrons
steaming from the cath-
ode are deflected down
by an electric field E and
up by a magnetic field
B. When the strengths
of the field are adjusted
so that the beam is un-
deflected, the velocity
of the electrons is determined to be E/B. The
charge-to-mass ratio is
determined by removing
the magnet and measur-
ing the deflection caused
by known electric field.

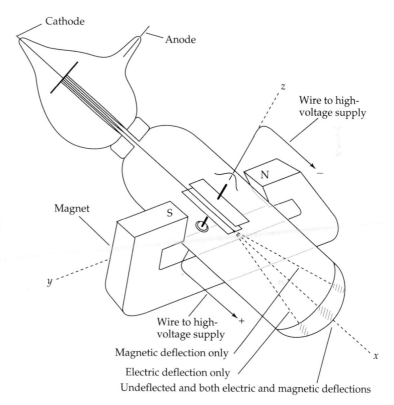

Thomson built the device shown in Figure 10.4, consisting of an evacuated glass tube with a narrow constriction. Two electrodes were inserted into one part of the tube, to which a high voltage (supplied by a Ruhmkorff coil) was applied. Cathode rays streamed off in all directions from the negative cathode. The constriction in the tube was surrounded by a metal ring that allowed a narrow beam of cathode rays to enter the larger portion of the tube. The beam struck the far end of the tube, producing a small fluorescent spot. Two additional electrodes were connected to parallel plates within the tube. This was the means by which Thomson subjected the cathode rays to an electric field. At the same time, he placed a magnet near the tube, thus subjecting the cathode rays to a magnetic field as well. Thomson could study the effect of electric and magnetic fields on the cathode rays by measuring the deflection of the glowing spot.

If cathode rays were charged particles, they must be deflected by both electric and magnetic fields. We learned in Chapter 8 that the force on a moving charge is

$$\mathbf{F} = q\,(\mathbf{E} + \mathbf{v} \times \mathbf{B}) \tag{2}$$

where **E** and **B** are the electric and magnetic fields at the position of the particle. However, when Thomson connected the parallel plates in his tube to a high-voltage source, he saw no lasting deflection but only "a slight flicker in the beam when the electric force was first applied." Noticing that flicker made the difference between a good scientist and a great one.

Thomson realized that gas molecules in the imperfect vacuum of his tube were ionized by the electric discharge. The ions migrated to the electrodes and almost instantaneously neutralized their electric charges. "The absence of deflection is due to the presence of gas—to the pressure [in the tube] being too high," wrote Thompson, "thus the thing to do was to get a much higher vacuum. This was more easily said than done." Thomson eventually got his better vacuum, and with it he showed that cathode rays are indeed deflected by electric as well as magnetic fields. This was a necessary condition for his view of the particulate nature of cathode rays to be correct, but it was not sufficient. If cathode rays consisted of rapidly moving charged particles, then surely the properties of the particles could be measured. What was the velocity v of the electrons in the tube? What is the electric charge $-e$ carried by an electron? What is the mass m of an electron? As Thomson later wrote:

> This result removed the discrepency between the effects of magnetic and electric forces on the cathode particles; it did much more than this; it provided a method of measuring v the velocity of these particles, and also e/m, where m is the mass of the particle and e is its electric charge.

Measuring the Charge-to-Mass Ratio of Electrons

Thomson subjected the cathode rays in his tube to electric and magnetic fields at the same time. Suppose the cathode rays are moving in the x direction. The parallel plates inside the tube, when electrified, produce a known electric field E in the upward z direction. The effect of this electric field, according to equation 2, is to drive the negatively charged cathode rays downward. An electromagnet placed outside the tube produces a known magnetic field B in the y direction. The effect of this field is to drive the electrons upward. Suppose that both fields extend over the same length l along the trajectory of the cathode rays. This situation was analyzed in Example 8.3. The electric and magnetic forces cancel one another if the following condition is met:

$$v = E/B$$

where v is the velocity of the charge. Thomson adjusted the strength of the magnetic field so the cathode rays were not deflected by the crossed fields. From the known values of the compensating fields, he deduced the velocity of particles making up the rays.

Once he measured v in this manner, Thomson turned off the magnetic field. The glowing spot on the tube was then driven down by the electric field. From the distance the spot was deflected, Thomson determined the ratio of the charge of the electron to its mass, the quantity e/m. This analysis is carried out as Example 10.2.

FIGURE 10.5
An electron passes through a region of length l in which there is an electric field E pointing up. The electron is deflected downward by a distance d.

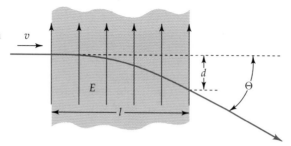

EXAMPLE 10.2
How Thomson Measured e/m of the Electron

Suppose that an electron with charge $-e$ and mass m is moving to the right, as shown in Figure 10.5. It passes through a region of length l in which there is an electric E field pointing up. If the electron is deflected downward by a distance d as it passes through the field, what is the value of e/m, its charge-to-mass ratio?

Solution
Since electrons are pushed down by an electric field pointing up, the charge of the electron is negative. The magnitude of the downward electric force is eE. The electrons accelerate down with the vertical acceleration eE/m during their traversal of the horizontal distance l. (They also fall with the acceleration of gravity g. In practice, however, g is negligible compared to eE/m, the acceleration due to the electric field.) We learned in Chapter 2 that the vertical displacement d of a uniformly accelerated body is $\frac{1}{2}at^2$, where a is the acceleration and t is the time interval over which it is accelerated. In Thomson's experiment, $t = l/v$ and $a = eE/m$. Thus:

$$d = \frac{eEl^2}{2mv^2} \qquad \text{or} \qquad \frac{e}{m} = \frac{2dv^2}{El^2}$$

The unknown quantity e/m is expressed in terms of the known quantities d, E, v, and l. Notice that the deflection of the electron, in this example and in Thomson's tube, determines neither the value of e nor the value of m, but only their ratio. Let's put numbers into this result that are similar to those in Thomson's experiment. Suppose that the electric field is $E = 4 \times 10^3$ volts/m and that it extends over a distance of $l = 2$ cm. Suppose too that the horizontal speed of the electrons is 10^7 m/s and that their deflection, after passing through the field, is $d = 1.4$ mm. We find that $e/m \simeq 1.8 \times 10^{11}$ coulombs/kg. The downward acceleration of the electron by the electric field is $eE/m \simeq 7 \times 10^4$ m/s^2. ∎

In another experiment, Thomson measured the charge-to-mass ratio of positively charged H^+ ions and found that it was about 1000 times smaller than the corresponding ratio for electrons. It seemed plausible, from Faraday's earlier studies of electrolysis, to assume that the magnitudes of charges of the electron and the ion were the same. From this, it followed that the hydrogen atom must be about 1000 times heavier than the electron. Thomson's measurements of charge-to-mass ratios were not precise; the actual ratio of masses is closer to 1837. The essential point, however, is that Thomson showed that electrons are much lighter than atoms. Thomson pressed onward and left precision experiments for others to perform:

> If e were the same as the charge of electricity carried by an atom of hydrogen—as was subsequently proved to be the case—then m, the mass of the cathode–ray particle, could not be greater than one thousandth part of the mass of an atom of hydrogen, the smallest mass hitherto recognized. These results were so surprising that it seemed more important to make a general survey of the subject than to endeavor to improve the determination of the exact value of the ratio of the mass of the electron to the mass of the hydrogen atom.

Thomson found that electrons could be produced from many different materials and in many different ways. He studied the charged particles produced when ultraviolet radiation strikes a metal surface (the photoelectric effect, to which we return in Chapter 11). He studied the charged particles produced when carbon or metal filaments are heated to incandescence (as in a lamp bulb). He "measured, by methods based on similar principles to those used for cathode rays, the value of e/m for the carriers of negative electricity in those cases, and found that it was the same as for cathode rays."

Although they were discovered in electric discharges, Thomson showed that his new particles were everywhere: Electrons are parts of the atoms that make up matter.* Thomson saw no escape from the following conclusions:

> That atoms are not indivisible, for negatively electrified particles can be torn from them by the action of electric forces, impact of rapidly moving atoms, ultraviolet light or heat.
>
> That these particles are all of the same mass, and carry the same charge of negative electricity from whatever kind of atom they may be derived, and are constituents of all atoms.
>
> That the mass of these particles is less than one thousandth part of the mass of a hydrogen atom.
>
> I at first called these particles corpuscles, but they are now called by the more appropriate name electrons. I made the first announcement of the existence of these corpuscles at a Friday Evening Discourse at the Royal Institution on April 29, 1897. There were very few people who believed in the existence of these bodies smaller than atoms. I was even told afterward by a distinguished physicist that he thought I had been pulling their legs. I was not surprised at this, as I myself had come to this explanation of my experiments with great reluctance, and it was only when I was convinced that the experiment left no escape from it that I published my belief in the existence of bodies smaller than atoms.

---

*A few years after Thomson's discovery of the electron, in 1902, the Curies deepened the mystery of radioactivity by proving that β rays consist of electrons that have somehow acquired energies many thousands of times greater than those in cathode rays.

*Measuring the
Charge of the
Electron*

Thomson knew what the charge of an electron should be. Faraday's studies of electrolysis in the 1830s, which we discussed in Chapter 8, showed that the charge carried by a singly ionized atom is

$$e \simeq \frac{96,500 \text{ C}}{N_A} \simeq 1.6 \times 10^{-19} \text{ coulombs} \tag{3}$$

where N_A is Avogadro's number. Thomson believed (correctly) that an ion is an atom that has lost or acquired one or more electrons, so that the electron has a charge given by equation 3. To confirm his suspicion, Thomson set out to measure directly the charge on an electron. His experimental technique involved the physics of fogs and clouds.

As the warm moist air of a summer day rises, its pressure decreases and water vapor condenses about dust particles or ions to form droplets. A billowing cloud is formed. Taking his lead from this phenomenon, Thomson passed X rays through damp and dust-free air in a flask to produce a large number of singly charged ions. When the pressure in the flask was reduced, water vapor condensed about each ion, forming a mist of charged water droplets. The number of droplets was found by studying the appearance of the mist. From the number of droplets and the measured electric conductivity of the mist, Thomson determined the charge on each droplet.

This difficult experiment was more easily said than done. Thomson did not obtain as precise a determination of e as he had hoped for. His published results ranged from two-thirds to twice the expected value given by equation 3. Thomson's methods were later refined by his younger colleague C. T. R. Wilson. Years later, Wilson invented the *cloud chamber*, an instrument by which rapidly moving charged particles can be "seen" as they leave visible trails of water droplets along their paths. We shall hear more about cloud chambers later on.

In 1906, the American physicist Robert Andrews Millikan, "being dissatisfied with the variability" of Thomson's results, repeated the experiment, "without obtaining any greater consistency." His failure made him try harder:

> This attempt, while not successful in the form in which it had been planned, led to a modification of the cloud method which seemed at the time, and which has actually proved since, to be of far-reaching importance. It made it for the first time possible to make all the measurements on individual droplets, and thus not merely to eliminate all of the questionable assumptions and experimental uncertainties involved in the cloud method of determining e, but, more important still, it made it possible to examine the properties of individual isolated electrons and to determine whether different ions actually carry one and the same charge. That is to say, it now became possible to determine whether electricity in gases and solutions is actually built up of electrical atoms, each of which has exactly the same value.

Millikan finally got a definitive result for the electron's charge by a straightforward procedure known to later generations of physics students as the Millikan oil-drop experiment. Its repetition in a laboratory course is often a rite of passage into the world of professional physics.

Millikan knew that a liquid, when it is forced through a narrow metal nozzle, forms a mist of tiny droplets that become electrically charged by the frictional forces making the spray. Each droplet can lose one or more electrons and thereby acquire a positive electric charge that is a small integer multiple of the charge of the electron, sometimes zero, but often e, $2e$, $3e$, and so on. Using oil because it does not readily evaporate, Millikan produced tiny electrically charged oil droplets with an atomizer. Instead of studying the properties of a cloud of many droplets, Millikan focused on the behavior of one droplet at a time.

An oil droplet was introduced into the space between two parallel horizontal plates, as shown in Figure 10.6. Its vertical position was monitored by an attentive observer with a microscope. At first, the plates were uncharged and gravity made the droplet fall. Unlike a falling brick, a falling oil droplet is very much affected by air resistance. It descends at a slow and constant speed that may be expressed in terms of the density of the oil, the mass of the droplet, and a known property of air called its viscosity. Consequently, a measurement of the speed of the falling droplet determines its mass M. (Of course, Millikan might have measured the size of the droplet and computed its mass from its density, but the foregoing procedure is far more accurate.)

The speed of the falling droplet was measured well before it struck the lower plate. As soon as v was determined, the plates were connected to a high-voltage supply. They became charged and produced a constant vertical electric field in the space between. The strength of the field was adjusted by changing the voltage until the droplet hung motionless in the air, with the electric force neatly balancing the force of gravity:

$$qE = Mg \qquad (4)$$

where q was the unknown charge on the droplet. Because E, M, and g were known quantities, Millikan could use equation 4 to determine q. Lo and behold! Every time that Millikan measured the charge of a

FIGURE 10.6
The Millikan oil-drop experiment. A tiny oil drop carrying a small electric charge remains suspended in midair if the electric force is adjusted to balance the weight of the drop. The charge on the oil drop is determined from its mass and the strength of the electric field.

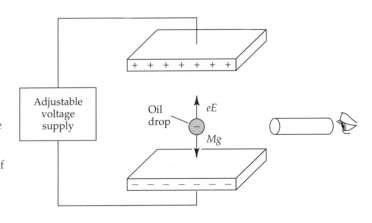

droplet, he obtained a value for q that was a small integer multiple of a fixed quantity, which he interpreted as e, the magnitude of the charge of the electron. His value for e agreed with equation 3, the electrolytically determined charge of an ion. Millikan had measured the charge of the electron.* A visit by the American poet Robert Frost to Millikan's laboratory inspired a poem that began

> *Did I see it go by,*
> *That Millikan mote?*
> *Well, I said that I did.*
> *I made a good try.*

By 1913, Millikan had measured e to a precision of 1 percent, and four years later to a precision of one part in a thousand. He also noted that the charge on a droplet would occasionally change when it captured an ion:

> I have observed, all told, the capture of many thousands of ions in this way, and in no case have I ever found one the charge of which did not have either exactly the value of the smallest charge ever captured or else a very small multiple of that value. Here, then, is direct unimpeachable proof that the electron is not a "statistical mean," but that rather the electrical charges found on ions all have either exactly the same value or else small exact multiples of that value.

So much for the notion of an electrical fluid. Electrons are indeed "atoms of electricity," all of them with precisely the mass m and charge $-e$. Both quantities are now known to a precision better than one part in a million:

$$e = (1.602\,177\,33 \pm 0.000\,000\,49) \times 10^{-19} \text{ C}$$

$$m = (9.109\,389\,7 \pm 0.000\,005\,4) \times 10^{-31} \text{ kg} \tag{5}$$

and it is absolutely certain that all electrons are identical to one another. Moreover, the electric charge of the neutral hydrogen atom is known to vanish to a precision of $\pm 10^{-21}\,e$, which means that the electric charges of the electron and the proton are very likely to be exactly equal and opposite. The numerical equality of two such disparate numbers cannot be a coincidence—it must reflect a fundamental principle of nature. Although today's theory suggests this equality, a decisive and convincing proof of the fact from first principles is still lacking.

---

*In a series of sensitive experiments carried out by William Fairbank and his collaborators at Stanford University in the 1970s, tiny magnetically levitated, tungsten-coated niobium balls replace oil drops. Based on these experiments, the scientists claimed to observe fractional charges of $\pm e/3$ and $\pm 2e/3$. Some physicists believed these results revealed isolated quarks, a contention contrary to the expectations of current theory. Because there has been no confirmation or successful repetition of this work, the consensus is that the experiment was flawed.

EXAMPLE 10.3
The Millikan
Oil-Drop Experiment

A spherical drop of oil with radius $r = 10^{-6}$ m carries charge e. The density of the oil is 0.8 g/cm^3. What electric field is needed to balance the force of gravity?

Solution

We first compute the mass of the droplet. Its volume is $4\pi r^3/3$, or 4.2×10^{-12} cm^3. Thus, its mass is $M \simeq 3.4 \times 10^{-15}$ kg. Equation 4 says that $E = Mg/e$. Using $g \simeq 10$ m/s^2 and e given by equation 5, we find $E \simeq 2 \times 10^5$ newtons per coulomb, or volts per meter. The electric field must point up.   ∎

The Light Year and the Electron Volt

Physical science deals with all features of the universe, however large or small. The metric units we have been using are just right to describe everyday things. Mammals are about a meter in size, give or take a factor of 10. The mass of this book is a kilogram or so, and as it falls to the floor, its kinetic energy is a few joules. Meters and kilograms are not the units of choice of astronomers, however—they use light years and solar masses. The masses of most stars are 1 solar mass, give or take a factor of ten, and there are 10 stars in the sky that lie less than 10 light years from Earth.

Chemists and physical scientists are often concerned with the properties of individual atoms or particles. We learned in Chapter 9 that in spectroscopy, the natural unit of length is the angstrom, or 10^{-10} m. The angstrom is approximately the diameter of an atom, and the atoms in a crystal are a few angstroms apart. To describe the physics of the microworld, the most favored atomic unit of energy is the *electron volt*. This is the energy unit we use most often in the remainder of this book.

Recall that the joule is the energy acquired by 1 coulomb of charge in traversing an electric potential difference of 1 volt. The electron volt (or eV) is defined to be the energy acquired by one electron in traversing an electric potential difference of 1 volt:

$$1 \text{ eV} \simeq 1.602\,177 \times 10^{-19} \text{ J} \tag{6}$$

The ratio of the electron volt to the joule is identical to that of the electron's charge to the coulomb, as given in equation 5. The following examples show how convenient it is to describe atomic phenomena in eV:

- The average kinetic energy of an air molecule at room temperature is about 1/40 eV.
- The combustion of one molecule of a fossil fuel releases a few eV.
- We shall learn in Chapter 11 that light shows particle as well as wave properties. The energy carried by a single particle of visible light is 2 or 3 eV.
- An energy of 13.6 eV is required to extract the electron from a hydrogen atom.

The different disciplines of subatomic physics are neatly separated by their relevant energy scales. Chemical and optical phenomena have to do with the structure of the outer parts of the atom, and involve energies of a few eV. X-ray phenomena are concerned with the inner structure of atoms, and involve energies of a few thousand eV, or KeV (thousand electron volts). The particles emitted from the nuclei of radioactive atoms have energies of a few million eV, or MeV (million electronic volts). And, most of what we know about the physics of elementary particles was learned by studying the collisions of particles with energies of a few billion eV, or GeV (billion electron volts).

EXAMPLE 10.4
The Electron Volt

In Example 10.2, the electrons in a beam move at a speed of 10^7 m/s. What is the kinetic energy of one of these electrons in joules and in electron volts?

Solution

The kinetic energy of an electron is $\frac{1}{2}mv^2$ if its speed is small compared to the velocity of light. (Otherwise, relativistic effects become important, as we shall see in Chapter 11.) In this example, the speed of the electron is about 1/30 the speed of light, and its kinetic energy is $0.5 \times 9.1 \times 10^{-31} \times 10^{14}$ or 4.55×10^{-17} J, where we have used the value of m given in equation 5. The energy of the electron is converted to eV by means of equation 6 to yield 284 eV. Electrons of this energy may be obtained by accelerating stationary electrons across a potential difference of 284 volts.   ∎

Exercises

1. Medical X rays typically have frequencies from 5×10^{18} to 5×10^{19} Hz. What is the corresponding range of X-ray wavelengths?

2. Show that Avogadro's number of electron volts is approximately equal to 100 kilojoules.

3. The atomic layers of a crystal lie parallel to its surface and are 5 angstroms apart. X rays of wavelength 1.8 angstroms impinge at an angle θ to its surface. For which values of θ between 0 and 90 degrees is there strong reflection?

4. What is the frequency of a gamma ray whose wavelength equals the radius of a hydrogen nucleus, 1.2×10^{-15} m?

5. Use Faraday's result (equation 3) and the measured charge of the electron (equation 5) to compute Avogadro's number.

6. In Example 10.2, the electron is deflected downward a distance d by the electric field E. After it passes the electric field, it moves in a straight line, but no longer horizontally. Show that the angle of deflection θ, indicated in Figure 10.5 satisfies the equation $\tan\theta = eEl/mv^2$.

7. Electrons in a TV tube are accelerated by passing through an electric potential difference of 2500 volts.
   a. What is the energy of each electron in eV as it reaches the screen?

   b. What is its speed at the screen?
   c. What fraction of the speed of light does it acquire?

8. What is the strength of the electric field, acting over a distance of 10 cm, that is needed to accelerate an electron from rest to a speed of $0.1c$?

9. In a modern X-ray tube, electrons are emitted at very low energies by a hot filament. They are accelerated by a constant electric field of $8 \times 10^5 N/C$. After being accelerated to their final velocity, they strike the anode and produce X rays.
   a. What is the acceleration of the electron in the electric field?
   b. Over what distance does the electric field extend if the electrons striking the anode have energies of 40 KeV?
   c. What is the current passing through the tube if 40 watts of X-ray power is produced? (Assume that all the energy of the electron beam is converted into X rays.)

10. When 1 kg of carbon is burned, 3.6×10^7 J of heat is produced. How much energy in joules is produced by burning 1 mole of carbon? How much energy in eV is released by the combustion of one carbon atom?

11. The heat of vaporization of water is about 2000 J/g. How much energy in eV is needed to evaporate a single water molecule?

12. Calculate the mean kinetic energy (kT) in eV of an air molecule at room temperature of 300 K. Calculate the mean kinetic energy of a particle at the center of the Sun, where $T \simeq 1.5 \times 10^7$ K.

13. Two electrons are separated by a distance of 10^{-10} m, about an atomic size. What is their electric potential energy in eV?

14. The solar constant is the power received by Earth per unit area when the Sun is overhead, and is about 1350 W/m². Sunlight consists of photons with a mean energy of 3 eV each. How many photons strike each square centimeter of Earth per second at midday?

15. An accelerator produces a beam of 1 MeV electrons. The current transported by the beam is 1 milliampere.
   a. What is the power of the beam in watts?
   b. How many electrons are produced in one second?

16. On a clear day, the atmospheric electric field points down with a magnitude of 100 volts per meter. A dust grain with a mass of 10^{-18} kg carries the charge of an electron. What is the magnitude and direction of its acceleration due to the sum of the gravitational and electric forces?

10.2 The Atomic Nucleus

Once electrons were discovered, physicists could begin to piece together a description of the atom. Surely electrons were one of the ingredients of

FIGURE 10.7
The saturnian and plum pudding models of atomic structure. In both models, the positive part of the atom bearing most of its mass was assumed to fill most of the atom's volume. The electrons orbit around the large positive charge in the saturnian model and they are embedded within the large positive charge in the plum pudding model.

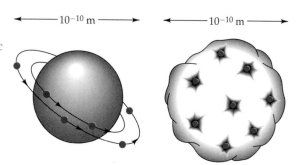

atoms, but they could not be the only constituent because: (1) electrons repel one another and (2) ordinary matter is electrically neutral. There had to be an additional positively charged portion of the atom.

Electrons were far too light to be the dominant component of the atom's mass. Furthermore, determinations of e/m for positive ions consistently gave results thousands of times smaller than for electrons, and never larger than that of the hydrogen ion. It followed that most of the mass of an atom consisted of its positively charged component or components. In addition, the atom contained a number of electrons sufficient to neutralize this charge. Consequently, it was both plausible and correct to assume that the negatively charged electrons of the atom were held to the massive positive component by the force of electrical attraction.

Because only a tiny part of the mass of an atom consisted of electrons, most physicists believed that the positive core of an atom made up the bulk of its measured volume as well. There were two popular models of atomic structure. Hantaro Nagaoka, in Tokyo, proposed the "saturnian model." In this model, electrons were supposed to move in circular orbits lying close to a large, massive, and spherical positive core, much like the rocks and dust that circle Saturn as its rings. At about the same time, Thomson put forward his "plum pudding model." Again, the positive charge was distributed throughout a region the size of the atom, but the electrons were electrically trapped within the positive charge, like the plums in an English pudding (Figure 10.7).*

As we shall see, neither of these models was close to the truth. In fact, most of the mass of an atom is concentrated in a tiny *nucleus*, whose electric charge is an integer multiple of e. The integer is denoted by Z and is called the *atomic number* of the atom. Because the atom as a whole is electrically neutral, Z is also the number of electrons in the atom. The electrons are in orbits about the nucleus somewhat like the planets of a

---

*Magnetize a number of needles and plunge them through corks. Float them in water with their north poles pointing up so they repel one another. Bring the south pole of a large bar magnet nearby. The floating magnets are drawn into a stable pattern consisting of several concentric rings! In Thomson's model, the mutually repelling electrons are like the floating magnets. The uniformly distributed positive charge of the atom provides the attractive force. Thomson hoped his model would yield stable atoms with properties periodic in the number of electrons. It didn't.

solar system. The concept of the atomic nucleus, which led to the modern picture of the atom, was introduced in 1911 by Ernest Rutherford.

Discovering the Atomic Nucleus

It seemed that the atom was not the "nice hard fellow" that Rutherford first imagined it to be. He had shown that energetic electrons can penetrate a significant thickness of matter. Scientists were beginning to realize that the space occupied by an atom is as empty as the interstellar void. In Manchester, Rutherford decided to use a beam of α particles as a probe to determine the structure of atoms. When a positively charged α particle penetrates an atom, it is repelled by the atom's positive core and deflected from its original path. Rutherford set out to measure how often, and by what angles, α particles would scatter from gold atoms. By this means, he could determine how the positive charge was distributed within the atom.

Rutherford's scattering experiments were performed collaboratively with Hans Geiger (who later designed the "Geiger counter," which electronically detects and counts individual charged particles) and an undergraduate named Eugene Marsden. A radioactive source supplied a beam of α particles with energies of 5 MeV or 5×10^6 eV. The beam was directed toward a thin sheet of gold foil. The scattered particles were detected by the tiny flashes of light they produced as they struck a zinc sulphide screen (Figure 10.8). According to Marsden:

> One day Rutherford came into the room where we were counting α particles. [He] turned to me and said, "See if you can get some effect of the α particles directly reflected from a metal surface." I do not think he expected any such result, but it was one of those hunches that perhaps some effect might be observed. To my surprise, I was able to observe the effect.

Geiger and Marsden published their results in 1909. "If the high velocity and mass of the α particle be taken into account," they wrote,

FIGURE 10.8
The Rutherford scattering experiment. A beam of α particles was directed at a sheet of gold foil. Deflected α particles were detected when they produced flashes of light at a zinc sulfide screen. Some of the particles were deflected by large angles, and some even bounced backward.

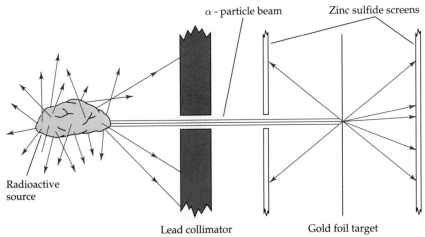

α - particle beam   Zinc sulfide screens

Radioactive source

Lead collimator   Gold foil target

"it seems surprising that some of the α particles [about one in 8000] can be turned in a layer of 6×10^{-5} cm of gold through an angle of $90°$, and even more." The α particles were traveling at speeds of 15000 km/s.

A BIT MORE ABOUT
A Particle and a Professor

Ernest Rutherford (1871–1937) was born and brought up in the New Zealand hinterland as one of 12 children of a jack-of-all trades. When the winner of a government fellowship to study abroad turned it down, Rutherford was made an alternate. His incredibly prolific scientific career began in 1895, when he became a junior collaborator of J. J. Thomson at the Cavendish Laboratory.

Soon after radioactivity was discovered, Rutherford found (in his words)

> that the uranium radiation is complex, and that there are present at least two distinct types of radiation—one that is very readily absorbed, which will be termed for convenience the α radiation, and the other of a more penetrating character, which will be termed the β radiation.

Another component was later called γ radiation. Alpha (α), beta (β), and gamma (γ) are the first three letters of the Greek alphabet. (In 1914, Rutherford showed that γ rays are electromagnetic waves with wavelengths even shorter than X rays.) Through most of his career, Rutherford focused on α rays. A colleague wrote that "The α particles were Rutherford's pets—and how he made them work!"

In 1899, Rutherford accepted a professorship at McGill University, in Montreal. "£500 is not so bad and as the physical laboratory is the best of its kind in the world, I cannot complain," he wrote to his mother. In Canada, Rutherford found that α radiation is deflected by magnetic fields and consists of heavy and fast-charged particles. What were they? Rutherford would eventually find out, but in 1903

he found a clue. He showed that helium is a by-product of radioactivity: When α particles came to rest in matter, they became helium atoms!

Despite their large mass and high velocities, α particles were often deflected by a few degrees when passing through sheets of matter. "Such results bring out clearly," Rutherford wrote, "that the atoms of matter must be the seat of very intense electrical forces." His continuation of this line of research in England would reveal the atomic nucleus, but most of Rutherford's effort at McGill was devoted to an extraordinary collaboration with Frederick Soddy, whose luscious fruit—the law of radioactive transformation—is savored in Chapter 13.

In 1907, Rutherford returned to England—to Manchester, where Dalton had founded atomic theory a century earlier. Within a year, he determined the nature of the α particle. "We may conclude," he wrote in 1908, "that an α particle is a helium atom, or, to be more precise, the α particle, after it has lost its positive charge, is a helium atom." More precisely yet, the α particle is a rapidly moving and positively charged helium nucleus ejected from a radioactive atom in the process of α decay. When it catches two electrons, it becomes a neutral helium atom. Don't fault Rutherford: He had not yet discovered the atomic nucleus. Rutherford returned to the Cavendish Laboratory in 1919 to succeed J. J. Thomson as its director. The event was celebrated in sophomoric song:

> He's the successor
> Of his great predecessor,
> And their wondrous deeds can never be ignored:
> Since they're birds of a feather,
> We link them both together,
> J. J. and Rutherford.

And yet, some of them bounced backward from the gold atoms—they emerged from the same side of the foil as they entered! Such a result was inconceivable in terms of Thomson's model of atomic structure, in which the positive charge of the atom was spread out over the atomic volume. The observation of large-angle deflections of α particles by gold atoms was more than surprising—it was mind-boggling. After two years of brooding, Rutherford concluded that these results ruled out all contemporary models of atomic structure. He concluded that the positive charge of an atom had to be concentrated within a tiny nucleus. The nucleus is thousands of times smaller than the atomic radius, yet it accounts for over 99.9 percent of its mass! Much later, Rutherford wrote:

> It was quite the most incredible event that has ever happened to me in my life. It was almost as incredible as if you fired a 15-inch shell at a piece of tissue paper and it came back and hit you. On consideration, I realised that this scattering backwards must be the result of a single collision, and when I made calculations I saw that it was impossible unless you took a system in which the greatest part of the mass of the atom was concentrated in a tiny nucleus.

Why Rutherford Was Amazed (Qualitative)

Earlier atomic models spread the positively charged and massive part of the atom over a large fraction of the atom's volume. The large-angle scattering of α particles was incompatible with this view. Such scattering could only be understood in terms of a nuclear or planetary model of atomic structure.

Let us explore an analogous problem. We know that peaches have pits. Suppose, however, that an inquisitive young child (who does not know) is given the challenge of determining the structure of the fruit on a tree growing next to a barn by shooting projectiles at it (Figure 10.9). The peaches are concealed by leaves and cannot be seen directly. She must learn the nature of the peach by performing remote experiments, like those Rutherford did.

She is given a BB gun whose projectiles are not energetic enough to pass through a peach. She shoots a great many BBs at the tree and observes their impacts on the barn wall. From the shadowed regions, she measures the size and shape of a peach but cannot discern its inner structure. It is like examining the gold foil with light, most of which is reflected or absorbed. Only a very small amount of light passes through the foil. To the easily deflected optical probe, gold atoms occupy almost all the space in the foil.

The child next acquires a 22-caliber rifle whose bullets readily pass through the flesh of the peach but ricochet by large angles when they strike the pit. Seeing the occasional large deflection of a bullet, she concludes that there must be something hard buried within the peach. By counting the number of scattered bullets, she finds that the pit is much smaller than the peach itself. She has accomplished all this without ever having "seen" a peach, let alone its pit. She shows promise as a physicist. In Rutherford's experiment, the α particles easily pass through

FIGURE 10.9
Finding the pit without
touching the peach.

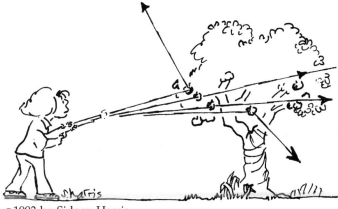

©1993 by Sidney Harris.

the soft outer part of the gold atoms. A very few encounter the tiny
and "hard" nucleus within the atom and are deflected by large angles.
Rutherford "found" the atomic nucleus in the very same way that our
aspiring scientist "found" the peach pit.*

Why Rutherford Was Amazed (Quantitative)

Let us reexamine the gist of the analysis that led Rutherford to his tiny
nucleus. The large-angle scatterings of α particles by gold atoms take
place because of the electrical force exerted on the projectiles by the
charged constituents of the target atoms. Gold atoms, with atomic weight
197, were known to contain about half that number of electrons. (The
correct number would later be determined to be 79). Atomic electrons
do not significantly affect the motion of α particles because they are
thousands of times lighter than the projectiles. The big bounces are
scatterings off the heavy, positively charged parts of atoms.

Suppose the positive charges within a gold atom are uniformly distributed throughout a sphere of radius R. Think of the projectile as a
much smaller and lighter object of charge $2e$. (In Thomson's plum pudding model, R was about the size of the atom, $\sim 10^{-10}$ m.) The force
between the α particle and the positive portion of the gold atom is at its
peak value when their separation equals R—that is, when the projectile
just skims the surface of the positive structure, as shown in Figure 10.10
(on page 440).

The projectile changes its direction because it is pushed sideways by

*Some 60 years after Rutherford's experiment, scientists were provided with much
more powerful subatomic probes: beams of particles from accelerators. Using electrons
with energies of 10 billion electron volts, three scientists at Stanford's giant accelerator
discovered the existence of hard and pointlike particles within the proton. Rutherford's
methodology, in a technologically more advanced context, led them to prove the existence
of quarks! For this discovery, Jerome Friedman, Henry Kendall, and Richard Taylor won
the 1990 Nobel Prize in physics.

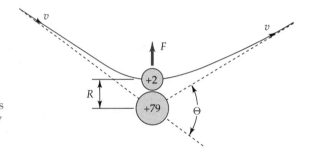

the electric force exerted by the target nucleus. Let R be the distance of closest approach of the α particle to the gold nucleus, as shown in Figure 10.8. If R is larger than the nuclear radius, the force on the α particle at the moment of closest approach is given by Coulomb's law: $F = k_c(2e)(79e)/R^2$, where $2e$ is the charge of the α particle and $79e$ is the charge of the nucleus of a gold atom. The force acts over a time interval during which the α particle is near the nucleus: $\Delta t \sim R/v$, where v is the speed of the projectile. The transverse momentum Δp acquired by the α particle as it passes the nucleus is approximately the product of these quantities: $\Delta p \sim F(R/v)$. If the α particle is to be deflected by a large angle, the change in its momentum must be comparable to its initial momentum:

$$\Delta p \sim p \qquad \text{or} \qquad 158\, k_c e^2/Rv \sim Mv$$

where M is the mass of the α particle. We solve this equation for R to find

$$R \sim \frac{158\, k_c\, e^2}{Mv^2} \qquad (7)$$

If an α particle is to bounce back from a gold atom, its mass and positive charge must be concentrated in a sphere whose radius is not less than R. We evaluate R by putting numbers into equation 7: $k_c = 9 \times 10^9\ \mathrm{N\,m^2/C^2}$, $e = 1.6 \times 10^{-19}$ C and Mv^2 (twice the kinetic energy of the α particle) $= 10$ MeV, or 1.6×10^{-12} J. We find $R \sim 2.3 \times 10^{-14}$ m.

The conclusion is astonishing—the observation of large-angle α scatterings demands that the positive charge of the gold atom is concentrated within a radius R that is thousands of times smaller than the radius of the gold atom itself. Pointlike electrons move in planetary orbits in the relatively vast empty spaces surrounding the tiny central nucleus. Atoms are mostly empty space!

The Mystery of the Nuclear Charge

Since atoms are electrically neutral, their nuclei must carry integer multiples of electric charge e. But what integer multiples? Thomson, Rutherford, and their contemporaries suspected that the nuclear charge Z is determined by its mass A, and that, in general, $Z \simeq \frac{1}{2}A$. This could not be an exact relation because, for hydrogen, $Z = A = 1$. Atomic weight seemed to be of primary importance, for it determined (or seemed to de-termine) the place of a chemical element in the periodic table. They had it backwards: The charge of the nucleus, not its mass, determines the num-ber of electrons in an atom, its chemical properties, and its place in the periodic table. This key fact was pointed out by a lawyer with an ama-teur's interest in theoretical physics—a Dutchman named van den Broek. By 1913, he formulated what came to be known as Broek's rule. It held that the nuclear charge Z equals the position of a chemical element in the periodic table. Thus, $Z = 1$ for hydrogen, $Z = 2$ for helium, and so on.

Harry Moseley established the truth of Broek's rule. When Moseley heard of von Laue's work on X-ray diffraction, he wrote to his mother, "Some Germans have recently got wonderful results by passing X rays through crystals and photographing them." He had found his life's work. When energetic electrons strike the metal anode in a vacuum tube, they produce X rays. Moseley set out to see how the wavelengths (or frequen-cies) of these X rays depend on the composition of the anode. Because of their tiny wavelengths, which are comparable to the size of the atom itself, perhaps a study of the X rays produced by an atom could reveal something about its inner structure. Moseley's technique depended on the wave nature of X rays: by measuring the angle by which they were diffracted by a crystal, he could deduce their wavelengths. Half a year later, having set up his laboratory in Manchester, the dutiful son once again wrote to his mother:

> We find that an X ray bulb with a platinum target gives out a sharp line spectrum of five wavelengths. Tomorrow we search for the spectra of other elements. There is here a whole new branch of spectroscopy, which is sure to tell one much about the nature of the atom.

Optical spectroscopy was discussed in Chapter 9. Each element has a characteristic spectrum of lines consisting of the frequencies of light their atoms can emit. Moseley created the discipline of X-ray spectroscopy. He showed that each element, when it is bombarded with energetic elec-trons, produces X-ray "lines" with characteristic frequencies. The most conspicuous X-ray line is called the K_α line. Moseley set out to find the relationship among the frequencies of the K_α lines produced by different elements.

A fortuitous visit by Niels Bohr to Manchester in July of 1912 focused Moseley's research vision. Niels Bohr was already formulating his rev-olutionary quantum rules of atomic structure. Bohr's recollection of this meeting indicates an ideal synergy between experimental and theoretical physics:

> I got to know Moseley really partly in this discussion about whether the nickel and cobalt should be in the order of their atomic weights.* Moseley

---

*Cobalt occupies the 27th place in the periodic table and nickel the 28th. The atomic mass of cobalt is 58.9. That of nickel is 58.7, less than that of cobalt! Which has the larger Z? That was the question.

FIGURE 10.11
Moseley designed and made an X-ray tube containing a series of carriages forming a trolley in which samples of different elements could be placed. By manipulating silk strings, he could change the target that was to be bombarded with electrons.

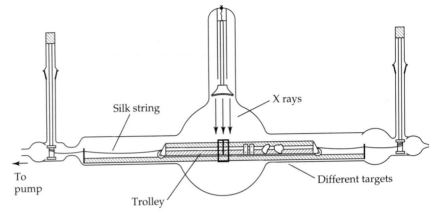

Silk string

X rays

To pump

Trolley

Different targets

asked what I thought about that. And I said, "There can be no doubt about it. It has to go according to the atomic number." And then he said, "We will try and see." And that was the beginning of his experiments. And then he did it at tremendous speed.

Moseley built a meter-long cathode ray tube in which he placed a tiny aluminum trolley carrying many targets of different chemical elements (Figure 10.11). As a silk fishing line was pulled, each target in turn was brought into the path of the cathode rays, where it emitted X rays characteristic of its atoms. What he would do next is explained in a letter to his sister:

> The method of finding the wavelengths is to reflect the X rays which come from a target of the element investigated. I have then merely to find at which angles the rays are reflected, and this gives the wavelengths. I aim at an accuracy of at least one in a thousand.

Moseley observed a startling regularity among the frequencies of X rays produced by different elements, which he attributed to the systematic variation of the nuclear charge of atoms. He introduced the concept of atomic number:

> It is evident that Q [the square root of the frequency of the K_α line of an element] increases by a constant amount when we pass from one element to the next, using the chemical order of the elements in the periodic system. Except in the case of nickel and cobalt, this is also the order of the atomic weights. While, however, Q increases uniformly the atomic weights vary in an apparently arbitrary manner, so that an exception in their order does not come as a surprise. We have here a proof that there is in the atom a fundamental quantity which increases by regular steps as we pass from one element to the next. This quantity can only be the charge on the central positive nucleus, the existence of which [Rutherford has proved.] We are therefore led by experiment to the view that Z [the nuclear charge] is the same as the number of the place occupied by the element in the periodic system. This atomic number is then for H 1, for He 2, for Li 3 for Ca 20, for Zn 30, &c.
>
> The similarity between the X ray spectra of different elements shows that these radiations originate inside the atom, and have no direct connexion with

the complicated light-spectra or chemical properties which are governed by the structure of its surface.

When an energetic cathode ray electron strikes an atom with $Z > 2$, it often dislodges one of its innermost electrons. Subsequently, an electron from the atom's second shell may take its place. This process is accompanied by the emission of a K_α X ray. Moseley discovered that the square root of the K_α X-ray frequency is a linear function of Z, as shown in Figure 10.12. Moseley deduced the following formula for these characteristic frequencies:

$$f = f_0 (Z - 1)^2 \left(\frac{1}{1^2} - \frac{1}{2^2} \right) \qquad K_\alpha \text{ X-ray frequency} \qquad (8)$$

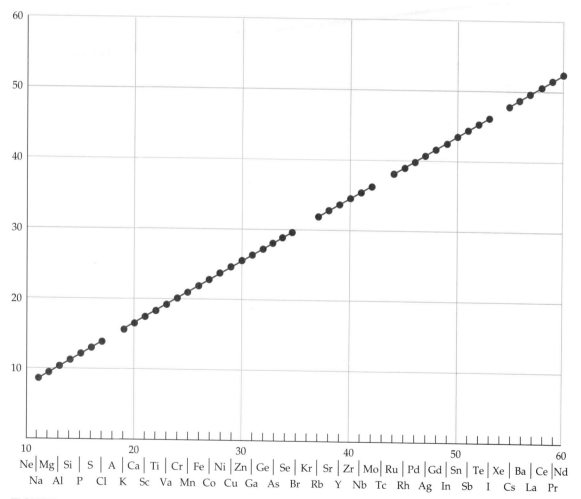

FIGURE 10.12 Moseley's experimental results. The square root of the frequency of the K_α X-ray line is plotted (in arbitrary units) against atomic number. Moseley showed that the charge of the atomic nucleus determines its place in the periodic table.

Equation 8 has been written in complicated form to facilitate its comparison with the first line of the Lyman series of hydrogen's spectral lines, as given by equation 9.20. The constant f_0 appearing in Moseley's formula is identical to the constant appearing in the Lyman series and given by equation 9.17. Bohr had just deduced the X ray spectra of the elements from his quantum model of the atom. Moseley's work offered one of its first confirmations:

> This numerical agreement between the experimental values and those calculated from [Bohr's] theory designed to explain the ordinary hydrogen spectrum is remarkable, as the wavelengths dealt with in the two cases differ by a factor of 2000.

The second of Moseley's two papers, entitled *The High-Frequency Spectra of the Elements*, was published in April 1914. From a systematic and accurate study of X-ray spectra, he came to the following conclusions, all of them correct:

1. Every element from aluminum to gold is characterized by an integer Z which determines its X ray spectrum. Every detail in the spectrum of an element can therefore be predicted from the spectra of its neighbors.
2. This integer Z, the atomic number of the element, is identified with the number of positive units of electricity contained in the atomic nucleus. [Consequently, Z is the number of electrons surrounding the nucleus of a neutral atom.]
3. The atomic numbers of all the elements from Al to Au have been tabulated on the [correct] assumption that Z for Al is 13.
4. The order of the atomic numbers is the same as that of the atomic weights, except where the latter disagrees with the order of the chemical properties.
5. Known chemical elements correspond with all the numbers between 13 and 79 except three. There are here three possible chemical elements still undiscovered. [Rhenium was found in 1923. Technetium and promethium are absent from Earth and had to be synthesized to be seen. The characteristic X rays of elements with $Z < 13$ were too long for Moseley to measure. See exercise 22.]
6. The frequency of any line in the X ray spectrum is approximately proportional to $A (Z-b)^2$, where A and b are constants. [As is equation 8 for the frequencies of the $K\alpha$ lines of different elements.]

The Classical Model of the Hydrogen Atom

One electron, of mass m and charge $-e$, is bound to a tiny but much heavier nucleus of charge $+e$. The electron moves with constant speed in a circular orbit around the nucleus (Figure 10.13). The binding energy W of the atom is the energy that must be supplied to extract the electron. It is known to be 13.6 eV. Let us use the laws of classical mechanics to find the radius r of the electron's orbit and its orbital frequency f, the number of times it circles the nucleus per second.

FIGURE 10.13
The nuclear atom. In
the classical picture of
an atom, electrons move
in circular orbits around
a much heavier nucleus.
The hydrogen atom con-
tains one electron, and its
singly charged nucleus is
called a proton.

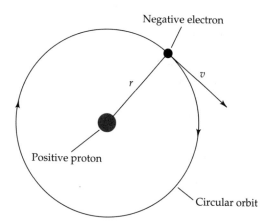

FIGURE 10.13
The nuclear atom. In
the classical picture of
an atom, electrons move
in circular orbits around
a much heavier nucleus.
The hydrogen atom con-
tains one electron, and its
singly charged nucleus is
called a proton.

Let v be the speed of the electron in its orbit. This problem is analogous to the problem of planetary motion, as discussed in Chapters 2 and
3. The binding energy W of a planet in an orbit is the negative of the
total energy of the planet, as given by equation 3.59. In the present case,
the potential energy of the electron is $-k_c e^2/r$, so its binding energy is

$$W = \frac{k_c e^2}{r} - \frac{m v^2}{2} \tag{9}$$

We also know that the mass of the electron times its centripetal acceleration equals the electric attraction of the nucleus:

$$\frac{m v^2}{r} = \frac{k_c e^2}{r^2} \tag{10}$$

Equation 10 tells us that the binding and kinetic energies are equal:

$$W = \frac{m v^2}{2} = \frac{k_c e^2}{2 r} \tag{11}$$

and from this result we may express r and v in terms of W. Since the
orbital frequency of the electron is $f = v/(2\pi r)$, we obtain:

$$r = \frac{k_c e^2}{2 W} \quad \text{and} \quad f = \frac{(2 W)^{3/2}}{2 \pi k_c e^2 \sqrt{m}} \tag{12}$$

To use equation 12 to evaluate r and f, W must be expressed in joules:
$W = 13.6 \times 1.6 \times 10^{-19}$ J. The results are

$$r \simeq 5.3 \times 10^{-11} \text{ m} \quad \text{and} \quad f \simeq 6.58 \times 10^{15} \text{ Hz} \tag{13}$$

By adjusting the binding energy of the classical hydrogen atom to
be its known value, we find that its radius, as given by equation 13,
agrees with the observed size of the hydrogen atom! In addition, the

orbital frequency of the electron is exactly twice the frequency f_0 that characterizes hydrogen's spectrum.

The Failure of the Classical Model

Niels Bohr, the great Danish physicist, agreed with Rutherford's classical picture of a hydrogen atom as an electron bound to a tiny, heavy, and positively charged nucleus. The electric attraction exerted by the nucleus keeps the electron in a circular orbit. Bohr realized that this model correctly relates the binding energy of the atom to its measured size, but he pointed out that in other respects the classical model of the hydrogen atom is a disaster.

In the first of a remarkable trilogy of papers Bohr published in 1913, he pinpointed two distinct but related problems with the classical model of the hydrogen atom. One has to do with atomic sizes. According to classical mechanics, the electron's orbit is a circle or an ellipse whose size determines the size of the atom. However, just as planetary orbits can be small (Mercury) or large (Pluto), so can electron orbits. Within classical physics, there is no explanation for why the size of the electron orbit is what it is. This is terrible. Because nothing in the theory determines the size of the hydrogen atom, some could be larger, some smaller. Yet it was known that all the atoms of any element are identical in size. Something was missing from the theory and (Bohr wrote) "it seems necessary to introduce in the laws in question a quantity foreign to the classical electrodynamics."

The second problem has to do with the stability of atoms. According to Maxwell's laws, an accelerating electron must radiate electromagnetic waves. When this is taken into account, the electron will no longer describe a fixed orbit but will "approach the nucleus describing orbits of smaller and smaller dimensions, and with greater and greater frequency; the electron gaining in kinetic energy as the whole system loses energy. This process will go on until the dimensions of the orbit are the same as those of the nucleus." The atom, if it obeyed the laws of classical physics, would collapse in a tiny fraction of a second. Bohr points out that the behavior of a classical atom is "very different from that of an atomic system occurring in nature. The actual atoms in their permanent state seem to have absolutely fixed dimensions and frequencies."

So far, my extracts from Bohr's paper simply reveal the trials and tribulations of a young theoretical physicist trying desperately to make sense out of Rutherford's model of the atom. But Bohr was about to pull a quantum rabbit out of a classical hat. Planck and Einstein, a few years earlier, had found the "foreign quantity" that had to be inserted into the laws of classical physics. It was h, Planck's quantum of action. Bohr would put it to good use. By supplementing classical physics with a few inexplicable rules, Bohr showed how all the known properties of the hydrogen atom (and many properties of larger atoms, such as Moseley's result) could be calculated from fundamental principles. Much later, the consistent theory that incorporated and explained Bohr's rules was formulated; it is called quantum mechanics.

Exercises

17. The radius of the hydrogen nucleus (the proton) is about 1.2×10^{-15} m, and the radius of the hydrogen atom is 5×10^{-11} m. If the proton were blown up to the size of the Sun, whose radius is 7×10^5 km, what would the size of the atom be? Would Earth lie inside? Would Pluto lie inside?

18. Using the information from exercise 17, determine the fraction of the atomic volume of hydrogen occupied by its nucleus.

19. There exist at least three other pairs of adjacent elements (besides cobalt and nickel) whose atomic weights are in the inverse order of their atomic numbers. Moseley didn't know about them because atomic weights were not yet known with sufficient precision. Find them with the aid of the periodic table in the Appendix.

20. The mass of the α particle is about 6.65×10^{-27} kg.
    a. What is the speed of a 5-MeV α particle?
    b. What is the energy of an α particle moving at a tenth the speed of light?

21. Calculate the wavelengths of the K_α X rays of Al ($Z = 13$) and Au ($Z = 79$) from equation 8.

22. Suppose that the K_α X-ray wavelengths of various elements are measured by determining the angles of strong reflection from a crystal. If the crystal layers are 4.5 angstroms apart, what is the element of lowest Z for which this procedure works?

23. Rutherford showed that the diameter of the nucleus of a gold atom is less than 5×10^{-14} m. Estimate the diameter of a gold atom from its atomic weight, its density (19.3 g/cm^3), and Avogadro's number.

24. The gold nucleus has charge $79e$ and radius 8×10^{-15} m. A helium nucleus has charge $2e$ and radius 2×10^{-15} m.
    a. Calculate the electric potential energy in MeV of an α particle that just touches a gold nucleus.
    b. What is the minimum energy in MeV of an incident α particle such that the two nuclei touch one another in a head–on collision?
    c. How closely can a 5-MeV α particle approach a gold nucleus?

25. What is the speed of the orbiting electron in a hydrogen atom according to classical physics? What is its kinetic energy in eV? Assume that its orbit is a circle with a radius of 5×10^{-11} m.

26. The nucleus of a fluorine atom has charge $+9e$. It is surrounded by nine orbiting electrons with negligible masses. It is a bit like our Sun and its nine planets. For the atom, the relevant force is electric while for the solar system it is gravitational. From the point of view of classical mechanics and aside from a difference in scale, in what way are these two dynamical systems different from one another.

Where We Are and Where We Are Going

Three discoveries of the 1890s allowed us to learn the inner structure of atoms in the early twentieth century. Electrons were found! Each one carries a negative charge $-e$ equal to the "atom of electricity" suggested by Faraday's experiments with electrolysis. They are the negatively charged constituents of atoms but are far too light to make up the bulk of an atom's mass.

Rutherford used α particles from radioactive elements as probes to show that most of the atom's mass is concentrated in its tiny positively charged nucleus. In fact, α particles themselves are rapidly moving helium nuclei. Moseley used X rays to show that nuclear charges are integer multiples of e. These integers run from $Z = 1$ (for hydrogen) to $Z = 92$ (for uranium). In each case, the nuclear charge is balanced by Z orbiting electrons to form a neutral atom of the chemical element with atomic number Z. Three sets of questions remained unanswered:

- What are the rules by which an atomic nucleus combines with electrons to form an atom? How do these rules explain the properties of chemical elements and, in particular, the success of the periodic table?

- What is radioactivity? Why are some elements radioactive and others not? What is the source of the energy released in radioactive processes?

- Are atomic nuclei elementary or are they made of simpler things? What are the nuclear constituents and how are they held together?

All of these questions have been answered, but the answers depended on two revolutions of twentieth-century physics. Many of the deepest held beliefs of classical physics had to be abandoned. Modern physics incorporates both quantum mechanics and the special theory of relativity. For things of our size and speed, its predictions are the same as those of classical physics. For atoms and their parts, it's a whole new story and it's told in the next two chapters.

11 Quantum Mechanics

When the history of this century is written, we shall see that political events—in spite of their immense cost in human lives and money—will not be the most influential events. Instead the main event will be the first human contact with the invisible quantum world.

Heinz R. Pagels, American physicist

Things on a very small scale behave like nothing you have any direct experience about. They do not behave like waves, they do not behave like particles, they do not behave like clouds, or billiard balls, or weights on springs, or like anything that you have ever seen.

Richard P. Feynman, American physicist

We wish to speak in some way about the structure of atoms, but we cannot speak about atoms in ordinary language.

Werner Heisenberg, German physicist

Niels Bohr (1885–1962) devised quantum-mechanical rules with which to explain atomic spectra. His concepts formed the basis of quantum mechanics. *Source: AIP Emilio Segre Visual Archives, Uhlenbeck Collection.*

Classical physics consists of Newton's theory of motion, Maxwell's theory of electrodynamics, the laws of thermodynamics, and the kinetic theory of gases. In the late nineteenth century, it seemed a solid foundation from which scientists could explain the behavior of matter. Not even the newly discovered electron or the unexpected phenomenon of radioactivity were thought to pose insuperable problems. The first inkling that something was dreadfully wrong with classical physics came from an unexpected direction. For decades, physicists had been searching for a mathematical description of the radiation produced by a hot body. In 1900, the German physicist Max Planck found the answer. In Section 11.1, we explain Planck's discovery and the impact it had on classical physics.

Section 11.2 recounts how Niels Bohr exploited the wave–particle duality of light to create a marvelously predictive model of the simplest atom, hydrogen. His quantum rules conflicted with classical physics. The entire logical structure of mechanics and electrodynamics required revision. Section 11.3 describes the emergence of a consistent theory of quantum mechanics. Finally, quantum mechanics is applied to the question of atomic structure in Section 11.4.

11.1 The Origins of Quantum Theory

Anyone sitting next to a hot radiator knows it radiates heat. If an object is heated further, it radiates light as well. Newton speculated about the origin of this radiation in 1704:

Do not all fix'd Bodies, when heated beyond a certain degree, emit Light and shine; and is not this Emission perform'd by the vibrating motions of their parts?

Newton was right. Radiation produced by a hot object—whether radiant heat or light—is called *thermal radiation* and results from collisions among chaotically moving molecules within the object. Radiators are not hot enough to glow. Their thermal radiation—radiant heat—consists mostly of electromagnetic waves with wavelengths much longer than those of visible light. Radiant heat lies in the infrared part of the electromagnetic spectrum. Thus, heat comes in two different forms: as the internal motions of the molecules within an object and as the radiant heat emitted by a hot object.

Both light and radiant heat consist of waves with a continuous range of wavelengths. The amount and the nature of thermal radiation depends on the temperature of the radiating object. As an object's temperature increases, the molecular velocities increase and molecular collisions become more violent. Consequently, the total power radiated by the hot object also increases, and its distribution according to wavelength changes.

Let's look at an example. A lamp bulb is equipped with a dimmer to control the applied voltage. When the voltage is low, the filament radiates heat but not light. At a higher voltage, the filament grows hotter and glows redly and dimly. As the voltage is turned up further, the temperature of the filament increases, as does the amount of radiant energy emitted. The nature of the radiated light changes as well: from dull red, it becomes bright orange, then yellow-white, then blue-white. The colors are those of stars: red stars are the coolest, blue-white stars the hottest; our Sun is a middling hot, yellow-white star.

Blackbody Radiation

Gustav Kirchhoff (whom we met in Chapter 9) was puzzled by the relation between the temperature of a body and the radiation it emits. The continuous spectrum of thermal radiation includes all the colors of the rainbow, as well as invisible infrared and ultraviolet radiation. Furthermore, there are lines in the spectra of most hot bodies such as stars or chemicals heated in a flame: dark lines due to the absorption of light of definite wavelengths by cool atoms and bright lines due to light emission by hot atoms.

Kirchhoff suggested how to produce purely thermal radiation from which the effects of individual atoms were removed. Consider an enclosed and heated cavity, such as the interior of an oven or kiln. The collisions of molecules of the oven walls produce electromagnetic waves that bounce around the enclosure and are repeatedly absorbed and emitted by these walls. The mechanisms for bright-line emission and dark-line absorption are both present at the same time. As a result, there can be no spectral lines in the radiation contained in the oven. Kirchhoff proved that the nature of this lineless thermal radiation (its intensity and its spectrum) cannot depend on the size, shape, or composition of the oven, or on the manner by which its walls are heated. Rather, thermal radiation is completely determined by the temperature of the enclosure in which it is produced. (In discussing thermal radiation, temperature T will be expressed relative to absolute zero in the Kelvin scale.)

Suppose a tiny hole is made in the oven wall. If an electromagnetic wave from somewhere outside the oven happens to enter the hole, it is unlikely to find its way out. For this reason, Kirchhoff called the hole in the oven a *blackbody* (Figure 11.1). Of course, some of the thermal radiation within the oven will stream out of the hole. If the oven is hot enough, the aperture will not be black but glowing. The emergent radiation (whose nature is determined by the temperature of the oven) is called *blackbody radiation.* The radiation emitted by a star, by the hot filament of an incandescent lamp bulb, by a person, or even by an ice cube is approximately the same as that of an ideal blackbody at the same temperature.

At the temperature of a red-hot branding iron, only a small part of the blackbody radiation streaming from the oven (or any hot object) will be in the form of visible light. At chicken-roasting temperature, the radiation will be less intense and will be mostly in the form of radiant heat. At room temperature, thermal radiation will be barely noticeable. The universe itself is in many ways like a gigantic oven at a temperature of 2.75 K. Nonetheless, its thermal radiation—the cosmic background radiation—has been observed and carefully measured.

FIGURE 11.1
An oven is maintained at temperature T. Blackbody radiation escapes through a small hole in the oven wall. The nature of this radiation posed a mystery that was solved by Planck's quantum hypothesis.

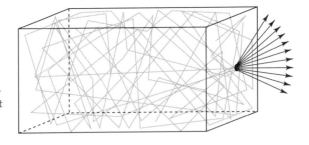

The Challenge Is Put!

The problem Kirchhoff posed but could not solve is simply stated. What is the mathematical function that expresses how much power, at different wavelengths, is present in the thermal radiation produced by an object at temperature T? In 1860, Kirchhoff wrote:

> It is a highly important task to find this function. Great difficulties stand in the way of its experimental determination. Nevertheless, there appear grounds for the hope that it has a simple form, as do all functions that do not depend upon the properties of individual bodies and which one has become acquainted with before now.

Physicists immediately set to work on Kirchhoff's problem. In 1879, Josef Stefan measured the power P in thermal radiation and found that it increased with the fourth power of the Kelvin temperature of the radiating object. Boltzmann deduced Stefan's experimental result from theoretical arguments. Their result is called the *Stefan–Boltzmann law*:

$$P = \sigma A T^4 \tag{1}$$

where T is the Kelvin temperature and A is the area of the object. The Stefan–Boltzmann constant σ was measured, but its value could not be calculated from basic physical principles. Equation 1 says that the power radiated by a blackbody increases by a factor of 16 when its Kelvin temperature is doubled. The observed value of the Stefan–Boltzmann constant is

$$\sigma \simeq 5.67 \times 10^{-8} \ \text{W/m}^2\,\text{K}^4 \tag{2}$$

with power measured in watts.

EXAMPLE 11.1
The Temperature of the Sun

The total power radiated by the Sun (the solar luminosity) is $L_\odot \simeq 3.83 \times 10^{26}$ W. Its radius is $R_\odot \simeq 7 \times 10^8$ m. Use equations 1 and 2 to estimate the solar surface temperature. Aside from its spectral lines, the radiation produced by a star is nearly the same as that produced by an ideal blackbody.

Solution

According to equation 1, $L_\odot = \sigma T^4 A$, where $A = 4\pi R_\odot^2$ is the Sun's area. Solving this equation for T, we obtain:

$$T = \left(\frac{L_\odot}{4\pi R_\odot^2 \sigma} \right)^{1/4} \simeq 5760 \ \text{K} \qquad \blacksquare$$

The Stefan–Boltzmann result was a partial answer to Kirchhoff's question. It described the power in thermal radiation, but it did not describe how the radiant energy is distributed among different wavelengths and why the average wavelength of thermal radiation decreases as the temperature increases. This property of thermal radiation was deduced by the German theoretical physicist, Wilhelm Wien (1864–1928). Let λ_{max} be the wavelength at which thermal radiation is most intense. Wien showed that the product of λ_{max} and the Kelvin temperature of a heated object is a constant. The constant was determined by experiment, and his relation is known as the *Wien displacement law*:

$$\lambda_{\text{max}} T = 2.90 \times 10^{-3} \ \text{m K} \tag{3}$$

Most of the electromagnetic waves in thermal radiation have wavelengths lying in the vicinity of λ_{max}, whose value varies inversely with the Kelvin temperature of the radiating object (Figure 11.2, on page 454). Wien's law explained the progression—from red to yellow to white—of lamp filaments or stars at different temperatures.

Careful experiments confirmed the validity of equations 1 and 3 in visible and infrared wavelengths. However, Kirchhoff's question was not yet fully answered. Suppose we measure the intensity of blackbody radiation as a function of wavelength at a particular temperature.

FIGURE 11.2
The Wien displacement law. In this logarithmic graph, the value of λ_{max} is shown against the temperature T of a body. The thermal radiation of the universe, which is at a temperature of about 3 K, lies in the microwave portion of the electromagnetic spectrum. An object at room temperature radiates energy in the infrared region. Molten iron radiates thousands of times more power and glows brightly, but most of its radiation is infrared. The Sun, at a temperature of 5760 K, emits most of its thermal radiation as visible light.

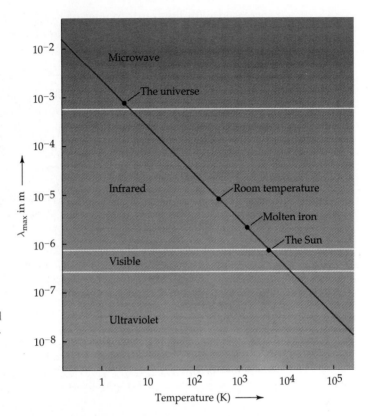

What is the shape of this curve? Let's be more precise. Let $\Delta\lambda$ denote a small interval of wavelength lying between λ and $\lambda + \Delta\lambda$. The radiant power emitted in this wavelength interval by a heated object of unit area is given by $\mathcal{J}_T(\lambda)\,\Delta\lambda$, where the function $\mathcal{J}_T(\lambda)$ characterizes blackbody radiation. The subscript T reminds us that the amount of radiation and its spectral distribution are determined by the temperature of the body. The shape of $\mathcal{J}_T(\lambda)$ was measured at different temperatures and is shown in Figure 11.3. The observations were compared with many theoretical models, but none of them agreed with the data. The Stefan–Boltzmann and Wien laws said something about $\mathcal{J}_T(\lambda)$, but they did not pin the function down. Classical physics could do no more.

EXAMPLE 11.2
Using Wien's Law

Calculate the value of λ_{max} for the radiation produced by the Sun (for which $T = 5760$ K), for a body at room temperature (300 K), and for a body with the mean temperature of the universe (2.7 K). For each case, in what part of the electromagnetic spectrum does λ_{max} lie?

Solution

Using Wien's displacement law, equation 3, we find that λ_{max} for the Sun is about 5×10^{-7} m (5000 angstroms), corresponding to yellow light. At room temperature, $\lambda_{max} \simeq 10^{-5}$ m, which lies in the infrared portion of the spectrum. At the mean temperature of the universe, $\lambda_{max} \simeq 1$ mm, corresponding to microwave radiation. ∎

FIGURE 11.3
The Planck distribution $\mathcal{J}_T(\lambda)$ of blackbody radiation at different temperatures. Notice that λ_{max} decreases as the temperature increases. The total radiated power grows with T^4.

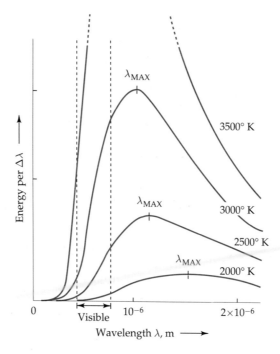

The Challenge Is Met!

Max Planck (1858–1947) struggled to explain the shape of the observed blackbody spectrum. Eventually, he found a mathematical expression for the spectral distribution function $\mathcal{J}_T(\lambda)$ that agreed with all available experimental data. However, a formula by itself is not enough—unless it can be shown to follow logically from the laws of physics. Planck realized that the known laws of thermodynamics, electromagnetism, and probability theory were insufficient. To explain his discovery, he had to make a bizarre and then unjustifiable conjecture lying outside the rules of classical physics. The quantum century began in 1900, when Planck was forced to the daring hypothesis that electromagnetic energy is not continuously transferred among atoms. Instead, atoms emit or absorb radiation in discrete bundles, or *quanta*. Emitting or absorbing electromagnetic radiation, according to Planck, has something in common with selling or buying milk, which comes in containers of 1, 2, 4, and 8 pints, but not in other size packages.

The energy ΔE carried by a single quantum must be an integer multiple of a certain quantity:

$$\Delta E = nhf \qquad \text{where} \quad n = 1, 2, 3, \ldots$$

where f is the frequency of the radiation. The universal constant h is now called Planck's constant, or the quantum of action, and is known with exquisite precision:

$$h \simeq 6.626\,075 \times 10^{-34}\,\text{J s} \simeq 4.1357 \times 10^{-15}\,\text{eV s} \qquad (4)$$

Because particle energies are measured in eV, we give h both in SI units and eV.

From his radical quantum hypothesis, Planck deduced the formula for thermal radiation. Planck was "pushing the envelope" of rational scientific inquiry. Much later, he wrote of his discovery that it was "an act of desperation. I had to obtain a positive result, under any circumstances and at whatever cost." His seminal paper of 1900 reads more professionally:

> One can evaluate—without knowing anything about a spectral formula or about any theory—the distribution of a given amount of [radiant] energy over the different colors of the normal spectrum.

Planck found the function that Kirchhoff dreamed of. The *Planck distribution* gives the intensity of thermal radiation at different wavelengths:

$$\mathcal{J}_T(\lambda) = \frac{2\pi hc^2}{\lambda^5} \frac{1}{e^{hc/kT\lambda} - 1} \tag{5}$$

where k is the Boltzmann constant and $e \simeq 2.718$ is the base of natural logarithms. Equation 5 passed every experimental test. Fifty years afterward, Planck wrote:

> Later measurements, too, confirmed my radiation formula again and again—the finer the methods of measurement, the more accurate the formula was found to be.

Equation 5 encompassed all that had been learned about thermal radiation. The Wien displacement law and the Stefan–Boltzmann law are its mathematical consequences. Furthermore—and this is the essence of what physics is about—Planck expressed the experimentally measured Stefan–Boltzmann constant in terms of fundamental constants of nature:

$$\sigma = \frac{2\pi^5 k^4}{15c^2 h^3} \tag{6}$$

Equation 5 exposed the soft underbelly of classical physics. Planck's hypothesis that light energy is exchanged in bundles hinted that the laws of Newton and Maxwell fail in the microworld. The radical revision of these laws is quantum mechanics, and Planck's tiny new constant blazed the path for the quantum revolution that left the older generation (and Planck himself) in the dust. Planck won his well-earned Nobel Prize in 1918, but youngsters—Einstein, Bohr, Heisenberg, and their friends—soon took over.

DO YOU KNOW

The Temperature of the Universe?

The Planck distribution is relevant to the universe itself, which was born long ago in the hot Big Bang. Nearly a million years afterward, it cooled to a temperature of 3000 K, low enough for neutral atoms to form. It suddenly became transparent and has been expanding and cooling ever since. Aside from its stars and galaxies, today's universe is like a blackbody, with a temperature $T = 2.7$ K. The peak of its spectrum lies in the microwave domain. This radiation, the last frozen remnant of the Big Bang, was first detected in 1964. In 1990, the COsmic Background Observer (COBE) satellite surged into orbit to measure the precise nature of the cosmic microwave background radiation. Figure 11.4 shows the measured spectrum along with a Planck distribution at a temperature that best fits the data. The two curves are identical within tiny experimental errors: our universe, the relic of a titanic explosion, is now just a cold blackbody with stars.

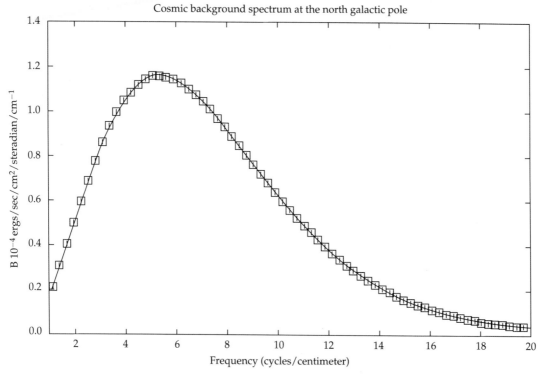

FIGURE 11.4 The cosmic background radiation, as observed by NASA's COBE satellite. The points show experimental observations. The smooth curve passing through all the points is a Planck distribution corresponding to a temperature of 2.74 K. The frequency is indicated as the number of wavelengths per centimeter.

*The Photon
Hypothesis*

Einstein turned his attention to thermal radiation in 1905. He knew that the Planck distribution was experimentally correct, but he also uncovered a subtle mathematical flaw in Planck's argument. If Planck had corrected this error, he would have been led to a result that did not agree with experiment. To obtain a correct derivation of the Planck distribution, Einstein had to strengthen the quantum hypothesis.

Planck demanded that light of frequency f be packaged in quanta whose energies are multiples of hf, but he never said or thought that light behaves like particles. Every physicist believed what Thomas Young had established a century earlier: "Radiant light consists in undulations of the luminiferous ether." Heinrich Hertz, who discovered radio waves, asserted that "The wave theory of light is a certainty." Einstein, however, was forced to adopt a revolutionary particle-like stance:

> The energy in a beam of light emanating from a point source is not distributed continuously over larger and larger volumes of space but consists of a finite number of energy quanta, localized at points of space which move without subdividing and are absorbed or emitted only as units. Light behaves in thermodynamic respect as if it consists of mutually independent energy quanta of magnitude $E = hf$

where E is the energy of one light quantum of frequency f. Not only is electromagnetic energy packaged in discrete quanta, Einstein realized, but it remains in discrete quanta as it propagates through space. That is, light behaves as if it consists of particles.

Einstein showed how the Planck distribution follows from his light-quantum hypothesis, but he was compelled to treat blackbody radiation as if it consisted of a gas of particles of light, or *photons.** To what extent is light made of particles rather than waves? Einstein's Nobel Prize-winning paper continued:

> If monochromatic radiation behaves as a discrete medium consisting of energy quanta of magnitude hf, then this suggests an inquiry as to whether the laws of generation and conversion of light are also constituted as if light were to exist as energy quanta of this kind.

The photon hypothesis was not just a mathematical artifice with which to derive the Planck distribution. In 1905, Einstein pointed out that it had direct and testable physical implications, which we now discuss.

*The Photoelectric
Effect*

Hertz had seen that a piece of metal is electrified by ultraviolet light. Thomson showed that electrons are driven from the metal by what is known as the *photoelectric effect.* When ultraviolet light falls on a sheet of metal, it is absorbed. Some of the incident energy is transferred to atomic

---

*The term *photon,* denoting the particulate avatar of light, was introduced in 1926 by the American physicist Gilbert Lewis. It took a while before the concept of light-as-particles was accepted. We use "photon" freely in place of Einstein's awkward "light quantum." We also use "light" lightly to refer to any electromagnetic radiation, including those with frequencies eyes do not see.

electrons, which are ejected as rapidly moving particles (photoelectrons) from the metal's surface (Figure 11.5a).

A certain fixed amount of energy W must be supplied to an electron to liberate it from a metal. W is known as the *work function*. Its value varies from metal to metal, but W is always a few electron volts, or eV. This energy could easily be supplied by electromagnetic waves. Careful experimental studies of the photoelectric effect had not yet been carried out, but the accepted view, based on the wave theory of light, led to the following predictions:

- The energy of individual photoelectrons should not depend critically on the frequency of the incident radiation. Once an electron had, bit by bit, accumulated enough energy, it would come popping out of the metal.

- If the intensity of the incident radiation (and, hence, the power delivered to the metal) were increased, both the number of photoelectrons and their energies should increase. If the intensity were very small, it would take time for the electrons to acquire enough energy to be ejected. Thus, there would be a perceptible delay between turning on the light source and observing photoelectrons.

Einstein analyzed the photoelectric effect differently. He interpreted the process in terms of light-as-particles rather than light-as-waves. Suppose that a photon of frequency f is absorbed by an electron in a metal and that it transfers its energy hf to the electron. If the photon energy is larger than the work function W, the electron is driven off. Conservation

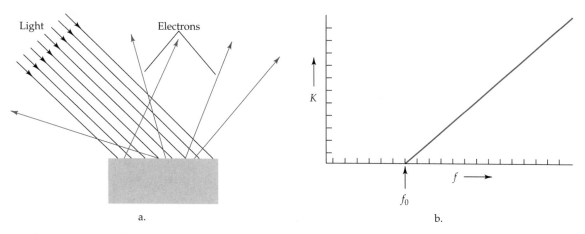

a.   b.

FIGURE 11.5 *a*. The photoelectric effect. When light strikes a metallic surface, electrons may be driven from it. The photon energy must exceed $W = hf_0$, the work function of the metal. The kinetic energy of the emitted electrons is $h(f - f_0)$. *b*. Millikan confirmed Einstein's predictions. The kinetic energy of the photoelectrons is plotted against the frequency of light. The slope of the line is equal to Planck's constant. Its x intercept is f_0. This experiment proves that light can behave as if it consists of a stream of particles, but it cannot be understood in terms of light as waves.

of energy tells us what the kinetic energy K of the emitted photoelectron must be:

$$K = hf - W \tag{7}$$

This simple formula is fraught with experimental consequences that are completely different from those of the wave theory of light:

- Below a critical frequency, $f_0 = W/h$ (which depends on the particular metal), no electrons are liberated, no matter how long you wait or how intense the radiation.
- The energy of each photoelectron is a linear function of the frequency of the light, $K = h(f - f_0)$, whose slope does not depend on the identity of the metal. The slope is Planck's constant h, which had been introduced into physics from altogether different considerations.
- The number of photoelectrons that are released—but not the energy of each electron—is proportional to the intensity of the light.

In 1905, experimental observations of the photoelectric effect were not sophisticated enough to distinguish between the classical expectations and Einstein's radically different predictions.

Waves vs. Particles

Einstein's theory of the photoelectric effect was vindicated in 1915 by Robert A. Millikan at the University of Chicago. (In Chapter 10, we saw how Millikan, a decade before, measured the charge of the electron.) Ordinary metals do not emit electrons unless they are struck by ultraviolet light. Millikan used targets made of sodium and other alkali metals because these metals produce electrons when visible light hits them. The targets were placed in a vacuum tube and exposed to light of different colors or frequencies. Millikan measured the kinetic energies of the emitted photoelectrons and found that equation 7 was precisely satisfied: A graph of the kinetic energy of the electrons versus the light frequency is a straight line (Figure 11.5b). Millikan measured the slope of this graph and thereby obtained a precise measurement of h. In 1948, he recalled:

> I spent ten years of my life testing that 1905 equation of Einstein's and contrary to all my expectations, I was compelled in 1915 to assert its unambiguous verification in spite of its unreasonableness, since it seemed to violate everything we knew about the interference of light.

Millikan proved that light, when it strikes a metal plate and ejects electrons, behaves as if it were a stream of individual particles, each with energy hf.

EXAMPLE 11.3
The Photoelectric Effect

A metal plate has a work function of $W = 5$ eV. If electromagnetic radiation with a wavelength of $\lambda = 2 \times 10^{-7}$ m falls on the plate, what is the energy of the emitted photoelectrons?

Solution   Equation 7 determines the electron energy in terms of the frequency of the incident light, which is c/λ or 1.5×10^{15} Hz. Using equation 4, we find that the energy of the incident photon is $hf = 6.2$ eV. Thus, the kinetic energy of the electron is $K = 6.2 - 5 = 1.2$ eV. ∎

To see what bothered Millikan, consider a two-slit diffraction experiment, such as that discussed in Section 9.3. In the experiment shown in Figure 11.6a, a beam of monochromatic light is sent through two slits. The screen is bright or dark depending on whether the waves coming from the two slits add up or cancel one another. An interference pattern consisting of bright and dark stripes appears on the screen. Let D be a dark point where the two waves cancel. When we obstruct the lower slit, the pattern changes. Half as much light falls on the screen and most points become darker than they were. However, places that were already dark, such as D, brighten because waves coming through the open slit are no longer being cancelled by those coming through the other (Figure 11.6b). If light were a beam of particles, the photons arriving at any point on the screen must have passed through one slit or the other. If one slit is closed, how could one part of the screen become brighter?

Diffraction (but not the photoelectric effect) is easily explained in terms of light as waves. The photoelectric effect (but not diffraction) is easily explained in terms of light as particles. In fact, neither wave nor particle is the right word for what light really is. J. J. Thomson (the electron's discoverer) wrote that the conflict between these two models of light was like "a struggle between a tiger and a shark; each is supreme in his own element, but helpless in that of the other."

Einstein treated light as particles so that he could describe blackbody radiation and the photoelectric effect. It is no wonder that Einstein's photon hypothesis went over like a lead balloon. Von Laue, whose fame

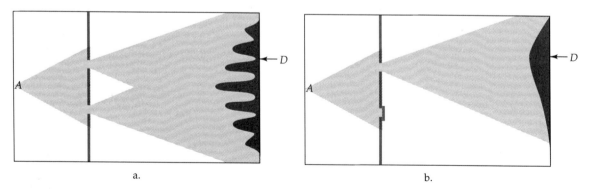

a.   b.

FIGURE 11.6 *a.* A source of light at A is allowed to pass through two slits. Where the waves passing through the slits add together, a bright line appears on the screen. Where they cancel out, a dark line appears. The shading indicates the intensity of light at the screen. *b.* In this case, one of the slits is covered. Half as much light falls on the screen. However, at point D, the intensity of light is greater than it was before. This phenomenon is easy to understand if light is a wave. It is puzzling from the point of view of light as particles.

rested on his discovery that X rays are waves, wrote to Einstein in 1907: "I would like to tell you how pleased I am that you have given up your light-quantum hypothesis." Millikan wrote in 1916: "Einstein's photoelectric equation appears in every case to predict exactly the observed results. Yet the physical theory of which it was designed to be the symbolic expression is found so untenable that Einstein himself, I believe, no longer holds it." No way! Einstein never abandoned his theory. He expressed reservations about his radical hypothesis in 1911,

> I insist on the provisional nature of this concept which does not seem reconcilable with the experimentally verified consequences of the wave theory

but he pursued its logical consequences doggedly. By 1916, he concluded that light was even more particle-like than he had first thought:

> If a bundle of radiation causes a molecule to emit or absorb an amount of energy hf, then a momentum hf/c is transferred to the molecule.

Einstein was not claiming that light and other forms of electromagnetic radiation consist of particles. Their wavelike properties were undeniable. Rather, Einstein argued that light behaved like particles under some circumstances—with each particle carrying a definite portion of energy E and momentum p. Under other circumstances, light behaved like waves—with definite frequencies f and wavelengths λ. Light is neither a particle nor a wave, but a little bit of each. The particle-like and wavelike properties of electromagnetic radiation are linked to one another by Einstein's relations:

$$E = hf = hc/\lambda \tag{8a}$$

$$p = hf/c = h/\lambda \tag{8b}$$

The energy of a photon is the product of Planck's constant and its frequency. The momentum of a photon is the quotient of Planck's constant by its wavelength. Its direction is that of the wave's propagation. Millikan's studies of the photoelectric effect in 1915 confirmed equation 8a, the relation between the energy of a photon and its frequency. Another eight years would pass before equation 8b was directly verified by laboratory experiments.

EXAMPLE 11.4
Photons of
Visible Light

Human eyes can detect electromagnetic waves ranging in wavelength from 4.0×10^{-7} to 7.5×10^{-7} m. What is the range in energy (expressed in eV) of the photons of visible light?

Solution

The energy of a photon is given by equation 8a as hc/λ. The quantity hc can be evaluated from equation 4 to be 1.24×10^{-7} eV m. We find that hc/λ ranges from 1.6 eV at the red end of the visible spectrum to 3.1 eV at the violet end.

The Compton Effect

In 1923, the American physicist Arthur Holly Compton (1892–1962) detected and measured the scattering of X rays by atomic electrons. In terms of photons (which are denoted by γ), the reaction he studied was

$$\gamma + e \longrightarrow \gamma + e$$

In particular, X rays with wavelength λ collided with atomic electrons in a carbon target. Because X-ray energies are much larger than the energies of atomic electrons, the process Compton studied was practically the same as the scattering of X rays by electrons at rest. Occasionally, the X rays would bounce directly backward from the electrons. Compton observed that their wavelength was changed in this process. The wavelength λ' of the back-scattered X rays was lengthened by an amount independent of the wavelength of the incident X rays:

$$\lambda' \simeq \lambda + 4.8 \times 10^{-2} \text{ angstrom}$$

This result could not be understood in terms of waves. However, it had a simple explanation in terms of the particle properties of electromagnetic radiation, as given by equations 8a and 8b.

The law of conservation of energy requires that the energy of the incident photon equal the sum of the energies of the scattered photon and the recoil kinetic energy E_e of the electron. According to equation 8a, we find

$$hc/\lambda = hc/\lambda' + E_e \tag{9}$$

This result tells us that λ' is larger than λ, but it does not tell us by how much because E_e is unspecified. The law of conservation of momentum, supplemented by the second of Einstein's predictions, equation 8b, gives the answer. The momentum of the incident photon must equal the sum of the momentum of the back-scattered photon and P_e, that of the recoil electron:

$$h/\lambda = -h/\lambda' + P_e \tag{10}$$

The energy and momentum of the electron may be expressed in terms of its mass m and velocity v. When this is done, equations 9 and 10 become two equations involving the two unknown quantities v and λ'. The derivation will be presented in Section 12.3, but the result for λ' is simply:

$$\lambda' = \lambda + 2\lambda_c \tag{11}$$

where λ_c is a universal constant that may be expressed in terms of the mass of the electron m, its charge e, and the velocity of light c:

$$\lambda_c \equiv \frac{h}{mc} \simeq 2.4 \times 10^{-2} \text{ angstrom} \tag{12}$$

Thus, Compton's experimental result follows from the particle interpretation of light. The quantity h/mc, with the dimensions of length, is known as the *Compton wavelength* of the electron. It is much smaller than the atomic size of about 1 angstrom.

When electromagnetic radiation scatters backward from electrons, its wavelength is increased by 2 Compton wavelengths, about 0.05 angstrom. If the incident radiation is visible light, the wavelength shift is tiny compared to its wavelength of thousands of angstroms. That's why Compton had to use X rays with wavelengths of 1 angstrom or so to detect the effect. He showed that photon–electron collisions satisfy the mechanical laws of energy and momentum conservation that apply to billiard ball collisions. Compton established the particle nature of light and its quantitative specification by equations 8a and 8b. Einstein was right again.

Kirchhoff's challenge led Planck to its solution via an ad hoc quantum hypothesis. Einstein's improved derivation of the Planck distribution forced him to revive the concept of light as particles. But light was known to consist of waves. How can a thing be both particle and wave? Today's physicists, tutored and trained in quantum theory, regard such questions as mere wordplay. Language evolves experientially—when a pebble falls in a pond the pebble is certainly the particle, the ripples the waves. Everyday laws of physics do not describe the microworld, so it should come as no surprise that everyday language is inappropriate to atomic phenomena: Light is what it is, neither particle nor wave. The microworld is correctly and consistently described by quantum theory, but not by the words and pictures characterizing events of human life. Such arguments satisfy most physicists, but they never satisfied Einstein, who wrote toward the end of his life:

> All these fifty years of pondering have not brought me any closer to answering the question, What are light quanta?

EXAMPLE 11.5
Compton Scattering

A photon scatters backward from a stationary electron and loses 10 percent of its energy. Calculate the wavelength λ and energy E (in eV) of the incident photon.

Solution

Because the energy of the scattered photon is $0.9E$, its wavelength is $\lambda' = 10\lambda/9$. Equation 11 tells us that $\lambda = 18\lambda_c$, or about 0.44 angstrom. The energy of the photon is hc/λ or $mc^2/18$. Thus $E \simeq 4.5 \times 10^{-15}$ J, or about 28 KeV, a typical X-ray energy. ∎

Exercises

1. An object is heated from 500 K to 600 K. Calculate, using equation 1, by what factor its radiant power increases.

2. Show that the right-hand side of equation 5 has dimensions of power over volume. Show that equation 2 determines the same dimensions for σ as equation 6.

3. Show that the dimensions of h are the same as those of: (a) momentum times length, (b) energy times time, (c) angular momentum mvr.

4. A photon with an energy of 10 KeV is scattered backward by an electron at rest. What is its energy afterward?

5. An electron in an atom has a speed ~ 1 percent of light. What is the product of its momentum and the atomic radius ($\sim 10^{-10}$ m) in units of h?

6. How much power is radiated by a blackbody at $T = 27°\,C$ with area $1\,m^2$? How much power is radiated by a blackbody with area $1\,mm^2$ at a temperature of $2727°\,C$?

7. Evaluate h in units of eV times seconds. Evaluate mc^2 (where m is the electron mass) in units of eV.

8. Potassium has a work function of 2.2 eV.
   a. What is the maximum wavelength λ of light that produces photoelectrons on striking potassium?
   b. If the incident light has $\lambda = 3 \times 10^{-7}$ m, what is the kinetic energy of each photoelectron?
   c. What value of λ yields photoelectrons with 1 eV of kinetic energy?

9. A beam of light with a power of 10 W falls on a metal surface whose work function is 3 eV. Each photon carries an energy of 4 eV.
   a. What is the energy of each photoelectron?
   b. How many photons are incident on the plate per second?
   c. If every photon produces a photoelectron, how many electrons are emitted from the metal surface in 1 second? The flow of electrons from the plate can be regarded as a current toward the plate. What is the value of this current in amperes?
   d. The power of the photoelectric current produced by the light beam is the product of the number of electrons released per second and their kinetic energy. What is this power in W?

11.2 Niels Bohr Explains the Hydrogen Atom

The first quantum thoughts, those of Planck and Einstein, concerned the nature of light not the structure of the atom. Circumstances took a dramatic turn when Niels Bohr linked the wave–particle nature of light to Rutherford's discovery of the nuclear atom.

Bohr accepted the photon hypothesis and applied it to the mystery of atomic structure. He began with a classical picture of the simplest atom, hydrogen, which was known to consist of a single electron bound to a much heavier, positively charged nucleus (a proton). For simplicity, he considered circular orbits of radius r. Bohr's classical analysis of this system was explained in equations 10.9–10.13. The rotational frequency f of the electron (the reciprocal of its period) and the radius r of its orbit was expressed in terms of its binding energy W:

$$r = \frac{k_c e^2}{2\,W} \quad \text{and} \quad f = \frac{(2W)^{3/2}}{2\pi k_c e^2 \sqrt{m}} \tag{13}$$

The results were tantalizing. When the binding energy of the electron is set equal to its known value, $W \simeq 13.6$ eV, the size of the hydrogen atom came out right. Furthermore, the frequency of light emitted when a proton captures an electron from afar is suggestively equal to $f/2$. However, equation 13 encountered two serious difficulties:

1. Nothing in classical mechanics fixes the radius of the orbit or, equivalently, the size of the atom. Equation 13 has solutions for any value of W, and hence for any value of r. Yet, all hydrogen atoms, under ordinary conditions, have exactly the same size.

2. An electron accelerates in its orbit. Maxwell's equations predict that any accelerated charged particle emits electromagnetic waves and loses energy. According to classical electromagnetic theory, electrons quickly spiral into their nuclei, causing atoms to collapse almost instantaneously.

To save the picture of the planetary atom, Bohr changed the rules! Like Planck and Einstein before him, he made guesses that could not be justified by any then-known laws of physics. To see how Bohr was led to his model of atomic structure, we retrace Bohr's arguments step by step, through a series of four brief extracts from his seminal papers of 1913. In several cases, details are presented just so you can see that the methods of science are deductive and mathematical, with just a bit of magic (or intuition) thrown in.

1. *Stationary States* "We are led to assume that [certain solutions to equation 13] correspond to states of the system in which there is no radiation of energy; states which consequently will remain stationary if not disturbed from outside."

Bohr postulated that some classical orbits are allowed, but most of them are forbidden. He called the allowed orbits *stationary states* of the electron. This assumption is contrary to the laws of classical mechanics! He declared that an electron in a stationary state does not radiate light. This assumption is contrary to the laws of classical electromagnetism!

Having rejected the laws of classical physics in the subatomic domain, Bohr had to fly blind. Bohr's inspired contributions to atomic theory must be judged by their resounding success, not by their logical foundation. A consistent mathematical approach to the problem of the hydrogen atom required the construction of an entirely new approach to mechanics rather than a series of patchwork adjustments to classical arguments. Bohr assumed that such a theory would someday arise (as indeed it did in 1926), and he set about to explore what its consequences would be.

It was one thing to say that only some electron orbits are allowed. To make further progress, Bohr had to figure out which of all possible classical electron orbits corresponded to stationary states of the atom. Bohr knew that Planck's constant h must play a key role in any theory of atomic structure. He knew that the dimensions of h are energy multiplied by time, the same as those of angular momentum. Surely, the angular momentum l of the electron must play a key role in atomic structure because l, like the binding energy W, is a constant of the motion for an electron moving around a proton. For a circular orbit, the angular momentum is:

$$l = mvr \tag{14}$$

in terms of the mass m, the speed v, and the radius r of the electron's orbit. What Bohr did next had nothing to do with any so-called scientific method—it was an act of genius. Again, in his words:

2. *The Quantization of Angular Momentum* "The angular momentum of the electron round the nucleus in a stationary state of the system is equal to an integer multiple of a universal value, independent of the charge on the nucleus."

At this point, Bohr made his most radical proposal. He conjectured that an electron cannot carry any amount of angular momentum. He assumed that the value of l must be an integer multiple of $h/2\pi$,

$$l = mvr = n\,(h/2\pi) \qquad n = 1, 2, 3, \ldots \tag{15}$$

and not any other value. This conjecture let Bohr determine the allowed electron orbits and their binding energies.

Each value of n (1, 2, 3, ...) was to correspond to a stationary state of the hydrogen atom. To determine the properties of these states, Bohr had to express W and r, given by equation 13, in terms of l. The details of this calculation are given in Example 11.6 but the results are

$$W = \frac{k_c^2\,e^4\,m}{2l^2} \quad \text{and} \quad r = \frac{l^2}{k_c\,e^2\,m} \tag{16}$$

By inserting the permitted values of l, as given by equation 15, into this purely classical result, Bohr found the possible binding energies W_n and radii r_n of the different quantum states of the hydrogen atom. He obtained

$$W_n = \frac{2\pi^2 m k_c^2 e^4}{n^2 h^2} \tag{17a}$$

$$r_n = \frac{n^2 h^2}{4\pi^2 m k_c e^2} \tag{17b}$$

where W_n and r_n are the binding energy and radius of the nth stationary state. Equations 17a and 17b describe the allowed electron orbits in a hydrogen atom, and thereby determine its spectrum. They summarize Bohr's model of the hydrogen atom. The rest of the story is an explanation of the significance of these results.

Bohr's key step was equation 15, the statement that angular momentum is quantized.* Everything else followed from an otherwise classical

---

*The orbital angular momentum of the electron in the ground state of the hydrogen atom is 0, not $h/2\pi$, as in Bohr's prescription.

model of the hydrogen atom. The physically realizable electron orbits are those in which l is an integer multiple of $h/2\pi$. The orbital angular momentum of each electron in an atom is a small integer multiple of this fundamental unit. The precise values of these orbital angular momenta play vital roles in the theory of atomic structure and determine the chemical properties of atoms.

The quantization of angular momentum is a general result applying to the motion of any object. Its consequences are ordinarily undetectable because the angular momentum of any macroscopic spinning object is enormous when expressed in units of $h/2\pi$: about 10^{30} for a spinning billiard ball and about 10^{10} for a wee whirling virus. The stationary states of things much larger than atoms are exceedingly closely spaced in energy and are virtually infinite in number. Quantum mechanics applies to the motions of all objects, but the description of large objects by classical mechanics is both simpler and adequate. This is an aspect of what Bohr called the *correspondence principle:* Quantum mechanics must reduce to classical mechanics when applied to motion in the everyday world. Its novel effects ordinarily pertain only to molecules, atoms, and subatomic phenomena. In other words, quantum mechanics is just as good as classical mechanics for everyday objects, but can describe the phenomena of the microworld as well.

EXAMPLE 11.6
Where Did Equation 16 Come From?

The purpose of this example is to show that Bohr, once he had made his radical hypotheses regarding stationary states and the quantization of angular momentum, proceeded deductively. Equation 16 is a consequence of classical mechanics. To see this, we recall equations 10.11:

$$W = \frac{m\,v^2}{2} \quad \text{and} \quad W = \frac{k_c\,e^2}{2\,r} \tag{18}$$

Solution

If the second equality of equations 18 is squared and divided by the first, the result is

$$W = \frac{k_c^2\,e^4}{2mv^2r^2}$$

The quantity appearing in the denominator is $2l^2/m$ according to equation 14. Making this substitution, we obtain the first of equations 16, an expression for W in terms of l in which r and v do not appear explicitly. The second equality of equation 18 may be written $r = k_c e^2/2W$. Replace W by its value in terms of l to obtain the desired expression for r in terms of l.   ∎

3. *The Ground State* "We see that the value of W is greatest if n has its smallest value 1. This case corresponds to the most stable state of the system, *i.e.,* to the state of highest binding energy, the breaking up of which requires the greatest amount of energy."

The binding energies of the allowed states of the hydrogen atom (or any other atom or molecule) may be put in descending order: $W_1 > W_2 > W_3 > \cdots$. The state with the highest binding energy is called the *ground state* of the atom. Under ordinary circumstances, atoms emit photons and soon find their way to their ground states. (At high temperatures, or when struck by an energetic particle, an atom may be excited to a state of higher energy. At even higher temperatures, electrons are knocked out of atoms and the atoms become ionized.)

A hydrogen atom is ordinarily found in its ground state, as described by equations 17a and 17b with n set equal to 1. The radius of the orbit of the ground state of hydrogen is the smallest of the allowed orbits. The radius of the electron's orbit in the ground state of hydrogen is given a special name: It is known as the *Bohr radius* r_B:

$$r_B = \frac{h^2}{4\pi^2 m k_c e^2} \simeq 5 \times 10^{-11} \text{ m} \tag{19}$$

By putting Planck's constant into the game via his quantization condition, Bohr computed the size of the hydrogen atom in terms of fundamental constants of physics: h, m, e, and k_c. He also obtained numerical results for the orbital frequency of the electron and its binding energy in the ground state:

$$f_1 \simeq 6 \times 10^{15} \text{ Hz} \qquad W_1 \simeq 13.6 \text{ eV} \tag{20}$$

of which he could proudly boast: "We see that these values are of the same order of magnitude as the linear dimensions of the atoms, the optical frequencies, and the ionization potentials."

Bohr knew that his quantum rules did not make a coherent theory, but he was confident that others would do that when the time was ripe.

When a hydrogen atom is struck by an energetic particle, it may gain energy and jump to an excited state with $n > 1$. The size of these states and their binding energies are determined by equations 17a and 17b and are shown in Table 11.1.

TABLE 11.1 The States of the Hydrogen Atom

Value of n	Orbital Radius	Binding Energy
1 (ground state)	a_1	W_1
2	$4a_1$	$W_1/4$
3	$9a_1$	$W_1/9$
...	...	...
n	$n^2 a_1$	W_1/n^2

Bohr believed that the integer n appearing in equations 17a and 17b completely characterized the possible states of the hydrogen atom. We shall learn in Section 11.4 that the quantum state of an atom is described by four *quantum numbers*, of which Bohr identified only one. Bohr's quantum number (somewhat redefined) determines the approximate binding energy of the electron in the hydrogen atom.

The most tightly bound state of the hydrogen atom, its ground state, corresponds to $n = 1$. States for which $n > 1$ are called *excited states*. When the electron in a hydrogen atom finds itself in an excited state, it quickly jumps to a smaller orbit, with a consequent decrease of its total energy. The excess energy appears as an emitted photon whose energy is determined by the initial and final values of n. For example, the electron may jump from the $n = 2$ state to the ground state and emit a photon of energy $W_1 - \frac{1}{4}W_1 = \frac{3}{4}W_1$. Conversely, an atomic electron can absorb a photon that has just the right energy to let the electron jump to an orbit with a larger value of n. Equation 17 determines the *energy levels* of the hydrogen atom and thereby the energies (or wavelengths) of the photons it can absorb or emit. With his expression for W_n (the negative of the total energy of the atom in its nth quantum state), Bohr computed the spectrum of hydrogen from purely theoretical arguments.

4. *The Spectrum of Hydrogen* "The amount of energy emitted by the passing of the system from a state corresponding to $n = n_1$ to one corresponding to $n = n_2$ is consequently:

$$W_{n_2} - W_{n_1} = \frac{2\pi^2 m k_c^2 e^4}{h^2}\left(\frac{1}{n_2^2} - \frac{1}{n_1^2}\right)$$

If we adopt Einstein's photon hypothesis and suppose that the amount of energy emitted is equal to hf, where f is the frequency of the radiation, we find:

$$W_{n_2} - W_{n_1} = hf$$

and from this:

$$f = f_0\left(\frac{1}{n_2^2} - \frac{1}{n_1^2}\right) \qquad f_0 = \frac{2\pi^2 m k_c^2 e^4}{h^3} \tag{21}$$

We see that this expression accounts for the law connecting the lines in the spectrum of hydrogen. If we put $n_2 = 2$ and let n_1 vary, we get the ordinary Balmer series. If we put $n_2 = 3$, we get the series in the infrared observed by Paschen. If we put $n_2 = 1$ and $n_2 = 4, 5, \ldots$ we get series respectively in the extreme ultraviolet and the extreme infrared, which are not observed but the existence of which may be expected."

The structure of the hydrogen spectrum, as schematically indicated in Figure 11.7, had already been found by experiment, as we described in Chapter 9 and in equations 9.16–9.20. Bohr deduced equation 21 from purely theoretical arguments, and it agreed precisely with what had been learned in the laboratory. The constant f_0, which had been experimentally measured to be

$$f_0 \simeq 3.29 \times 10^{15} \text{ Hz}$$

was determined by equation 21 in terms of fundamental constants of nature. Using Einstein's relation between the energy of a photon and its frequency, we may rewrite Bohr's result in terms of the photon energies that a hydrogen atom can emit or absorb:

$$E = \left(\frac{1}{n_2^2} - \frac{1}{n_1^2}\right) \text{Ry}, \qquad \text{where Ry} = hf_0 \qquad (22)$$

The symbol Ry in equation 22 signifies the *rydberg*, a unit of energy approximately equal to 13.6 eV. It is equal to Bohr's W_1 and to the energy needed to ionize a hydrogen atom in its ground state.

FIGURE 11.7
The circles indicate how the size of Bohr's stationary orbits increase with the quantum number n. Transitions to the ground state from excited states produce the Lyman series of ultraviolet lines. Transitions to the $n = 2$ state from more highly excited states produce the Balmer series. The binding energy of the nth stationary state is 13.6 eV divided by n^2, and the radius of the classical electron orbit is the Bohr radius multiplied by n^2.

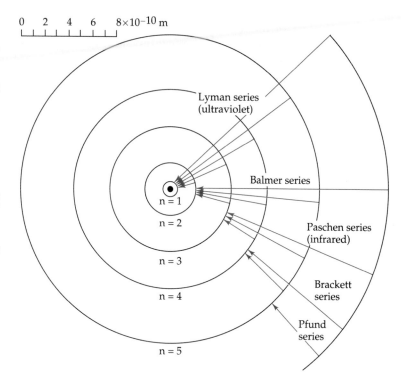

What Bohr Did and Didn't Do

The explanation of the hydrogen spectrum was Bohr's masterstroke: There had to be something to a theory, however vague its foundations, that had such quantitative power. Bohr analyzed hydrogen because it is the simplest of all atoms—and he got the right answer. His approach applied just as well to *hydrogenic atoms*, which are ions with just one electron, such as He^+ and Li^{++}. (In this connection, he showed that the mysterious Pickering series of lines, seen in certain stars, are the spectral lines of ionized helium).

Furthermore, the notion of stationary states of definite energies was applicable to any atom, not just hydrogen. Suppose that the binding energies of the quantum states of a multi-electron atom are E_1, E_2, E_3, $\ldots$ The atom can absorb a photon and jump to a state of lower binding energy, or it can emit a photon and fall to a state of higher binding energy. Any photon with which it interacts must have energy given by $E_i - E_j$ for some i and j. Every atom or molecule is thereby characterized by a set of fundamental frequencies, $f_i = E_i/h$. Any photon that can be absorbed or emitted by the system must have a frequency given by a difference between two of these frequencies. This property of atomic and molecular spectra, the Ritz combination principle (which we discussed in Chapter 9) was noticed before Bohr invented his theory—it is one more aspect of nature that Bohr explained from basic principles and brilliant insights.

Bohr strove to compute the spectra of atoms with more than one electron, but he failed. None of his tricks yielded the binding energy of the helium atom, let alone those with more than two electrons. The German mathematical physicist Arnold Sommerfeld wrote, in 1923, that "all attempts made hitherto to solve the problem of the neutral helium atom have proved to be unsuccessful." There were two good reasons Bohr's quantum rules could not be extended to deal with anything more complicated than hydrogen:

1. Bohr's analysis of hydrogen was prescriptive rather than deductive. Fiddling with classical theory, as Bohr unabashedly did, revealed an element of truth. However, the theory underlying his quantum rules—quantum mechanics, which we discuss in the remainder of this chapter—had not yet been discovered.

2. When the laws of quantum mechanics were put on a firm logical foundation, the calculation of the properties of an atom with many electrons remained a formidable mathematical problem. Even today, when the theory of atomic structure is complete and powerful super-computers are available, a precise calculation of the optical spectrum of the iron atom—a system with 26 mutually interacting electrons—is impossible to carry out. It's a miracle that scientists learned nature's ground rules. Knowing them doesn't answer all of the questions, but it was a good start.

Exercises

10. Show that equations 17 follow from equations 15 and 16.

11. A hydrogen atom in its ground state absorbs a photon and ends up in the $n = 2$ state. Calculate the wavelength of the photon in angstrom units. What is the energy of the photon in eV?

12. According to Bohr's model, what is the smallest energy in eV that may be absorbed by a hydrogen atom in its ground state? What is the largest energy the atom may absorb without being ionized?

13. What is the energy in eV of a photon emitted by a hydrogen atom in each of the following transitions? In each case, specify whether the photon corresponds to the ultraviolet, visible, or infrared portion of the electromagnetic spectrum.
    a. From the $n = 3$ state to the $n = 1$ state?
    b. From the $n = 4$ state to the $n = 2$ state?
    c. From the $n = 5$ state to the $n = 3$ state?

14. What is the radius of a hydrogen atom in the $n = 10$ state? What is its binding energy?

15. A *muon* is a singly charged elementary particle with a mass ~ 200 times that of the electron. It is bound to a proton forming a hydrogen-like atom in its ground state. What is its binding energy?

16. The spectrum of singly ionized helium is described by equation 22, but with Ry replaced by 4 Ry. The photon energies corresponding to $n_2 = 4$ and $n_1 > 4$ comprise the Pickering series. Show that alternate lines of the Pickering series correspond to the same photon energies as those of hydrogen's Balmer series.

17. Suppose a particle existed with the same mass as an electron but twice its charge. What would be its binding energy (in eV) to a proton in its ground state?

11.3 Quantum Mechanics Becomes a Theory

By the early 1920s, physicists were becoming accustomed to the dual nature of electromagnetic radiation. They recognized that light displays wave or particle properties under different circumstances. In 1923, a French graduate student named Louis de Broglie (1892–1987) added a surprising twist: What if all particles also had wave properties? He wrote in his doctoral dissertation: "Because photons have wave and particle characteristics, perhaps all forms of matter have wave as well as particle characteristics." De Broglie asserted that the association of particles and waves is universal. Under the right conditions, electrons (or any other particles) should behave as if they were waves. He conjectured that equation 8b, Einstein's relation between the momentum of a photon and its wavelength, applies equally well to a particle. The wavelength de Broglie associated with an electron of mass m and velocity v was

$$\lambda = h/p = h/mv \qquad (23)$$

where $p = mv$ is the momentum of a particle.

Einstein's immediate reaction to matter waves was encouraging: "I believe it is a first feeble ray of light on this worst of our physics enigmas." Feeble it may have been, but de Broglie's brainstorm could be tested by experiment. He predicted that wavelike electrons must be diffracted by the regular array of atoms in crystals.

It was already known that X rays behave like waves and could be made to diffract from crystals. According to de Broglie, electrons should behave the same way. C. J. Davisson and L.H. Germer were working at the Bell Telephone Laboratories when they discovered electron diffraction. In their experiment, a beam of electrons was directed toward the face of a nickel crystal. If electrons behaved like waves, they must be diffracted by the crystal just as X rays are. Davisson and Germer set out to prove this and found that electrons did indeed behave in the manner de Broglie expected (Figure 11.8). They proved that electrons diffract, and thereby showed their wavelike nature.

In particular, when the electron energy was 54 eV (corresponding to a speed of approximately 1.5 percent of the speed of light), Davisson and Germer observed a narrow beam emerging from the crystal at an angle of about 50°. From this result, and from the known spacing of atoms in a nickel crystal, they found that the de Broglie wavelength of the electrons was in perfect agreement with equation 23. "These results," they concluded, "are highly suggestive of the ideas underlying the theory of quantum mechanics, and we naturally inquire whether the wavelength which we thus associate with a beam of electrons is in fact the h/mv of L. de Broglie."

FIGURE 11.8
A beam of electrons, all with the same energy, is directed at a nickel crystal. Davisson and Germer observed diffracted beams emerging from the crystal at certain angles. They concluded that electrons behave like waves and confirmed de Broglie's formula for their wavelength.

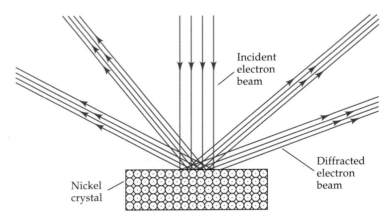

EXAMPLE 11.7
De Broglie Waves

What is the de Broglie wavelength of an electron with kinetic energy of 54 eV (that is, 9×10^{-18} J)?

Solution

The kinetic energy of the electron is $\frac{1}{2}mv^2$. Setting this equal to 9×10^{-18} J and using $m \simeq 9 \times 10^{-31}$ kg, we find $v^2 \simeq 2 \times 10^{13}$ (m/s)^2 or $v \simeq 4.4 \times 10^6$ m/s. Thus, the momentum of the electron is $p = mv \simeq 4 \times 10^{-24}$ kg m/s. Using $h \simeq 6.6 \times 10^{-34}$ kg m^2/s from equation 4, we find that the de Broglie wavelength of a 54 = 1eV electron is $\lambda \simeq 1.6 \times 10^{-10}$ m.   ∎

Meanwhile, in an incredible spurt of discovery from 1926 to 1927, the theory underlying Bohr's quantum rules was born. Quantum mechanics, as the theory is now known, seemed at first to be a double birth. The German physicist Werner Heisenberg (1901–1976) invented what was called "matrix mechanics" at about the same time that the Austrian physicist Erwin Schrödinger (1887–1961) devised "wave mechanics." Because these were soon recognized to be two equivalent mathematical approaches to the same underlying theory, we shall focus on Schrödinger's approach. However, it was Heisenberg who first pointed out that the measurement of one quantity may disturb the system under observation, and thereby change the values of previously known quantities. That is, he showed that a certain amount of uncertainty is built into the laws of quantum mechanics.

Heisenberg's Uncertainty Relations

To measure a quantity, we must "look at it." That is, we must examine it with an external probe of some sort. The uncertainty implicit in quantum mechanics follows from the wave particle duality. Whatever probe is used to measure the position of an object must itself have both wave and particle properties.

Suppose we try to measure both the position and speed of a tiny object with a powerful microscope. The probe consists of the light that illuminates the object. The wavelike nature of light limits the accuracy with which the position of the object may be measured to a precision no better than the size of the probe, which in this case is the wavelength of the light being used. (This is another way of saying that objects smaller than 1 wavelength leave no shadows because of diffraction.) Thus, the uncertainty Δx in the position of the object is:

$$\Delta x \geq \lambda$$

No problem—in principle, we can use light of arbitrarily small wavelength, thereby making Δx as small as we wish. There is no intrinsic limit to how well the position of an object may be measured.

However, light behaves like a particle when it interacts with what is being observed. To be seen, the object must be struck by a photon, which transfers its momentum h/λ to the object. The direction of the initial photon is unknown, so the measurement changes the momentum $p = mv$ of the object by an unknown amount. The momentum uncertainty Δp produced by the observation depends on the wavelength of the light. If λ is small, the photon has a large momentum and gives a large kick to the object. If λ is large, it produces a smaller momentum uncertainty. When geometric factors are taken into account, we find:

$$\Delta p \geq \hbar/2\lambda, \quad \text{where} \quad \hbar = h/2\pi \simeq 1.05 \times 10^{-34} \text{ J s} = 6.6 \times 10^{-16} \text{ eV s} \quad (24)$$

Because the combination $h/2\pi$ appears so often, physicists abbreviate it as $\hbar$ (h bar). Multiplying together the uncertainties in position and momentum, we obtain the quantum-mechanical constraint on what can be known about the object. This constraint is the most familiar instance of the Heisenberg uncertainty relations:

$$\Delta p \, \Delta x \geq \hbar/2 \quad \text{or} \quad \Delta v \, \Delta x \geq \hbar/2m \qquad (25)$$

According to equation 25, the better the position is measured, the larger is the uncertainty of the momentum. Conversely, the better the momentum is measured, the larger is the uncertainty of the position. The position and the momentum (or velocity) of a body cannot both be determined with arbitrary precision.

Our derivation of the uncertainty relation used light to make the measurement. Can we do better with another probe? Can an electron microscope make more delicate measurements and evade Heisenberg's result? But electrons also have wavelengths! The same argument, applied to any conceivable subatomic probe, yields the same constraint on what can be measured and what cannot. The uncertainty relation follows from the universal wave-particle duality. We could beat it with particles that are really particles or waves that are really waves, but quantum theory forbids such things. There is no escape from quantum uncertainty. However, Example 11.8 illustrates why the uncertainty relation (and, hence, quantum mechanics) is relevant to atoms and molecules but not to larger things like apples and oranges.

EXAMPLE 11.8
The Uncertainty
Relations

The position of an object of mass m is known with a precision of $\pm\Delta x$. Determine the minimum uncertainty in its velocity for the following cases:

  a. A bacterium with mass $m = 5 \times 10^{-15}$ kg is located to a precision equal to its own size of 1 micron; that is, $\Delta x = 10^{-6}$ m.
  b. A hydrogen atom ($m \simeq 1.7 \times 10^{-27}$ kg), where $\Delta x = 10^{-6}$ m.
  c. An electron ($m \simeq 9.1 \times 10^{-31}$ kg), is located to a precision of an atomic diameter, $\Delta x = 10^{-10}$ m.

Solution

  a. From equations 24 and 25, we find that $\Delta v \simeq 5.3 \times 10^{-35}/(m\,\Delta x)$, where the answer is in meters per second if mass in measured in kilograms and distance in meters. Thus, we find $\Delta v \simeq 10^{-14}$ m/s. This is a very tiny uncertainty in velocity. Quantum mechanics is not needed to describe the motions of things as large as microbes.
  b. Using the same procedure as in part a, we find $\Delta v \simeq 3$ cm/s. As a result of this velocity uncertainty, the uncertainty in the position of the atom grows rapidly. One second after the observation is made, the atom may have moved one way or another by a few centimeters.
  c. Quantum mechanics becomes very important in this case. The minimum velocity uncertainty of the electron is $\Delta v \simeq 6 \times 10^5$ m/s. ∎

The central issue in classical mechanics is the determination of the trajectory of a moving object. If the position and velocity of the object are known at $t = 0$, what are its position and velocity at later times? However, the uncertainty relation says that the precise position and velocity of an object cannot be known at the same time. Because the initial conditions cannot be known, the trajectory of an object (such as the orbit of an electron) cannot be determined. In fact, quantum mechanics tells us that there is no such thing as a precise trajectory. Although trajectories are undeniably useful to describe everyday phenomena like flying insects and falling bricks, the concept of an absolutely well-defined trajectory does not exist in quantum theory. The smaller an object is, the less meaningful it is to say that it travels in a definite path through space.

If we must abandon the notion of trajectory, how do we describe the motion of a small object like an electron? The mathematical construct that incorporates everything that can be known about a particle is called its *wave function*. In quantum mechanics, we cannot speak of the position of an object as a function of time. Instead, we speak of the evolution of the wave function with time.

EXAMPLE 11.9
Quantum Baseball

In an imaginary universe, $\hbar$ is much larger than it is in ours. How large must it be for a pitcher to be unable to throw a strike with confidence? The mass of the ball is 150 g, the pitcher to batter distance is 20 m, the radius of the strike zone is 0.3 m, and the speed of the pitch is 40 m/s.

Solution

If a strike is to be thrown, two conditions must be met. The uncertainty in the ball's position when pitched must be less than the radius of the strike zone:

$$\Delta x < 0.3 \text{ m}$$

The uncertainty relation tells us that the error in the transverse velocity of the baseball cannot be less than $\Delta v = \hbar/(2m\Delta x)$. The time of flight of the ball from the mound to the batter is $t = 0.5$ s. Because of its velocity uncertainty, the ball will swerve from its intended location in the strike zone by $t\Delta v$. If the pitch is to be a strike,

$$\hbar t/(2m\Delta x) < 0.3 \text{ m}$$

Since both of these inequalities must be satisfied to throw a strike, their product must also be satisfied:

$$\hbar t/2m < 0.09 \text{ m}^2, \quad \text{or} \quad \hbar < (0.09)(0.30)/(0.5) = 5.4 \times 10^{-2} \text{ J s}$$

In our universe, where $\hbar \simeq 10^{-34}$ J s, America's pastime is as classical as could be. Planck's constant would have to be more than 30 powers of 10 larger than it is for quantum mechanics to spoil the game. ∎

The Quantum-
Mechanical
Wave Function

A wave, according to classical physics, is a disturbance propagating through a medium. Ocean waves are fluid motions that result in a pattern of fluctuations in the height of the ocean surface. The complex and changing pattern is described by a function $h(x, y, t)$ whose value specifies the departure of the water surface from sea level at the point (x, y) and at the time t. The laws of fluid mechanics let us determine the future shape of the wave from its behavior in the past.

In the same fashion, the de Broglie (or matter) waves associated with an electron are decribed by a quantum-mechanical wave function $\Psi(\mathbf{r}, t)$. This function differs in several important respects from the functions describing classical waves.

- The function h describing a water wave can be positive (a crest) or negative (a trough). However, the quantum-mechanical wave function is usually complex. That is, Ψ assigns a *complex number* to each point in space and time. Complex numbers were first introduced by mathematicians as solutions to such equations as $x^2 + 1 = 0$. They are a convenient mathematical artifice (but not a necessity) for engineers and classical physicists, but they are intrinsic to quantum theory. Even though the results of measurements are always real numbers (such as angles of deflection or dial readings), the wave functions of actual objects, and the mathematical procedures leading to observable results, almost always involve complex numbers.

- The wave function $\Psi(\mathbf{r}, t)$ of a particle does not represent a disturbance propagating through a medium. Instead, it expresses all that can be known about the motion of the particle and all that may be said about its future behavior. That is, Ψ describes the result to any measurement that may be performed on the particle, such as finding its position, velocity, or kinetic energy. In classical mechanics, the trajectory of the particle is specified by a function $\mathbf{r}(t)$ giving the position of the particle at all times. However, in quantum mechanics, there is no trajectory, only a wave function. Furthermore, the results of observations of the particle are not necessarily determined by its wave function. In many cases, the most we may know are the likelihoods (or probabilities) of different experimental outcomes.

The wave function Ψ of a particle has a value at each point in space. Somehow, it contains the information of where the particle is likely to be found. The German-born physicist Max Born (1882–1970), the grandfather of Olivia Newton-John, explained how it does this. Ψ may be a complex number, but the quantity $|\Psi(x)|^2$, the square of the magnitude of the wave function at the point x, is a nonnegative number. Born interpreted this quantity as a *probability density*. It is a measure of the likelihood that the object is at the point x.

Let Ψ be the wave function of an electron and imagine a tiny volume v centered about the point x and within which Ψ is more or less constant. The quantity

$$|\Psi(x)|^2 \, v$$

is the probability that the electron is found in the volume v. A larger volume V, in which Ψ does vary, is broken up into many tiny regions. The probabilities that the electron is in each one are added together to give the probability of finding the electron in V. In short, the electron is likely to be found where $|\Psi|^2$ is large, and is unlikely to be found where it is small. Born's interpretation of the square of the absolute value of the wave function as a probability density made good sense: "We had never dreamt that it could be otherwise," said Bohr.

The stationary states of a hydrogen atom have definite energies and are described by time-independent wave functions. An isolated hydrogen atom in its ground state remains in that state indefinitely. It makes no sense to think of the electron "moving in an orbit" because nothing is changing inside the atom. The wave function of the electron Ψ is centered about the nucleus of the atom, as in Figure 11.9. The electron is unlikely to be found far from the proton, whose electrical attraction binds the electron and confines its wave function. Thus, Ψ is large near the proton and rapidly falls off at distances greater than the Bohr radius. The wave function does not change with time—the electron is in a stationary state, and nothing much happens to it unless the atom gets hit by something else. $|\Psi(x)|^2$ measures how likely it is for the electron to be found at point x.

FIGURE 11.9
The wave function of the hydrogen atom in its ground state. $\Psi(r)$ is shown as a function of r, the distance from the electron to the proton, in units of Bohr radii. It has a maximum value at $r = 0$ and decreases as r increases. At a distance of $r = 3r_B$, which is not shown in the figure, it has fallen to 5 percent of its central value.

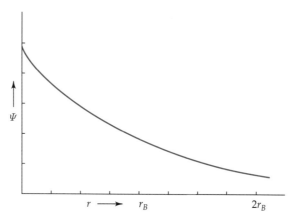

Figure 11.10 (on page 480) shows the likelihood for the electron in the ground state of the hydrogen atom to be in various regions. The chances are about 1 in 3 that it will be found within a Bohr radius of the proton, and about 99 in 100 that it lies closer than 4 Bohr radii. There is a chance—tinier than you can imagine—that the electron is a foot away from the proton! But it makes no sense to think about the electron moving in any particular orbit about the proton.

We may measure the position of the electron in the atom with arbitrary precision. Sometimes it will be here, sometimes there. Most of the time it will turn out to have been within a few Bohr radii of the proton. But the measurement disturbs the atom and changes its wave function. As a result of the measurement, the electron may jump to an excited state: Its wave function may change.

We may measure just as well the momentum of the electron in a hydrogen atom with arbitrary precision. If the atom as a whole is at rest, the average momentum of the electron is zero. Its uncertainty in

FIGURE 11.10
This figure shows the probabilities of finding the electron at various distances from the proton in the ground state of the hydrogen atom. The probability that the electron will be found closer to the proton than 1 Bohr radius is 32 percent. The probability that it will be further than 3 Bohr radii is only 6 percent.

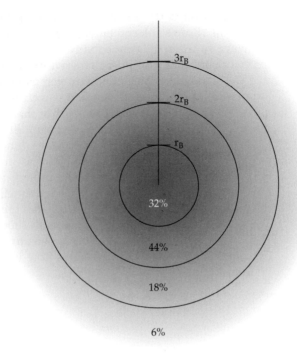

position is $\Delta x \simeq r_{B}$, so its uncertainty in momentum is $\Delta p \simeq \hbar/2r_{B}$. Each measurement will give a different answer, but most of the results will be comparable to Δp. Once again, a precise measurement of p would change the wave function and possibly ionize the atom.

The Superposition Principle

A peculiarly shaped wave function of a particle is shown in Figure 11.11. It is small everywhere except near the points $x = 5$ and $x = -5$, where the probability distribution is concentrated. A measurement of the position of the particle will give a result that is either near $x = 5$ or near $x = -5$. The likelihood of either result is 50 percent. The particle is said to be in a superposition of two states at two different locations, but its location cannot be determined until it is measured.

FIGURE 11.11
This wave function is large near $x = 5$ and $x = -5$. The particle has a 50 percent chance of being found near either point.

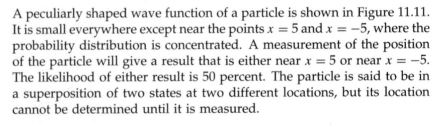
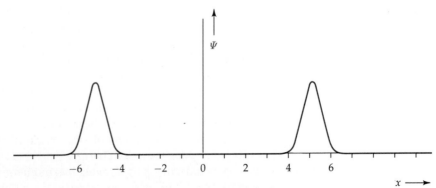

A similar situation pertains to the two-slit experiment described in Section 11.1. Since electrons are just as wavelike as photons, this experiment could be performed with photons or with electrons. Suppose that electrons are sent, one at a time, through two closely spaced slits, A and B. The wave function of each electron at the screen consists of two parts: Ψ_A describes the portion of its wave function that passes through slit A, and Ψ_B describes the portion that passes through slit B. The wave function of the electron is the sum of its parts:

$$\Psi = \Psi_A + \Psi_B$$

At some points on the screen, Ψ_A and Ψ_B have the same sign and add up constructively. $|\Psi|^2$ is large, so these are likely destinations for the electron. At other points on the screen, Ψ_A and Ψ_B have opposite signs and add together destructively. $|\Psi|^2$ is small, so these are unlikely destinations. In time, as more and more electrons fall on the screen, an interference pattern of bright and dark lines builds up. The pattern can literally be seen if the electrons impinge on a phosphor screen making light. It cannot be said of any particular electron that it passed through slit A or slit B. The essence of quantum mechanics lies in the statement that the wave function of each electron, as it arrives at the screen, is the sum of two terms: the wave coming through slit A and the wave coming through slit B.

What if a particle detector is positioned to determine through which slit each electron passes? This act of measurement changes the wave function of the electron and necessarily destroys the knowledge of its sign (or more accurately, its complex phase). We can no longer know whether the two waves, at a point on the screen, will add up or cancel out. In this case, the electrons that fall on the screen will not produce an interference pattern. This is another instance of how the act of measurement can change the wave function of what is being observed.

What about Particles That Move?

Thus far, we have not shown how quantum mechanics describes motion. Consider the case of an object in uniform motion. No forces act on a body of mass m moving along the x-axis with velocity v. If the object starts at the origin at $t = 0$, its classical trajectory is $x(t) = vt$. However, we cannot specify both the position and velocity at $t = 0$. Suppose that its probability distribution $|\Psi(x, 0)|^2$. at $t = 0$ resembles the drawing in Figure 11.12a (on page 482). It is spread out over a small distance Δx, corresponding to the uncertainty in the initial position of the body. According to the uncertainty relation, equation 25, the velocity of the body is uncertain by at least $\Delta v = \hbar/(2m\,\Delta x)$. Thus, the body is not moving at velocity v, but at a velocity that is most likely somewhere between $v + \Delta v$ and $v - \Delta v$. Because its velocity is uncertain, the uncertainty in its position grows larger as time passes. Its wave function spreads out, as indicated in the subsequent parts of Figure 11.12.

FIGURE 11.12
Part *a* shows the ini-
tial probability distribution of a free electron
moving to the right. Its
uncertainty in position
is Δx. At subsequent
times, shown in parts *b*,
c, and *d*, the wave function moves uniformly to
the right, but it broadens
as it moves.

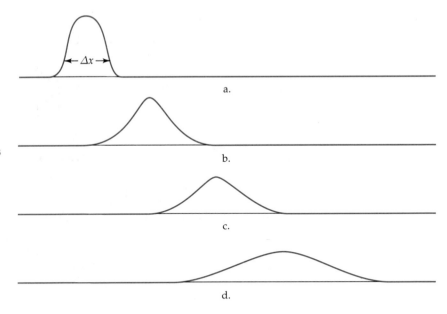

a.

b.

c.

d.

If the object we have been considering is macroscopic in size, the quantum-mechanical spreading of its wave function is such a small effect as to be entirely negligible. The wave function of the microbe discussed in Example 11.8 starts off with a position uncertainty of 10^{-6} m and an obligatory velocity uncertainty of 10^{-14} m/s. Its wave function widens, but it would take several years to double in size. The concept of a precise trajectory makes sense for macroscopic objects, where quantum mechanics reduces to the commonsense laws of classical mechanics.

Schrödinger's Cat

Quantum mechanics normally comes into play only for tiny things like electrons or atoms. However, it also applies to radioactive decay. The radioactive isotope ^{117}Te has a half-life of one hour. Each atom has a 50 percent chance of decaying in an hour. It is impossible to predict when the atom will decay. In fact, there is no measurable difference between an atom that is about to decay and one that will live until doomsday.

Let Ψ_1 be the wave function of a ^{117}Te atom and Ψ_2 be the wave function of its decay products. Suppose we put one such atom in a sealed box. Initially, its wave function is Ψ_1. A week later, its wave function is almost certainly Ψ_2. At in-between times, the wave function is a linear superposition of Ψ_1 and Ψ_2. But when we open the box after one hour, we will find either an intact atom or a decayed atom. The act of performing the measurement will have altered the wave function so as to force it to choose one alternative or the other. In 1935, Erwin Schrödinger described how the essentially quantum-mechanical uncertainty of radioactive nuclei can be thrust upon an unsuspecting cat:

© 1993 by Sidney Harris.

> One can even construct quite burlesque cases. A cat is shut up in a steel container, together with the following diabolical apparatus (which one must keep out of the direct clutches of the cat): In a Geiger tube there is a tiny mass of radioactive substance, so little that in the course of an hour perhaps one atom of it disintegrates, but also with equal probability not even one; if it does happen, the counter responds and through a relay activates a hammer that shatters a little flask of prussic acid. If one has left this entire system to itself for an hour, then one will say to himself that the cat is still living, if in that time no atom has disintegrated. The first atomic disintegration would have poisoned it. The Ψ-function of the entire system would express this situation by having the living and the dead cat mixed or smeared out (pardon the expression) in equal parts. An uncertainty originally restricted to the atomic domain has been tranformed into a macroscopic uncertainty, which can then be resolved through direct observation.

Is the system really described by a superposition of two wave functions, one of a live cat, the other of a corpse? Does the observation force the wave function into one or the other of these states? What does the cat think of all this? Are there parallel universes with cats alive and dead? These questions are nettlesome and often discussed. There are almost as many answers as there are licensed quantum mechanics. What I do know is that quantum mechanics lets me compute the probabilities for the outcomes of experiments, including Schrödinger's felinocidal fantasy.

The Schrödinger Equation

Schrödinger's contributions to quantum mechanics were not limited to killing imaginary cats. He also discovered an equation that describes all

Erwin Schrödinger.
Source: The Bettmann
Archive.

the phenomena we have been discussing: stationary states, de Broglie waves, conservation of probability, the uncertainty relations, spreading wave packets, and almost everything else.

The first of Schrödinger's six great papers establishing the theory once known as wave mechanics was written during a two-week vacation at Christmas, 1925. According to a recent biography,*

> Erwin wrote to "an old girl friend in Vienna" to join him in Arosa, while Anny [his wife] remained in Zürich. Efforts to establish the identity of this woman have so far been unsuccessful. Like the dark lady who inspired Shakespeare's sonnets, the lady of Arosa may remain forever mysterious. We know that she was not Lotte or Irene [or] Felicie. Whoever may have been his inspiration, the increase in Erwin's powers was dramatic, and he began a twelve-month period of sustained creative activity that is without parallel in the history of physics.

Schrödinger's first paper on wave mechanics, written from his love nest in the Swiss Alps, begins with a bang:

> In this communication, I wish first to show in the simplest case of the hydrogen atom that [Bohr's] rules for quantization can be replaced by another requirement, in which mention of "whole numbers" no longer occurs. Instead the integers occur in the same natural way as the integers specifying the number of nodes in a vibrating string. The new conception can be generalized, and I believe it touches the deepest meaning of the quantum rules.

Schrödinger's triumph was to find the equation for the wave function of an object moving in response to a force. Applying his equation to an electron subject to the electrical attraction of a proton, he deduced Bohr's quantum rules for the hydrogen atom.

Schrödinger started from a classical equation long known to describe sound waves or the vibrations of a metal rod. Wherever the wavelength appeared in the equation, he replaced it by de Broglie's expression h/mv. After several logical twists and turns, he obtained what has become known as the *Schrödinger equation*:

$$\nabla^2 \Psi + \frac{2m}{\hbar^2}(E - V)\,\Psi = 0 \qquad (26)$$

where E is the total energy of the electron, V is its potential energy, and m is the electron mass. ∇^2 stands for a mathematical operation that indicates the curvature of the wave function.

Equation 26 has two kinds of solutions. E can have any positive value. Any such solution corresponds to an electron not caught by the proton—after all, the electron could be in the room next door. A negative value of E corresponds to a bound state of the electron with binding energy $W = -E$. Schrödinger showed that his equation had solutions only for certain negative values of E—namely, the values given by

---

*Walter Moore, *Schrödinger, Life and Thought* (Cambridge: Cambridge University Press, 1989), 195.

equation 17, which Bohr found: $E = -W_n$. These solutions to Schrödinger's equation are the electron wave functions. Bohr's magic gave way to Schrödinger's method.

Schrödinger obtained Bohr's result for the spectrum of hydrogen without making further hypotheses. He computed the explicit wave functions for each of the stationary states and confirmed Bohr's guess that the angular momentum is quantized: It was no longer an ad hoc hypothesis, but a consequence of equation 26. For the $n = 1$ ground state, Schrödinger found that $l = 0$ (rather than Bohr's result, $l = 1$). For the $n = 2$ state, he found two possible values of the angular momentum: $l = 0$ or 1. In general, in the nth quantum state, there is a solution to the equation for any non-negative integer value of l less than n.

Heisenberg developed an equally powerful alternative version of quantum mechanics. His theory, which we shall not describe, involved matrices rather than a differential equation. Both Schrödinger and Heisenberg, from different mathematical starting points, created a consistent theory of quantum mechanics that encompassed Bohr's intuitive system of quantum rules.

The Schrödinger equation applies to many other situations in physics. His second paper presented quantum-mechanical solutions to an assortment of classical problems, such as the vibrations and rotations of a molecule. In later papers, he explored the effects of electric fields on atomic spectra and examined various scattering problems. There seemed to be no limit to the versatility of his new equation and to the power of wave mechanics.

Quantum Mechanical Tunneling

Some things possible classically are impossible quantum mechanically. For example, because of the uncertainty relations, we cannot determine the initial position and velocity of a particle. We cannot define a precise orbit for the motion of an atomic electron. The converse is also true: many things that are impossible classically can (and do) happen quantum mechanically by a process called *tunneling*.

Suppose that the process $A \rightarrow B$ is exothermic; that is, it releases energy. A may be an upright cereal box and B may be the same cereal box that has fallen over. Or, A may be this book, and B may be this book as it suddenly and spontaneously begins to burn. Or, A may be a proton and a deuteron in the Sun, and B the He^3 nucleus resulting from their fusion. In all these cases, energy is released by the action. However, there is an obstacle in the path from A to B. The box must tilt before it falls, and it takes energy to do this. The fire must be ignited. The proton does not have enough energy to penetrate the repulsive electric barrier and come close enough to the deuteron to fuse with it. If any of these processes are to take place spontaneously, the law of energy conservation must be momentarily repealed! The box will not fall over in any imaginable time, and the book is unlikely to burn of its own accord. Tunneling is a quantum-mechanical phenomenon and does not happen in the everyday world. However, the classically forbidden fusion process does take place in stars, for otherwise they could not shine.

To illustrate quantum-mechanical tunneling, consider a tiny ball placed in a well (Figure 11.13). If the ball is at rest, it cannot climb over the barrier and escape. According to quantum mechanics, it does escape! It tunnels through the barrier, converts its gravitational potential energy to kinetic energy, and leaves the scene of its crime to roll happily away. Schrödinger's equation lets us calculate the wave function Ψ of the ball in terms of its mass and the shape of the well. Most of the wave function lies inside the well, but a small tail of Ψ extends into the barrier and beyond. There is a small probability of the ball's being found outside the well. For any macroscopic situation, the probability of escape is absurdly small. However, an α particle in a nucleus is analogous to a ball in a well. Although the α particle does not have enough energy to penetrate the nucleus, it can tunnel through the nucleus if the overall reaction is exothermic. Quantum-mechanical tunneling is the mechanism behind α decay and underlies the radioactivity of uranium and many of its daughter nuclei. The half-life for α decay is billions of years for uranium, days for radon, and mere seconds for astatine.

FIGURE 11.13
A ball bearing is trapped in a well between two humps. Its wave function is concentrated within the well, but some of it pokes outside. Thus, there is a small chance that the ball may be found outside of the well. That is, it can quantum-mechanically tunnel through the well to appear outside the barrier.

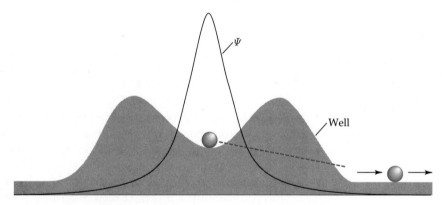

What Does It All Mean?

In the fall of 1926, Schrödinger presented his work to the great master, Bohr, in Copenhagen. Heisenberg, who was also present, recounted their fiery interaction:

> *Schrödinger:* It is claimed that the electron in a stationary state of an atom first revolves in some sort of an orbit without radiating. There's no explanation given of why it should not radiate; according to Maxwell theory, it must radiate. Then the electron jumps from this orbit to another one and thereby radiates. Does this transition occur gradually or suddenly? If it occurs gradually, it's not comprehensible how this can give sharp frequencies for spectral lines. If the transition occurs suddenly, one must ask how the electron moves in the jump. Why doesn't it emit a continuous spectrum, as electromagnetic theory would require? And what laws determine its

motion in the jump? Well, the whole idea of quantum jumps must simply be nonsense.

Bohr: Yes, you are right in what you say, but that does not prove that there are no quantum jumps. It only proves that we can't visualize them—that the pictorial concepts we use to describe the events of everyday life do not suffice to represent also the process of a quantum jump. That is not so surprising when one considers that the processes with which we are concerned here cannot be the subject of direct experience.... .

Schrödinger: If we are still going to have to put up with these damn quantum jumps, I am sorry that I ever had anything to do with quantum theory.

Bohr: But the rest of us are very thankful for it, and your wave mechanics in its mathematical clarity and simplicity is a gigantic progress over the previous form of quantum mechanics.

The discussion went on and on until poor Erwin developed a fever and had to be nursed back to health by the master's wife. Bohr had the best of the argument, and Schrödinger never fully accepted the quantum theory he helped to create. Nothing in his later works rivals his one great burst of creativity. He tried to understand the meaning of quantum mechanics, life, and the universe.* Schrödinger chased after women and the will-o'-the-wisp unified field theory that fells so many physicists, but he succeeded only in the former endeavor.

Exercises

18. What is the de Broglie wavelength of the following:
    a. An electron with a kinetic energy of 10^4 eV, such as may be found inside a TV tube?
    b. A bacterium with a mass of 10^{-15} kg moving at a velocity of 1 micron per second?
    c. An air molecule in this room moving at 300 m/s?

19. Suppose the Davisson-Germer experiment were performed with a beam of 81 eV electrons. The diffraction pattern coincides with that seen in the scattering of X rays with wavelength equal to the de Broglie wavelength of the electrons.
    a. What is the de Broglie wavelength of an 81 eV electron?
    b. What energy X rays (in eV) produces the same diffraction pattern?
    c. What energy protons (in eV) would produce a similar pattern? (The proton mass is 1.67×10^{-27} kg.)

20. The position of a free electron is known to a precision of $\Delta x = 0.5$ mm. What is the uncertainty in its velocity?

---

*Schrödinger's short monograph, *What Is Life?* inspired a generation of molecular biologists.

21. The location of a hydrogen atom is known to a precision of $\Delta x = 1$ cm. What is the uncertainty of its velocity? If no forces act on the atom, what is its position uncertainty one hour later?

22. An excited atom emits a 3.3 eV photon and falls to a lower energy state. The energy of the photon is the difference in energy between the two states. However, the time at which the photon is emitted is uncertain by an amount called the "lifetime of the excited state," Δt. Quantum mechanics relates Δt to the uncertainty ΔE of the emitted photon: $\Delta t \Delta E > \hbar/2$. Suppose that the lifetime of the state is 10^{-12} s. What is the energy uncertainty of the emitted photon in eV? What is the width in angstroms of the spectral line of this photon?

11.4 The Structure of Atoms

Schrödinger's equation solved the mystery of the hydrogen atom and was successfully applied to the behavior of small molecules. However, it did not seem to apply to the structure of atoms containing more than one electron. Two other concepts remained to be discovered: the *spin* of the electron and the *Pauli exclusion principle*. When these two concepts were put together with Schrödinger's equation, all the secrets of atomic structure came tumbling out of the theory. In particular, the successes of the periodic table were explained in terms of the properties of electrons and atomic nuclei.

Quantum Numbers

The measurable properties of a quantum-mechanical system are called *observables*. Consider the possible wave functions of the hydrogen atom. Some observables can be measured without disturbing the atom. Others (like the position or momentum of the electron) cannot. To classify the wave functions and understand their properties, we must identify all of their simultaneously knowable attributes, or *quantum numbers*. One of these is the binding energy of the atom, as specified by Bohr:

$$W_n = \mathrm{Ry}/n^2 \qquad n = 1, 2, 3 \ldots \qquad \text{where } \mathrm{Ry} \simeq 13.6 \text{ eV}$$

The integer n is one of four measurable quantities that characterize the hydrogen atom. Because it determines the energy of the atom, it is called the *principal quantum number*.

What other quantum numbers of the hydrogen atom can be measured without changing its wave function? The second quantum number of a hydrogen atom is the magnitude l of its angular momentum. A measurement of l can be carried out without affecting the atom: W and l, unlike position and momentum, are compatible observables. Bohr was correct in his belief that angular momentum plays a special role in atomic structure.

*Angular
Momentum*

Something more about a hydrogen atom wave function may be specified besides its principal quantum number n and the magnitude of its angular momentum l. The third quantum number involves the direction of the angular momentum vector $\mathbf{l}$, which points along a direction in space and has three Cartesian components: l_x, l_y, and l_z. Each of these components is an observable, but no two components of $\mathbf{l}$ may be measured at the same time. For example, a measurement of l_x destroys all knowledge of l_z. The quantum numbers characterizing the states of the hydrogen atom consist of (1) the principal quantum number n, (2) the magnitude of its angular momentum l, and (3) the component of its angular momentum in any one direction, say l_z.

As Bohr guessed and Schrödinger showed, the angular momentum of any object is quantized. The magnitude of l is restricted to be an integer multiple of $\hbar$. The same is true of its components. Not only is the magnitude of the angular momentum quantized, but so also is the direction in which it can point. If the magnitude of an electron's angular momentum is l, its component l_z can take on any any one of $2l + 1$ integer values ranging from $-l$ to $+l$. The value of l_z is the third quantum number in the Schrödinger picture of the hydrogen atom. All three of the quantum numbers: n, l, and l_z are integers.

The $n = 1$ ground state of the hydrogen atom is unique because its angular momentum is 0. There is only one such wave function, and it corresponds to

$$n = 1, \qquad l = 0, \qquad l_z = 0$$

The principal quantum number alone does not specify the wave function if $n > 1$. For the case of $n = 2$, the angular momentum can be either $l = 0$ or $l = 1$. However, n and l alone do not quite pin down the electron wave function. Although there is only one $n = 2$ wave function with $l = 0$, there are three different wave functions corresponding to $l = 1$, depending on the orientation of the angular momentum. These three states correspond to $l_z = -1$, $l_z = 0$, and $l_z = +1$. The pattern of quantum numbers of the $n = 2$ wave functions is

$$n = 2 \quad \begin{cases} l = 0 & l_z = 0 \\ l = 1 & \begin{cases} l_z = +1 \\ l_z = 0 \\ l_z = -1 \end{cases} \end{cases}$$

Each of these wave functions corresponds to a state with the same binding energy. A hydrogen atom in any one of them may "jump" to the ground state and emit a photon of energy 3 Ry/4. The four quantum states with $n = 2$, and their spectral lines, are said to be *degenerate*. However, the degeneracy is removed if the atom is placed in an electric or magnetic field pointing in the z direction. The binding energy of the electron is affected by the field in a way that depends on which way the angular momentum points. What was a single line in the spectrum of the atom becomes several nearby but separate lines.

Atomic physicists denote electronic wave functions with a curious notation by which s stands for $l = 0$, p stands for $l = 1$, d stands for $l = 2$, and f stands for $l = 3$. The letters once described spectral lines: They derive from the words *sharp, principal, diffuse,* and *fine.* The next letters in the series are g, h, and so on. Thus, the ground state wave function* is denoted by $1s$. The $n = 2$ states of the hydrogen comprise one $2s$ state and three different $2p$ states. The $n = 3$ wave functions comprise one $3s$ state, a triplet of $3p$ states, and a quintuplet of $3d$ states. A similar notation is used to denote the electron configurations in atoms with many electrons.

The Pauli Principle, Electron Spin, and the Periodic Table

Two essential ingredients were missing from the theory of atomic structure based on Bohr's rules and Schrödinger's equation. Even before Bohr invented his quantum rules, model builders sought atomic structures in which the electrons distribute themselves in concentric shells so as to reproduce the periodicity of Mendeleev's table. The inertness of the rare gases corresponds to atoms with particularly stable electronic configurations. However, the Schrödinger equation allows all of the electrons in an atom to fall into the lowest quantum state. If this were the whole story, there would be no reason for electrons to distribute themselves in concentric shells.

Wolfgang Pauli put forward an ad hoc solution to the puzzle. The *Pauli exclusion principle* states that no two electrons can occupy the same quantum state at the same time; that is, they cannot have the same set of values for their quantum numbers. At the same time, Pauli introduced a fourth quantum number to supplement n, l, and l_z. Pauli assumed that every atomic electron has a *two-valued attribute* intrinsic to itself. Let us tentatively denote these two somehow different forms of the electron as $e\uparrow$ and $e\downarrow$. Pauli wrote: "I cannot give a more precise reason for this rule."

The explanation was found in 1925 by two Dutch theorists, Sem Goudsmit and George Uhlenbeck. The angular momentum of an electron in a stationary state consists of two parts: (1) its *orbital angular momentum,* which is always an integer in accord with the Schrödinger equation, and (2) an intrinsic angular momentum, or electron *spin.* The electron not only revolves about the nucleus but it also rotates about its own axis: like Earth in its orbit, but not quite. The spin angular momentum of an electron is always $s = \frac{1}{2}$ in units of $\hbar$. According to the quantum rule, the z component of the electron spin must be either $s_z = +1/2$ or $s_z = -1/2$, never more, never less, and never in between. What we called $e\uparrow$ is an

---

*Chemists use the word *orbital* to designate an electronic wave function.

electron spinning in one direction, while $e\!\downarrow$ is an electron spinning in the opposite direction.*

Spin was one missing ingredient of quantum mechanics, the exclusion principle the other. With these additions, the discipline of nonrelativistic quantum mechanics was complete. The chemical and physical properties of the elements and the ways in which they interact with one another—the whole of the sciences of chemistry and atomic physics—are completely determined by three things:

1. The existence of atomic consituents: heavy nuclei and light electrons with spin,
2. Which exert electromagnetic forces on one another
3. In accordance with the laws of quantum mechanics.

With the help of the Pauli exclusion principle and the notion of electron spin, we can assemble what we have learned about quantum states and quantum numbers to understand the arrangement of the elements in the periodic table.

$Z = 1$ The hydrogen atom in its $n = 1$ state has one electron in the lowest energy state. Its electron configuration is denoted by $1s$.

$Z = 2$ The helium atom has two electrons in the $1s^2$ configuration. Both $e\!\uparrow$ and $e\!\downarrow$ occupy the lowest quantum state. The first shell of electrons is filled, and the element is chemically inert.

$Z = 3$ The lithium atom has three electrons. Two fill the lowest shell, but the third must be put into a $2s$ orbital to begin a second shell: $1s^2 2s$. The outer electron is easily lost, making lithium a very reactive alkali metal.

$Z > 3$ As we proceed to larger atoms, more electrons are put into the second shell. Both $e\!\uparrow$ and $e\!\downarrow$ can occupy each of the four distinct $2s$ and $2p$ orbitals, so there is room for precisely eight electrons in the second shell. Fluorine (F), with $Z = 9$, has a single vacancy in its second shell. Its electron configuration is $1s^2\, 2s^2\, 2p^5$. Nirvana for fluorine is finding one more electron to fill its $2p$ shell. A fluorine atom can borrow the extra electron of a lithium atom to form a stable molecule of LiF. Neon, with $Z = 10$, has a complete second shell of electrons and is another inert gas. Sodium $Z = 11$ is an active alkali metal like lithium: It has one extra $3s$ electron in its third shell. So it goes.

As we continue on to larger values of Z, the electrons in the ground states of these atoms sequentially fill up the unoccupied quantum states. Oxygen ($1s^2 2s^2 2p^4$) has two vacancies in its outermost shell, hence it has

*If $e\!\uparrow$ and $e\!\downarrow$ are states with $l_z = \pm\frac{1}{2}$, how do we describe states whose spin is oriented sidewise, say with $s_x = \pm\frac{1}{2}$? The answer is implicit in the idea of superposition. An electron with sidewise spin is described by the combinations $e\!\uparrow \pm e\!\downarrow$.

a valence of -2. Halogens have one vacancy, and alkali metals have an outer shell with just one electron and a valence of $+1$.

The rare gases such as helium and neon are chemical elements with complete outer shells of electrons. They are chemically inert because they have neither extra electrons to donate to other atoms nor vacancies to accept such donations. The particular sequence in which shells are filled is shown in Table 11.2.

TABLE 11.2   Closed Shells of the Inert Gases

Inert Gas	Z	Last Complete Shell
Helium	2	$1s^2$
Neon	10	$2s^2 2p^6$
Argon	18	$3s^2 3p^6$
Krypton	36	$3d^{10} 4s^2 4p^6$
Xenon	54	$4d^{10} 5s^2 5p^6$
Radon	86	$4f^{14} 5d^{10} 6s^2 6p^6$

Exercises

23. The ^{11}B nucleus has spin $s = \frac{3}{2}$. What are the possible values of s_z?

24. The ^{10}B nucleus has spin $s = 3$. What are the possible values of s_z?

25. Lithium and sodium are the first two alkali metals in the periodic table. From Figure 11.14, determine the atomic number of the third

FIGURE 11.14
The electronic struc-
ture of several atoms
is shown. Colored cir-
cles denote electrons
and open circles indi-
cate vacancies in the
outermost valence shell.
Helium and neon are
examples of chemically
inert rare gases: Their
valence shells are filled.
Lithium and sodium are
both chemically active
alkali metals with va-
lence 1. The alkali metals
are atoms with a single
electron in their valence
shells. Fluorine is the
first halogen in the periodic table. Halogens are
very reactive nonmetals
with one vacancy in the
valence shells of their
atoms. The next halogen
is chlorine, with $Z = 17$.

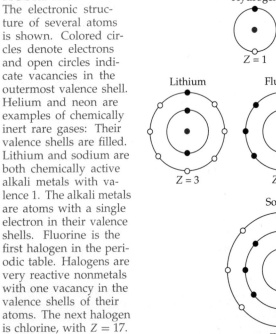
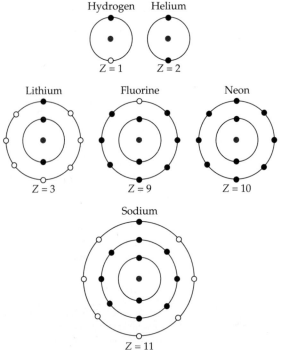

alkali metal, potassium. Draw a picture indicating its electron shells. What is its electron configuration?

26. A quantum-mechanical system has angular momentum J, which is either an integer or half an integer. Neighboring values of J_z are one unit of $\hbar$ apart and $|J_z| \leq J$. Show that there are $2J + 1$ allowed values of J_z.

27. In the progression from xenon to radon in the periodic table, the $4f$, $5d$, $6s$, and $6p$ electron shells are filled. Show how this explains why the difference of their atomic numbers is 32.

28. The quantum states of a hydrogen atom with principal quantum number n can have orbital angular momentum l from 0 to $n - 1$. Show that there are n^2 different wave functions with binding energy Ry/n^2.

29. The wave functions of the hydrogen atom with $n = 3$ can have $l = 0$, or $l = 1$, or $l = 2$. What are the values of l and l_z for each of the nine distinct states?

Where We Are And Where We Are Going

We have come to a turning point in our historical development of the physical sciences. With a quantitatively correct theory of atomic structure in hand, the tools became available to develop a deep understanding of the nature of matter, and most physicists turned their attention to the exciting new horizons suddenly revealed: the physics and chemistry of macroscopic forms of matter. Quantum mechanics provides a triumphant description of the phenomena of atomic physics, molecular physics, atmospheric physics, chemistry, the biological sciences, material sciences, plasma physics, low-temperature physics, and the physics of the solid and liquid states. Beginning in the 1920s, scientists learned at long last how to cope with many of the questions an inquisitive child might ask:

> *Why copper red in sky so blue?*
> *Why snowflake fall when else be dew?*
> *Why diamond does as soot's black dust,*
> *Or soap make bubble, iron rust?*
> *What be diagnostic doctor scanner,*
> *Laser, sonar, radar, maser?*
> *Should fossil fuel or solar powers*
> *Warm our quartzless quantum hours?*
> *Fluoro chloro hydrocarbon,*
> *Jargon, jargon, jargon, jargon:*
> *Do global warming ozone hole*
> *And acid rain leave room for soul?*

At this promising juncture we take our leave of atomic theory and non-relativistic quantum mechanics so that we may pursue our journey into the heart of the nucleus and beyond. As quantum theory scored success after success, a few scientists remained fascinated by the fundamental problems that remained unsolved. We make the great leap inwards to the questions that early quantum theorists could not address. What is radioactivity? How is the atomic nucleus built of simpler things? What powers the Sun? What is the ultimate nature of matter, space, and time?

12 The World According to Einstein

Einstein once told me in the lab: "You make experiments and I make theories. Do you know the difference? A theory is something that nobody believes except the person who made it, while an experiment is something that everybody believes except the person who made it"

H. F. Mark, German chemist

If Einstein's theory should prove to be correct, as I expect it will be, he will be considered the Copernicus of the 20th century.

M. Planck, German physicist

How often have I said to you that when you have eliminated the impossible, whatever remains, however improbable, must be the truth?

Arthur Conan Doyle, British author

Albert Einstein created both the special theory of relativity (which replaced Newton's theory of motion) and the general theory of relativity (which replaced Newton's theory of gravity). However, his Nobel Prize was awarded for his explanation of the photoelectric effect in terms of quantum mechanics. *Source:* UPI/Bettmann.

Without the careful experiments of Coulomb, Ampère, Faraday, and their kindred spirits, there could never have been a Maxwell or an Einstein, nor could we have learned so much about the world and its wonders. Einstein realized that space and time are not independent quantities, but are linked together in a seemingly complex fashion. The true relation between them appears at first sight to be weird and counterintuitive. To the modern scientist, however, the relation is elegant and even inevitable. The consequences of relativity are not apparent in our daily lives because the velocities of everyday objects are tiny compared to the velocity of light. However, Einstein's theories have been verified by experiments. They are essential to our understanding of everything from elementary particles to the universe itself. Section 12.1 describes the first experimental hint that something was lacking in our understanding of space and time. Section 12.2 explains the consequences of relativity for measurements of positions, times, and velocities. Section 12.3 describes how the concepts of mass, energy, and momentum must be revised, and how the conservation of mass and energy are combined into a single law. Einstein's general theory of relativity, which supplanted Newton's theory of gravity, is introduced in Section 12.4.

12.1 The Velocity of Light Is the Velocity of Light

Waves are disturbances in elastic media by which energy is conveyed from one place to another. Maxwell proved that light consists of electromagnetic waves. Like thunder claps and ocean swells, light waves travel hither to yon. Unlike sound and water waves, light crosses the voids of space, from heavenly bodies to Earth. We see the Sun, but we do not hear it. It was natural for scientists to believe that a real and rigid elastic medium they called the *ether* permeated all space. An electromagnetic wave was regarded as a mechanical disturbance that propagates through the ether, just as a sound wave consists of pressure fluctuations that propagate through the air. The hypothetical ether was invisible and impalpable—its only role was to be the medium for the passage of light. In the early nineteenth century, Thomas Young, an advocate of the wave theory of light, said that the ether passes through matter "like the wind through a grove of trees."

Maxwell computed the speed of light through the vacuum from the results of electrical and magnetic experiments. This velocity is universally called c and is approximately 2.998×10^8 m/s. It was originally interpreted as the velocity of light waves relative to the ether, if such a thing existed. Based on the assumption that light travels at the speed c relative to the ether and the fact that Earth speeds around the Sun at $v = 10^{-4}c$, scientists concluded that there must be an "ether wind" relative to an observer on Earth, whose direction continually changes as Earth rotates about its axis (Figure 12.1). According to the classical interpretation of space and time, light moving in the direction of the ether wind should travel at the speed $c + v$; light moving against the ether wind should travel at the speed $c - v$. Thus, the velocity of light, as viewed by an observer on Earth, should depend on its direction.

Two American physicists, Albert Abraham Michelson (1852–1931) and Edward Williams Morley (1838–1923) set out to measure the relative velocity of Earth and the ether. On April 17, 1887, Morley wrote to his father: "Michelson and I have begun a new experiment to see if light travels with the same velocity in all directions. I have no doubt we shall get decisive results." It does and they did.

FIGURE 12.1
Earth moves at an orbital speed of $\sim 10^{-4}c$. If the ether is at rest relative to the Sun, Earth is moving through the ether. The ether wind blows from the west at midday and from the east at midnight.

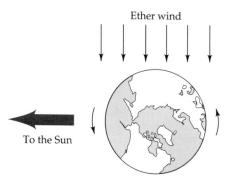

Ether wind

To the Sun

The Michelson–
Morley
Experiment

In Chapter 9, we described how Fizeau and Foucault performed precision measurements of the velocity of light by sending a beam in a round trip to a distant mirror and back. This procedure was adapted by Michelson and Morley to compare the velocity at which light travels in different directions. To understand what they did, let's look at an analogous situation.

Imagine a river 1 km wide flowing at $v = 1.25$ km/h. A man and a woman—both of whom swim at constant speeds of 3.25 km/h relative to the water—engage in a contest. Starting at one shore, the woman must swim to the nearest point on the opposite shore, turn around, and return to the starting point. Meanwhile, the man must swim 1 km upstream and 1 km back (Figure 12.2). What is the time it takes the swimmers to complete their respective 2-km circuits?

To reach the designated point across the river, the woman swims at an angle such that the upstream component of her velocity is 1.25 km/h, so as just to compensate for the river's flow. The transverse component of her velocity (her rate of progress across the river) is $\sqrt{3.25^2 - 1.25^2}$, or 3 km/h. She swims across the river in 20 min and back in another 20 min. The round trip takes 40 min. More generally, if her swim speed is c, the river current is v, and the river width is L, the time $T_\perp$ required for the transverse round trip is

$$T_\perp = \frac{2\,L}{c} \frac{1}{\sqrt{1 - v^2/c^2}}$$

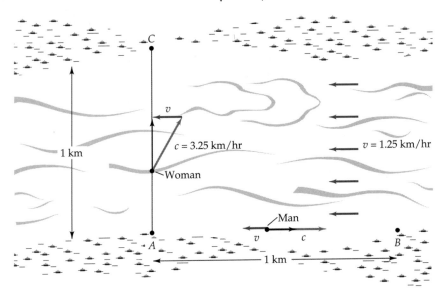

FIGURE 12.2
A woman and a man compete head to head in a 2-km swim meet. The man swims against the current from A to B, and returns with the current. The woman swims from A to C (directly across the river) and back. Although both racers swim at the same distance at the same water speed, the woman wins the race.

When traveling upstream, the man progresses at a speed equal to his swimming speed minus the speed of the river, or 2 km/h. The first leg takes 30 min. Returning with the current, his rate of progress is his swimming speed plus the speed of the river, or 4.5 km/h. The second leg takes 13.3 min. In this case, the round trip takes 43.3 min. The man loses the race by 3.3 min. More generally, the time $T_\parallel$ required for a round trip along the river is

$$T_\parallel = \frac{L}{c+v} + \frac{L}{c-v} = \frac{2L}{c} \frac{1}{1 - v^2/c^2}$$

where c is the speed of the swimmer and v is the speed of the river. For any values of v and c, the longitudinal round trip takes longer than the transverse round trip.

In the Michelson–Morley experiment, light beams were sent on round trips in two perpendicular directions. The beams play the roles of the swimmers. The hypothetical flow of the ether relative to Earth plays the role of the river current. Because v is very much smaller than c, the difference between $T_\parallel$ and $T_\perp$ is given by a simple approximate formula

$$T_\parallel - T_\perp \simeq (v/c)^2 L/c \qquad (1)$$

The time difference between the two round trips depends quadratically on the small quantity v/c.

Michelson and Morley used a partially transparent mirror to split a single beam of light into two beams traveling perpendicular to one another. The beams were then reflected from distant mirrors, as shown in Figure 12.3. According to equation 1, the round trip from A to B should take longer than the round trip from A to C. If the distance between the beam-splitting mirror and each of the distant mirrors were $L = 11$ m and the ether-wind velocity were $v = 10^{-4}c$, then the difference in the

FIGURE 12.3
The Michelson–Morley experiment. A beam of light was split in two at A by a semitransparent mirror. One beam traveled a distance L to B, where it was reflected back to A. The other beam traveled the same distance to C, where it was reflected back to A. The two beams recombined at A and their interference pattern was observed. The hypothetical ether wind flowed in the direction from A to B. When the interferometer was rotated by 90°, the interference pattern was expected to change. It didn't.

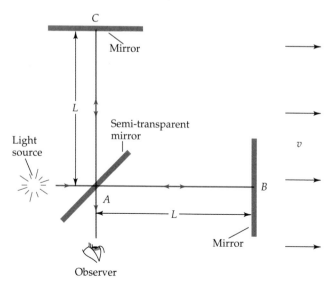

light travel times for the two round trips would be $\Delta T = T_{AB} - T_{AC} \simeq 4 \times 10^{-16}$ s. This time difference is far too short to be measured directly, so Michelson and Morley used an indirect approach—they observed the interference of the two beams with a device called an *interferometer*. The waves returning from the two mirrors were recombined to form a single beam. Because the two mirrors were imperfectly aligned, the two beams interfered differently at different points. Where they added together, a bright band resulted. Where they cancelled out, a dark band resulted. The overall pattern of parallel fringes was viewed through an eyepiece.

The interferometer was placed on a stone slab sitting in a bath of mercury so that it could be turned around. The interference pattern seen in one orientation was compared to the pattern seen in another. When the two beams were recombined, the interference pattern was expected to differ considerably from the original pattern to account for the effect of the ether wind. Instead, Michelson and Morley were amazed to discover that the pattern did not change at all. On August 17, 1887, Michelson wrote to the eminent physicist Lord Rayleigh:

> The experiment on the relative motion of the earth and the ether has been completed and the result was decidedly negative. It follows that the relative motion of the earth and the ether is less than one-sixth of earth's velocity.

Other physicists repeated the Michelson–Morley experiment with the same negative result. No relative velocity between the ether and Earth could be detected. In 1892, H. A. Lorentz wrote to Lord Rayleigh in despair:

> I am totally at a loss to clear away this contradiction, and yet I believe that if we were to abandon [the wave theory of light], we should have no adequate theory at all. Can there be some point in the theory of Mr. Michelson's experiment which has yet been overlooked?

What to Do?

Nothing had been overlooked. The experimental result was correct. The relative velocity of Earth and the ether was and is undetectable. Many tentative explanations of the Michelson–Morley experiment were put forward, including the following:

- Maybe the ether moves along with Earth in its orbit, like molasses pulled along by a spoon. In that event, there would be no relative motion between Earth and the ether on the surface of Earth. The Michelson–Morley result would be explained! However, the "ether drag" hypothesis is incompatible with astronomical observations. If Earth dragged the ether along, there could be no aberration of starlight (see Chapter 9). Yet, aberration had been observed and measured. This explanation had to be abandoned.

- Maybe the universal value c for the velocity of light should be interpreted as the velocity of light relative to its source. In this case, the motion of the ether relative to Earth is irrelevant to the Michelson–Morley experiment. The two beams would travel at the velocity c relative to the light source and the paradox would be resolved! However, this hypothesis was also in conflict with astronomical observations. Pairs of stars were observed to rotate about one another. Sometimes one component of the double star would move toward us (so that its light would travel faster), and sometimes it would move away from us (so that its light would travel more slowly). Each of the stars would periodically disappear or appear in triple images! No such effect had ever been seen. This explanation had to be abandoned.

• The Irish physicist George Francis Fitzgerald came up with an alternative explanation in 1899:

> The [Michelson–Morley] result seems opposed to other experiments showing that the ether in the air can be carried along only to an inappreciable extent. The only hypothesis that can reconcile this opposition is that the length of material bodies changes, according as they are moving through the ether or across it, by an amount depending on the square of the ratio of their velocities to the velocity of light.

The Michelson–Morley could be explained, but only at the cost of a contrived and entirely ad hoc hypothesis. To Fitzgerald, the ether was a real thing, and time and space were absolute. However, Fitzgerald had hit upon a germ of the truth: Moving bodies do shrink along their direction of motion.

In fact, there is no ether wind. The ether, which had been an integral part of science for centuries, does not exist. The correct solution to the puzzle posed by the Michelson–Morley experiment is called the *special theory of relativity*, and it was created by Albert Einstein (1879–1955) in 1905. However, the Michelson–Morley experiment was not the inspiration for the theory of relativity. Einstein claims that he came upon his theory by wondering what it would be like to travel at the speed of light—something that neither he nor anyone else will ever do. As we shall see in the next section, the special theory of relativity is a radical revision of our understanding of the nature of space and time. Light propagates through the vacuum at exactly the same speed relative to all observers, no matter how fast, or in what directions, they are moving. That is, "the velocity of light is the velocity of light is the velocity of light."

Because the vacuum velocity of light c is the same to all observers, it is truly a universal constant. Since 1983, the SI unit of length, the meter, has been defined in terms of c. It is the distance traveled by light in $(1/299\,792\,458)$ s. Thus, the distance light travels in 1 s through a vacuum has been assigned the precise value of 299 792.458 km. The velocity of light c in a vacuum is no longer a measurable quantity but is defined to be:

$$c = 2.99792458 \times 10^8 \text{ m/s}$$

EXAMPLE 12.1
Thought Experiments

When velocities are comparable to c, the correct description of motion differs from the classical description. But the velocity of light is enormous in everyday terms. To visualize the effects of relativity, it is convenient to imagine situations that are difficult or impossible to realize: *thought experiments*. As an example of a thought experiment, imagine three observers: (1) an officer in a patrol car cruising north at a road speed V; (2) a truck driver whose semi is further north and is hurtling southward at velocity $-V$; and (3) a pedestrian on the sidewalk nearby. As the two vehicles approach one another, the patrol car flashes its blinker and sounds its siren (Figure 12.4). The observations discussed below are performed before the two vehicles pass one another.

a. Suppose that the velocity of sound in still air is 300 m/s and that $V = 200$ m/s. What is the velocity of the siren signal relative to each of the observers?

b. Suppose that $V = 2c/3$. What is the velocity of the blinker signal relative to each of the observers?

Solution

a. Sound waves traveling north from the moving siren travel at 300 m/s relative to the air (which is at rest relative to the ground). Relative to the patrol car (which moves at 200 m/s in the same direction), the signal speed is 100 m/s. The truck travels at the opposite speed relative to the air, and the siren signal moves at 500 m/s relative to it. To the pedestrian, the speed of the siren signal is 300 m/s.

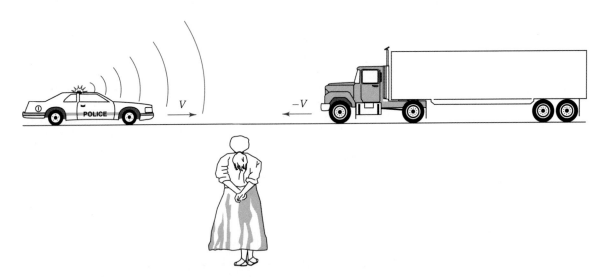

FIGURE 12.4 The police car is moving at ground velocity V while flashing its lights and sounding its siren. The truck approaches the police car at ground velocity $-V$. Each of three observers—the officer in the car, the truck driver, and a pedestrian—measures the velocities of the police siren and blinker.

b. The Michelson–Morley experiments and others proved that the velocity of light is the same when it is measured by any observers. To the police, the truck driver, and the pedestrian, the blinker signal travels at the same velocity c.   ∎

Exercises

1. From the value of the astronomical unit and the length of the year, show that the orbital speed of Earth is approximately $10^{-4}c$.

2. What is the speed of the fastest moving visible object you have ever seen, read, or heard about? Work out the ratio of this velocity to the speed of light.

3. Imagine living south of a cliff. On a windless day, your shout returns as an echo from the cliff precisely 4 s later.
   a. One day, a strong north wind is blowing. The time delay of the echo increases to 4.04 s. At what fraction of the speed of sound is the wind blowing?
   b. The wind changes its direction but not its speed. When it blows from the east, what is the time delay of the echo?

4. This exercise deals with the race described on pages 498–499.
   a. Suppose that the woman swam at a water speed of 3.25 km/h, as in the exercise, but the man managed to swim a little bit faster so as to achieve a tie. What was his swimming speed?
   b. Suppose that the woman erred. She swam both ways across the river in a direction perpendicular to the current. Afterward, she had to swim upstream to return to her starting point and complete the race. How long did her swim take?

5. Answer the questions in Example 12.1 after the two vehicles have passed one another. In this case, the relevant light and sound waves are those moving to the south.

6. Two stars of the same mass rotate about their center of mass with speeds of $v = 10^{-4}c$ relative to Earth. They are 100 light-years from Earth. Suppose that light travels at c relative to its source, but at $c \pm v$ relative to Earth depending on whether the source is approaching or receding. How much longer would light take to get to us from one of the stars when it is receding from Earth than when it is approaching?

12.2 The Special Theory of Relativity—Space and Time

Two principles form the basis of the special theory of relativity: one is due to Galileo, the other to Einstein. They are: (1) The laws of physics are identical to all uniformly moving observers, and (2) the velocity of light in the vacuum, c, is independent of the motion of the source and the motion of the observer. Although these two postulates may seem simple and reasonable, they give surprising results, as Banesh Hoffman (an American collaborator of Einstein's) pointed out:

> Watch closely. It will be worth the effort. But be forewarned. As we follow the gist of Einstein's argument we shall find ourselves nodding in agreement, and later almost nodding in sleep, so obvious and unimportant will it seem. Beware. We shall by then have committed ourselves and it will be too late to avoid the jolt; for the beauty of Einstein's argument lies in its seeming innocence.

Even the "simple" matter of how to combine velocities must be reconsidered in the light of these principles. Before exploring the consequences of the two principles of relativity, we first define the ground rules.

Measurements by Different Observers

Special relativity examines the relation between measurements of events performed by different observers who are in relative motion. An *event* is something that happens at a definite time and a certain place: the absorption of a photon by an atom, the decay of a radioactive nucleus, the shooting of John Lennon. An event is located in space and time by four numbers: three spatial coordinates (latitude, longitude, and altitude often do) and the time at which it took place.

The space–time locations of various events are measured by *observers*. Different observers move at uniform velocities relative to one another. Observers are equipped with identical clocks and rulers with which to measure times and distances. Each observer makes measurements relative to a fixed Cartesian coordinate system (Figure 12.5).

Imagine that a train moves at constant speed v along perfectly straight tracks. Thought experiments are carried out by two idealized observers: Harpo, who is on the ground near the tracks, and his twin sister Oprah, who is aboard the train. Both observers have the superhuman ability to

FIGURE 12.5
The space and time coordinate systems of two observers. Oprah's system (primed) is moving at velocity v along the positive x-axis of Harpo's system (unprimed). The spatial axes of the two systems are similarly oriented. Harpo and Oprah use identical clocks that are synchronized at the moment their spatial axes coincide.

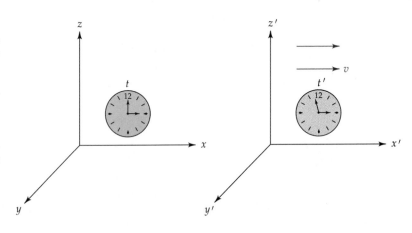

measure times and distances with absolute precision. Harpo measures the locations of events relative to himself. That is, he uses a coordinate system and a clock with respect to which he is at rest. His spatial coordinates are x (in the direction of the train), y (horizontal, but perpendicular to v) and z (vertical). His clock reads the time of an event as t. Oprah's coordinates x', y', and z' are aligned with Harpo's. His clock reads time t, hers reads t'. The origins of the two systems are chosen to coincide. An event at $x = y = z = t = 0$ coincides with event at $x' = y' = z' = t' = 0$.

Events in Different Coordinate Systems

Oprah observes various events and records their positions and times in her coordinate system. For example, she sees a squirrel jump at the position (x', y', z') and at the time t'. Harpo observes the same events and records their positions and times. In his coordinate system, the squirrel jumps at the position (x, y, z) and at the time t. How are the primed quantities related to the unprimed quantities? This is the central question underlying the special theory of relativity.

Let's work out the relations between x', y', z', t' and x, y, z, t according to conventional (prerelativistic) modes of thought. Oprah and Harpo will agree on the time at which a given event takes place. Each observer sees the squirrel jump at the same time; that is $t' = t$. According to classical physics, the time at which an event takes place has an absolute significance. The observers would agree about the times at which events happen.

Furthermore, they agree about the positions of events in directions perpendicular to their relative velocity. Suppose that the squirrel jumped from a branch 3 m high on a tree 10 m from the railroad track according to Harpo. That is $y = 3$ and $z = 10$. Oprah would certainly agree. She finds $y' = 3$ and $z' = 10$.

Of course, the twin's measurements of positions in the direction of motion of the train would not agree. The observers are in relative motion along the x-axis. Suppose that Harpo measures the x component of the jumping squirrel to be 100. The result that Oprah gets depends on where she is when the squirrel jumps. Oprah finds that the x' coordinate of the jumping squirrel is $100 - vt$. What we have just said about the squirrel jump applies just as well to the position and time of any event. We may express the relationships between the two coordinate system in terms of four equations:

Changing Coordinates According to Galileo

From Oprah's Coordinate System to Harpo's Coordinate System	From Harpo's Coordinate System to Oprah's Coordinate System	
$x' = x - vt$	$x = x' + vt'$	
$y' = y$	$y = y'$	
$z' = z$	$z = z'$	
$t' = t$	$t = t'$	(2)

If Oprah is at rest in the train at $x' = a$, Harpo measures her position to be $x = a + vt$. If a bird flies along the tracks at speed u, its position as a function of time is $x = b + ut$, according to Harpo. Its position, according to Oprah, is $x' = b + (u - v)t'$. She sees the bird move at speed $u - v$. This result is just as we would expect, but it is wrong!

Equations 2 are wrong because they contradict Einstein's fundamental principle. Think of a beam of light moving in a direction opposite to the train. If light travels at c relative to Harpo, the classical rules say that light travels at $v + c$ relative to Oprah. But we have accepted as a fundamental principle that light travels at the same speed relative to all observers! That is why we abandon equations 2 and search for the correct relations between Harpo's coordinates and Oprah's coordinates. The replacements for equations 2, the *Lorentz transformations,* are the essence of special relativity. We shall be led to them through a series of imaginary, impractical, and undoable thought experiments.

The First Experiment: Distance Measurement

First, let's see what happens when two moving observers try to measure a distance that is perpendicular to their relative velocity.

Imagine that Oprah and Harpo each hammers a long nail through one end of a meterstick. Oprah thrusts her meterstick through the open window of her train compartment along her y'-axis. She holds it firmly with the nail protruding upward. Harpo extends a similar meterstick toward the tracks (in the negative y direction) with its nail pointing down (Figure 12.6). As Oprah passes her brother, each nail scratches

FIGURE 12.6 The first experiment shows that distances transverse to two moving observers are the same to each. The protruding nails of each ruler scratch the other as they pass. The symmetry of the situation demands that the distances (d and d') from the nail to the scratch on each ruler be equal.

the other meterstick. Each of these occurrences is an event, and the distance between the two events is indelibly recorded on each meterstick. Furthermore, both observers agree that the two events occurred at the same time. Harpo measures the distance between the nail and the scratch on his meter stick to be d. Oprah measures the distance to be d'. It had better be that $d = d'$ because the situation is entirely symmetrical between the two observers. Transverse distances are the same, as measured by either observer. The second and third of equations 2 are perfectly correct:

$$y = y' \quad \text{and} \quad z = z'$$

Thus, distances perpendicular to their relative velocity are the same according to both observers. What could be less surprising? In this case, the "intuitively obvious" relations coincide with the correct relativistic relations.

The Second Experiment: Simultaneity

Two events are said to be *simultaneous* if they are observed to take place at the same time. Are two events that are simultaneous to one observer also simultaneous to another? Here's how our imaginary twins discovered the surprising answer.

During a thunderstorm, Harpo sees two lightning bolts strike at exactly the same time: One hits the front of the train, the other strikes the rear. Each bolt leaves a singe mark on the track (Figure 12.7, on page 508). Harpo knows that he was standing precisely halfway between the two marks and that light from each of the bolts reached his eyes at exactly the same time: $t_1 = t_2 = L/2c$, where L is the distance between the singe marks (and the length of the train, according to Harpo). According to his analysis of the data, both bolts hit the train simultaneously at $t = 0$.

While Harpo is watching from the ground, Oprah is standing precisely at the center of the train as it travels to the right at speed v. She also sees the lightning bolts. However, because she is moving to the right, the light pulse from the right reaches her eyes before the light pulse from the left. Thus, Oprah concludes, one lightning bolt struck the front of the train before the second lightning bolt struck the rear! She did not see the two events happen at the same time! Simultaneous events to Harpo (events for which $t_1 = t_2$) are not necessarily simultaneous to Oprah ($t_1' \neq t_2'$). Conversely, if the two lightning strokes were simultaneous to Oprah ($t_1' = t_2'$), they cannot have been simultaneous to Harpo ($t_1 \neq t_2$).

You cannot get out of this by saying that Harpo is "at rest" and Oprah is "moving." One inertial coordinate system is no better than any other. Suppose that two spatially separated events A and B are simultaneous to one observer. There are other inertial coordinate systems in which A precedes B, and still others in which B precedes A. The concept of simultaneity is relative. There is no coordinate-system-independent meaning to the statement that two events happen at the same time, unless they also happen at the same place.

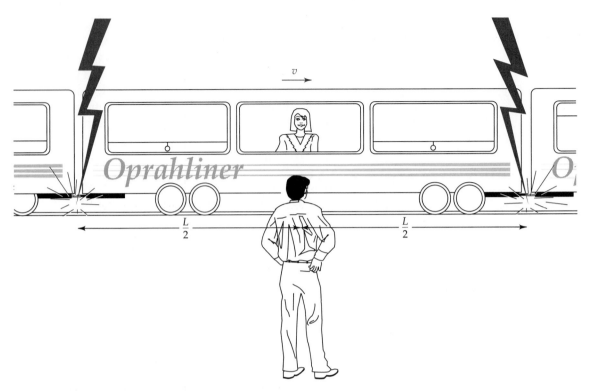

FIGURE 12.7 The second experiment. Lightning strikes the front and back of the moving carriage at the same time according to Harpo. Oprah disagrees. Because she is moving to the right, she sees the forward flash occur before the rear flash.

Why do we not notice that simultaneity is relative? Put numbers into our results and you will see. Trains, planes, missiles, satellites, and Earth itself move at tiny fractions of c, and we cannot detect a difference between classical and relativistic mechanics. You've heard this story before: We don't see quantum effects because h is so small. In the subnuclear world, however, things are small enough and fast enough to require a theory taking both c and h into account.

The Third Experiment: Time Dilation

If two observers can't agree about whether two events are simultaneous, what will they say about the time between events? As it turns out, the time between two events as measured by Oprah is different from the time between the same events as measured by Harpo. To see how we are forced to this surprising result, let's use idealized "photon clocks" consisting of two mirrors mounted one above the other a distance d apart. A single photon bounces back and forth between the mirrors. The clock ticks when the photon strikes the upper mirror and tocks when it strikes the lower mirror (Figure 12.8a). Identical photon clocks are placed on the train and on the ground. The two clocks are synchronized at $t = t' = 0$.

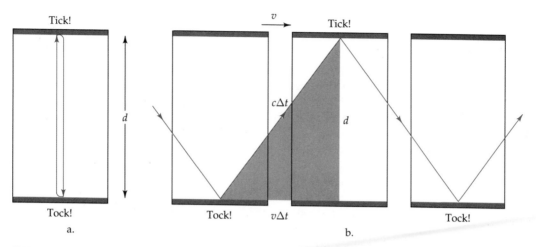

FIGURE 12.8 The third experiment. *a.* Oprah and Harpo are provided with identical "photon clocks." The photon in the clock travels back and forth (at velocity *c*) between two mirrors. The clock ticks as the photon bounces off the upper mirror and tocks when the photon bounces off the lower mirror. Time intervals are measured by the number of intervening ticks. The time between successive ticks of Oprah's clock is $\Delta t' = 2d/c$, according to Oprah. *b.* According to Harpo, to whom Oprah's clock is moving, the photon must traverse a distance longer than *d* between a tick and a tock. Oprah's clock appears to Harpo to run slow. The situation is symmetrical: According to Oprah, it is Harpo's clock that runs slow compared to her own. Both conclusions are correct.

According to the result of the first experiment, Oprah and Harpo agree that the distance between the mirrors of each clock is *d*. Let's focus on Oprah's clock: She deduces that the time between a tick and the next tock (or a tock and the next tick) of her clock is

$$\Delta t' = d/c \tag{3}$$

Here's the key question: What is the time interval Δt between a tick and a tock of his sister's clock as measured by Harpo? Oprah's clock is moving at velocity v relative to Harpo. From his standpoint, the photon in her clock takes a diagonal path to get from one mirror to another. Between tick and tock, the clock moves a distance $v\Delta t$, (as shown in Figure 12.8b). This distance forms the base of a right triangle whose height is d. Its hypotenuse is the distance $c\Delta t$ that the photon travels between tick and tock. The speed of a photon (to Harpo or anyone else) is c. Applying the Pythagorean rule, Harpo finds $(c\Delta t)^2 = (v\Delta t)^2 + d^2$. This equation is readily solved for Δt:

$$\Delta t = \frac{d}{c} \frac{1}{\sqrt{1 - (v/c)^2}} \tag{4}$$

Harpo observes that the time Δt between the tick and tock of Oprah's clock is longer than $\Delta t'$, its value for his own identical clock. Harpo sees his sister's clock running slower than his own. When Harpo's clock reads t, he observes his sister's clock to read t', where

$$t' = t\sqrt{1 - v^2/c^2} \tag{5}$$

If the train were moving at 60 percent of the speed of light, Oprah's clock would read $t' = 0.8\,t$ according to Harpo. It would lose ten minutes per hour compared to his own clock!

This argument is perfectly symmetrical between the two observers. Oprah observes that the time between the tick and tock of Harpo's clock is longer than it is for her own. According to Oprah, Harpo's clock runs more slowly than her own. Each observer sees a moving clock ticking more slowly than an identical stationary clock. This is the effect known as *time dilation*. Nothing in the argument depends on the particular clocks being used. Not only are moving clocks slowed down, but so are all physical and biological processes. The photon clock can be redesigned to tick at the same rate that Oprah's heart beats. When she holds the clock close to her chest, the ticks of the clock and the beats of her heart can be made to coincide in time and in position. Her heartbeats and the ticks of her clock are coincident events. Harpo observes his sister's heart to be slow by the same factor that her clock is slow.

To illustrate the consequences of relativistic time dilation, we imagine a hand grenade containing a clock that detonates the grenade precisely 1 s after it is thrown (Figure 12.9). The clock is at rest relative to the grenade and measures time in the *rest frame* of the grenade: the coordinate system in which the grenade is at rest. However, the rest frame of the grenade is not the reference frame of the soldiers in the field. If the grenade is thrown at velocity v, how far D does it travel before it blows up? According to prerelativity physics, the time interval is frame-independent: 1 s to the grenade is 1 s to the soldiers. Let's work in *light-seconds*, or ls, the distance light travels in 1 s. (The distance to the Moon is a bit more than 1 ls.) Classically, D is the product of v and 1 s. Thus, if $v = 0.5c$, the grenade travels 0.5 ls before it explodes. If $v = 0.9\,c$, it travels 0.9 ls, and so on. Because (as we shall see) v cannot exceed c, the grenade cannot travel more than 1 ls before exploding. These are the classical results, and they are wrong.

Let's attack the problem correctly and relativistically. The clock in the bomb is moving at velocity v relative to the ground. Equation 5 says that 1 s in the rest frame of the grenade corresponds to a longer

FIGURE 12.9
The detonator of the grenade is set to explode at fixed time t (in its rest frame) after it is thrown. If it is thrown at velocity v, it travels the distance $D = \gamma\, vt$ before exploding. Only for v very much smaller than c can D be approximated by the classical formula $D = vt$.

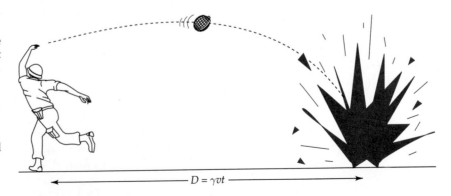

$D = \gamma vt$

time in the battlefield. The moving grenade's lifetime is extended by the multiplicative factor

$$\gamma = 1/\sqrt{1 - v^2/c^2} \qquad (6)$$

The relativistic parameter γ will be convenient and important to us. Notice that it is greater than 1 except for $v = 0$. The result for D, according to relativity theory, is the product of v and γs. The values of D for various grenade speeds are given in Table 12.1.

There is no limit to how far the grenade can go before exploding! It's only a question of how close to the speed of light it can be thrown. At any human speed, however, the effect of time dilation is undetectable.

TABLE 12.1   The Relativistic Hand Grenade

Grenade Speed	Distance Traveled
$0.3\,c$	0.31 ls
$0.5\,c$	0.58 ls
$0.8\,c$	1.33 ls
$0.9\,c$	2.06 ls
$0.99\,c$	7.02 ls
$0.999\,c$	22.3 ls

EXAMPLE 12.2
Cosmic Ray Muons

The *muon* is a teeny analog to the relativistic hand grenade. It is an unstable elementary particle with a half-life of 2 microseconds: Muons typically decay in about that time in their rest frames. Classically, such a particle cannot travel further than about c times 2 microseconds, or 600 m, before decaying. However, cosmic rays striking atoms in the upper atmosphere produce muons that reach the ground! How can this be? The muons are moving almost at the speed of light relative to Earth. How far does a muon travel before decaying if it travels at 99 percent of light speed?

Solution

At 0.99 c, $\gamma \simeq 7$ according to equation 6. Thus, the muons travel about 4 km before decaying, long enough to traverse the atmosphere. Some muons move at 99.9 percent of c and travel over 20 km before decaying or making a collision. To particle physicists, time dilation and the special theory of relativity are parts of everyday life. ∎

The Fourth Experiment: Length Contraction

Time dilation is related to another consequence of relativity: the contraction of moving bodies. Oprah takes the same photon clock used in the third experiment and turns it around so that the photon bounces back and forth along, rather than perpendicular to, the x-axis. This doesn't affect the operation of the clock. The time between consecutive ticks is unchanged—it is $2d/c$ to Oprah and $2d\gamma/c$ to Harpo.

Putting the cart before the horse, let's allow for the possibility that the distance between the mirrors is not the same as measured by different observers. Oprah's clock is at rest relative to Oprah, who measures the distance between its mirrors to be d. Oprah's clock is in motion relative to Harpo, who measures the distance to be d' (Figure 12.10).

Let's trace the path of the photon as Harpo sees it. At $t = 0$ it strikes the aft mirror and makes a tick. It gets to the fore mirror, which is receding from it at speed v, at time t_1. The distance traveled by the photon, ct_1, equals the distance to the fore mirror at $t = t_1$, which is $d' + vt_1$. From this equality (and a perfectly analogous one for $t_2 - t_1$, the time needed for the photon to return from the fore mirror to the aft mirror), we find

$$t_1 = \frac{d'}{c - v} \qquad t_2 - t_1 = \frac{d'}{c + v}$$

Adding t_1 to $t_2 - t_1$, we find the time between consecutive ticks of Oprah's clock as seen by Harpo:

$$\Delta t = 2d'\,c/(c^2 - v^2)$$

The orientation of a clock cannot affect its performance. This result must coincide with the period of Oprah's clock as seen by Harpo when it is turned the other way, as given by equation 4. Thus:

$$\frac{2d'\,c}{c^2 - v^2} = \frac{2d}{c\sqrt{1 - v^2/c^2}} \qquad \text{or} \qquad d' = d/\gamma$$

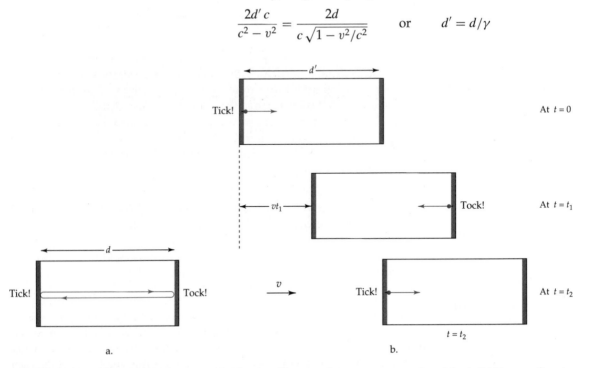

FIGURE 12.10 *a.* Oprah's clock is turned sidewise. The time between consecutive ticks is $2d/L$ according to Oprah, as it is when the clock is upright. *b.* Three views of the clock in motion, as seen by Harpo. He sees its length to be d'. The clock ticks at $t = 0$ when the photon begins its voyage to the front mirror. At $t - t_1$, the photon bounces off the front mirror. The clocks ticks again at $t - t_2$, when the photon returns to the rear mirror.

Consistency demands that Harpo sees Oprah's clock to be shortened along the direction of its motion by the factor $1/\gamma$. Everything in the train (including Oprah) and the train itself appears similarly foreshortened to Harpo. Rapidly moving objects appear to contract along their direction of motion (Figure 12.11). This effect was foreseen by Lorentz and Fitzgerald in an ether model, but it is an integral part of Einstein's special theory of relativity. If L is the length of an object in its rest frame, then the length of the object according to an observer moving at velocity v along its length is

$$L' = L\sqrt{1 - v^2/c^2} \qquad (7)$$

Suppose that the train is 100 m long according to Oprah. That is, its *rest length* is 100 m. The train will appear to be shorter than it will to an observer in motion relative to the train. If the train is moving at a speed of half the velocity of light, Harpo will measure its length to be $100\sqrt{1 - 0.25}$, or about 87 m. However, because Harpo in moving relative to Oprah, she sees her brother to be much skinnier than he is in person.

FIGURE 12.11
The Lorentz contraction. Moving bodies appear shorter than they are in their rest system. An object moving at $0.6c$ is reduced to 80 percent of its rest length. At $0.8c$, it is reduced to 60 percent.

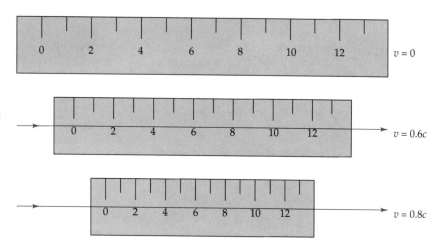

The Fifth Experiment: The So-Called Twin Paradox

Oprah takes a vacation, during which she inadvertently challenges the theory of relativity. She boards the spaceliner *Albert* to visit the nearest stellar attraction—the triple-star system α-Centauri, lying 4.4 light-years from Earth. The *Albert* cruises at a constant speed of $0.8c$ to its destination, then comes about and returns home at that same speed. According to Harpo, the round trip should take 11 years, as indeed it does—for him.

However, when Oprah comes home to Harpo, she is now his younger sister, not his twin. Shipboard, only 6.6 years have passed! Harpo quickly puzzles it out. His sister was in uniform motion relative to

© 1993 by Sidney Harris.

him. For half the trip, she was speeding at 0.8c outbound. For the other half, she was speeding home at −0.8c. In both cases, $\gamma = 1/\sqrt{1 - 0.64} = 1/0.6$. Throughout her voyage—except for the short time she spent alien-watching and turning around—the clocks on her ship, including her body clock and those of her shipmates, were running slow compared to Earth-bound clocks. The 11 years he had been waiting at home were only $11/\gamma$ or 6.6 years to Oprah.

Oprah argues the contrary. She says that the situation is symmetrical. Harpo's clocks were in uniform motion relative to hers, so they were the slow ones. Oprah's hour aboard is but 36 minutes to Harpo at home. Less time passed on Earth than on the *Albert*. According to her understanding of the theory of relativity, she has become the elder sibling. (She attributes her youthful appearance to years without orange juice or sunshine.)

There can be only one correct answer: Each sibling knows for sure how much subjective time has passed. One of them made an error in computing the other's age. It only *seems* like a paradox. In their train experiments, Oprah's clocks were slow to Harpo. So also were his clocks slow to her. But the spaceliner differs from the train because it had to turn around! The situation was not symmetrical. Harpo's reference frame was inertial (or in uniform motion), but Oprah's reference frame was not always so. To return, she had to spend a period of time decelerating from +0.8c to −0.8c. Oprah is correct to say that Harpo's clocks are ticking more slowly than hers as long as the liner cruises with no change of velocity. But, the special theory of relativity cannot be naively applied in an accelerated coordinate system, such as the *Albert* while it is changing its course.

Oprah found her error when she remembered the relativistic length contraction. From the vantage point of the *Albert* at its cruising speed of $0.8c$, the distance between Earth and α-Centauri is shorter than 4.4 light-years. According to equation 7, the distance she must travel is only $4.4\sqrt{1-0.64}$, or a mere 2.64 light-years. Consequently, her outward passage took only 2.64/0.8 or 3.3 years, and the return took the same. Harpo was right after all. The stay-at-home brother is now 4.4 years older than his errant sister. The special theory of relativity is paradox free and right as rain.

The Lorentz Tranformations

The experiments we examined can pin down the form of the Lorentz transformations. Rather than deduce them, I will tell you what they are. In terms of the useful parameter γ defined by equation 6, they are not so formidable:

Changing Coordinates According to Galileo

From Oprah's Coordinate System to Harpo's Coordinate System	From Harpo's Coordinate System to Oprah's Coordinate System
$x' = \gamma(x - vt)$	$x = \gamma(x' + vt')$
$y' = y$	$y = y'$
$z' = z$	$z = z'$
$t' = \gamma(t - vx/c^2)$	$t = \gamma(t' + vx'/c^2)$ (8)

These formulas replace their nonrelativistic expressions, equations 2. Let's use them to recapture our earlier results:

- *Time dilation:* Oprah sits in a deck chair located at $x' = a$ at $t' = b$. She leaves the chair a time interval T later, at $x' = a$ and $t' = b + T$. To find the times of these events according to Harpo, we use equation 8 to express t in terms of t' and x'. The first event takes place at $t = \gamma(b - va/c^2)$. The second event takes place at $\gamma(b + T - va/c^2)$. The difference between these expressions is simply γT. The quantity γ is greater than 1: the time she spends in the chair, according to Harpo, is longer than T.

- *Lorentz contraction:* To Harpo, the length of the train is the distance between its front and rear when measured at the same value of his own clock time, t. Suppose the train extends from $x' = 0$ to $x' = L$ in Oprah's frame. Using the formula $x' = \gamma(x - vt)$, we find, for the rear of the train, $x = vt$, and, for the front, $x = vt + L/\gamma$. The difference between these, L/γ, is the foreshortened length of the train according to Harpo.

- *The relativity of simultaneity:* Harpo sees lightning strike each end of the train at the same time. To him, the events are located at

$$(t = 0;\ x = 0) \qquad \text{and} \qquad (t = 0;\ x = L/\gamma)$$

where L is the *rest length* of the train and γ takes into account its relativistic contraction. Oprah's coordinates for these same events, obtained from equations 8, are

$$(t' = 0;\ x' = 0) \qquad \text{and} \qquad (t' = -vL/c^2;\ x' = L)$$

As we have seen before, the events are not simultaneous to her. The Lorentz transformations can be used to solve problems without the need to analyze and interpret the exchange of light signals between events and observers.

How to Add Velocities

Let's use the Lorentz transformations to answer the following question. Oprah (on the train moving at speed v relative to the ground) shoots a bullet forward. Its muzzle velocity is u and its trajectory (according to Oprah) is $x' = ut'$. What is its speed, which we call w, according to Harpo? We must express x' and t' in terms of Harpo's coordinates x and t. If we use the old-fashioned (and wrong) Galilean transformation, the trajectory becomes $x = (u + v)\,t$; the bullet's speed to Harpo is $u + v$. To do it right, we use equations 8 to express x' and t' in terms of x and t:

$$\gamma\,(x - vt) = u\,\gamma\,(t - xv/c^2) \quad \text{and hence} \quad x = (u + v)\,t\,/(1 + u\,v/c^2)$$

The true value of the slug's speed to Harpo is

$$w = \frac{u + v}{1 + uv/c^2} \tag{9}$$

This is the relativistic formula for the addition of velocities.

If u and v are small compared to c, the denominator in equation 9 is 1 plus something very small. For everyday (or even astronomical) velocities, the classical formula $w = u + v$ suffices. Earth revolves about the Sun at the velocity $v \simeq 10^{-4}c$, while the Sun moves relative to the Galaxy at $u \simeq 10^{-3}c$. If the directions of u and v were the same, the classical result for Earth's velocity relative to the Galaxy would be wrong by only 1 part in 10^7. Table 12.2 illustrates the use of equation 9 for larger velocities, such as are encountered in particle and nuclear physics.

TABLE 12.2   The Addition of Large Relative Velocities

v/c	$\oplus$	u/c	$=$	w/c
0.1		0.1		0.198
0.1		0.5		0.566
0.1		0.9		0.917
0.5		0.5		0.800
0.5		0.9		0.969
0.9		0.9		0.995
0.99		0.99		0.99995

The relativistic sum of any two velocities u and v smaller than the speed of light is itself smaller than the speed of light. Imagine a body moving at a speed $v < c$ relative to the ground. Catch it up, then toss it forward with the additional velocity v. Keep doing this. It will go faster and faster, but it can never be made to travel at the speed of light.

EXAMPLE 12.3
Adding Velocities

A vacationer is on a ship cruising at $0.5c$ relative to the ocean. He runs forward at a speed of $0.5c$ relative to the deck. While running, he throws a ball at a speed of 0.6c straight ahead. What is the speed of the ball relative to the ocean?

Solution   Applying classical (but incorrect) reasoning, we would find that the runner's speed is c and the ball's speed is $1.6c$ relative to the ocean. Using equation 9, we find that the runner's speed, in units of c, is actually $(0.5 + 0.5)/1.25$, or $0.8c$. To obtain the ball's speed relative to the ocean, we combine the ball's speed relative to the runner ($0.6c$) with the runner's speed relative to the ocean ($0.8c$) to obtain $(0.6+0.8)/1.48$, or about $0.95c$. ∎

Exercises

7. Suppose that the two lightning flashes in the second experiment were simultaneous to Oprah. Show that they could not have been simultaneous to Harpo.

8. The relativistic hand grenade explodes precisely 1 s (in its rest frame) after it is thrown. At what speed must it be thrown if it travels a distance of 1 light-second (relative to the thrower) before exploding?

9. An astronaut chooses one of two identical and synchronized watches before being launched into orbit. The shuttle speed in a low-altitude

DO YOU KNOW
About Tachyons?

One may imagine three kinds of motion, depending on the speed of a body:

1. Motion at a velocity u that is less than c
2. Motion at light speed, where $u = c$
3. Motion at a velocity u that is greater than c

Electrons, atoms, baseballs, and planets are type 1 objects. They travel at *subluminal* velocities. Photons are particles of type 2. They travel at the speed of light. Nothing in the special theory of relativity forbids the existence of particles of type 3, which travel at *superluminal* velocities. Physicists refer to such hypothetical particles as *tachyons*.

Suppose a body is viewed by a second observer moving at a subluminal velocity v. He will observe the body moving at the speed w given by equation 9. The following results can be established:

1. A subluminal object to one observer is subluminal to all observers.
2. An object traveling at light speed to one observer travels at light speed to all observers.
3. A superluminal object to one observer is superluminal to all observers.

Experimenters have searched everywhere for tachyons but have never found them. This is fortunate because theoretical physicists cannot imagine how to construct a consistent quantum theory in which tachyons exist.

polar orbit is $2.65 \times 10^{-5}c$. After the year-long mission is completed, the astronaut lands safely and returns home.

  a. About how many times did the spaceship circle Earth?

  b. Explain why his perfect watch was about a hundredth of a second behind the watch that stayed home (use $\gamma \cong / + v^2/2c^2$).

10. A meterstick is moving along the direction of its length. It appears to be 99 cm long. What is its velocity?

11. For what value of v/c does a moving body appear to be halved in length?

12. For what value of γ does a muon travel 60 km within its half-life of 2×10^{-6} s?

13. The velocity of light in still air is less than c by 82 km/s. Let S be an observer moving through the air at the speed $c/3$. According to S, by how much is the velocity of light through the air less than c? (Consider two cases: light moving along the direction of motion and light opposite to it.)

14. While Oprah is on her outbound journey, the *Albert's* captain observes its sister ship, the *Marie Curie*, returning from its jaunt to α-Centauri and to be on a collision course 1 light-year away. Both liners' speeds are $0.8c$, but in opposite directions.

  a. How rapidly are the ships approaching one another according to observers on the *Albert* or on the *Marie Curie*?

  b. If no evasive action is taken, how much time will pass on the *Albert* before the impact?

  c. The *Albert* has the right-of-way. Its captain radios the *Marie Curie* of the problem. How far away is the *Marie Curie* from observers in the *Albert* when the message is received?

15. Spaceliner *Albert* must decelerate from $v = 0.8c$ to $v = -0.8c$. Suppose it does so at twice the acceleration of gravity (or -20 m/s) according to Harpo on Earth. According to him, how long does the *Albert* take to reverse its direction? Express your answer in months.

16. Suppose that A moves at velocity u relative to B, and that B moves at velocity v relative to C, and that C moves at velocity w relative to D. The three velocities are in the same direction. Show that V, the velocity of A relative to D, is given by:

$$V = \frac{u + v + w + uvw/c^2}{1 + (uv + vw + wu)/c^2}$$

12.3 The Special Theory of Relativity—Energy and Momentum

The conservation laws for momentum and energy are among the most basic concepts in mechanics. They must be obeyed in all inertial coordinate systems. However, the special theory of relativity changes the relationship among velocities measured by different observers. To maintain the conservation laws, we must modify our understanding of mass, momentum, and energy. First of all, we must clarify what is meant by the mass of an object or a system of objects. According to classical mechanics, the mass of an object is the ratio of the force acting on it to the acceleration it experiences. The relativistic definition of mass is exactly the same, with one proviso: The mass of an object or a system of objects is to be determined in its center-of-mass, or rest, system.

According to classical mechanics, the mathematical expressions for the kinetic energy of an object and its momentum are

$$\text{Classical mechanics:} \quad E = mv^2/2, \quad \mathbf{p} = m\mathbf{v} \tag{10}$$

There is a firm upper limit to v. The special theory of relativity allows a massive object to have a velocity arbitrarily close to c, but its velocity cannot exceed or even reach c. According to equations 10, E cannot exceed $mc^2/2$ and p cannot exceed mc. Something is wrong because there is no limit to the amount of kinetic energy or momentum a moving object may have. The object may always be struck from behind so as to increase both E and $\mathbf{p}$. When velocities are comparable to c, the classical expressions for energy and momentum must be replaced by appropriate relativistic expressions.

Relativistic Momenta

According to the special theory of relativity, the momentum of a freely moving object with velocity $\mathbf{v}$ is

$$\text{Relativistic momentum:} \quad \mathbf{p} = \gamma\, m\, \mathbf{v} \tag{11}$$

This expression looks deceptively simple because its complexity is hidden in the factor of γ defined by equation 6. Does the replacement of the classical formula by the relativistic one mean that everything we learned about classical physics is to be forgotten? Not at all! The new expression simply extends the range of validity of the classical expressions to a new domain of large velocities.

The relativistic formula for momentum coincides with the classical expression except for the factor γ. If v/c is a very small number, we may use an approximate formula for γ:

$$\gamma \simeq 1 + \frac{1}{2}\,(v/c)^2 \tag{12}$$

Under ordinary circumstances, the first term suffices and we may use $\mathbf{p} = m\mathbf{v}$. In celestial mechanics, where planetary speeds are typically $v \simeq 10^{-4}c$, equation 12 errs by only a few parts per billion. Relativity might make a difference of seconds per century in the predicted time of an eclipse, for example. For velocities that are small compared to c, the old and new formulas for momentum give approximately the same results; as v approaches c, however, γ (and consequently p) increases without limit. Under this circumstance, equations 10 can no longer be used (Figure 12.12).

FIGURE 12.12
The momentum of a moving body. The classical (and incorrect) formula for the momentum of an object with velocity v is mv. It is shown here as a straight line vs. v/c. The correct relativistic expression is $p = \gamma mv$. This is shown as a curved line. For small v/c, the two expressions agree. As v approaches c, however, the momentum of the body (but not mv) increases without limit.

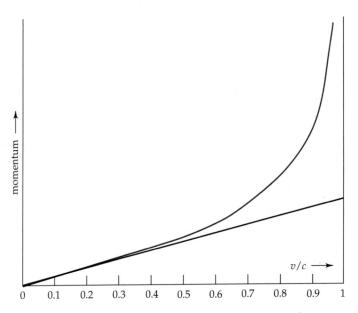

EXAMPLE 12.4
Classical vs.
Relativistic Momenta

At the same velocity, the classical expression for momentum, $p = mv$, is smaller than the correct relativistic expression. For what values of v/c does the classical formula underestimate the momentum of an object by 0.1 percent, 1 percent, and 10 percent?

Solution

According to equation 11, the ratio of a body's momentum to mv is γ. To answer this question, we must determine the values of v for which $\gamma = 1.001$, 1.01, and 1.1, respectively, where $\gamma = 1/\sqrt{1 - v^2/c^2}$. For the various cases, we obtain:

$$\gamma = 1.001, \quad \text{or} \quad 1 - v^2/c^2 \simeq 1.002, \quad \text{or} \quad v \simeq 0.045\, c$$

$$\gamma = 1.010, \quad \text{or} \quad 1 - v^2/c^2 \simeq 1.020, \quad \text{or} \quad v \simeq 0.14\, c$$

$$\gamma = 1.100, \quad \text{or} \quad 1 - v^2/c^2 = 1.21, \quad \text{or} \quad v \simeq 0.46\, c$$

∎

Relativistic Energies

The relativistic expression for the energy of an object also differs from its classical expression. The total energy of an object of mass m and velocity v is

$$\text{Total energy:} \quad E = \gamma mc^2 \tag{13}$$

This formula has a curious and important property: The total energy of an object at rest does not vanish. If $v = 0$, then $\gamma = 1$ and the total energy of the object is

$$E_0 = mc^2 \tag{14}$$

E_0 is called the *rest energy* of an object of mass m. This is a famous formula that you may have seen before. It suggests that energy and mass are somehow interconvertible. We shall soon see in what sense this is true.

If an object is moving, γ is greater than 1. Not surprisingly, the total energy of an object moving at velocity v is greater than its rest energy. The energy of motion of an object is the difference between its total energy and its rest energy: Kinetic energy $= E - E_0$. If v is small compared to c, we may use equation 12 to evaluate γ and obtain an approximate formula for E:

$$E \simeq mc^2 + \frac{1}{2}mv^2$$

Thus, the kinetic energy of a slowly moving object is approximately $\frac{1}{2}mv^2$, as in classical mechanics. However, as v approaches c, the factor γ increases without limit. The energy of a rapidly moving body may be arbitrarily large. If v is comparable to c, the classical expression for kinetic energy cannot be used (Figure 12.13).

An important relation between the energy and the momentum of an object is found by subtracting the square of its rest energy from the square of its total energy:

$$E^2 - E_0^2 = (\gamma^2 - 1)m^2c^4 = \gamma^2 m^2 v^2 c^2 = p^2 c^2$$

This result may be used to express the energy of an object in terms of its momentum, or the magnitude of its momentum in terms of its energy:

$$E = \sqrt{m^2c^4 + p^2c^2} \qquad p = \sqrt{E^2/c^2 - m^2c^2} \tag{15}$$

FIGURE 12.13
The kinetic energy of a moving body. The classical formula for kinetic energy is $\frac{1}{2}mv^2$. It is shown as the lower curve. The correct expression $[(\gamma - 1)mc^2]$ is shown as the upper curve. For small v/c, the two expressions agree. As v approaches c, however, the kinetic energy (but not $\frac{1}{2}mv^2$) increases without limit.

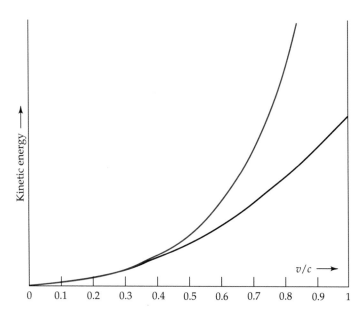

Furthermore, the rest energy of an object may be expressed in terms of its energy and momentum:

$$mc^2 = \sqrt{E^2 - p^2c^2} \tag{16}$$

The energy E and momentum p of the object depend on the frame of reference of the observer, but the right-hand side of this equation does not: It is the square of the rest energy of the body. The rest energy (or the mass) of a moving object can be determined by measurements of its energy and its momentum.

EXAMPLE 12.5
Energy

Consider the three cases $v = 0.1c$, $v = 0.5c$, and $v = 0.9c$. At each of these velocities, calculate (a) the ratio of the energy of an object to its rest energy, and (b) the ratio R of the kinetic energy of the object to its classical value.

Solution

a. The ratio of energy to rest energy is γ. Thus, the answers to this question are

If $v = 0.1c$, then $\gamma = 1/\sqrt{1 - .01} \simeq 1.00504$

If $v = 0.5c$, then $\gamma = 1/\sqrt{1 - .25} \simeq 1.155$

If $v = 0.9c$, then $\gamma = 1/\sqrt{1 - .81} \simeq 2.294$

Notice that the energy of an object moving at 90 percent of the speed of light is over twice its rest energy.

b. The ratio of kinetic energy to $\frac{1}{2}mv^2$ is $2(\gamma - 1)c^2/v^2$. Making use of the values for γ just determined, we obtain for the desired ratios:

If $v = 0.1c$, then $R = 2(.00504)/.01 \simeq 1.008$

If $v = 0.5c$, then $R = 2(.155)/.25 \simeq 1.24$

If $v = 0.9c$, then $R = 2(1.294)/.81 \simeq 3.20$

At $v = 0.1c$, the classical formula for kinetic energy is correct to within 1 percent. The error increases to 24 percent at half light speed. Beyond that point, the classical formula is useless. ■

Conserving
Energy and
Momentum

Equations 11 and 13 are the relativistic expressions for energy and momentum. Two fundamental conservation laws are satisfied in any process by which objects come together, collide, or fall apart, such as

$$A + B \rightarrow C, \qquad A + B \rightarrow C + D, \qquad A \rightarrow B + C$$

Conservation of momentum says that the sum of the momenta of the initial objects is equal to the sum of the momenta of the final objects.

Conservation of energy says that the sum of the energies of the initial objects is equal to the sum of the energies of the final objects. However, there is no independent law of conservation of mass. In particular, the sum of the masses of the reactants is not always equal to the sum of the masses of the products.

According to the special theory of relativity, neither mass nor kinetic energy is independently conserved in reactions. Total energy is conserved, but the total energy of each participant includes both its rest energy and its energy of motion. However, a quantity analogous to mass is not only conserved but is independent of the state of motion of the observer. This quantity is called the *invariant mass* of a system and is closely related to its *invariant energy*. We will use important concepts later on.

Imagine a system of two objects with energies E_1 and E_2 and momenta $\mathbf{p}_1$ and $\mathbf{p}_2$. Their total energy is $E_1 + E_2$ and their total momentum is $\mathbf{p}_1 + \mathbf{p}_2$. The values of each of these quantities depend on the state of

---

A BIT MORE ABOUT
Photons

The photon is a special kind of particle: It is massless and travels at speed c to all observers. For photons, equations 11 and 13 make no sense at all: $m = 0$ and $\gamma = \infty$. What are we to do? The answer lies in equation 15 with $m = 0$. The relation $E = pc$ holds for any massless particle, such as the photon. This result is implicit in Einstein's view of photons. We learned in Chapter 11 that the energy of a photon of wavelength λ is $E = hc/\lambda$ and its momentum is h/λ. These relations imply that $E = pc$ for photons. Let's apply relativistic mechanics to the Compton scattering problem. What is the shift of a photon's frequency when it scatters backward from a stationary electron of mass m?

A photon with momentum p collides with an electron at rest and bounces backward with momentum $-p'$. The total energy of the system of two particles is $E = mc^2 + pc$

and the total momentum is p. Conservation of momentum says that the electron's momentum after the collision is $p + p'$. The only remaining condition is conservation of energy:

$$mc^2 + pc = \sqrt{m^2c^4 + (p + p')^2c^2} + p'c$$

Bring the term $p'c$ to the left and square both sides of the equation. We find

$$(mc + p - p')^2 = m^2c^2 + (p + p')^2$$

or

$$2mc(p - p') = 4pp'$$

or

$$1/p' - 1/p' = 2/mc$$

and finally

$$\lambda' - \lambda = 2h/mc$$

The last equality expresses the fact that the wavelength of the back-scattered photon is increased by $2\lambda_c$, where $\lambda_c = h/mc$ is the Compton wavelength of the electron.

motion of the observer. However, the quantity

$$(E_1 + E_2)^2 - |\mathbf{p}_1 + \mathbf{p}_2|^2 \, c^2 = M^2 c^4 \tag{17}$$

is independent of the frame of reference. The quantity M is interpreted as the invariant mass of the system of two particles. M is necessarily larger than the sum of the two individual masses. The quantity Mc^2 is interpreted as the invariant energy of the system of particles. These quantities are called invariants because (1) their values do not depend on the state of motion of the observer and (2) they remain unchanged in the course of any reaction between the two objects. Using equation 17, we may express the invariant mass M of a system of two moving bodies as

$$M = \sqrt{(E_1 + E_2)^2/c^4 - |p_1 + p_2|^2/c^2} \tag{18}$$

If the bodies collide, their invariant mass does not change. The sum of the individual masses of the particles produced can be equal to or less than M, but it cannot exceed M.

How Mass Can Become Energy

In the process of nuclear fission, a uranium nucleus comes apart into two smaller nuclei and a few neutrons. For simplicity, let's consider a simpler situation. A nucleus of mass M is initially at rest. It splits into two fragments, each with mass m. Because the fissioning nucleus was at rest, the momenta of the fragments ($\mathbf{p}_1$ and $\mathbf{p}_2$) are equal and opposite:

$$\mathbf{p}_1 = \gamma m \mathbf{v}, \qquad \mathbf{p}_2 = -\gamma m \mathbf{v}$$

The conservation of energy says that the initial energy (Mc^2) equals the sum of the final energies:

$$Mc^2 = 2\gamma mc^2 \qquad (\text{or} \quad \gamma = M/2m)$$

The left-hand side of this equation is the rest energy of the initial nucleus. The right-hand side is the invariant mass of the two-body system it becomes. If the fission reaction is to take place, the original nucleus must be more massive than the sum of its decay products. The amount of kinetic energy released is c^2 times the amount of mass that is lost:

$$\text{Kinetic energy release} = (M - 2m) \, c^2$$

The fraction of mass lost in a typical fission process is 10^{-3}, corresponding to $\gamma = 1.001$ for fission products of equal mass. Each of the two fragments emerge from the reaction with velocities of about 4.5 percent c.

Our discussion applies to many other phenomena than nuclear fission. Suppose an object at rest comes apart into two other objects: $A \longrightarrow B + C$. The process could be:

- A grenade exploding into two fragments
- A molecule decomposing, such as: $N_2O_3 \rightarrow NO + NO_2$
- An atom in an excited state emitting a photon

- A nucleus fissioning into two smaller nuclei or decaying through α decay
- An elementary particle decaying into two particles, such as: $\pi \to \mu + \nu$.

In each of these processes, both energy and momentum are conserved. In particular, the mass of the object that comes apart is necessarily greater than the sum of the masses of the objects into which it comes apart. If the process is to take place, the mass of the parent must exceed the sum of the masses of the daughters. That is:

$$A \longrightarrow B + C \quad \text{implies} \quad M_A \geq M_B + M_C$$

The loss in mass in the course of a reaction, when multiplied by c^2, is the kinetic energy released and carried off by the decay products. That is, the sum of the kinetic energies of B and C is equal to $M_A c^2 - (M_B + M_C)c^2$.

Lavoisier was wrong: The sum of the masses of the products is a bit less than the sum of the masses of the reactants in a reaction that releases energy. It is a bit more in a reaction that consumes energy. The rest energy of an object represents the amount of energy that is bound up in the form of mass. Suppose we blindly apply equation 14 to a quarter-pound hamburger ($m = 150$ g). The result is 10^{16} J, which is the amount of energy released by a whopper of a nuclear bomb. Yet, the nutritional energy of the burger is a mere 10^6 J. This means that our guts can extract only a very tiny fraction (about 10^{-10}) of the rest energy of food. Most of the rest energy of matter is in the form of atomic nuclei. Under special circumstances, we can get at some of it. Large nuclei can be induced to split apart (by nuclear fission) or small nuclei can be coaxed to coalesce (by nuclear fusion). When this happens, a larger fraction of the rest energy of matter becomes accessible to us—as much as 1 part per 1000. And conversely, the total mass of the waste products is only 99.9 percent of the fuel that was consumed. Thus, the law of conservation of mass must be reconsidered: It is just not true in its original form.

How Energy Can Become Mass

Consider a reaction that requires energy to take place, such as the decomposition of water into hydrogen and oxygen:

$$2H_2O + \text{Energy} \longrightarrow 2H_2 + O_2$$

The energy added to the water becomes mass. If 1000 tons of water are decomposed, the mass of the gaseous products is about 0.3 g heavier than 1000 tons. This is a negligible effect—chemists may safely ignore the relativistic departures from conservation of mass. However, the conversion of energy to mass is of critical importance in particle physics.

Let A and B be two energetic particles with mass m and equal and opposite velocities. They can collide with one another to form a system whose mass is larger than $2m$:

$$A + B \longrightarrow C$$

For example, the collision can result in the production of four particles at rest, each with mass m. Conservation of energy tells us that $2\gamma mc^2$ (the initial energy) equals $4mc^2$ (the rest energy of the final particles). It follows that $\gamma = 2$. If the particles have equal and opposite velocities, their speeds must be $v = 0.87c$. The situation is quite different if particle A is moving and particle B is at rest. If the collision is to produce a system with invariant mass equal to $4m$, the invariant mass of particles A and B must be at least that great. In this case, equation 18 can be used to show that particle A must travel at a velocity of 99 percent that of the speed of light.

One of the most important experiments in high-energy physics was the discovery of the Z boson, to be discussed in Chapter 15. This particle is like the photon, but its mass is that of a large atom. The Z boson is produced in a collision of two particles, and its mass is about 100,000 times the sum of the masses of the colliding particles! In this process, energy is converted to mass with a vengeance.

Two of the bulwarks of classical physics are the laws of conservation of mass and of energy. The theory of relativity tells us that these laws are not independent of one another. In fact, neither energy nor mass is absolutely conserved. When a ton of coal is burned, the mass of the

---

DO YOU KNOW
About Antimatter?

When energy is extracted from a system, its mass decreases. A spent battery has a mass that is a tiny bit smaller than the same battery when it is charged. Nuclear processes are more efficient: They can convert 10^{-3} of the mass of uranium into energy. There are two ways by which all (or a large fraction) of the energy inherent in matter may be extracted.

Later on, we shall introduce you to *antimatter*, which is a kind of matter not normally found on Earth. It exists only in the sense that it can be made in tiny quantities by means of the collisions of high-energy particles. When matter encounters antimatter, the rest energy of both can be converted into electromagnetic energy. Antimatter could supply an endless amount of energy. However, astronomers are certain that there is hardly any antimatter in the solar system or in the stars. We can and do manufacture antimatter particle by particle, but its energy cost far exceeds its energy content. The total amount of antimatter produced by the laboratories of the world would fit on a pinhead.

Physicists have found another way to extract the rest energy from ordinary matter. Some stellar explosions result in the creation of *black holes*—objects with the mass of a star but the size of a mountain. When ordinary matter falls into a black hole, about half of its rest energy is consumed by the black hole, but the rest emerges as heat and light. If we had a convenient nearby black hole, we could extract limitless energy by throwing our garbage into it, thereby solving two problems at the same time. Unfortunately, we have neither found a nearby black hole nor can we make one.

carbon dioxide produced is about 1 milligram shy of the mass of carbon and oxygen consumed. Mass and energy are convertible currencies, with certain restrictions. According to the special theory of relativity, the two laws become one: the conservation of mass-energy. Loosely speaking, energy may be transformed into mass, or mass into energy.

Exercises

17. A proton has a kinetic energy that is equal to its rest energy.
    a. What is the value of γ?
    b. At what fraction of the velocity of light is it moving?

18. A proton is traveling at a speed of $c/2$.
    a. What is the value of γ?
    b. What is the ratio of its kinetic energy to its rest energy?

19. The Sun radiates 4×10^{26} W of power. By how much does the mass of the Sun decrease in 1 s?

20. Show that the invariant mass of a system of two objects cannot be smaller than the sum of their rest masses. (*Hint:* Because the result is independent of the choice of reference frame, you may assume that one object is at rest.)

21. How much heavier (in eV/c^2) is a hydrogen atom in its $n = 2$ state than one in its ground state?

22. The combustion of one ton of carbon releases 3.4×10^{10} J of energy in the form of heat. How much lighter in grams are the products than the reactants?

23. An object of mass m and total energy γmc^2 collides with a stationary object of the same mass.
    a. Show that the invariant mass of the two bodies is $m\sqrt{2(1+\gamma)}$.
    b. If the collision results in the production of a single object of mass $4m$, show that $\gamma = 7$.

12.4 Einstein Overthrows Newton—The General Theory of Relativity

What's so special about the special theory of relativity? There are many possible coordinate systems. They may spin around or be accelerated in one direction or another. The special theory of relativity offers a framework for the description of motion in a special class of coordinate systems: inertial, or uniformly moving, systems. Einstein was not satisfied with his theory of 1905. He sought and found a more general framework, one that is totally independent of the choice of coordinate system. His generalization of the special theory of relativity is known as the *general theory of relativity*.

The special theory of relativity follows from the requirement that the laws of physics—including the value of c, which follows from Maxwell's equations—are the same to all uniformly moving observers. According to Einstein's general theory of relativity, the laws of physics are the same to all observers, whether inertial or accelerated. Consider an observer in a sealed elevator accelerating "upward" at 10 m/s^2 in deep space, far from any gravitating bodies. He experiences an upward force exerted by the floor of the elevator on his feet, just as if the accelerator were sitting at rest on Earth. There is no way to tell the difference between a constant acceleration and a constant gravitational force. Acceleration is a perfect mimic of gravity. A theory that is valid in all coordinate frames must describe gravity as well. It took Einstein until 1915 to construct such a scheme. As he later wrote, the tale began in 1907, when

> there occurred to me the happiest thought of my life, in the following form. The gravitational field has only a relative existence, because for an observer falling freely from the roof of a house there exists—at least in his immediate surroundings—no gravitational field. Indeed, if the observer drops some bodies then these remain relative to him in a state of rest or of uniform motion, independent of their particular chemical or physical nature. The observer has the right to interpret his state of motion as "at rest." The known matter independence of the acceleration of fall is a powerful argument for the fact that the relativity postulate has to be extended to coordinate systems which, relative to each other, are in non-uniform motion.

In Einstein's view of general relativity, dynamics is replaced by geometry. Space itself is warped by the nearby presence of a massive body. General relativity is a very mathematical discipline. It deals with the geometry of an intrinsically curved four-dimensional space–time universe. The equations of general relativity determine the geometry of space–time that is produced by a distribution of masses. Just as Newton's analyses forced him to develop the mathematics of calculus and differential equations, Einstein had to learn, and in some cases create, new branches of mathematics called tensor calculus and differential geometry. We shall not attempt to explicate his theory in any detail. However, we can get a feeling for the complexity of curved space–time by recalling the problems that mapmakers have in presenting the shapes and sizes of the continents and oceans on a flat map.

We live on the curved surface of Earth. You cannot cut out a portion of the globe, say the United States, and spread it on the page of an atlas without distortion. As the British representative of the Nyasa-Tanganyika Boundary Commission put it a century ago:

> If the area to be represented bear a very small ratio to the whole surface of the sphere, the matter is easy. If it is larger, then the curvature begins to be sensible. The sphere cannot be opened out into a plane like a cone or a cylinder; consequently, in a plane representation of configurations on a sphere it is impossible to retain the desired proportions of lines or areas

or equality of angles. But though one cannot fulfil all the requirements, we may fulfil some by sacrificing others.

Whatever the problems mapmakers have, the geometry of curved surfaces is simple compared to the geometry of four-dimensional space–time. The curvature at a point on any surface is entirely characterized by a single number called the radius of curvature. In four dimensions, fully twenty different numbers are needed to specify the geometry at any point! Nonetheless, we can begin to appreciate Einstein's identification of gravity in terms of geometry with a two-dimensional toy model. Think of the universe as a perfectly flat and frictionless sheet of rubber. A massive body (the Sun) is placed at one point. It distorts the sheet about it just as the real Sun distorts three-dimensional space. A second smaller body placed on the rubber sheet quite naturally describes a curved orbit about the larger body, just as planets trace ellipses in the curved space about the Sun (Figure 12.14).

Einstein completed his general theory of relativity in 1915. It is a powerful and useful generalization of the special theory of relativity in at least three senses:

1. General relativity is a theory of gravity that replaces Newton's theory. Its predictions differ slightly from those of its predecessor. Einstein realized that his theory accounted for a discrepancy that had bedeviled generations of astronomers. As Einstein put it, the theory quantitatively explains "the secular rotation of the orbit of Mercury discovered by Le Verrier without the need of any special hypothesis. For a few days I was beside myself with joyous excitement."

2. General relativity predicts the existence of new phenomena. Among them are the bending of light by gravity and the possibility of star-like objects from which light cannot escape—*black holes*. The bending

FIGURE 12.14
A weight placed on a flat and frictionless rubber sheet causes a deformation of the sheet— a change in its two-dimensional geometry. A smaller weight rolling along the sheet describes a curved trajectory. In a similar way, the Sun affects the geometry of three-dimensional space, thereby causing planets and comets to describe curved trajectories.

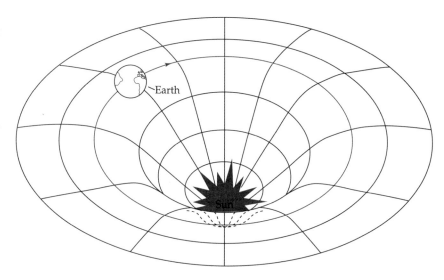

of light by gravity has been observed, and most astronomers are convinced that black holes exist in the centers of some galaxies.

3. General relativity provides the framework from which cosmologists can explain the birth and evolution (and even the ultimate fate) of the universe. We return to this subject in the concluding chapter of this book.

Few scientists were convinced when Einstein put forward his new theory. The mathematics was difficult and unfamiliar, and the experimental consequences of the theory seemed to be very few. However, many experiments have confirmed Einstein's predictions.

The Three Classic Tests of General Relativity

No matter how compelling or elegant it is, a theory of physics must be subject to experimental verification or it differs little from medieval theology. At the time of its invention, Einstein suggested three ways in which his theory differed in its consequences from Newton's.

1. *The Precession of the Perihelion of Mercury* The orbit of Mercury is not a perfect ellipse. It does not quite close upon itself, but moves as shown in Figure 12.15. Its axis turns about the Sun once in about 20,000 y. Most of this effect is due to the perturbing influences of the other planets, but a tiny residual precession cannot be explained by Newtonian theory. Today, the effect is measured to an accuracy of half a percent and it agrees perfectly with the prediction of the general theory of relativity.

2. *The Bending of Light by the Sun* The first test of general relativity was "retrodictive" rather than predictive: It resolved a long-standing problem. More impressive was a prediction of a new and altogether unexpected effect—the effect of gravity on the trajectory of light. Einstein's theory demands that starlight, when grazing the surface of the Sun, be bent by 1.74 seconds of arc—the angle subtended by your pinkie nail at a distance of two miles (Figure 12.16). Of course, you cannot ordinarily see the stars when the Sun is in the sky, so empirical proof had to wait for a total eclipse. Einstein's confidence in his new theory seemed

FIGURE 12.15
The precession of Mercury. Mercury's approximately elliptical orbit precesses about the Sun as shown. The long axis of the ellipse completes a turn about the Sun in about 20,000 years. Most of this effect was explained by the gravitational pull of other planets, but a small anomaly (43 seconds of arc per century) remained. The resolution of this discrepancy was the first triumph of Einstein's general theory of relativity.

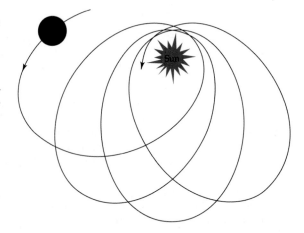

FIGURE 12.16
The bending of light by the Sun. According to the general theory of relativity, starlight is deflected by 1.74 seconds of arc when it passes nearby the Sun. This effect was first observed during a total eclipse of the Sun in 1919. It was the second experimental triumph of Einstein's theory of gravity.

unbounded: "I do not doubt any more the correctness of the whole system whether the observations of the solar eclipse succeed or not." Yet, when World War I endangered the eclipse expedition, he wrote: "Only the intrigues of miserable people prevent the execution of the last, new, important test of the theory."

Einstein's prediction was confirmed by the 1919 British eclipse expedition, but to a precision of only 20 percent. More recently, radio astronomers measured the bending of radio waves by the Sun. Their results agree with general relativity to a precision of 1 percent. A new and even more sensitive test was proposed by Irwin Shapiro in 1964. It turns out that light passing nearby the Sun is not only bent but delayed. The Viking space probes carried transponders with which the travel time of a signal sent from Earth to Mars and back was measured. The time is affected by the presence of the Sun, and the data accords with general relativity to an accuracy of 0.1 percent.

In 1979, the gravitational bending of light appeared in an entirely new and exciting arena. The British science journal *Nature* rhapsodized about the startling discovery of identical twin quasars:*

> The twin quasar is the first known example of the gravitational lens effect. This has long been predicted if a massive galaxy should happen to be almost on the line of sight between us and a quasar. The twin consists of two very similar objects only 6 arc seconds apart. Walsh, Carswell, and Weymann found almost identical spectra for both objects, and were astonished to find that they had identical red shifts [meaning that they are at the same distance from us]. The twins were so identical that they could only reasonably be

*Quasars or quasi-stellar objects are the most distant objects that can be seen. A quasar is relatively small (perhaps the size of our solar system), but it generates more energy than a galaxy consisting of billions of stars.

accounted for as two images of the same object, exactly as would be expected if a massive galaxy lay on the line of sight.

The discovery of the gravitational lens comes only a few months after the excitement of the orbital changes in the binary pulsar, which demonstrated the existence of gravitational waves. Einstein has indeed been well celebrated in his centenary year.

Galaxies are not only lens shaped, but thanks to general relativity, we know that they behave as lenses as well. Many more twin images caused by gravitational lenses have since been found. Their study is a way of studying the distribution of matter in the universe. The bending of light by gravity, which was once a difficult to perform test of general relativity, has become a powerful astrophysical tool.

3. *The Effect of Gravity on Clocks* According to general relativity, clock rates are affected by gravity. The higher the gravitational potential, the slower the clock. The effect is tiny and was not measured until 1959. As shown in Figure 12.17, one of two identical clocks is placed in a cellar, 75 feet below the other. Gravity causes it to lose a second in ten million years! Robert Pound and Glenn Rebka detected the effect at Harvard in the very building in which I sit! Their clocks were gamma-ray sources placed at the top and bottom of a tower. They observed that the frequency of the lower source was lower by just the predicted amount. Einstein's theory passed its third test to an accuracy of 1 percent. Once again, there have been dramatic experimental improvements since. In 1976, an atomic clock placed in a rocket provided a test of the gravitational redshift to an accuracy of 0.02 percent.

FIGURE 12.17
The gravitational slowing of time. Two identical clocks are synchronized. One is placed in the attic, the other in the basement. Because of the difference in gravitational potential, the attic clock ticks more rapidly than the basement clock, which falls behind. In billions of years, the effect will be as large as this figure indicates. Nonetheless, the gravitational slowing of clocks has been observed and measured.

A recent review article commemorates the 75th anniversary of Einstein's construction of the general theory of relativity:*

> On the 75th anniversary of the genesis of General relativity, we find that the theory has held up under extensive experimental scrutiny. The question then arises, why bother to continue to test it? One reason is that gravity is a fundamental interaction of nature and as such requires the most solid empirical underpinning we can provide. Another is that all attempts to quantize gravity and to unify it with the other forces suggests that gravity stands apart. The more deeply we understand gravity and its observable implications, the better we may be able to confront it with the other forces.
>
> Finally, and most importantly, the predictions of general relativity are fixed; the theory contains no adjustable constants, so nothing can be changed. Thus every test of the theory is a potentially deadly test. A verified discrepancy between observation and prediction would kill the theory, and another one would have to be substituted in its place. Although it is remarkable that this theory, born 75 years ago out of almost pure thought, has managed to survive every test, the possibility of finding a discrepancy will continue to drive experiments for years to come.

Einstein would smile.

Where We Are and Where We Are Going

The last two chapters described the "tools of the trade" of modern physics: quantum mechanics and the theory of relativity. Things are not as they seem: The rules obeyed by rapidly moving particles are entirely different from the rules we intuit from everyday experience. Ordinarily, classical physics works admirably and agrees with our sensibilities. But that is just because we are large and slow on the atomic scale. The world we live in is intrinsically relativistic and quantum mechanical. The laws of classical physics can describe neither the particles of which matter is made nor the complex fashion by which they interact with one another. Using quantum theory and relativity, we are now poised to descend into the subatomic world, where we shall explore the deepest secrets of nature.

The first frontier appeared early in the twentieth century: It was the atomic nucleus and its mystifying radioactive transformations. Nuclear physics is the subject of Chapter 13. Although quantum mechanics sufficed to give us a quantitative understanding of atomic structure, the puzzles posed by nuclear phenomena could only be solved in terms of relativity and quantum mechanics. Aside from technological developments (such as nuclear power, medicine, and weaponry) the new discipline of nuclear physics solved the mysteries of how stars are born, how they evolve, and how they created the matter of Earth.

*C. M. Will, *Science* 250 (1990): 770–775.

13 Inside the Nucleus

The practical utilization of the abundant store of energy locked up in every atom of matter is a problem which only the future can answer. Remember, at the dawn of electricity, it was looked on as a mere toy.

Henri Becquerel, French physicist

Anyone who looked for a source of power in the transformation of the atoms was talking moonshine.

Ernest Rutherford, British physicist

"Men have forgotten a truth," said the fox. "You are responsible forever after for anything you tame."

A. de Saint-Exupéry, French author

Pierre and Marie Curie in their laboratory. The husband and wife team, as well as their daughter and son-in-law, made major contributions to the new science of radioactivity. Madame Curie is most well known for her discovery and isolation of radium. *Source: The Granger Collection.*

By 1930, physicists and chemists agreed that the basic laws governing atomic and molecular phenomena were known. Quantum mechanics explained the chemical and optical properties of atoms and provided a logical framework from which a description of the properties of bulk matter could be deduced. However, the structure and behavior of the atomic nucleus was still unknown. This chapter describes the development of nuclear physics from the discovery of radioactivity to the construction of the first nuclear reactor. Section 13.1 introduces the fundamentals of nuclear science and describes its beginnings: how it was learned that radioactivity changes one chemical element into another while satisfying a statistical law of decay. Elements were found to come in isotopes with different masses, some of which might be radioactive, others not. Section 13.2 tells of the discovery of the neutron and how it led to the quantitative science of nuclear physics. Section 13.3 applies the law of energy conservation to the nuclear processes of radioactivity, fission, and fusion. We learn how stars shine and how to mimic them on Earth.

13.1 What Is Radioactivity?

The discovery of radioactivity in 1897 posed a deep problem for scientists. Energy is produced by radioactive materials, but nobody understood the mechanism by which it is produced. Radioactivity is now known to be a nuclear process, but the constituents of the nucleus (neutrons and protons) were not identified until 1932. This section focuses on a remarkable collaboration between the physicist Ernest Rutherford and the chemist Frederick Soddy. The collaboration took place at McGill University in Canada, years before Rutherford returned to England to discover the nucleus. Before we begin, let us sacrifice mystery to clarity and review some of the properties of nuclei and nuclear transformations discovered long after Rutherford and Soddy blazed the trail.

Isotopes

Atomic nuclei are made of two kinds of particles with about the same masses. A given nuclear species contains Z positively charged *protons* and N electrically neutral *neutrons*. The sum $N + Z$ is called A, the *atomic mass number*. A is approximately equal to the atomic weight of the atom expressed in atomic mass units. The chemical properties of an atom are determined by its atomic number Z, the nuclear charge. Neon is the tenth element in the periodic table, and every neon nucleus contains ten protons. Different atoms of the same element often contain different numbers of neutrons. These different species are called *isotopes*. Two isotopes of the same element have the same value of Z but different values of A. Isotopes were named and discovered by Soddy as a by-product of his collaboration with Rutherford—they didn't know about neutrons and could not measure atomic weights, but they found chemically identical samples of the same element with different radioactive properties.

The link between isotopes and atomic weight arose in 1913, when J. J. Thomson (the discoverer of the electron) separated neon atoms into two components that were chemically identical but had different atomic weights. Neon, whose atomic weight is 20.2, consists of a mixture of isotopes, with $A = 20$ and $A = 22$. One isotope contains 10 neutrons, the other 12; these are denoted as $^{20}\mathrm{Ne}_{10}$ and $^{22}\mathrm{Ne}_{10}$, or more simply, as $^{20}\mathrm{Ne}$ and $^{22}\mathrm{Ne}$. Neither form of neon is radioactive: Thomson found the first stable isotopes.

Francis Aston, a young researcher at the Cavendish Laboratory with Thomson and Rutherford, spent his career identifying and measuring isotopes. He found 212 of the 281 naturally occurring stable isotopes. Since Aston's time, thousands of radioactive isotopes have been found. Some occur naturally, but most of them are synthetic. Early investigators of radioactivity were baffled by the fact that one isotope of a heavy element may be intensely radioactive, another much less so, and a third not radioactive at all.

Atomic Weights

Atomic weights are defined relative to the isotope $^{12}\mathrm{C}$, the mass of whose atom is defined to be 12 atomic mass units, or amu:

$$1 \text{ amu} = 1.66 \times 10^{-27} \text{ kg} = 931.5 \text{ MeV}/c^2 \tag{1}$$

Some elements have atomic weights that depart significantly from integers. (The case of chlorine was discussed in Chapter 6.) One reason for this is that elements occur as mixtures of different isotopes. The atomic weights of pure isotopes are always nearly whole numbers. For example, the atomic weight of ^{1}H (the common isotope of hydrogen and the lightest atom) is 1.0080, while that of ^{238}U (the common isotope of uranium and the heaviest natural atom) is 238.05. Isotopic atomic weights are not exactly whole numbers because nuclei are made up of protons and neutrons, whose masses are not identical. The neutron is about 10^{-3} amu heavier than the proton. Neutrons and protons are referred to generally as *nucleons*.

We learned in Chapter 12 that the mass of an unstable system exceeds the sum of the masses of its decay products. Conversely, the mass of a nucleus is less than the sum of the masses of its constituent nucleons. The mass difference, multiplied by c^2, is called the *binding energy* of the nucleus. For example, the mass of the α particle (the nucleus of ^{4}He) is less than twice the sum of the proton and neutron masses by 0.03 amu. This mass difference corresponds to a binding energy of about 28 MeV. Thus, it takes 28 MeV of energy to decompose an α particle into its component nucleons. Conversely, 28 MeV are released if two neutrons combine with two protons to form an α particle.

The ABC of Radioactivity

Natural radioactivity consists of three distinct components: α rays, β rays, and γ rays. α rays consist of energetic helium nuclei, β rays consist of energetic electrons, and γ rays are photons with energies of several MeV. Each type of radiation is produced by a nuclear transformation. Suppose that the radioactive parent nucleus has atomic number Z and mass number A.

α. In the process of α decay, the parent nucleus emits an α particle (two neutrons and two protons bound together). The decaying nucleus is transformed into a "daughter" nucleus whose atomic number is $Z - 2$ and whose mass number is $A - 4$.

β. In the process of β decay, the parent nucleus emits an electron. In this case, the daughter nucleus has mass number A, but its atomic number increases to $Z + 1$ so that electric charge is conserved.

γ. In the process of γ decay, a radioactive nucleus in an excited state emits a photon, thereby falling to a less excited state or to its ground state. Neither A nor Z changes. The process of γ decay is analogous to the emission of a photon by an excited atom. In the atomic process, the photon energy is usually a few eV. In the nuclear case, the photon energy is usually a few MeV—a million times larger!

The Discovery of Radium

Let's return to the beginning of the twentieth century, when little of what we just described was known. Marie Sklodowska, the daughter of a Warsaw physics teacher, came to Paris in 1891 to study physics. She met her first love and husband-to-be in 1894: the promising young physicist Pierre Curie. She met her second love in 1896: Becquerel's newly found phenomenon of radioactivity.

By April 1898, Madame Curie showed that thorium is more radioactive than uranium and some of its ores are more radioactive yet. "This fact is very remarkable," she wrote, "and leads one to believe that these minerals contain an element much more active than uranium." By July, the couple isolated a substance that was 400 times more active than uranium which, they believed, "contains a not-yet observed metal which we propose to call polonium, after the country of origin of one of us." The Curies suspected an even more radioactive element to lurk in pitchblende. After months of arduous chemical extractions, on the day after Christmas, 1898, they isolated the intensely radioactive element radium. They found (and deserved) instant fame as well. The magnitude of their task (and the intensity of radium's activity) can be seen from the fact that there is only about 150 mg of radium in a ton of pitchblende.

Pierre won a professorship at the prestigious Sorbonne University in Paris while Marie taught at a girl's high school. Lecturing, research, and motherhood left her little time to complete her thesis, but by 1903 she did. Later that year, she and Pierre shared the Nobel Prize in physics with Becquerel. Pierre was killed while crossing a street in 1906. His widow succeeded him at the Sorbonne as its first female professor in 600 years. Five years afterward, just after she isolated metallic radium and determined its chemical and physical properties, she won a second Nobel Prize—this time in chemistry—for her discoveries and studies of radium and polonium.

Madame Curie showed that radioactive materials produce a prodigious amount of energy per atom—millions of times more than any chemical process. About 200 W of power are generated by 1 kg of radium. The energy comes pouring out in apparently endless quantity, as in the fairy tale about the never-ending penny. And, nothing we can do affects it! Where did this energy come from? If the radium's activity were eternal, the principle of conservation of energy would have to be abandoned. In fact, the energy of radioactivity results from nuclear transformations, radium loses half its strength in about 1600 years, and energy is always conserved.

A Chemist and a Physicist Disagree

In 1899, Rutherford joined McGill University's physics department. While the Curies were busy with polonium and radium, Rutherford showed that α rays are helium nuclei and β rays are electrons. Frederick Soddy became a demonstrator in chemistry at McGill in 1900. Like the chemists of his day, Soddy regarded atoms as primal and fundamental entities. However, physicists—especially Rutherford—believed that atoms were structured and mutable systems made of simpler things. A debate was held in the fall of 1901. According to Rutherford, the physicists "hope to demolish the chemists." Soddy began the confrontation:

> My object has been to show that matter is possessed of very positive attributes which leads us to consider the atom as the material entity. I think the onus of proving that [radiation such as electrons and radioactivity] is really a form of matter rests on the new school, and until it is shown that

it is affected by gravity, or otherwise possesses features distinct from the ether, there will be no necessity for chemists to modify the Atomic Theory.

Possibly Professor Rutherford may be able to convince us that matter as known to him is really the same matter as known to us, or possibly he may admit that the world in which he deals is a new world demanding a chemistry and physics of its own, and in either case, I feel sure chemists will retain a belief and reverence for atoms as concrete and permanent identities, if not immutable, certainly not yet transmuted.

Soddy was interested in, but ignorant of, the latest news about radioactivity and the electron. Rutherford reminded the assembled chemists that electrons were particles of known charge and mass and were produced by radioactivity:

Electrons have been produced by the agency of light, heat, and the electric discharge. In some cases they are spontaneously emitted. Becquerel has shown that the radioactive uranium and radium give out some rays. These rays were found to be analogous in all respects to high velocity cathode rays.

The differences between the scientists and their disciplines were irreconcilable: Rutherford's atom had electrical substructure and could produce mystifying radiations; Soddy's atom was static and indivisible. Rutherford and Soddy decided to resolve their differences by working together. Instead of arguing with one another, they would appeal to the ultimate authority, nature. By 1903, their experiments transformed the chemists' world to conform to the new physics of radioactivity. A revolutionary picture of radioactive disintegration emerged whose central principles are these:

- *Radioactivity is spontaneous transmutation!* When a radioactive atom decays, it emits one or more energetic particles. At the same time, it is transformed into an atom of a different chemical element.
- *Radioactivity is a random process!* It is impossible in principle to predict when a given radioactive atom will decay. All that may be foretold is the probability for it to decay in the next second, and this probability remains absolutely constant throughout the life of the atom. Radioactive atoms eventually decay, but they do not age or change meanwhile.

Let's look more closely at each of these principles.

Radioactivity as Spontaneous Transmutation

In 1899, Rutherford and Robert Owens noticed a curious feature of radioactivity. Unlike the rock-steady activity of uranium, thorium's activity was affected by a draft of fresh air! There seemed to be two distinct components to the radiation thorium produced: a steady portion and a mysterious "radioactive emanation." Whether the emanation was a material substance or some new ineffable fluid was unknown. Rutherford and Soddy attacked this puzzle. They showed that the emanation associated with thorium could not be permanently removed—if thorium were cleansed of its emanation, the emanation soon reappeared. They proved that "the power of giving an emanation is really a specific property of thorium," and not of any other chemical contaminant.

Could the emanation be ordinary air that was somehow activated by radioactivity? To test this hypothesis, they put a thorium sample in a flask filled with CO_2 but saw no change in its behavior. The emanation was extracted from thorium and treated with powerful chemical reagents and extreme heat. It could not be induced to form chemical compounds. It was, therefore, an inert gas like neon or krypton, but it was radioactive! They had shown that the radioactive decay of thorium produces a new chemical element: radon. When realization dawned, Soddy later recalled, he blurted out: "Rutherford, this is transmutation: the thorium is disintegrating and transmuting itself into an inert gas of the argon family!"*

Rutherford and Soddy realized that the daughter of a radioactive parent is often radioactive itself, as is the granddaughter, and so on until the radioactive product is a stable atomic species. The cascade forms a *radioactive decay chain* from the parent to the stable end product, which is an isotope of lead for all naturally occuring isotopes of thorium or uranium.

Most thorium on Earth consists of the very long-lived isotope ^{232}Th. It decays by emitting an α particle to become a shorter-lived radioactive isotope of radium. A sequence of four more radioactive transformations yields an isotope of radon: Rutherford's emanation. After ten radioactive decays thorium is converted into stable lead. Thorium ores found on Earth contain tiny amounts of all nine of its radioactive daughters, granddaughters, and so on. Some live for years, some only for seconds, but as atoms of each generation decay they are replaced by the decays of the preceding generation. Because the average ^{232}Th atom lives for many billion years, the activity of the ore has come to equilibrium: Its radiation is constant in time, as are the relative abundances within it of thorium's even more radioactive descendants.

Radioactivity as a Random Process

The concept of radioactivity as a random process is both simple and astonishing. I'll first explain what it means, then describe how it was found out. Compare the life expectancies of a radioactive atom and a person. The median lifespan of an American at birth is about 75 years. As a person ages, his or her life expectancy decreases: perhaps it is 35 years at age 50 and less than a decade at age 90. The likelihood of death in the next year increases with age once one survives infancy. Much the same is true of shoes, tires, and light bulbs: The older they

---

*Rutherford and Soddy called the thorium emanation Thorium-X. In fact, it was ^{220}Rn, a very short-lived radon isotope. Uranium's decay produces a much longer lived (and thus, harder to detect) radon isotope, ^{222}Rn. Ramsay (the codiscoverer of argon) and Gray isolated this gas in 1908 and measured its atomic weight. They called it niton. The word *radon* was introduced in 1923 for the 86th element in all its isotopic avatars.

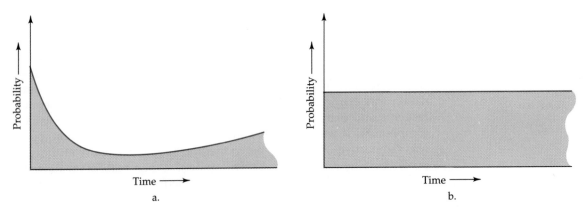

FIGURE 13.1 *a*. In everyday life, the probability that something will happen usually depends on time. This curve could show the likelihood of your awakening with a fever versus your age. Or it could show the likelihood of your car failing to start: It is higher for a new car until the bugs are straightened out. The probability of motor trouble rises as the car ages. *b*. The probability that a given radioactive atom will decay does not change with time. All atoms of a given isotope are identical to one another; atoms do not wear out (like cars) or improve with age (like wine).

are, the sooner they wear, blow, or burn out (Figure 13.1). Not so for the radioactive atom! Its life expectancy and the chance of its decaying in the next minute never change. Radioactive decay is a random event with a never-changing probability of occurrence. Unlike people and their artifacts, there is no difference between a newborn radioactive atom and an identical one that has existed for eons: Both have exactly the same chance to live another day.

The constancy of a radioactive atom's decay probability is expressed as an exponential law for the most likely number of survivors in a large sample of radioactive atoms (Figure 13.2). If the number of atoms of a certain radioactive material is N_0 at $t = 0$, and if these atoms are not being replenished, the most likely number remaining at a future time t is given by

$$N(t) = N_0\, e^{-t/\tau} \qquad \text{or equivalently} \qquad N(t) = N_0 \exp(-t/\tau) \qquad (2)$$

In every time interval τ, the number of survivors falls by a factor of $1/e$, which is approximately 0.368. The decay law specified by equation 2 applies to α, β, and γ processes, but it is not limited to radioactive decays. An atom in an excited state lives for a while before decaying to its ground state with the emission of a photon; it satisfies equation 2 with τ equal to a fraction of a nanosecond. A nucleus in an excited state behaves the same way. So do unstable elementary particles.

The exponential formula can be used to determine fraction F of radioactive atoms decaying in a time interval Δt that is small compared to τ:

$$\text{If } \Delta t \ll \tau \qquad \text{then } F = \Delta t/\tau \qquad (3)$$

The interval τ is an intrinsic attribute of a radioactive atom called its *lifetime* for decay, or *mean* life. Identical radioactive atoms may decay sooner or later, but their average life expectancy is τ. The mean life is

FIGURE 13.2

The exponential decay law. There are N atoms of a radioactive isotope at $t = 0$. The half-life of the isotope is 2 days and its lifetime is about 2.9 days. Notice that the number of surviving atoms falls to half its value every 2 days.

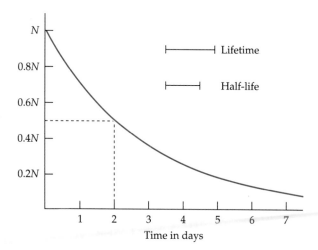

closely related to, but not quite the same as, the atom's *half-life*, which we now define. Many who are not scientists are unfamiliar with exponential functions. For this reason, it is convenient and useful to introduce the notion of a half-life $\tau_{1/2}$ of a radioactive species:

$$\tau_{1/2} \equiv \tau \log_e 2 \simeq 0.69\,\tau \tag{4}$$

The half-life is the period of time over which approximately half of the atoms in a sample of radioactive material decay. Half-lives can very short (about a minute for ^{220}Rn) or very long (14 billion years for ^{232}Th).

The radioactive isotope ^{220}Rn has a half-life of about one minute. When will a particular atom decay? The laws of physics cannot answer this question: It is a quantum-mechanical unknowable. All that can be known is a statistical estimate of how many atoms will decay in a given interval. Suppose we start with 6 atoms of ^{220}Rn. After one minute passes, the most likely result is that 3 atoms will have decayed. However, the chance that this will happen is only 31 percent. The chances are better than 2 to 1 that either more or fewer than 3 atoms will decay in the first half-life. There is even a 1 in 64 likelihood that none will decay, and an equal likelihood that all 6 of the atoms will decay!

However, if we start with 6 million atoms, the chances are better than even that 3 million plus or minus a few thousand will decay in one minute. The likelihood that fewer than 2,990,000 (or more than 3,010,000) will decay is very tiny. The more atoms we start with, the better equation 2 describes the actual behavior of $N(t)$. The intrinsic inaccuracy of the radioactive decay law—its probabilistic nature—was an early intimation of quantum-mechanical indeterminacy.

To discuss radioactivity quantitatively, we must define what is meant by the activity of a sample. The appropriate unit is named in honor of Madame Curie. The *curie* is an activity corresponding to 3.7×10^{10} radioactive disintegrations per second. Its value was chosen to be approximately the number of atoms decaying per second in 1 g of radium.

WORKING WITH HAHNIUM, WHICH HAS A HALF-LIFE OF 35 SECONDS

LET'S START. IS IT OVER? WHAT HAPPENED?

THERE IT IS. THERE IT GOES. I'M LEAVING.

I SNEEZED, AND I MISSED IT. LET'S EAT. WHAT SHOULD WE HAVE?

FAST FOOD. FAST FOOD.

©1993 by Sidney Harris.

EXAMPLE 13.1
Half-Lives

The half-life of ^{226}Ra is 1600 y.

a. By what fraction does the activity of a sample of radium decrease per decade?

b. Show that the radioactivity of 1 g of radium is about 1 curie.

c. The decay of each radium atom releases 38 MeV of energy. Show that 1 kg of radium produces about 200 W of power.

Solution

a. We determine the radioactive lifetime from equation 4 to be $\tau = 1600/0.69$, or 2319 y, so the fraction decaying in 10 y is 10/2319, or 4.3×10^{-3}. The activity is reduced to 99.57 percent of its original value.

b. The atomic weight of radium is 226, so 1 g contains $N_A/226 \simeq 2.7 \times 10^{21}$ radium atoms. Its radioactive lifetime is 2319 y, or 7.3×10^{10} s. The number of atoms that decay in 1 s is approximately $(2.7 \times 10^{21})/(7.3 \times 10^{10})$, or about 3.7×10^{10}, which is 1 curie, according to its definition.

c. From the definition of the curie, 1 kg of radium generates power at the rate

$$(3.7 \times 10^{13})(3.8 \times 10^{7}) \simeq 1.41 \times 10^{21} \text{ eV/s}$$

Because 1 eV $= 1.60 \times 10^{-19}$ J, the power produced by 1 kg of radium is 226 W.   ∎

A BIT MORE ABOUT
The Thorium Decay Chain

Table 13.1 shows the entire pattern of thorium's decay chain, involving a sequence of six α decays and four β decays. Similar decay chains describe the sequential decays beginning with ^{238}U and ^{235}U.

The half-life of ^{232}Th is about three times the age of Earth, and the stable end product of its decay chain is the most common isotope of lead. The nine radioactive descendants of ^{232}Th are found only in thorium ores. None of them is long lived; their half-lives range from a fraction of a microsecond to several years, and their presence on Earth results from thorium decay. None of the radioactive ancestors of ^{232}Th have extremely long half-lives either. Consequently, none of them are found on Earth today. The longest-lived predecessor of ^{232}Th is the plutonium isotope ^{244}Pu$_{94}$, whose half-life is 80 million years. Earth's primordial plutonium decayed long ago, but the plutonium in the universe is continually

TABLE 13.1 Thorium unto the Tenth Generation

Isotope	Old Name	Decay	Half-Life
^{232}Thorium$_{90}$	Thorium	α	1.4×10^{10} y
^{228}Radium$_{88}$	Meso-Thorium$_1$	β	5.76 y
^{228}Actinium$_{89}$	Meso-Thorium$_2$	β	6.13 h
^{228}Thorium$_{90}$	Radiothorium	α	1.91 y
^{224}Radium$_{88}$	Thorium-X	α	3.66 d
^{220}Radon$_{86}$	Thoron	α	55.6 s
^{216}Polonium$_{84}$	Thorium A	α	0.15 s
^{212}Lead$_{82}$	Thorium B	β	10.64 h
^{212}Bismuth$_{83}$	Thorium C	β	60.6 m
^{212}Polonium$_{84}$	Thorium C'	α	3×10^{-7} s
^{208}Lead$_{82}$	Thorium D	None	Stable

being replenished by the explosions of distant supernovae. All of the uranium and thorium on Earth was produced long before Earth was formed, by ancient supernovae. Tiny quantities of ^{244}Pu have been detected within meteorites, but much larger amounts of its lighter isotope ^{239}Pu are produced by nuclear reactors and used as a source of energy.

How Rutherford and Soddy Found the Radioactive Decay Law

Consider an idealized situation involving two radioactive species. Suppose that the nucleus X α decays to another nucleus Y with a very long lifetime, $\tau_X = 10^{10}$ y, and that Y thereafter β decays to the stable species Z with a short lifetime, $\tau_Y = 10^{-2}$ y, or about 3 days.

$$X \xrightarrow{\alpha} Y \xrightarrow{\beta} Z$$

Let N_X be the number of X atoms in the sample and N_Y the number of Y atoms. If there were a millimole of X atoms, then N_X would be 6×10^{20}. The fraction of X atoms that decay each year is only 10^{-10}. Therefore, N_X is an enormous number that remains approximately constant during the course of the experiment.

Consider a small interval of time Δt. According to equation 3, the number of Y atoms created during that interval via the $X \to Y$ process is $N_X \Delta t / \tau_X$. (If Δt were 1 day, this number would be about 2×10^8.) The number of Y atoms decaying to Z atoms in the same time interval is

$N_Y \Delta t/\tau_Y$. After some time has passed, things come to a stand off: The rates of Y production and Y decay become equal to one another:

$$Y's \text{ produced} = Y's \text{ decaying} \quad \text{or} \quad N_X/\tau_X = N_Y/\tau_Y$$

and the number of $Y's$ takes on a constant value: $N_Y = (\tau_Y/\tau_X) N_X$. In our example, $N_Y \simeq 6 \times 10^8$.

The rate of Y decays in the original sample is measured and found to be steady in time. Afterward, the sample is chemically separated into two parts. The first extract contains all the Y atoms, and the second extract contains all the X atoms. The overall β activity is not changed just because the atoms have been segregated into two different samples. Because, at first, there are no $Y's$ in the second sample, all of the β radioactivity occurs in the first sample.

What happens later on? The $Y's$ in first sample decay and there are no $X's$ to replenish them. The activity in the sample gradually falls according to

$$(N_Y)_2 = N_Y \exp -(t/\tau_Y) \qquad \text{for sample 1} \qquad (5)$$

Although the second sample had no $Y's$ at the start, the $Y's$ are gradually replenished by the decays of the X nuclei. Its activity rises according to

$$(N_Y)_1 = N_Y [1 - \exp - (t/\tau_Y)] \qquad \text{for sample 2} \qquad (6)$$

As the activity of sample 1 decreases, the activity of sample 2 increases. However, the sum of the two activities remains constant:

$$N_Y = (N_Y)_1 + (N_Y)_2 = \text{Constant}$$

Curves showing the time dependence of the activity of the two samples are the hallmark of the radioactive decay law. Figure 13.3 shows the actual results obtained by the experiments of Rutherford and Soddy.

FIGURE 13.3
The actual data of Rutherford and Soddy. The solid circles are the measured activities of the sample in which the β decaying nuclei were concentrated. The open circles are those of the sample in which the progenitors of the β decaying nuclei were concentrated. The sum of the two activities is constant, as it was in the unseparated sample. Smooth curves drawn through the data points agree with the theoretical result, equations 5 and 6. The curves intersect at a time equal to the half-life of the β-active nuclei.

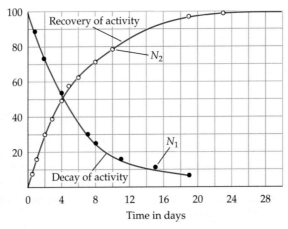

How Isotopes Were Discovered

As the radioactive decay chains of uranium and thorium were explored, substances turned up that had identical chemical properties but that differed in their radioactive properties. Soddy wrote in 1911:

> We have in these examples no mere chemical analogues, but chemical identities. Chemical homogeneity is no longer a guarantee that any supposed element is not a mixture of several different atomic weights, or that any atomic weight is not merely a mean number. The constancy of atomic weight, whatever the source of the material, is not a complete proof of its homogeneity. The absence of simple numerical relationships between the atomic weights becomes a matter of course and not for surprise.

Soddy coined the word *isotopes* (meaning same form) to describe the varieties of an element with different half-lives and decay properties. The five radioactively distinct substances once called uranium-X, thorium, ionium, radiothorium, and radioactinium are chemically alike. They are the thorium isotopes ^{234}Th, ^{232}Th, ^{230}Th, ^{228}Th, and ^{226}Th.

Soddy was wrong to believe in the constancy of atomic weights of naturally occurring elements, whatever their source. With today's sensitive analytic procedures, tiny systematic variations in isotopic composition can be detected and measured. Isotopic studies of minerals, meteorites, and moondust tell us much of what we know about the origin and evolution of the solar system. Isotopic studies of ancient human bones tell us about our origins. Precise measurements of the ^{234}U/^{230}Th ratios at burial sites in the Near East suggest that early human societies coexisted with Neanderthals about 100,000 years ago. The ^{13}C/^{12}C ratio in bones is affected by diet. Recent measurements of this ratio in pre-Columbian relics are helping to determine how the cultivation of corn (maize) spread northward from Mexico to Native American societies.

EXAMPLE 13.2
Carbon Dating

^{14}C is present in air because it is made by cosmic rays striking atoms in the atmosphere. However, this radioactive isotope of carbon decays with a half-life of 5730 years. Living plants and animals are continually exchanging their atoms with those in the air or in their food. The decay of ^{14}C in their bodies is balanced by its replacement with newly formed ^{14}C. Consequently, the relative abundance of the rare ^{14}C isotope compared to its stable cousin ^{12}C remains constant in living things. When a creature dies, it no longer breathes or transpires, so its ^{14}C is no longer replenished. Its age is determined by comparing its ^{14}C abundance with that of contemporary organic matter. The Shroud of Turin was alleged to date from the time of Christ. It was subjected to carbon dating at three different laboratories. They all agreed that the shroud is less than 600 years old. How was this done?

Let $R(t)$ be the ratio of ^{14}C to ^{12}C abundance in the shroud as a function of its age t. Its value is: $R(t) = R_0 \exp(-t/\tau)$, where $\tau \simeq \tau_{1/2}/0.69 \simeq 8300$ y. R_0 is the abundance ratio in living things or in things recently deceased, such as your cotton bedsheets. If t were 600 y, R would be $0.93 R_0$. If t were 2000 y, it would be $0.79 R_0$. The experiments found the former value, thereby proving that the shroud was manufactured in the fifteenth century. ∎

Atomic Physics vs. Nuclear Physics

Rutherford and Soddy found that radioactivity is a process of atomic transformation. An atom emits an electron (in β decay) or an alpha particle (in α decay) and becomes an atom of another element. Each radioactive species is characterized by a half-life and decays according to the exponential law. Elements may have several isotopes with different masses and radioactive properties but virtually identical chemical properties. In the course of unraveling the mystery of radioactivity, neither the Curies nor Rutherford nor Soddy knew that it is a nuclear phenomenon. They were puzzled because the energies of α and β rays (which are both of nuclear origin) were so much larger than those of photons emitted by excited atoms.

Indeed, radioactivity and other nuclear phenomena form a discipline far removed in distance and energy scales from the physics of the atom as a whole. The size of a nucleus is to the size of an atom as an inch is to a mile. The binding energies of nuclear particles are measured in MeV while those of valence electrons (relevant to chemical and optical phenomena) are measured in eV. They are to one another as the energy of a charging elephant is to that of a fleeing mouse. At the 1911 Solvay Conference, Mme. Curie noted that the phenomena of heat, light, elasticity, magnetism, and chemistry depend only on the outer structure of atoms, whereas,

> radioactive phenomena form a world apart, without any connection with [the above phenomena.] It seems, therefore, that radioactive phenomena originate from a deeper region of the atom, a region inaccessible to our means of influence and probably also to our means of observation, except at the moment of atomic explosions.

The innermost structure of the atom is more observable than she thought. Rutherford spied the nucleus just a few months earlier by scattering α particles from it. A few years later, he engineered the first artificial nuclear reaction:

$$\alpha + {}^{14}N_7 \longrightarrow {}^{17}O_8 + p$$

By bombarding nitrogen with α particles, he turned some of the atoms into oxygen. Although this feat of modern alchemy was of only academic interest, the relevance of nuclear phenomena to everyday life was demonstrated two world wars later at Hiroshima and Nagasaki, and later yet, by the lives saved by nuclear medicine.

Atoms are neither elementary nor immutable: They are built of simpler things. Atomic physics deals with the electrical interactions of Z electrons tethered to a tiny, massive, and positively charged nucleus. The atom is characterized by its quantum states of definite energy, or *energy levels*. Figure 13.4a shows part of the *energy-level diagram* of atomic sodium. The nucleus is another arena in which quantum mechanics holds sway. Nuclei are neither elementary nor immutable. They are built of simpler things. Nuclear physics deals with the interactions among Z protons and N neutrons bound together as an atomic nucleus by the nuclear force. Every composite system is characterized by its quantum states of definite energy. Figure 13.4b shows part of the energy-level

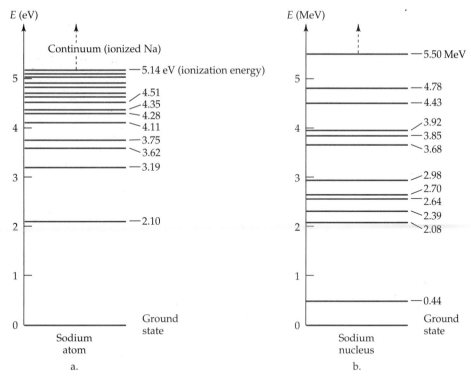

FIGURE 13.4 *a.* Energy-level diagram of the sodium atom. The energies of some of the quantum states of the sodium atom are indicated in eV relative to its ground state. If more than 5.14 eV is supplied to the atom, one or more electrons are driven off. *b.* Energy-level diagram of the sodium nucleus. The energies of some of the quantum states are indicated in MeV relative to its ground state. If more than 8.9 MeV is supplied to the nucleus, one or more nucleons are driven off.

diagram of the sodium nucleus. On the face of it, the patterns of atomic and nuclear levels are not very different. However, the energy scales differ by a factor of a million. When an atom in an excited state jumps to its ground state, it emits a photon with a few eV. When a nucleus in an excited state decays, it emits a photon with a few MeV in the radioactive process of γ decay.

Radioactivity is a nuclear phenomenon. Whether a particular nucleus is radioactive and how it decays are problems in nuclear physics. However, what the energetic particles do afterward is a problem in atomic physics. For example, an α particle, as it passes through matter, loses energy by collisions and leaves a trail of ions in its wake. When it comes to rest, it catches two electrons and becomes a neutral helium atom. Atomic and nuclear phenomena are disparate, but a working knowledge of both is needed by physicians, engineers, generals, politicians, and scientists: to treat cancer or radiation sickness, to diagnose disease, to remove radioactivity from our dwellings, to design safe nuclear plants, to dispose of nuclear wastes, to study the neutrinos coming from the Sun, to negotiate test ban treaties, to prevent nuclear weapon proliferation, and much more. A responsibly concerned citizen should know something about the atom and its nucleus.

------◆··◆------

Exercises

1. Into what isotopic species does ^{14}C β decay? Into what isotopic species does ^{238}U α decay?

2. An excited state of a ^{77}Se nucleus at rest γ decays to its ground state, which is 1.7×10^{-4} amu less massive. What is the approximate energy of the γ ray in KeV? (The recoil of the nucleus may be neglected.)

3. Tritium has a half-life of about 12 years. It is manufactured mostly for military purposes. If we stopped its production when we had a stockpile of 100 tons, how much would be left 36 years later?

4. Suppose that 10^{-12} of the atoms in a gram of carbon are the radioactive isotope ^{14}C. How many of them decay per hour? What is the activity of the sample in curies?

5. What is the radioactive decay chain leading from $^{246}Cm_{96}$ to $^{238}U_{92}$? What is the radioactive decay chain leading from $^{206}Hg_{80}$ to $^{206}Pb_{82}$?

6. If 10^9 atoms of ^{126}I, whose half-life is about 13 days, are placed in a test tube, approximately how many of them decay per second? About how many of the original atoms remain after 130 days? At that time, how many of them decay each second?

7. ^{159}Dy decays into stable ^{159}Tb with a half-life of 146 days. At an unknown time in the past, 1 mg of this radioisotope was sealed in a tube and placed in a time capsule. When it was opened, it was found to contain only 31.25 μg of ^{159}Dy. How long ago was the time capsule buried?

8. Natural potassium now contains 1.2×10^{-4} of radioactive ^{40}K with a half-life of 1.2×10^9 y. That is, for every 100,000 potassium atoms, 12 would be ^{40}K. When Earth was formed 4.8×10^9 y ago, how much ^{40}K was there in natural potassium?

9. The half-life of ^{238}U is approximately the age of Earth, while that of ^{235}U is about a seventh as long. Today, the relative abundance of the lighter isotope is 0.72 percent. What was it when Earth was born?

10. The three most abundant radioactive species on Earth each undergo a complex chain of radioactive decays, ultimately becoming different stable isotopes of lead:

$$^{238}U_{92} \rightarrow {}^{206}Pb_{82}, \qquad ^{235}U_{92} \rightarrow {}^{207}Pb_{82}, \qquad ^{232}Th_{90} \rightarrow {}^{208}Pb_{82}$$

In each case, how many α and β decays are involved in the chain?

11. An isotope of the synthetic element fermium decays by a sequence of five α emissions to yield long-lived thorium. Part of its decay pattern is shown here along with several half-lives:

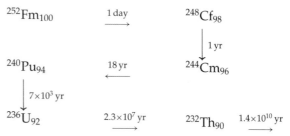

Suppose we start with a pure sample of the fermium isotope. After a month passes, almost all of the fermium has decayed and the sample is mostly californium. After four years have passed, it is mostly curium. After a century, it is mostly plutonium. After a million years, it is mostly uranium. After a billion years, it is mostly thorium. Explain why this is.

13.2 The Structure of the Nucleus

Rutherford discovered the atomic nucleus in 1911 and proposed the name *proton* for the simplest nucleus, that of hydrogen. Two atomic building blocks were known: light, negatively charged electrons and heavy, positive protons. The masses of most nuclei were nearly integer multiples A of the proton mass. Their charges were the integers Z. Many physicists imagined the nucleus to be made up of A protons (to make the mass come out right) and $A - Z$ electrons (to fix the nuclear charge). The fact that some radioactive nuclei emit energetic electrons when they β-decay seemed to confirm this view.

Electrons occurred in two positions in this model: Some were inside the nucleus while others lay outside, in Bohr's quantum states. This scheme led to a conflict with quantum mechanics. The uncertainty principle says that an electron, when confined to a nucleus 10^{-15} m in size, has a momentum uncertainty so large that it cannot remain bound to the nucleus. Furthermore, nuclei and electrons behave like tiny magnets, yet the magnetic properties of nuclei were known to be thousands of times smaller than electronic magnetism. How could this be if there were electrons inside the nucleus?

Today we know there are no nuclear electrons. Nuclei are made of protons and neutrons. The idea of the neutron was introduced when Rutherford suggested in 1920 that "it may be possible for an electron to combine much more closely with the H-nucleus. The existence of such [particles] seems almost necessary." He soon he coined the word *neutron* to describe what he imagined to be a tightly bound system consisting of a proton and an electron. However, the neutron turned out to be a particle in its own right. It is neither more nor less elementary than a proton. The English physicist James Chadwick (1891–1974) set out to find Rutherford's particle: "I think we shall have to make a real search for the neutron," he wrote. He found it in 1932, almost a decade later.

The Discovery of the Neutron

Some years before, a penetrating form of radiation had been observed to emerge from beryllium foil that was bombarded by energetic α particles from a polonium source. The unknown radiation consisted of electrically neutral particles that, when striking a target of hydrogen gas, would eject protons with energies up to 5 MeV (Figure 13.5). In fact, the new radiation consisted of neutrons, but the Joliot–Curies (Mme. Curie's daughter and son-in-law) were loathe to introduce a new particle just to explain their data. They believed that the radiation produced in the beryllium consisted of energetic photons. If this had been true, the photons would have needed energies of 50 MeV to do the trick, as we show in Example 13.3. But an α particle from polonium has a mere 5 MeV of kinetic energy! By what magic were photons of such great energy being produced?

EXAMPLE 13.3
Photon–Proton Collisions

Show that photons must have energies of 50 MeV if, upon striking hydrogen nuclei (protons) at rest, they can eject them with energies of 5 MeV.

Solution

Let the proton mass be m and photon energy be E_γ. (The photon momentum is E_γ/c.) A proton at rest is kicked hardest when the photon is deflected backward. When the photon bounces backward, the proton acquires a momentum of $2E_\gamma/c$ and a velocity of $v = 2E_\gamma/mc$. Its energy is $\frac{1}{2}mv^2$, or $2E_\gamma^2/mc^2$. If we set this equal to 5 MeV and recall that the rest energy mc^2 of the proton is nearly 1000 MeV, we find that $E_\gamma = 50$ MeV. ∎

FIGURE 13.5
Schematic diagram of Chadwick's experiment. A radioactive substance A produces 5 MeV α particles, a beam of which passes through an aperture in a lead plate B and strikes a thin beryllium foil C. Collisions of α particles with beryllium nuclei produce 5 MeV neutrons, which strike a target D consisting of hydrogen gas, or one of six other elements. Elastic impacts of neutrons drive the target nuclei toward the particle detector E. At first, Chadwick used a Geiger counter to measure the velocities of the recoiling nuclei. Later, he used a cloud chamber. Chadwick's results proved that the particles produced by the α particles were neutrons.

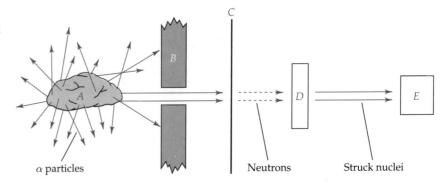

α particles   Neutrons   Struck nuclei

Chadwick believed the beryllium radiation consisted of neutral particles with roughly the proton's mass—neutrons—with energies of about 5 MeV. In a collision of a moving body with a stationary body of the same mass, all the energy can be transferred, just as in the impact of the cue ball with the 8 ball. Chadwick tested his hypothesis by replacing the hydrogen in the target with other gases. In each case, he showed that the particles ejected are nuclei of the target element. By measuring the energies of the ejected nuclei, he found that the neutron hypothesis explained the data but the photon hypothesis did not. In his paper announcing the discovery of the neutron, he wrote:

These results are very difficult to explain on the assumption that the radiation from beryllium is [photons.] The difficulties disappear, however, if it be assumed that the radiation consists of particles of mass 1 and charge 0, or neutrons. The capture of the α particle by the ^{9}Be nucleus may be supposed to result in the formation of a ^{12}C nucleus and the emission of a neutron. From the energy relations of this process the velocity of the neutron emitted in the forward direction may well be 3×10^7 m/s. The collisions of this neutron with the atoms through which it passes give rise to the recoil atoms, and the observed energies of the recoil atoms are in fair agreement with this view.

Although photons of 50 MeV or neutrons of 5 MeV can equally well explain the recoils seen off hydrogen, the photon hypothesis was eventually rejected because it predicted smaller recoil energies of heavier atoms than had been observed. Chadwick had discovered the neutron and the nuclear reaction by which it was produced:

$$\alpha + {}^9\text{Be}_4 \longrightarrow {}^{12}\text{C}_6 + n \tag{7}$$

EXAMPLE 13.4
Photons or Neutrons?

Two hypotheses could explain the ejection of 5 MeV protons by the beryllium radiation: (1) the radiation consisted of 50 MeV photons, and (2) it consisted of 5 MeV neutrons. Let the target be switched from hydrogen to nitrogen, whose atomic weight is 14. What are the energies of the ejected nitrogen nuclei under each hypothesis?

1. The nitrogen nuclei are struck by 50 MeV photons. If we change m to $M = 14m$ in the analysis of Example 13.3, the result for the maximum recoil energy becomes 5 MeV/14, or 0.36 MeV.

2. The nitrogen nuclei are struck by 5 MeV neutrons. This is an exercise in classical mechanics. A cue ball of mass m and velocity v strikes head-on a stationary ball of mass M. The recoil velocity V of the struck ball is $V = 2mv/(m + M)$. Its kinetic energy is $\frac{1}{2}MV^2$, which can be worked out to be $4mM/(m + M)^2$ (or about 0.25) times the kinetic energy of the cue ball. For the problem at hand, the cue ball is a 5 MeV neutron. The recoil energy of the nitrogen nucleus under the neutron hypothesis is 1.25 MeV. ∎

The Nuclear Force

Once the neutron was found, scientists realized that nuclei are built from neutrons and protons. What force holds these nucleons together in a nucleus? Gravity is far too weak and the electric force between protons is repulsive. A third fundamental force of nature was required: the *strong nuclear force* with the following properties:

• The nuclear force must give rise to a strong attraction between nucleons. The strength of the nuclear force is revealed by a comparison between nuclear and atomic energies. The energy levels of a nucleus are typically a few MeV apart, while those of an atom are only a few eV apart. It takes 16 MeV of energy to extract a proton from a ^{12}C nucleus, but only 13.6 eV to steal an electron from a hydrogen atom.

• The nuclear force between two nucleons must be effective only at very short distances. Nucleons hardly affect one another when they are separated by a few nuclear radii. Furthermore, experiments show that the volume of a nucleus is proportional to A, the number of constituent nucleons. Unlike the atom, in which electrons are spread out over a large volume of space, the nucleons in a nucleus touch one another. Experiments show that the radius of a nucleus is given by

$$R \simeq (1.3 \times 10^{-15} \text{ m}) A^{1/3} \qquad (8)$$

Consequently, the nuclear volume is approximately equal to A times the volume of a single nucleon. The nucleus is a quantum-mechanical system, but it is sometimes useful to think of it as a bunch of tiny marbles stuck together with glue or as an approximately spherical but deformable droplet of liquid.

In the 1930s, the mystery of the nuclear force was still unsolved, but there was one tantalizing clue. Neutrons and protons are nearly equal in mass:

$$m_p \simeq 938.272 \text{ MeV}/c^2 \qquad m_n \simeq 939.566 \text{ MeV}/c^2$$

The fact that light nuclei are often made up of equal numbers of each particle suggested that the relation between them is more profound than a mere coincidence in mass. Of the first eight elements in the periodic table, seven have stable isotopes with $Z = N$. None of the 15 lightest nuclear species have more than one excess neutron. However, heavier elements show a systematic increase in the neutron-to-proton ratio. ^{40}Ca$_{20}$ is the heaviest stable nucleonic democracy; thereafter, neutrons predominate. For ^{238}U, the n/p ratio is almost 1.6. Although the strong nuclear force treats neutrons and protons similarly, the weaker electric force does not. Protons repel each other a little bit, but lots of protons repel each other a lot. This is why larger nuclei have a neutron excess.

Heisenberg concluded that the nuclear force between neutrons is identical to the nuclear force between protons. The connection is even deeper, however. Under appropriate circumstances, the nuclear force between a proton and a neutron is identical to that between neutrons or protons. Apart from their tiny mass splitting and the differences in their electromagnetic properties, neutrons and protons are interchangeable and indistinguishable particles. The intrinsic similarity between protons and neutrons is mathematically formulated as *isospin symmetry* and explains why groups of particles or nuclei appear in multiplets with different electric charges yet nearly equal masses. The proton and neutron is an example of a *isotopic doublet*. We shall encounter other isotopic multiplets when we study elementary particles in Chapter 15.

Getting to Know the Nuclei

The angular momentum of every quantum-mechanical system must be an integer or half-integer multiple of $\hbar$. In particular, electrons have a built-in angular momentum (or spin) of one-half unit. So do protons and neutrons! In the case of the hydrogen atom, the total angular momentum is a sum of three terms: the orbital angular momentum of the electron, and the intrinsic angular momenta of the electron and the proton. Not only are nucleons spin 1/2 particles, but they satisfy the Pauli principle, just as electrons do: No more than two protons can occupy the same quantum state—one with spin up, the other with spin down. The same is true for neutrons.

Nucleons in nuclei sequentially fill up quantum shells just as electrons do in atoms. The lowest quantum state of the atom (because it has zero orbital angular momentum) has room for only two electrons of opposite spin. It's the same for the lowest quantum state of the nucleus, which can accommodate no more than two protons and two neutrons. Let's see how these ideas apply to the smallest nuclei, those with $A < 9$:

A = 1   The proton is the nucleus of the ordinary hydrogen atom ^{1}H. The neutron, since it is electrically neutral, does not form an atom. Furthermore, it is an unstable particle that β decays with a half-life of about 10 min to become a proton, an electron, and (as we shall see in Section 13.3) an antineutrino.

A = 2   Neither two protons nor two neutrons stick together. The only $A = 2$ nucleus is the deuteron, which consists of one proton and one neutron. This stable particle has spin 1 and is the nucleus of ^{2}H, the heavy, rare, and stable isotope of hydrogen.

A = 3   Two protons and one neutron form the nucleus of a rare stable isotope of helium, ^{3}He. Conversely, two neutrons and one proton form the nucleus of tritium, or ^{3}H, a third isotope of hydrogen. These are the only $A = 3$ nuclei. Tritium β decays into ^{3}He with a half-life of about 12 years. Each $A = 3$ nucleus contains two identical nucleons whose spins must be opposite to one another and thus cancel one another. That is why both nuclei have spin 1/2, the spin of the odd nucleon. The symmetry between proton and neutron is reflected in the nuclei they form: ^{3}He and ^{3}H form an isotopic doublet just as nucleons do.

A = 4   Again, there is only one such state: the α particle, which is the nucleus of ^{4}He. Its nucleus has the structure

$$\{p\uparrow \; p\downarrow \; n\uparrow \; n\downarrow\}$$

This exceptionally well-bound nucleus has complete shells of protons and neutrons, just as its atom, helium, has a complete shell of electrons. Because its binding energy is so large, α is the particle of choice to be ejected in many radioactive decays.

A = 5   There are no stable or long-lived nuclei made up of five nucleons. (Actually, $A = 5$ nuclei may be created in the laboratory, but they survive for only 10^{-21} s.)

A = 6, 7   The only exemplars are the two stable isotopes of lithium. ^{6}Li$_3$ has one proton and one neutron in the second nuclear shell and its spin is 1. ^{7}Li$_3$ has one additional neutron and a spin of 3/2.

A = 8   There are no stable or long-lived nuclei made up of eight nucleons. (The longest lived $A = 8$ nuclei are ^{8}B$_5$ and ^{8}Li$_3$, with half-lives of almost 1 s.) The gaps at $A = 5$ and 8 are crucial to the history of the universe and the evolution of stars. If there were such nuclei, much of the nuclear fuel in the universe would have been consumed in the first few minutes after the big bang. It might not have been possible for stars to evolve! Aside from these exceptions, there is always at least one (but never more than three) stable nuclear species corresponding to every value of A less than 210, but there are no larger stable nuclei.

The periodic properties of the elements result from the Pauli principle: There is room for only so many electrons in each quantum s'ıell. Atoms with complete electron shells are inert gases. Quantum theory tells us which elements these are

$$\text{Inert gases}: \quad Z = 2,\ 10,\ 18,\ 36,\ 54,\ \text{and}\ 86$$

Because protons and neutrons also satisfy the Pauli principle, something analogous takes place for atomic nuclei. Neutrons and protons form shells in the nucleus just as electrons form shells in atoms. Certain *magic numbers* of neutrons or protons correspond to completed shells of neutrons or protons:

$$\text{Magic numbers}: \quad N \text{ or } Z = 2,\ 8,\ 20,\ 28,\ 50,\ 82,\ \text{and}\ 126$$

Nuclei with N or Z equal to magic numbers are particularly stable. All known radioactive decay chains end up at stable isotopes of lead, whose Z is a magic 82, or at the unique stable form of bismuth, whose N is a magic 126. No isotopes of lead and hardly any nuclei with $Z < 82$ are α radioactive. Nuclei with complete shells are called magic number nuclei and are of particular interest to nuclear scientists because they shed light on the details of the nuclear force.

The Binding Energy of Nuclei

The special theory of relativity says that things stuck together have less mass than the same things pulled apart. When coal is burned, each carbon atom combines with an oxygen molecule to form a CO_2 molecule. About 4 eV of energy is released. The mass of the CO_2 molecule is $4 \text{ eV}/c^2$ less than the sum of the masses of the reactants. Because the rest mass of CO_2 is about $4 \times 10^{10} \text{ eV}/c^2$, the fraction of the mass of the reactants converted into energy is a mere 10^{-10}.

Nuclear processes release far more energy per reaction than atomic processes because the glue holding nuclei together is more powerful than the electric force binding electrons to their atoms. The binding energy of a nucleus is the mass of its component nucleons minus the mass M of the nucleus itself:

$$\text{Binding energy} = (Z\,m_p + (A - Z)\,m_n - M)c^2$$

It is the energy needed to take the nucleus apart into its constituent nucleons. The binding energy of most nuclei is nearly 1 percent of their rest energies. For example, the sum of the masses of 92 protons and 146 neutrons is about 2 amu more than the mass of the ^{238}U nucleus.

The binding energy per nucleon is a measure of the stability of a nucleus. Let us define δ to be the binding energy of a nucleus divided

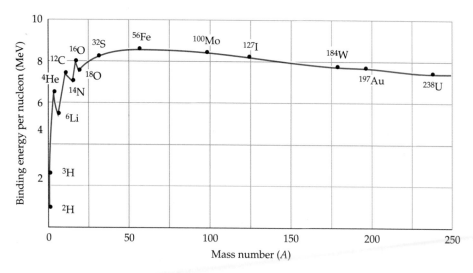

FIGURE 13.6 The curve of binding energy. The binding energy of a nucleus is the energy required to separate it into its constituent nucleons. This curve shows the binding energy per nucleon δ plotted against atomic mass number A. The larger is δ, the more tightly bound is the nucleus.

by A. The precise value of δ varies erratically from nucleus to nucleus, but its general trend is shown in Figure 13.6 as a function of A, the atomic mass number. The larger the value of δ, the more tightly bound is the nucleus. The best bound nuclei of all are the midsized ones near iron in the periodic table. Table 13.2 cites a few numerical examples.

TABLE 13.2 The Binding Energy per Nucleon of Selected Nuclei

Nucleus	δ (in MeV)	Nucleus	δ (in MeV)
^{2}H	1.1	^{28}Si	8.45
^{3}H	2.82	^{56}Fe	8.79
^{3}He	2.57	^{132}Xe	8.43
^{4}He	7.07	^{235}U	7.56
^{12}C	7.68	^{239}Pu	7.59

It takes energy to undo binding. Conversely, energy is released whenever a system becomes more bound. The principle of energy conservation determines whether a nucleus is radioactive (and unstable) or not radioactive (and stable) It tells what nuclei can be used as nuclear fuel, both in the stars and in power plants on Earth. In Section 13.3, we describe how a uranium nucleus can be induced to *fission* (or split apart) into fragments that are better bound than it is by about 1 MeV per nucleon. Consequently, the sum of the rest energies of the fission products is about 200 MeV less than that of the original nucleus (whose rest energy is about 200 GeV). A fission reactor converts about 1/1000 of the mass of its fuel into useful energy, but a conventional power plant ekes out merely 1 part in 10^{10}. Nuclear fission releases ten million times more energy per gram of fuel than combustion.

Exercises

12. Suppose that a 5 MeV neutron makes an elastic collision with a stationary ^{4}He nucleus. What is the largest possible recoil energy of the helium nucleus?

13. How much energy is released in MeV if two ^{2}H nuclei combine together to form one ^{4}He nucleus? (*Hint:* Use Table 13.2 to determine the total binding energies of one ^{4}He nucleus and of two ^{2}H nuclei.) How much energy is released in joules if 1 g of ^{2}H is converted into helium by this fusion process?

14. The atomic mass unit corresponds to a rest energy of 931.5 MeV. In these units, $M(\alpha) = 4.0026$, $M(^9\text{Be}) = 9.0122$, $M(^{12}\text{C}) = 12.0000$, and $M(n) = 1.0087$. In reaction 7 (a nuclear reaction by which neutrons are produced), which is greater: the total kinetic energy of the reacting nuclei, or the total kinetic energy of the product nuclei? What is the difference, per reaction, in MeV?

15. How much energy is required (in MeV) to disassemble one ^{12}C nucleus into its constituent nucleons?

16. Neutrons are readily absorbed by boron nuclei by the process

$$n + {}^{10}\text{B} \longrightarrow {}^{11}\text{B} + \gamma$$

where γ is an energetic photon or γ ray. The atomic mass of ^{10}B is 10.013 amu, that of ^{11}B is 11.009 amu, and that of the neutron n is 1.009 amu. What is the energy of the γ ray in MeV?

17. Use the data in Exercise 14 to determine the energy (in MeV) released if three α particles combine together to form a carbon nucleus?

13.3 Nuclear Processes and the Conservation of Energy

Which nuclei are radioactive and by which process do they decay? How much energy is carried off by an α particle or a β particle? How does the Sun extract nuclear energy from its substance? What must be done to extract useful nuclear energy from uranium or other nuclear fuels? The answers to these questions have to do with the masses of different nuclei and their relation to one another.

To Alpha-Decay or Not to Alpha-Decay?

Alpha radioactivity is characteristic of large nuclei. With a few exceptions, it occurs only for elements with $Z > 82$—that is, for elements beyond lead in the periodic table. *Whether* a given nuclear species α-decays has a simple answer: Those that can, do. Radioactive α decay takes place whenever it is energetically possible for a nucleus to do so—that is, whenever the mass of the products is less than the mass of the decaying nucleus. *When* an energetically permitted α decay takes place can only be answered statistically in terms of a half-life. The half-lives of radioactive nuclei can often be calculated by nuclear physicists, but the quantum-mechanical calculations are technical and difficult. Let's stick to the whether question.

Let $M(A, Z)$ be the mass of an isotope with atomic number Z and mass number A and $M(4, 2)$ be the mass of a ^{4}He atom. The decay

$$(A, Z) \longrightarrow (A-4, Z-2) + \alpha$$

takes place if

$$M(A, Z) - \{M(A-4, Z-2) + M(4, 2)\} = \Delta > 0$$

The lost mass reappears as the kinetic energy Δc^2 shared by the daughter nucleus and α particle.

How is the energy divided between the α particle and the daughter nucleus? The available energy for such decays is rarely more than 6 MeV, while the rest energy of the α particle is about 4000 MeV. Thus, the α particle is nonrelativistic. Let the decaying atom be at rest and m and v be the mass and velocity of the emitted α particle. Let M and u be the mass and recoil velocity of the daughter nucleus.*

Momentum conservation says: $m v + M u = 0$
Energy conservation says: $\frac{1}{2} m v^2 + \frac{1}{2} M u^2 = \Delta c^2$

Solving the momentum equation for u, we find $u = -(m/M)v$. Inserting this result into the energy equation, we obtain an equation for v:

$$\frac{1}{2} mv^2 + (m/M)\frac{1}{2}mv^2 = \Delta c^2$$

where the first term is the α energy and the second is the energy of the recoiling daughter. The daughter nucleus receives only a small fraction $m/(M + m)$ of the decay energy.

EXAMPLE 13.5   The smallest naturally occurring atom that α-decays is a common iso-
Alpha Decay Energies   tope of the uncommon element neodymium, ^{144}Nd$_{60}$. Its half-life is an astonishing 5×10^{15} years, a million times the age of Earth! Its decay produces a 1.8 MeV α particle. If the decaying atom is at rest, what is the recoil velocity of the daughter?

Solution   The atomic weight of the daughter nucleus is about 140 and the atomic weight of helium is about 4. Thus, the recoil kinetic energy of the daughter is 4/144 times 1.8 MeV, or 50 KeV. The rest energy Mc^2 of the daughter is about 130 GeV. Using the relations

$$Mu^2/2 = (u^2/c^2)Mc^2/2 = (u^2/c^2)(6.5 \times 10^{10} \text{ eV}) = 5 \times 10^4 \text{ eV}$$

we find that $u^2 \simeq 0.8 \times 10^{-6} \, c$ or $u \simeq 10^{-3} \, c$, which is a leisurely 300 km/s. The α particle is faster by a factor of 36; it whizzes off at about 10,000 km/s.

---

*The nuclear masses m and M are slightly smaller than the corresponding atomic masses $M(4, 2)$ and $M(A-4, Z-2)$. They do not include the electron masses. The Z electrons of the original atom play no role in the nuclear process, they are distant spectators.

Beta Radioactivity

Alpha decay is a relatively straightforward process involving the conflict between electromagnetism and the nuclear force. A nucleus that can α decay hangs around for a long time (typically a half-life) until the α particle finds itself outside of the nucleus. In other words, it quantum-mechanically tunnels through the repulsive electric barrier. Beta decay, on the other hand, is an intrinsically slow process involving the fourth force of nature: the *weak nuclear force*. If a nucleus contains too many neutrons to be stable, it can change one of its neutrons to a proton by emitting an electron in the process of β decay. (In Chapter 14, we shall discover that a nucleus with too many protons to be stable has an analogous remedy: It can emit a *positron*, the positively charged antiparticle of the electron.) The history of our understanding of β decay is full of surprising twists that have led us to today's remarkably correct and concise theory of nuclear and subnuclear phenomena. In 1913, Bohr presented two convincing arguments that β decay is a nuclear phenomenon:

1. The energies of β rays are very much greater than the kinetic energies of even the innermost electrons of large atoms.

2. Different isotopes of the same radioactive chemical element have different lifetimes and produce different β-ray energies. Thus, β decay has nothing to do with the chemical properties of an atom and everything to do with the internal constitution of its nucleus.

In the process of β decay, a negatively charged electron is emitted by a nucleus of charge Z. Since electric charge is conserved, the parent nucleus becomes a nucleus of charge $Z + 1$. The principle of energy conservation says that the decay process

$$(A, Z) \longrightarrow (A, Z+1) + e^-$$

takes place if the atomic masses satisfy

$$M(A, Z) - M(A, Z+1) = \Delta > 0$$

It would seem as if the electrons emitted by a specific isotope should all have precisely the same energies, as do the α particles from a specific α decay. To conserve energy, the electron should carry off the energy Δc^2. However, in this case, nature does not oblige.

Chadwick, who would later discover the neutron (as described in Section 13.2), was barely 22 in 1913, when he was awarded a fellowship to study abroad. He chose to work with Geiger in Berlin. Just weeks before the outbreak of World War I, he proved that the electrons produced in β decay do not all have the same energy. They can have any energy ranging from zero to Δc^2. The electron energies coming from the β decays of identical nuclei are not all the same. They display a continuous spectrum of energies, as shown in Figure 13.7. Weeks after his discovery, the war began and Chadwick was confined for its duration. Rutherford summarized the situation in 1915:

> The general evidence shows that β radiation gives a continuous spectrum due to β particles of all possible velocities. It is thus necessary that each atom [of a given nuclear species] does not emit an identical β radiation.

FIGURE 13.7
The electron energy spectrum from the β decay of ^{32}P. If energy were to be conserved, every electron would be expected to have an energy of 1.71 MeV. However, observations showed that there was continuous distribution of electron energies, all less than 1.71 MeV. Pauli postulated the existence of a light and neutral particle, the neutrino, which carried off the missing energy and preserved the law of energy conservation.

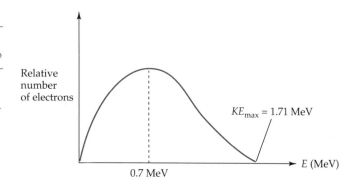

It was truly mystifying. Not only did β ray energies vary, but energy appeared to be disappearing into thin air. By 1929, Bohr was ready to sacrifice the law of energy conservation:

> We have no argument for upholding [the law of energy conservation] in the case of β ray disintegrations. The features of atomic stability responsible for the existence and properties of atomic nuclei may force us to renounce the very idea of energy balance.

Rutherford responded, "I will wait and see before expressing an opinion," while Dirac (whose contributions to physics are discussed in Chapter 14) declared that "I should prefer to keep rigorous conservation of energy at all costs." As it turned out, Bohr was wrong, Rutherford was cautious, and Dirac was wise.

The Neutrino

Wolfgang Pauli solved the problem of missing energy by postulating the existence of a new and unseen particle: the *neutrino*. In December 1930, he wrote a letter to a gathering of scientists interested in radioactive phenomena:

> Dear Radioactive Ladies and Gentlemen:
> I have come upon a desperate way out. To wit, the possibility that there could exist in the nucleus electrically neutral particles which I shall call neutrinos,* which have spin 1/2 and satisfy the exclusion principle. The mass of the neutrinos should be not larger than 0.01 times the proton mass— the continuous β spectrum would then become understandable from the assumption that in β decay a neutrino is emitted along with the electron, in such a way that the sum of the energies of the electron and the neutrino is constant.
>   I admit that my way out may not seem very probable a priori since one would probably have seen the neutrinos a long time ago if they exist. But

*He actually called them *neutrons;* we call them *neutrinos* or, more precisely, *electron antineutrinos.* Two years after Pauli's speculation, the neutral counterparts of protons were discovered; they are called neutrons. Neutrons and protons are the heavy and strongly interacting constituents of nuclei. Neutrinos are very light (perhaps even massless) particles that do not partake in the strong force. Trillions of them each second come from the Sun, pass all the way through Earth, and pierce your body at night. They are very weakly interacting particles and quite harmless.

only he who dares wins. One must therefore discuss seriously every road to salvation.—Thus, dear radioactive ones, examine and judge—unfortunately I cannot appear personally in Tübingen since a ball which takes place in Zürich makes my presence here indispensable. Your most humble servant, W. Pauli

Pauli was doubly right: A dance is more fun than a physics conference and a light and neutral spin 1/2 particle is emitted along with an electron in nuclear β decay. The available energy is shared between these two particles. Sometimes the electron gets the lion's share, sometimes the neutrino. The result is a continuous spectrum of electron energies. The idea made a lot more sense once Chadwick discovered the neutron. We may think of β decay as the following process taking place in the interior of an atomic nucleus:

$$n \longrightarrow p + e^- + \bar{\nu}$$

where the symbol $\bar{\nu}$ designates an *antineutrino*. (In Chapter 14, we introduce antimatter and learn that Pauli's particle is called an antineutrino rather than a neutrino.) The law of energy conservation was saved at the price of introducing a brand-new particle that nobody had seen.*

Neutrinos were observed a quarter century later by an experiment performed at a nuclear reactor near Savannah, Georgia. Its discoverers sent a telegram to Pauli in 1956: "We are happy to inform you that we have definitely detected neutrinos." Since then, the physics of neutrinos and antineutrinos has become a discipline in itself. Neutrinos from many sources on Earth and in the heavens are observed and their properties are studied. Scientists have even detected neutrinos coming from a supernova 160,000 light-years away.

In α decay, an α particle is ejected from a large nucleus if such an act is energetically permitted. No particles change their identities. Beta decay is different. Neither the electron nor the antineutrino are present in the original nucleus—they are created at the moment of decay. At the same instant, a neutron is transformed into a proton. Scientists knew that photons are created (that is, emitted) or destroyed (that is, absorbed), but the study of β decay revealed that other particles can be created or destroyed as well. The acts of particle creation and destruction underlie today's theoretical picture of fundamental physics.

The Discovery of Nuclear Fission

Yesterday's discoveries, like X rays and electrons, became today's tools. The neutron, too, became a powerful tool with which to explore the nucleus. Unlike α particles, which are repelled by electrical forces, neutrons are electrically neutral and can strike at the heart of a charged nucleus.

The Italian (later, American) physicist Enrico Fermi and his group, working in Italy, studied what happens when various chemical elements

---

*If a spin 1/2 neutron decayed into a spin 1/2 proton—a spin 1/2 electron and nothing else—there would be no way to conserve angular momentum. The neutrino hypothesis saves the law of conservation of angular momentum as well.

are bombarded with neutrons. They published over 300 papers on the subject from 1934 to 1938, showing that dozens of new radioactive materials are produced. In most cases, neutron absorption is followed by electron emission via β decay, yielding an element one higher in the periodic table:

$$n + (A, Z) \longrightarrow (A+1, Z) \longrightarrow (A+1, Z+1) + e^- + \bar{\nu}$$

What would happen if uranium were bombarded by neutrons? If the pattern held, a *transuranic element* with $Z > 92$ should be formed. Fermi discovered the presence of several new radioactive species in neutron-activated uranium. Their chemical properties were not those of uranium, nor of any of the elements from $Z = 82$ to $Z = 92$. Fermi was convinced he had synthesized element number 93, but he was wrong!* The German chemist Ida Noddack (who was the codiscoverer of the stable element rhenium) found a flaw in Fermi's reasoning. In 1934, she wrote:

> It is conceivable that the nucleus breaks up into several large fragments which would, of course, be isotopes of known elements but would not be neighbors of uranium.

Fermi would have to rule out all the known elements before he could be sure that he had made a transuranic element. Though she did not use the term, Noddack was the first scientist to point out the logical possibility of nuclear fission.

The radioactive by-products produced from uranium by neutron bombardment have different lifetimes and a variety of chemical properties. Otto Hahn, Lise Meitner, and Fritz Strassman, working in Nazi Germany, attempted to identify these materials. Meitner, an Austrian Jew, was forced to leave the team and emigrate to Sweden when Hitler took over Austria in the spring of 1938, just before the group's discovery of fission. A few months later, Hahn wrote to Meitner of the remarkable conclusion to their research. The decay products of neutron-activated uranium included the relatively small barium ($Z=56$) nucleus!

> Perhaps you can suggest some fantastic explanation [Hahn wrote]. We understand that it really can't break up into barium. So try to think of some other explanation.

Meitner discussed Hahn's letter with her nephew Otto Frisch, who recalls:

> But how can one get a nucleus of barium from one of uranium? We walked up and down in the snow trying to think of some explanation. Could it be that the nucleus got cleaved right across with a chisel? It seemed impossible that a neutron could act like a chisel, and anyhow, the idea of a nucleus as a

---

*Fermi was the first scientist to create an artificial chemical element. He synthesized technetium ($Z = 43$), a radioactive element that does not occur in nature. However, he did not find a transuranic element. Neptunium ($Z=93$) and plutonium ($Z=94$) were first produced with the Berkeley cyclotron in 1940. Americium ($Z = 94$) and curium ($Z = 95$) were discovered at Chicago in 1944. Today, all of the transuranic elements from $Z=93$ to $Z=109$ have been synthesized.

solid object that could be cleaved was all wrong; a nucleus was much more like a liquid drop. Here we stopped and looked at each other.

A few years earlier, Bohr and John Wheeler had argued that a large nucleus should behave like a droplet of liquid. When struck by a particle, it could become lopsided. If it were struck hard enough, it could split up into two smaller droplets (Figure 13.8). Frisch and Meitner pointed out to Bohr that his theory could explain nuclear fission. Frisch writes:

> I had hardly begun to tell him about Hahn's experiments and the conclusions that Lise Meitner and I had come to when he struck his forehead with his hand and exclaimed, "Oh, what idiots we have been. We could have foreseen it all! This is just as it must be!" And yet even [Bohr], perhaps the greatest physicist of his time, had not foreseen it.

FIGURE 13.8
Nuclear fission according to the liquid drop model. *a.* An energetic neutron strikes a uranium nucleus. *b.* The neutron is absorbed, producing an excited and deformed nucleus that *c.* comes apart into two fission fragments and a few neutrons. Nuclear energy is released by this process.

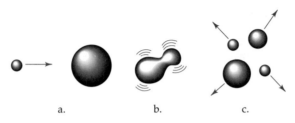

a.   b.   c.

We learned in Section 13.2 that δ, the binding energy per nucleon, is about 1 MeV larger for midsized nuclei than it is for uranium. If the uranium nucleus could be cajoled to split, a large amount of energy would be released. Neutrons are the keys to the nuclear energy of uranium. When an energetic neutron is absorbed by a ^{238}U nucleus, an excited state of ^{239}U is formed. It behaves like an unstable liquid drop. It wiggles about to get rid of its excess energy, and after a short time splits into two (usually unequal) droplets and a few energetic neutrons. Among dozens of possible decay schemes, it may do the following:

$$n + {}^{238}U_{92} \longrightarrow {}^{239}U_{92} \longrightarrow {}^{90}Kr_{36} + {}^{146}Ba_{56} + 3n \tag{9}$$

The fission fragments fly off with large kinetic energies. The Hungarian-born physicist Leo Szilard pointed out that if an element could be found that emitted two neutrons after absorbing one, then a chain reaction would ensue that could liberate useful amounts of energy (Figure 13.9). He even went so far as to patent his idea!

Chain Reactions

Because of its high neutron-to-proton ratio, uranium could fission into two smaller fragments and, most importantly, a few leftover neutrons. The fission neutrons could be captured by other uranium nuclei, which would fission and produce even more neutrons. The result could be a self-sustaining *chain reaction* (Figure 13.9). If it were carefully controlled, this process could produce useful power. If it were allowed to proceed without limit, it would produce a tremendous explosion. Szilard saw the military implications of the new physics and advocated secrecy early in 1939. Fermi was unconvinced, as the following discussion between him and Isadore Rabi reveals:

FIGURE 13.9
Chain reaction. A neutron strikes a uranium nucleus causing it to split into two smaller nuclear fragments and a few neutrons (two, in this figure). Each neutron causes another uranium nucleus to fission. As more and more uranium nuclei fission, a considerable quantity of energy is released. In a nuclear reactor, a balance is established between neutron production and neutron loss so that useful power may be extracted. Similar phenomena of exponential growth describe the multiplication of microbes in spoiled food or the spread of epidemics as each infected individual spreads the disease to several others.

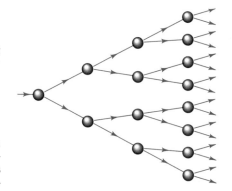

R. "Look, Fermi, I told you what Szilard said and you said 'Nuts!' and Szilard wants to know why you said 'Nuts!'"
F. "Well ... there is the remote possibility that neutrons may be emitted in the fission of uranium and then of course perhaps a chain reaction can be made."
R. "What do you mean by a remote possibility?"
F. "Well, ten percent."
R. "Ten percent is not a remote possibility if it means that we may die of it. If I have pneumonia and the doctor tells me that there is a remote possibility that I might die, and it's ten percent, I get excited about it."

Only a few months before, French and American scientists proved that the fission of a uranium nucleus releases somewhere between two and four neutrons. Szilard revised the likelihood of establishing a chain reaction to be above 50 percent. There was a good chance that a controlled fission reactor—or an uncontrolled bomb—could be built. German scientists discovered fission, and Germany would certainly try to build one. War seemed imminent and Szilard and his colleagues felt that America should get the bomb first. He and Eugene Wigner drafted the famous letter to President Roosevelt that Einstein, another new immigrant, would sign:

> Some recent work by E. Fermi and L. Szilard leads me to expect that the element uranium may be turned into a new and important source of energy in the immediate future. Certain aspects of the situation which has arisen seem to call for watchfulness and, if necessary, quick action on the part of the Administration. I believe therefore that it is my duty to bring to your attention the following facts and recommendations:
>
> In the course of the last four months it has been made probable—through the work of Joliot in France as well as Fermi and Szilard in America—that it may become possible to set up a nuclear chain reaction by which vast amounts of power would be generated. Now it appears almost certain that this could be achieved in the immediate future.
>
> This new phenomenon would also lead to the construction of bombs, and it is conceivable—though much less certain—that extremely powerful bombs of a new type may thus be constructed. A single bomb might very well destroy the whole port together with some of the surrounding territory.

In view of this situation, you may think it desirable to have some permanent contact maintained between the Administration and the group of physicists working on chain reactions in America.

The first thing was to demonstrate a chain reaction. While ordinary uranium fissions when struck by energetic neutrons, the neutrons released by the fission process are too slow to cause additional fission reactions. You can't have a chain reaction in pure ^{238}U. However, the rare isotope ^{235}U fissions when struck by slow neutrons. A nuclear reactor could be built from ordinary uranium, which contains 0.72 percent ^{235}U, but it would have to be large and cumbersome. Nonetheless, Fermi and his coworkers set about to build such a device under the stands of the University of Chicago's football field. They got it to work on the frigid afternoon of December 2, 1942:

> The clicks [of the neutron counter] came more and more rapidly, and after a while they began to merge into a roar; the counter couldn't follow anymore. That was the moment to switch to the chart recorder. Everyone watched in the sudden silence the mounting deflection of the recorder's pen. It was an awesome silence. Everyone realized the significance of that switch; we were in the high intensity regime. Again and again, the scale of the recorder had

The uncontrolled chain reaction taking place within a nuclear weapon produces an explosion. *Source:* Photo Researchers.

to be changed to accommodate the neutron intensity which was increasing more and more rapidly. Suddenly Fermi raised his hand. "The pile has gone critical," he announced. No one present had any doubt about it.

By that time America was at war, its troops fighting in North Africa and Guadalcanal. American physicists knew that their demonstration of a chain reaction would lead to the construction of a powerful bomb—by one side or the other.

Soon after the war, scientists, politicians, and the public expected that nuclear energy would provide a virtually endless supply of cheap energy. Many nuclear power plants were built in the United States and elsewhere. Today, 75 percent of the electric power in France is obtained from the fission of uranium. In the United States, however, the construction of nuclear power plants has stopped.

One of the factors retarding the growth of the nuclear power industry has to do with the safe disposal of nuclear waste. The neutron-to-proton ratio in uranium is much larger than in smaller, stable nuclei. For this reason, fission fragments are neutron-rich and highly radioactive. The spent fuel of a nuclear reactor contains all of the fission by-products and is chock full of highly dangerous materials—it cannot be disposed of casually.

Nuclear Fusion—
The Power
Behind the Sun

Midsized elements have the highest binding energies per nucleon. Fission reactors provide an environment where neutrons are slowed down and contained. They initiate and maintain a controlled chain reaction

---

DO YOU KNOW ABOUT
The World's First Fission Reactor?

Fermi didn't know it, but nature had beat him to the punch by building her own nuclear reactor long before there were birds, bees, and battleships on Earth. ^{235}U has a much smaller half-life than ^{238}U. When our planet was young, the fraction of ^{235}U in uranium was larger than it is now and it was easier to build a reactor. The subterranean conditions were just right 1.7 billion years ago in what is now Gabon, Africa, where there is a large and rich uranium deposit. At the time, the ^{235}U abundance was 3 percent. Water flowing through the ore acted as a natural *modera-*

tor to slow down the neutrons. A chain reaction took place. Over the course of millions of years, more nuclear energy was produced there than is produced by a large commercial reactor over its career.

French scientists discovered the relic reactor by measuring isotopic abundances in that region. The Gabon ores have a markedly smaller $^{235}U/^{238}U$ ratio than do other uranium ores. Some of the fissionable material had fissioned long ago! Most of the fission products that were once dangerously radioactive are still where they were when they were created eons ago. If nature can do it, then surely we may find a safe and secure haven in which to stash our toxic nuclear wastes.

by which large uranium nuclei are induced to break apart into better bound smaller ones. A glance at the curve of binding energy (Figure 13.6) shows that there is another way to get at nuclear power. Energy is released if small nuclei can be induced to coalesce to form larger and stabler nuclei by the process known as *nuclear fusion*.

Fusion is not a theoretical fancy: It is the secret of sunshine! By the early twentieth century, Arthur Eddington realized that the source of stellar power had to be subatomic. In the 1920s, physicists speculated that at the high temperatures in a stellar core, nuclei could collide and react with one another, releasing energy. They suggested that *thermonuclear* reactions were the mechanism of the solar furnace. Hans Bethe, in the 1930s, proposed a scheme by which stars could produce their energy. It earned him a Nobel Prize, although the Sun's stellar furnace doesn't quite follow his original scenario.

About half of the Sun's radiant energy is produced by the reaction

$$^3\text{He} + {}^3\text{He} \longrightarrow {}^4\text{He} + p + p \tag{10}$$

According to Table 13.2, the total binding energy of two ^{3}He nuclei is $2(3 \times 2.57) = 15.4$ MeV while that of ^{4}He is $4 \times 7.07 = 28.3$ MeV. The fusion of two ^{3}He nuclei described by reaction 10 liberates 13 MeV of energy. The total power produced by the Sun is $L_\odot \simeq 4 \times 10^{26}$ W. Consequently, it must perform this trick about 10^{38} times each second. All in all, the Sun is a most inefficient power plant. If we divide its power output $L_\odot$ by its mass $M_\odot \simeq 2 \times 10^{30}$ kg, we find that it generates only 0.2 watt per ton of solar material. A modern power plant, nuclear or otherwise, does a million times better in terms of power per pound! The alert student should pose two incisive questions at this point:

1. The nuclear force has a very short range—the two helium nuclei must touch one another if they are to fuse. However, ^{3}He nuclei are positively charged and repel one another electrically. How do they come close enough together for the fusion reaction to operate?

   Answer   The temperature in the center of the Sun is 15 million degrees Kelvin, which means that the average kinetic energy of a particle is 1 KeV. The high temperature ensures that at least in some collisions, helium nuclei have enough energy to quantum-mechanically tunnel through the potential barrier and fuse together. But it's not easy—the average ^{3}He nucleus spend an average of 100,000 years banging around the Sun before it manages to fuse. Some day, we may run out of fossil fuels and uranium. What then? We need to acquire a safe and nonpolluting source of power, and the answer may be controlled nuclear fusion. For decades, scientists have been trying to create the necessary pressures and temperatures to sustain the fusion reaction:

$$^2\text{H} + {}^3\text{H} \longrightarrow {}^4\text{He} + n$$

in a laboratory setting. Hydrogen isotopes are the fuel of choice. Because their electric charge is least, the coulomb barrier to fusion is lowest with these isotopes. At present, controlled fusion remains a distant goal. Uncontrolled fusion, where the hellish environment necessary for nuclear reactions to proceed is provided by a fission bomb, is a dangerous reality—it's called the hydrogen bomb.

2. Where does the Sun get its fuel? ^{3}He is a very rare isotope of helium, both on Earth and in the Sun. Furthermore, whatever primordial ^{3}He the Sun once had was consumed long ago.

Answer The Sun manufactures its fuel from hydrogen by means of nuclear reactions that proceed at an intrinsically slow rate: They depend on the intervention of the *weak nuclear force,* the agent underlying β decay, and the last of the four fundamental forces of nature. The Sun burns slowly and peacefully because the weak force is weak and the supply of ^{3}He for reaction 10 is provided at a slow and steady rate. To understand the working of the Sun, we must take into account each of the four forces of nature. Gravity provides the high-pressure, high-temperature environment. Electrical repulsion provides an essential control mechanism to ensure that the Sun acts more like a reactor than a bomb. Finally, the weak force allows the Sun to convert its protons into neutrons so that they may be assembled into nuclei. We'll learn more about stellar energy production in Chapter 14, after I introduce you to positrons, neutrinos, antineutrinos, and the weak nuclear force.

Exercises

18. If the fission of a ^{235}U nucleus releases 200 MeV, how many kilowatt-hours of energy reside in 1 g of such a nuclear fuel?

19. The β decay of ^{14}C produces ^{14}N whose atomic mass is 14.00307 amu. The maximum electron energy is 0.16 MeV. What is the atomic mass of ^{14}C?

20. An isotope of fermium decays with a half-life of 1 s by spontaneous nuclear fission. Suppose the reaction is

$$^{246}Fm_{100} \longrightarrow \, ^{150}Nd_{60} + \, ^{96}Zr_{40}$$

The mass of the Fm nucleus exceeds the sum of the masses of the fission products by about 250 MeV/c^2. A mole of the isotope decays in this fashion. What is its mass in kilograms? Show that the total mass of the reaction products is approximately 99.9 percent of the original mass. How much energy is released in joules?

21. The nucleus of ^{8}Be$_4$ decays with a half-life of 7×10^{-17} s into two α particles, each with 9 MeV of kinetic energy.
    a. What is the velocity of each α particle?
    b. Use Table 13.2 to determine the binding energy per nucleon of the ^{8}Be$_4$ nucleus.

22. The ^{210}Bi$_{83}$ nucleus can either β-decay or α-decay. In the former case, it becomes ^{210}Po$_{84}$, releasing 1.2 MeV of energy. The polonium

nucleus α-decays to $^{206}Pb_{82}$, releasing 5.3 MeV. In the latter case, the bismuth nucleus becomes $^{206}Tl_{81}$, releasing 5 MeV. How much energy is released when the thallium nucleus β-decays to $^{206}Pb_{82}$?

23. Tritium and $^{14}C_6$ are both β radioactive nuclei. To what nuclear species do they decay?

24. How much energy in joules is produced by the fusion of a mole of ^2H with 1 mole of ^3H to form 1 mole of ^4He and 1 mole of neutrons?

25. Consider an idealized scenario for a chain reaction. We begin with a mole of fissionable atoms. Each fission process releases two neutrons, each of which causes another nucleus to fission. The chain reaction begins when one nucleus fissions at $t = 0$. At $t = 1$ (in arbitrary and very small units of time), the fission neutrons cause 2 more nuclei to fission. At $t = 2$, 4 more nuclei fission, at $t = 3$, 8 more nuclei fission, and so on. At what time have all of the nuclei fissioned? If each fission reaction releases 200 MeV of energy, how much energy in joules is produced by the chain reaction?

Where We Are and Where We Are Going

The investigation of radioactivity led to a powerful theory of the structure of the nucleus. Scientists determined the nuclear reactions taking place within stars that make them shine. The story of stars, from their birth as clouds of gas to their sometimes peaceful and sometimes violent deaths, was unraveled by studying the behavior of the smallest imaginable particles.

Radioactive decay obeys a statistical and quintessentially quantum-mechanical law: In a sample of absolutely identical atoms, some will decay today, but others may live for millenia. The conversion of mass to energy in the process of radioactive decay is a relativistic phenomenon. Both quantum mechanics and the special theory of relativity are needed to describe nuclear phenomena. However, the theory of quantum mechanics, as it was originally conceived, was not fully compatible with the theory of relativity. The first consistent physical theory that was both quantum-mechanical and relativistic was developed in the 1930s and 1940s. It is called *quantum electrodynamics*, or QED, and it describes the interactions among charged particles and photons. Chapter 14 introduces you to the strange phenomena associated with QED: the existence and observation of antimatter, and the creation and annihilation of elementary particles.

14   Elementary Particles

Quantum mechanics solved the riddles of atomic and nuclear structure. The Schrödinger equation explained the fundamental processes underlying chemistry, material science, geology, and biology. It described nuclei well enough for scientists to work toward safe and secure nuclear power and for doctors to make use of radioisotopes and nuclear magnetism. The physical principles underlying today's technological society have been known for 50 years, but the search for the ultimate building blocks of matter blazes onward. Many "elementary particles" have been found. What are they, what laws do they obey, and how are they related to neutrons and protons? How and why do stars shine? Why

Is matter eternal or will it eventually disappear by a process analogous to radioactivity? An experiment performed in a salt mine deep under Lake Erie proved that the half-life of matter is greater than 10^{31} y. A tank containing 8000 tons of pure water was surrounded by sensitive electronic "eyes" that could detect the decay of a single nucleon. No events were seen, but an even larger experiment is now being deployed in Japan.
Source: IMB Collaboration/Courtesy, L. Sulak.

are there stars and galaxies, and what happens when stars explode and galaxies collide? Where did the chemical elements come from? How did the universe begin and how will it end? What are the limits of human knowledge? We are headed toward a unified theory of all physical laws and have partial answers to all but the last question. The synthesis of quantum mechanics and special relativity is called quantum field theory. Particles and forces are not independent concepts: Both are included in the concept of quantum fields. Relativistic quantum mechanics describes atoms and electrons to ten-decimal-place precision. Its success has led physicists toward a consistent theory of all the particles and forces of nature, but the road has been full of surprises.

As we plunge deeper into the structure of matter, we shall learn about particles with funny names that are not parts of atoms. Many new particles were discovered by physicists studying the interactions of cosmic rays, which are energetic particles traveling between the stars that occasionally strike Earth. Positrons were discovered in 1932; pions, muons, and "strange particles" a few years later. These particles have been the keys to our understanding of the structure of matter. (Today, positrons are used for medical imaging and pions for radiation therapy.) Section 14.1 describes the prediction and discovery of antimatter and the development of relativistic quantum mechanics. *Quantum electrodynamics,* the theory of electrons and light, has served as a paradigm for today's theory of all elementary particle phenomena. Section 14.2 applies what we have learned to the properties of subnuclear particles. Section 14.3 is an optional digression about the new science of neutrino astronomy.

14.1 The Marriage of Quantum Mechanics and Relativity

Quantum theory, as originally put forward, explained the gross features of the hydrogen spectrum but not its details. Furthermore, the Pauli exclusion principle and the existence of electron spin were not consequences of the theory. They had to be put in by hand if atomic structure were to be understood. Both difficulties reflect the failure of the Schrödinger equation to be consistent with the special theory of relativity.

Because the speed of an atomic electron is about $0.01c$, relativistic effects on atoms are small and can often be neglected. Nonetheless, to obtain a complete and precise description of atomic phenomena, physicists needed a replacement for the Schrödinger equation. Schrödinger himself, in his seminal papers of 1926, had searched for a relativistic equation for the electron but in the end settled for a nonrelativistic approximation. A relativistic generalization of the Schrödinger equation, the *Klein–Gordon equation*, was later developed to describe the behavior of relativistic particles without spin. Particles of this kind were discovered in the 1940s and their behavior is described by the Klein–Gordon equation. A different kind of equation, however, had to be found to deal with spin 1/2 electrons.

Pauli devised a clever way to treat the spinning electron. In Chapter 11, we learned that a measurement of the component of electron spin along any axis must yield the value $\hbar/2$ (an electron spinning "up") or $-\hbar/2$ (an electron spinning "down"). Pauli introduced two wave functions to describe the electron: ψ_{up} for an electron spinning up and ψ_{down} for an electron spinning down. These were combined to form a unified two-component wave function:

$$\Psi = \begin{pmatrix} \psi_{up} \\ \psi_{down} \end{pmatrix}$$

If $\psi_{down} = 0$, the electron spin points up. If $\psi_{up} = 0$, its spin is down. If neither component vanishes, Ψ describes an electron spinning in another direction, where $|\psi_{up}|^2$ is the probability that its spin points up, and $|\psi_{down}|^2$ is the probability that its spin points down. In the absence of a magnetic field, the components of the electron's wave function each satisfy the Schrödinger equation. However, magnetic forces can change the direction of the electron's spin and thereby mix up the two components.

The Klein–Gordon equation is consistent with relativity, but it cannot describe spin. The Pauli equation describes the effect of electromagnetic fields on a slowly moving electron, but it is not relativistic. In 1928, the English physicist Paul Adrien Maurice Dirac (1902–1984) found the relativistic equation for a spin 1/2 electron, the *Dirac equation*. Electron spin is built into Dirac's equation from the start. Furthermore, the equation predicted the existence of *antimatter* and led to the realization that particles could be created and destroyed.

The Dirac Equation

The Dirac equation can be described as the square root of the Klein–Gordon equation. However, the square roots of equations (like those of negative numbers) involve new mathematical concepts. Dirac was led to a wave function of the electron with four components that automatically included a description of electron spin. Maxwell's equations say that a spinning electric charge produces a magnetic field, so a spinning electron should behave like a tiny magnet. Dirac's quantum-mechanical equation predicted the strength of the electron's built-in magnetism—its magnetic moment. The equation applies both to the details of atomic spectroscopy and to the scattering of high-energy electrons.

Some of the solutions of the Dirac equation described the observed behavior of electrons. Other solutions, however, seemed to describe electrons in states of negative energy. If such states existed, one would expect an electron in a positive-energy state to jump to a negative-energy state by emitting a photon. It could then jump to a state of even lower energy by emitting another photon, and so on and on. The world would blow up in a cascade of photons! Dirac, who often expressed the sentiment that his equation was smarter than he was, pondered:

> The problem of the negative-energy states puzzled me for quite a while. The main method of attack to begin with was to try to find some way to avoid the transitions to negative-energy states, but then I approached the question from a different point of view. I was reconciled to the fact that the negative-energy states could not be excluded from the mathematical theory, or so I thought, so let us try to find a physical explanation for them.

To do this, Dirac invoked the Pauli exclusion principle. He proposed that the vacuum is not empty, but instead is chock-full of negative-energy electrons. He likened the universe to a gigantic atom of an inert gas with all of its quantum shells filled by negative-energy electrons. According to the exclusion principle, none of the negative-energy electrons can do anything because all the negative-energy states are occupied to begin with.

Suppose that one of the particles in Dirac's sea of negative-energy electrons is struck by an energetic particle and driven to a positive-energy state. A vacant space or "hole" would be left in Dirac's "sea." Because a hole is the absence of a negatively charged particle, the hole should behave as if it were a particle of positive charge. Dirac wondered whether protons could be holes in an otherwise full sea. In that case, a single equation could treat both protons and electrons, which were then all of the known elementary particles. Dirac soon saw his error. His equation does describe two different kinds of particles, but they are not electrons and protons. The Dirac equation implies the existence of a particle that is not ordinarily found on Earth—the positron.

Dirac was wrong about the sea of negative-energy electrons as well. His equation was correct, but its interpretation is simpler and more subtle than he first thought. As the theory of relativistic quantum mechanics developed, physicists realized that the hypothesis of a vacuum filled with negative-energy electrons was a superfluous scaffold, like the ether had been a generation before. American physicist Robert J. Oppenheimer (1904–1964), who later became the leader of America's atomic weapons program, found that the extra solutions of Dirac's equation do not describe negative-energy states of negatively charged electrons. Rather, they describe positive-energy states of positively charged electrons. These new particles would later be found and given the name *positrons*. Furthermore, Oppenheimer showed, positrons could not be protons, which are almost 2000 times heavier than electrons. He proved that the electron and positron masses must be the same. By 1930, Dirac, Oppenheimer, and the small band of particle theorists agreed that the four components of the Dirac wave function corresponded to electrons and positrons spinning up or down: $e^- \uparrow$, $e^- \downarrow$, $e^+ \uparrow$ and $e^+ \downarrow$. At the time, nobody had seen a positron, and hardly any experimental physicists took the wild-sounding theoretical ideas of Dirac and Oppenheimer seriously.

A BIT MORE ABOUT
Cosmic Rays

The differences between relativistic and non-relativistic quantum mechanics are most striking for collisions of energetic particles with velocities near c. Until the advent of particle accelerators, the heavens were the only source of relativistic particles: They were called *cosmic rays* and were discovered by the Austrian physicist Victor F. Hess. In 1912, he wrote: "The results to my [balloon] observations are best explained by the assumption that a radiation of very great penetrating power enters our atmosphere from above." In the 1920s, Millikan used a cloud chamber to study these peculiar radiations. He confirmed that cosmic rays came from outer space and sometimes penetrated to Earth's surface and below. He believed they were γ rays (energetic photons) released by the nuclear fusion of hydrogen atoms in interstellar space and described them as the birth cries of newly formed atoms. Millikan was wrong. Nuclear fusion takes place deep within stars, not in space. Fusion makes stars shine but does not produce cosmic rays. Later experiments showed that primary cosmic rays (those incident on the atmosphere) are charged particles consisting of energetic protons with a smattering of larger nuclei. Their source lies somewhere outside the solar system but inside the Milky Way. Cosmic ray particles may have been accelerated by interstellar magnetic fields or perhaps are the relics of ancient stellar catastrophes. However important cosmic rays have been in the history of physics, their origin is not yet fully understood.

When cosmic rays collide with nitrogen or oxygen nuclei in the atmosphere, they produce many different kinds of particles. For decades, cosmic ray collisions were the only tiny window into the world of high-energy phenomena. Positrons were discovered among the debris of cosmic ray interactions, as were many other so-called elementary particles.

The Discovery of Positrons

Positrons were first observed in 1932 by the American physicist Carl David Anderson (1905–1991). Anderson studied cosmic rays at the insistence of Millikan, his research advisor. To view the rays, Anderson built a cloud chamber (which he later described to the media as "nothing much but a sealed tube full of water vapor under low pressure") surrounded by a large magnet. Cosmic rays traversing the chamber left tracks of water droplets. The horizontal magnetic field bent the particle trajectories into helices that turned one way if the particle was positively charged, the other way if negatively charged. In his first experiment, Anderson observed as many particles turning one way as the other. It seemed that there were as many negatively charged particles among the sea-level cosmic rays as there were positively charged particles. Were they electrons and protons, or were they something else?

EXAMPLE 14.1
The Motion of a Charged Particle in a Magnetic Field

A charged particle moves horizontally in a region of constant vertical magnetic field **B** (Figure 14.1). Its mass is m, its charge is q, and its velocity is **v**. Recall from Chapter 9 that the particle experiences a horizontal magnetic force $F = qvB$ in the direction perpendicular to **v**. Its effect is to change the direction of **v** but not its magnitude. Consequently, the particle describes a circular orbit at constant speed v.

a. Express the radius of the orbit R in terms of q, B, and the momentum of the particle ($\mathbf{p} = m\mathbf{v}$, if the particle is nonrelativistic).

b. In Anderson's experiment, the strength of the magnetic field was 7.5 tesla. Most of the cosmic ray particles Anderson observed had momenta from 30 MeV/c to 300 MeV/c. If the particles were moving horizontally, what were the radii of their circular orbits?

Solution

a. The centripetal acceleration of a particle moving in a circle of radius R at speed v is v^2/R. Setting the magnetic force qvB equal to the mass m of the particle times its acceleration, we obtain $mv^2/R = qvB$, or $R = mv/qB$. Thus, we obtain the desired relation: $R = p/qB$. This formula is correct for relativistic particles if the relativistic formula for momentum is used: $\mathbf{p} = m\mathbf{v}/\sqrt{(1 - v^2/c^2)}$.

FIGURE 14.1
A positively charged particle moves at speed v through the shaded region, where there is a magnetic field B pointing into the page. The particle experiences a force F perpendicular to its velocity. Therefore, it moves in a circle. The magnetic force is qvB, where q is the charge of the particle. The orbital radius R is proportional to the momentum of the particle and inversely proportional to the magnetic field strength.

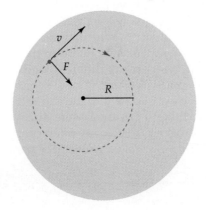

b. We are given the particle momenta in MeV/c. Because 1 MeV = 1.6×10^{-13} J, it follows that 1 MeV/c = $(1.6 \times 10^{-13})/(3 \times 10^8)$ kg m/s. If the particle is singly charged, we put $q = e = 1.6 \times 10^{-19}$ C into the result of part a to obtain

$$R = \frac{p \text{ (in kg m/s)}}{eB} = \frac{1}{300} \frac{p \text{ (in MeV/}c)}{B}$$

Thus, the radius of the circular orbit of a 30 MeV/c particle is about 1.3 cm, and that of a 300 MeV/c particle is about 13 cm. ∎

When particle momenta are large, the tracks left by protons or electrons in the chamber are indistinguishable. However, heavy particles with small momenta leave more ions in their wakes than light ones, thereby making broader tracks. Anderson focused on the tracks left by slow particles that bent the most in the magnetic field. He could tell which of these were made by heavy protons and which were made by less massive electrons. In 1982, Anderson reminisced:[*]

> Practically all of the low-velocity cases involved particles whose masses seemed to be too small to permit their interpretation as protons. The alternative explanations were that these particles were either electrons moving upward or some unknown lightweight particles moving downward. In the spirit of scientific conservatism, we tended at first to the former interpretation."

The track of a positive particle going down looks the same as that of a negative particle going up. To prove that he had seen a new kind of particle, Anderson had to find a way to determine the direction of a particle leaving a track. His act of genius was to divide the cloud chamber into two parts with a metal partition. As we saw in Example 14.1, the smaller the momentum of a particle, the more it curves in a magnetic field. When a charged particle passed through the partition, it lost energy, slowed down, curved more, and thereby revealed its direction of motion, as shown in the photo.

The first observed positron was traveling up rather than down, as most cosmic rays travel. It entered the bottom of Anderson's cloud chamber with an energy of 63 MeV, passed through a 6-mm lead plate, and emerged with an energy of 23 MeV. Notice how much more curved the trajectory becomes afterward. *Source:* Photo by C.D. Anderson/Courtesy AIP Emilio Segre Visual Archives.

[*]*The Birth of Particle Physics,* ed. L. M. Brown and L. Hoddison (Cambridge, England: Cambridge University Press, 1983), 139–140.

A lead plate was inserted across the center of the chamber in order to ascertain the direction in which these low-velocity particles were travelling and to distinguish between upward-moving negatives and downward-moving positives. It was not long after that a fine example was obtained in which a low-energy lightweight particle of positive charge was observed to traverse the plate, entering the chamber from below and moving upward through the lead plate. Ionization and curvature measurements clearly showed this particle to have a mass much smaller than that of a proton and, indeed, a mass entirely consistent with an electron mass.

Anderson saw many other positively charged, low-mass particles and proved they were positrons. Dirac's predicted particle had been found, but Anderson had not been guided by Dirac's theory:

It has often been stated that the discovery of the positron was a consequence of its theoretical prediction by Dirac, but this is not true. The discovery of the positron was wholly accidental. Despite the fact that Dirac's relativistic theory of the electron was an excellent theory of the positron, and despite the fact that the existence of this theory was well known to nearly all physicists, including myself, it played no part whatsoever in the discovery of the positron.

At other times, Anderson remarked: "I was too busy operating this piece of equipment to read [Dirac's] papers," and in any case, "their esoteric character was not in tune with most of the scientific thinking of the day."

Where did the positrons come from? Not from outer space. Primary cosmic radiation (that is, cosmic rays that have not yet struck the atmosphere) consists of protons and larger nuclei. Positrons are the secondary or tertiary by-products of primary collisions with atomic nuclei in the upper atmosphere. A cosmic ray proton strikes a nucleus and produces many particles, each of which makes more collisions and even more particles. Some of these cascading particles are fragments of struck nuclei, but most of them are brand-new particles.

In everyday experience, and even within the atom and its nucleus, particles move slowly compared to c, and their kinetic energies are much smaller than their rest energies. Cosmic rays, however, often have kinetic energies far greater than their rest energies, energies so large that new particles are created by their collisions. If an energetic photon strikes a proton or an electron at rest, it cannot create a single electron or positron because such a reaction would violate charge conservation. But the collision can produce a pair of particles consisting of an electron e^- and a positron e^+:

$$\gamma + p \longrightarrow p + e^- + e^+ \tag{1a}$$

$$\gamma + e^- \longrightarrow e^- + e^- + e^+ \tag{1b}$$

The creation of particles in high-energy collisions is the gist of high-energy physics.

EXAMPLE 14.2
Pair Production

The mass of the electron is $m = 0.511$ MeV/c^2. The mass of the initially stationary proton is $M = 938$ MeV/c^2.

a. What is the minimum photon energy E for reaction (1a) to take place?

b. What is the minimum photon energy for reaction (1b) to take place?

Solution **a.** The invariant energy (see Chapter 11) of the initial state is $\sqrt{(E + Mc^2)^2 - E^2}$. The invariant energy of the final state is at least the sum of the rest energies of the proton (Mc^2), the electron (mc^2), and the positron (another mc^2). Because the invariant energy of an isolated system cannot change, reaction (1a) can take place if

$$\sqrt{(E + Mc^2)^2 - E^2} \;>\; Mc^2 + 2mc^2$$

$$\text{or} \quad (E + Mc^2)^2 - E^2 \;>\; (M^2 + 4mM + 4m^2)c^4$$

$$\text{or} \quad 2Mc^2 E \;>\; 4mc^4 (M + m)$$

$$\text{or} \quad E \;>\; 2mc^2(1 + m/M)$$

We use $mc^2 = 0.511$ MeV and $m/M = 0.0005$ to obtain an explicit answer: $E \simeq 1.02$ MeV.

b. The preceding analysis applies to this case as well. All we need do is replace M by m in the formula for E. We obtain $E = 4mc^2 \simeq 2.04$ MeV. The minimum photon energy is about twice as large for reaction (1b) as for reaction (1a). ∎

What happens to positrons once they are made? A physics lab (and any other place) contains loads of electrons but hardly any positrons. A newborn positron is surrounded by hordes of electrons. If the positron is moving slowly, it can combine with an electron to form a structure much like an atom, called a *positronium*, in which the positron plays the role of the proton. This curious system is short-lived because the electron and positron soon annihilate each other. With a half-life of a microsecond or less, the positronium "atom" disappears and is replaced by two or three photons. The annihilation reactions may be written:

$$e^+ + e^- \longrightarrow \gamma + \gamma \qquad \text{or} \qquad e^+ + e^- \longrightarrow \gamma + \gamma + \gamma \qquad (2)$$

Despite its short lifetime, the positronium "atom" has been carefully studied and has provided some of the most sensitive tests of quantum electrodynamics.

Antiparticles

The positron is called the *antiparticle* of the electron. The relation is reciprocal: The electron is the antiparticle of the positron. Soon after Anderson discovered the positron, Dirac realized his theory "might be applied to protons. This would require the possibility of existence of negatively charged protons forming a mirror-image of the usual positively charged ones." In fact, relativistic quantum theory demands the existence of an antiparticle corresponding to every kind of particle. The masses of any particle and its antiparticle (like those of an electron and positron) are exactly the same and their electric charges are equal and opposite.

The negatively charged antiparticle of the proton p is called the antiproton $\bar{p}$. Its discovery is discussed in Chapter 15. The antiproton mass has been measured to be the same as the proton mass to an accuracy of a few parts per billion. Neutral particles also have antiparticles. The photon is its own antiparticle, but the neutron's antiparticle is a different particle called the antineutron $\bar{n}$. Nucleons and antinucleons, like electrons and positrons, annihilate each other on contact.

Particles and antiparticles have equal masses and opposite electrical and magnetic properties. They share other properties as well. The antiparticle of a stable particle is stable. If a particle is unstable, so is its antiparticle, and both have the same mean lives. Moreover, particles and antiparticles behave the same way in a gravitational field (that is, antimatter falls down, not up). All these predictions have been, and are being, tested by precise experiments.

Collisions between energetic electrons and positrons often proceed via the annihilation and subsequent creation of new particles. For example, the process

$$e^+ + e^- \longrightarrow \mu^+ + \mu^-$$

where μ is a particle about 200 times heavier than the electron, is routinely seen at accelerator laboratories. This process, and others like it, are correctly and completely described by the theory of *quantum electrodynamics.*

The substance of Earth is made up, by definition, of particles: electrons, protons, and neutrons. Positrons are occasionally produced by natural radioactivity or by cosmic rays. Antiprotons and antineutrons are routinely produced in collisions of energetic particles at accelerator laboratories. However, if the antiparticles synthesized by all the accelerators on Earth were gathered together, they would fit on a pinhead. In principle, antiparticles can be assembled into bulk antimatter. (In fact, the antihelium atom has been synthesized. It consists of two positrons bound to a nucleus made of two antineutrons and two antiprotons.) Scientists once thought that distant galaxies might be made of antimatter, but now it is known that they are not. In the early history of the universe, matter and antimatter played equivalent roles, but most of the matter and antimatter of the universe has been annihilated. For reasons that are just beginning to be understood, the remnant of this process—the stuff of stars, planets, and galaxies—is all in the form of matter. (What is called "matter" is a matter of words. It would be silly, but scientists could have defined electrons and protons as antimatter, and positrons and antiprotons as matter.)

Quantum Electrodynamics

The first offspring of relativistic quantum mechanics were the precise description of atomic structure and the prediction of the positron. And they were just the beginning. Physicists learned that electron-positron pairs were created in energetic collisions, and positrons were annihilated by electrons. The Dirac equation could describe the motions of electrons and positrons, but how could it take into account the creation and destruction of particles? The answer lay in the reinterpretation of the wave function as an operator that could create or destroy particles. What had been wave functions evolved into *quantum fields,* and a new mathematical formalism was developed to handle them. The first quantum field theory dealt with electrons, positrons, and photons, and their electromagnetic interactions. It was called *quantum electrodynamics,* or *QED,* and was developed principally by Richard Feynman, Julian Schwinger, and the Japanese physicist Sin-itiro Tomonaga in the 1940s. The triumphs of QED have made it a paradigm for the construction of today's more ambitious theories.

The creation and destruction of elementary particles is described graphically by Feynman diagrams.* These diagrams describe the processes that take place en route from an initial state to a final state, but they also stand for specific mathematical operations by which the details of these processes may be worked out: their probability of occurrence, how energy and momentum are shared among the particles, and so on. Feynman diagrams depict electrons as arrows pointing from left to right and positrons as arrows pointing from right to left. Photons are denoted by undirected wavy lines. Time runs from left to right: Particles and antiparticles in the initial state enter from the left. Particles and antiparticles in the final state exit to the right.

The simple diagrams shown in Figure 14.2 describe *being* not *becoming. A* is a point in space-time—then and there—from which a particle comes. *B* is another point in space-time—here and now—to which the particle goes. In Figure 14.2a, a solid arrow directed from *A* to *B* denotes the flight of an electron. Of course, there is no well-defined trajectory in quantum theory and the diagram must not be taken literally. Figure 14.2b is the same diagram with the direction of the arrow reversed. This diagram could be thought of as an electron traveling backward in time from *B* to *A*. What it really shows is a positron traveling from *A* to *B*. The wavy line in Figure 14.2c shows the flight of a photon from *A* to *B*. It is shown as an undirected wavy line rather than an arrow because the photon is its own antiparticle.

Electrons, positrons, and photons move freely through space. They satisfy Newton's first law: Their momenta remain constant. Until they interact! But there are no forces as such in diagrammatic language: no fields and no action-at-a-distance. Interactions are represented by means of more complex diagrams than those of Figure 14.2. The next most

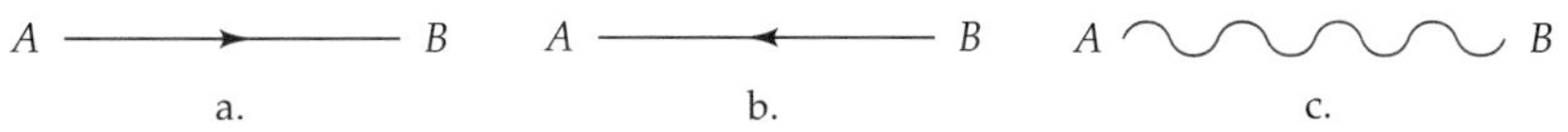

FIGURE 14.2
The ingredients of Feynman diagrams. Quantum electrodynamics describes the interactions of three kinds of particles: electrons, positrons, and photons. Solid arrows indicate electrons or positrons, depending on whether they point to the right or to the left. Wavy lines indicate photons.

Electron	Positron	Photon
$A \longrightarrow B$	$A \longleftarrow B$	$A \sim\!\sim\!\sim\!\sim B$
a.	b.	c.

*Feynman's small book, *QED: The Strange Theory of Light and Matter* (Princeton, NJ: Princeton University Press, 1985), makes a valiant attempt to explain quantum field theory in simple terms. It is recommended reading.

complicated diagrams, shown in Figure 14.3, portray the six fundamental acts by which particles interact with one another:

1. An electron may emit a photon.
2. An electron may absorb a photon.
3. A positron may emit a photon.
4. A positron may absorb a photon.
5. An electron and positron may annihilate into a photon.
6. A photon may create an electron-positron pair.

Energy and momentum conservation imply that none of these acts can take place as real physical processes. An electron cannot simply emit a photon: The invariant mass of the electron and photon necessarily exceeds that of the initial electron, and the invariant mass of an isolated system cannot change. Similarly, an electron or positron cannot simply absorb a photon. Nor can an electron-positron annihilate into or be created by a single photon. The fundamental acts of QED are forbidden!

Quantum mechanics comes to the rescue. Energy and momentum conservation may be set aside for a moment, as long as they are respected when the process comes to an end. (It's rather like buying a car on a very short-term loan—the actual time over which there may be a violation of the conservation laws is a tiny fraction of a second.) Several unphysical acts may be put together to produce a physically realizable process. Becoming is the result of a concatenation of two or more fundamental and forbidden acts.

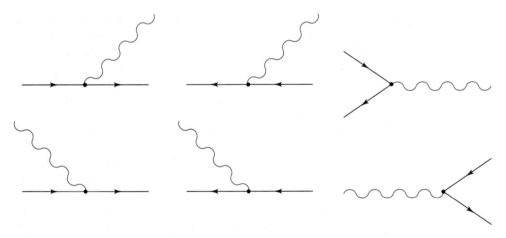

FIGURE 14.3 The fundamental acts of becoming in quantum electrodynamics. These simple acts are assembled into more complex diagrams to describe physically possible processes.

In Figure 14.4, two fundamental acts are wedded together to produce a diagram in which two electrons enter and two emerge. The physical process is the scattering of an electron by another electron, and the dia-

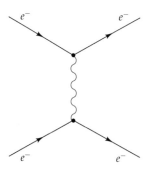

FIGURE 14.4
Two fundamental acts are linked together to form a two-act diagram describing the scattering of one electron by another: $e^- + e^- \rightarrow e^- + e^-$. The wavy line segment denotes a photon that is emitted and absorbed in the course of the process. It is a "virtual photon" in the sense that it acts to mediate the force between the two electrons.

gram is the simplest (and most important) contribution to the scattering process. In classical physics, we would say that one electron's trajectory is affected by the electric field of the other. In quantum mechanics, we would try to solve the Schrödinger equation for one electron in the electric field of the other. In QED, we say that one or more *virtual photons* may be exchanged, thereby transferring energy and momentum between the electrons.

It is not meaningful to ask if a photon is really exchanged. What really matters are the particles coming in and those leaving. All the rest is a computational artifice, a diagrammatic representation of a calculational procedure having nothing to do with "what really happened." Not even the time sequence of virtual processes is important. It doesn't matter whether the virtual particle is emitted by one particle and absorbed by the other, or vice versa. The probability of a particular scattering event is obtained by adding the contributions of the relevant diagrams and squaring the result.

Figure 14.5 shows three apparently distinct diagrams describing the scattering of a photon by an electron. In words, they can be described as follows:

a. An electron absorbs the incident photon and later emits the final photon.

b. An electron emits the final photon and later absorbs the incident photon.

c. The incident photon creates an electron-positron pair. The initial electron annihilates the positron producing the final photon.

Diagrams (14.5a) and (14.5b) represent distinct contributions to the Compton scattering process:

$$\gamma + e^- \longrightarrow \gamma + e^-$$

Changing the (irrelevant) time order of the two interaction points in Figure 14.5c by pulling the electron line straight makes it coincide with Figure 14.5b. They are simply different ways of drawing the same contribution to the photon-electron scattering process.

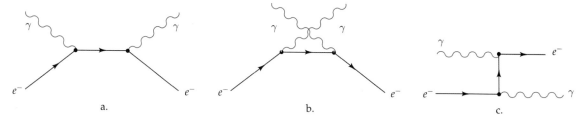

a.   b.   c.

FIGURE 14.5 Three diagrammatic contributions to the scattering of a photon by an electron: $\gamma + e^- \rightarrow \gamma + e^-$. If the electron line is straightened, diagram *c* becomes equivalent to diagram *b*. There are only two distinct contributions to this process in two acts.

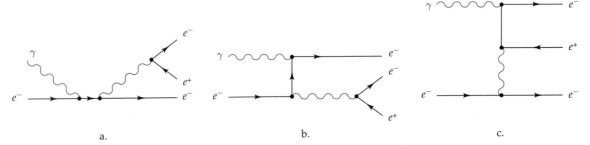

a.   b.   c.

FIGURE 14.6 Three Feynman diagrams contributing to the process by which an electron-positron pair is created in a collision between a photon and an electron.

The three diagrams in Figure 14.6 are each composed of three fundamental acts. They contribute to the pair-production reaction

$$\gamma + e^- \longrightarrow e^- + e^- + e^+$$

The photon strikes an electron and produces an electron-positron pair. QED offers a precise description of this process and any others involving the production, annihilation, or scattering of electrons, photons, and positrons. All possible processes involving electrons, positrons, and photons may be represented by Feynman diagrams.

The Feynman diagrams discussed so far are the simplest contributions to various physical processes. To obtain greater precision, scientists had to take into account more complex diagrams, such as the contribution to electron-electron scattering shown in Figure 14.7. In the original formulation of relativistic quantum mechanics, however, this procedure led to a paradox. When the indicated calculations were performed, the results came out to be infinity when they should have led to small and sensible corrections to the lowest-order calculations. Infinite answers don't make sense—something about the theory was very sick. The Austrian-born American physicist Victor Weisskopf put it this way:

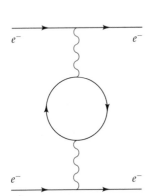

FIGURE 14.7
A complicated Feynman diagram contributing to electron-electron scattering, to which the diagram shown in Figure 14.4 is only a first approximation. To obtain precise results, many diagrams must be taken into account.

The appearance of infinite magnitudes in QED was noticed in 1930. Because they occurred only when a certain phenomenon was calculated to a higher order of accuracy than the lowest one in which it appeared, it was possible to ignore the infinities and stick to the lowest-order results which were good enough for the experimental accuracy of that period.

As technology developed and accurate experiments were performed, theoretical physicists needed more precise calculations to compare with experiments. They had to find out what was wrong with the theory. The first occasions in which QED runs into trouble correspond to simple diagrams such as that in Figure 14.8, an electron interacting with itself. The net effect of this diagram is to change the mass of the electron. It describes the *self-energy* of the electron due to its own electromagnetic field. However—and this was the principal stumbling block to the creation

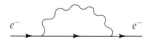

FIGURE 14.8
A Feynman diagram contributing to the electron's self-energy, the effect on its mass caused by its own electromagnetic field.

of QED—this diagram generates an infinite contribution to the electron mass. Of course, the electron mass is not infinite: It is about 10^{-30} kg.

This seemingly intractable problem was overcome in the 1940s by a procedure known as *renormalization*, by which the infinities of the theory are, so to speak, swept under the rug. The observed mass and charge of the electron are put into the theory at the start, and the results of calculations are expressed in terms of these quantities. When that is done, the infinities are transformed into answers to questions that should never have been asked, such as: What would the electron mass have been if electromagnetism did not exist? Because we cannot turn off electromagnetism, this is a philosophical question, not a physics question. When meaningful and measurable quantities are considered, QED always gives correct and finite answers.

Since QED was formulated, enormous theoretical and experimental advances have been made. For example, the magnetic moment of the electron is given as $\mu_0 = e\hbar/2m_e$ by the Dirac equation. Physicists using supercomputers have added up the corrections to this quantity due to hundreds of Feynman diagrams. Its predicted value μ_{th} is known to ten decimal places. Its measured value μ_{ex} has been determined with similar precision. The theoretical and experimental results agree:

$$\mu_{th} = 1.001\ 159\ 652\,\mu_0 = \mu_{ex}$$

This is one of many instances of the predictive power of QED. What more can one expect of a theory?

DO YOU KNOW
Whether the Photon Is Massless?

QED says the photon is massless, but nature is the ultimate authority. Physicists must determine the consequences of a nonzero photon mass and compare them to what is seen. According to Maxwell's equations, electromagnetic effects decrease gently with the distance between interacting objects. For example, electric force varies with $1/r^2$. If the photon had mass m_γ, Maxwell's equations would fail and electromagnetic effects would fall precipitously at separations beyond $d = \hbar/m_\gamma c$. The strongest limit on the photon mass was described by the Russian physicist G.B. Chibisov in 1976:

Virtually all physicists believe that the photon rest mass is exactly zero. On the other hand, there is no doubt that experiment has the last word in this important question. All experiments made to determine the photon rest mass give only upper limits for the mass. In particular, the photon mass may be zero. The best limit, obtained from the analysis of the mechanical stability of magnetized gas in the galaxies, shows that the photon's rest mass is at least 32 powers of ten less than the electron's. Do we really have to continue to infinity the succession of these upper limits in order to convince ourselves that the photon rest mass is zero. The answer is no.

Chibisov concluded that $m_\gamma < 3 \times 10^{-27}$ eV$/c^2$ or, equivalently, that the range of the electromagnetic interaction exceeds 10,000 light-years! For all intents and purposes, we may regard photons as massless particles.

QED is a relativistic quantum theory of electromagnetism. It tells everything anyone could want to know about the interactions of electrons, positrons, and photons, but it cannot describe nuclear particles or nuclear forces. The challenge to theoretical physicists was clear: to construct a theory of the strong and weak nuclear forces that was comparable to QED in power, consistency, and elegance. In Section 14.2, we approach this challenge by applying the language of Feynman diagrams to a variety of nuclear and subnuclear processes.

Exercises

1. Millikan knew that some cosmic ray particles at sea level travel upward. Explain how they can do so. (*Hint:* They cannot have passed through Earth from the antipodes.) How did the horizontal lead plate in Anderson's chamber help him to tell particles moving up from those moving down?

2. The rest energy of the muon ($\mu^\pm$) is about 106 MeV. Use the analysis of Example 14.2 to determine the minimum photon energy that can produce a pair of oppositely charged muons in the reaction:

$$\gamma + p \longrightarrow p + \mu^+ + \mu^-.$$

3. The Superconducting Supercollider being built in Texas will accelerate protons to momenta of 20 TeV/c, or 2×10^7 MeV/c. It is in the form of a ring 90 km in circumference. If it were exactly circular, and if magnets were placed along the entire periphery, how powerful would the magnets have to be to keep the protons in their orbits? (*Hint:* The answer is close to the strength of Anderson's magnet.)

4. Figure 14.6 exhibits three distinct Feynman diagrams contributing to $\gamma + e^- \to e^- + e^- + e^+$. Draw a fourth diagram for this process that cannot be deformed into any of the others.

5. Describe an experiment you can do that proves that the mass of the photon is less than a trillionth of the mass of the electron. (*Hint:* If the photon has mass m, the range of the electric force is $\hbar/mc$.)

6. Draw Feynman diagrams describing the following processes:
   a. $e^+ + e^- \longrightarrow \gamma + \gamma$
   b. $e^+ + e^- \longrightarrow \gamma + \gamma + \gamma$
   c. $\gamma + e^- \longrightarrow \gamma + \gamma + e^-$

14.2 Particles and Their Interactions

As quantum electrodynamics was being developed, other physicists turned to nuclear processes: to the mysteries of the strong and weak nuclear forces. The first suggestion as to the nature of the strong nuclear force came from the Japanese physicist Hideki Yukawa. We have seen that electromagnetic forces are mediated by the exchange of massless photons. In 1934, Yukawa proposed that nuclear forces result from the exchange of particles of a sort that had not yet been imagined, let alone seen in the laboratory.

Yukawa's hypothetical particle was massive so that it would produce a short-range nuclear force.* The relation between the range of a force and the mass of the exchanged particle lies in Heisenberg's uncertainty relation. The typical momentum of the exchanged particle is mc, so the position uncertainty associated with the virtual particle—and the range of the force—is $\hbar/mc$, the Compton wavelength of the particle. Yukawa chose his particle's mass to match the known range of the nuclear force, which is about the size of the proton (1.2×10^{-15} m). His conjectured particle—the proposed agent of the nuclear force, now called the *pion*—had to have a mass of about 150 MeV/c^2. The pion was found, but as we shall find later in this section, there was a surprise ending to the tale of its discovery.

The first hints about the nature of the weak nuclear force came from the study of nuclear β decay. If a nucleus contains too many neutrons for stability, one of its neutrons can become a proton with the simultaneous production of an electron and an antineutrino. We now turn to other processes governed by the weak force.

Positrons and Beta Decay

A high-energy collision can result in the creation of a pair of particles consisting of an electron and a positron. Pair production results from the electromagnetic force and is described by quantum electrodynamics. However, positrons can also be created by the weak nuclear force. It is an historical accident that positrons were first found in cosmic rays because they are produced by natural processes taking place on Earth.

In 1933, the Joliot-Curies (Madame Curie's daughter and son-in-law) were studying the impacts of α particles on atomic nuclei. When they irradiated a sheet of aluminum foil with α particles from a radioactive source, they observed positrons streaming from the foil. At first, the experimenters thought the positrons resulted from the disintegration of protons because, in their own (incorrect) words, "the proton is complex and results from the association of a neutron and a positron." This view is as untenable as its converse, once held by Rutherford, that a neutron is a combination of a proton and an electron. Protons and neutrons are both spin 1/2 particles and are equally elementary. However, the positrons did not emerge instantaneously—the foil continued to emit them long after it was removed from the radioactive source. The positron emission satisfied an exponential decay law with a half-life of a few minutes.

The Joliot-Curies concluded (correctly) that α bombardment of aluminum produced a new type of radioactive nucleus:

$$\alpha + {}^{27}\text{Al}_{13} \longrightarrow {}^{30}\text{P}_{15} + n$$

---

*Yukawa hoped that his new particle would generate both the strong and weak nuclear forces. It doesn't.

The mode of decay of ^{30}P was different from anything seen before. The decaying nucleus emitted a positron rather than an electron by a process akin to β decay. Consequently, the daughter nucleus lies one step below the parent in the periodic table:

$$^{30}P_{15} \longrightarrow {}^{30}Si_{14} + e^+ + \nu$$

The neutrino hypothesis, and its experimental confirmation, were discussed in Section 13.3. Ordinary β decay (which is more properly called β^- decay) results in the production of an electron and an antineutrino, but the decay of ^{30}P (an instance of β^+ decay) produces a positron and a neutrino. The Joliot-Curies did not mention the neutrino because it was still a highly speculative notion. They shared the 1935 Nobel Prize in chemistry for their discovery of a new form of radioactivity. The Curie women and their spouses earned a total of five Nobel gold medals.

The process of β^+ decay involves the conversion of a proton in a nucleus to a neutron and the simultaneous creation of a positron and a neutrino. Here is what happens inside the nucleus in the two varieties of β decay:

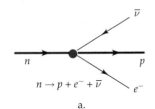

$$\text{In } \beta^- \text{ decay:} \quad n \longrightarrow p + e^- + \bar{\nu} \qquad Z \rightarrow Z+1 \qquad (3a)$$

$$\text{In } \beta^+ \text{ decay:} \quad p \longrightarrow n + e^+ + \nu \qquad Z+1 \rightarrow Z \qquad (3b)$$

The light neutral particle produced with an electron in β decay is defined to be an antineutrino $\bar{\nu}$, and that made with a positron is defined to be a neutrino ν. Both β processes change the identity of a nucleon and create a particle-antiparticle pair. The $e^- - \bar{\nu}$ pair produced by β^- decay or the $e^+ - \nu$ pair produced by β^+ decay is not present in the nucleus to begin with—it is created by the weak nuclear force.

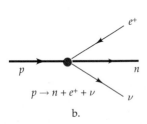

The processes of β^- and β^+ decay, as described by equations 3, can be portrayed as Feynman diagrams involving two kinds of arrows (Figure 14.9). Neutrons and protons are similar particles: Both are heavy spin 1/2 particles and are subject to the strong nuclear force. The word *nucleon* refers to either of them. In the diagrams, a nucleon is depicted by a colored arrow directed to the right.

Neutrinos and electrons are also similar particles: Both are light spin 1/2 particles and neither is subject to the nuclear force. The word *lepton* refers to electrons, neutrinos, and particles of their ilk. A black arrow directed to the right depicts a lepton, and a black arrow directed to the left depicts an antilepton. The effect of the weak force is shown as a blob into which a lepton line and nucleon line enter, and from which each emerges. In the process, one unit of electric charge is exchanged between the nucleon and the lepton.

FIGURE 14.9
a. The process of β^- decay. A neutron in a nucleus becomes a proton. At the same time, an electron and an antineutrino are created. *b.* The process of β^+ decay. A proton in a nucleus becomes a neutron. At the same time, a positron and a neutrino are created.

Electron Capture

In Section 14.1, we saw how the lines of a diagram describing one process may be turned around to produce a diagram describing another process. The diagrams shown in Figure 14.10 are variants of Figure 14.9b in which a lepton line has been brought from the right (corresponding to a final

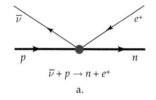

$$\bar{\nu} + p \rightarrow n + e^+$$

a.

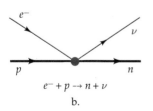

$$e^- + p \rightarrow n + \nu$$

b.

FIGURE 14.10
The Feynman diagrams shown here are obtained by displacing one or another of the lepton lines in Figure 14.9b. *a.* This diagram describes the process $\bar{\nu} + p \rightarrow n + e^+$. *b.* This diagram describes the process $e^- + p \rightarrow n + \nu$.

particle) to the left (corresponding to an initial particle). Figure 14.10a describes a process by which an antineutrino strikes a proton to become a neutron and a positron.

$$\bar{\nu} + p \longrightarrow n + e^+ \tag{4}$$

This reaction led to the first observations of antineutrinos.

Figure 14.10b describes a reaction by which a proton encounters an electron to form a neutron and a neutrino. This process takes place in some atoms, giving rise to a third form of β radioactivity called *electron capture*. A proton-rich nucleus consumes an orbital electron to become a nucleus with one less proton and one more neutron:

$$\text{In electron capture:} \quad e^- + p \longrightarrow n + \nu \quad Z + 1 \rightarrow Z \tag{5}$$

Electron capture is a radioactive process like β^+ decay in which A remains constant but Z decreases by one unit. Many isotopes, such as the naturally occurring but rare isotope ^{40}K, decay by capturing an electron. Electron capture induces the same nuclear transformation as β^+ decay, but requires less energy because an electron is consumed in the former process while a positron must be created in the latter.

DO YOU KNOW

How the Neutrino Was Discovered?

Reaction 4 describes the process by which the American physicists Frederick Reines and Clyde Cowan first detected antineutrinos in 1956. Working at a fission reactor in Georgia, they designed an experiment to detect antineutrinos produced by the nuclear reactions taking place within a nuclear reactor. They placed a large instrumented tank of *liquid scintillator* close to the core of the power-generating reactor at Savannah River, Georgia. The fluid in the tank had three important properties:

1. It converted γ rays into visible photons that were detected electronically.

2. It was rich in hydrogen to act as a target for

reaction 4 and to slow down the neuron released by the reaction.

3. It contained cadmium to absorb the neutron. (When a neutron is absorbed, the cadmium nucleus emits several detectable γ rays with a total energy of about 9 MeV).

When an $\bar{\nu}$ from the reactor interacted with a proton in the tank, the resulting positron encountered an electron. The positron and electron annihilated into back-to-back γ rays, each with the rest energy of an electron, about 0.5 MeV. The unambiguous signal of reaction 4 is the detection of the γ rays from positron annihilation, followed immediately by the detection of lower-energy γ rays coming from neutron capture. Cowan and Reines observed these signals and found the neutrino a quarter of a century after Pauli invented it.

To Beta-Decay or Not to Beta-Decay?

Nuclear stability is determined by the masses of the parent and daughter atoms. The standard of mass is the ^{12}C atom (including its six atomic electrons) with a mass of 12 amu, where 1 amu $\simeq$ 931.5 MeV/c^2. We refer to an atom with Z electrons and atomic mass number A as (A, Z) and to its mass as $M(A, Z)$.

Consider two neighboring isotopes with the same A but one step apart in the periodic table. Under what circumstances can (A, Z) β^--decay into $(A, Z + 1)$? The nuclear process involves the transformation $n \rightarrow p + e^- + \bar{\nu}$. The final state has just the right number of protons, neutrons, and electrons to constitute the $(A, Z + 1)$ atom. The mass of the neutrino is known to be very tiny. Thus, the β^- process takes place if

$$M(A, Z) > M(A, Z+1) \qquad \text{for } \beta^- \text{ decay} \qquad (6)$$

Similar reasoning determines the necessary condition for $(A, Z + 1)$ nucleus to capture an electron:

$$M(A, Z+1) > M(A, Z) \qquad \text{for electron capture} \qquad (7)$$

Because different isotopes never have exactly the same mass, one of the two inequalities, 6 or 7, must be satisfied. We have established an interesting and important result—two neighboring isotopes cannot both be stable. At least one of the isotopes $(A, Z+1)$ and (A, Z) must be radioactive.

It often happens that $M(A, Z+1)$ is sufficiently larger than $M(A, Z)$ to permit the β^+ reaction to compete with electron capture. In this case, the reaction is $p \rightarrow n + e^+ + \nu$. The β^+ process may take place if

$$M(A, Z+1) > M(A, Z) + 2m \qquad \text{for } \beta^+ \text{ decay} \qquad (8)$$

Testing the Law of Parity Conservation

The study of β decay in the 1950s led to the extraordinary discovery of *parity violation*, meaning that the fundamental processes of nature are not the same when viewed in a mirror. This result may not seem surprising. After all, your body is not the same when viewed in a mirror. Your heart is on the left and your liver on the right, but in a mirror, the locations are reversed. Biological molecules display a specific handedness as well. However, the left-right asymmetry of living things was probably an evolutionary accident. Life could have evolved differently—mirror-reflected men and women would have been just as good as we are.*

---

*One in a million of us is born with the wrong handedness. The condition is known in medicine as *situs invertus*. Although their hearts are on their right side and their livers on the left, these individuals can be perfectly healthy. The molecules in their bodies, however, have the same handedness as ours.

In the early 1950s, most physicists were certain that all physical laws were mirror symmetric. The notion was codified as the law conservation of parity and was once regarded as a sacred cow. The strong nuclear force conserves parity. So does the electromagnetic force. However, nobody bothered to ask whether the weak force conserves parity until 1956, when two young Chinese-American physicists at Columbia University, Tsung Dao Lee and Chen Ning Yang, pointed out that the emperor of parity had no clothes:

> The conservation of parity is usually accepted without question... There is actually no a priori reason why its violation is undesirable. As is well known, its violation implies the existence of a right-left asymmetry. We have seen in the above some possible tests of this asymmetry. These experiments test whether the present elementary particles exhibit asymmetrical behavior with respect to the right and the left.

Within months, Chien-Shiung Wu (known as Madame Wu) and her collaborators at Columbia University performed the test Lee and Yang proposed. They prepared a sample of radioactive ^{60}Co in which the nuclear spins were arranged to point in the same direction. The group discovered that the β rays from the ^{60}Co decays emerged preferentially along the nuclear spin direction (Figure 14.11).

FIGURE 14.11
Madame Wu and her collaborators placed a sample of radioactive ^{60}Co in a cryostat, where it was brought to a very low temperature, and an intense magnetic field caused the nuclear spins to line up with one another. Electrons produced by decaying cobalt atoms were detected and counted. The number of detected electrons when the field pointed down was about twice the number as when the field pointed up. Thus, the electron direction in β decay is correlated to the direction of the nuclear spin.

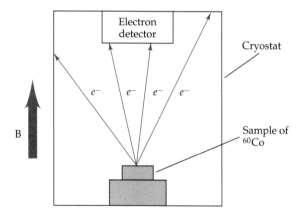

The specification of a direction of rotation entails a choice of handedness. If the cobalt nuclei spin in the direction of the curled fingers of the *right* hand, their spins—and the emitted electrons—point in the direction of the thumb. This shows that the β decay process is not right-left symmetric because the mirror image of a right hand is a left hand (Figure 14.12). Madame Wu and her colleagues proved that parity is violated by the weak nuclear force.

FIGURE 14.12
How the Wu experiment disproves the hypothesis of mirror symmetry. Both *a* and *b* show ^{60}Co nuclei with their spins pointing up along the fat arrows—like the outstretched right thumb when the fingers curl in the direction of nuclear rotation. Experiment revealed that case *a*, where the electron emitted by the decaying nucleus points away from its spin direction, is more probable than case *b*, where the electron points along the nuclear spin. In *c*, we see that the reflected image of case *a* is identical to case *b*. If the experiment were observed in a mirror, case *b* would have been more likely than case *a*. This experiment (and many more) showed that the laws of physics are not mirror symmetric.

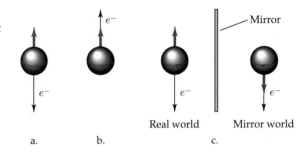

Real world   Mirror world

a.   b.   c.

Another deeply held belief of physicists fell in the 1960s: the law of microscopic time-reversal symmetry. In the everyday world, time has a well-defined direction. Eggs are eaten, break, rot, or become chickens, never the other way around. The apparent asymmetry of time is not a mystery. Things like neat desks and healthy bodies find more ways to lose their proper order than to keep intact. We need cleaning services and doctors because of the second law of thermodynamics, as explained in Section 5.4. However, the equations governing mechanics and electromagnetism are unchanged under reversal of the arrow of time.

Having lost their cherished mirror symmetry, physicists still clung to the principle of microscopic time-reversal symmetry. They felt certain that the basic laws of microphysics run the same forward or backward in time. Until 1964. At that time, Val Fitch and James Cronin observed a tiny time-reversal violating effect in the behavior of elementary particles. Not only does the weak force violate parity conservation, but it displays a preferred direction in time! This tiny effect is of little practical importance to us now, but it may explain why there is matter in the universe!

DO YOU KNOW

Why We Are Made of Matter, Not Antimatter?

Matter is made of protons, neutrons, and electrons rather than of their antiparticles. Long ago, when the universe was very hot, things were very different. The universe contained a hundred trillion times as many nucleons as it does today. There were an approximately equal number of antinucleons as well. As the universe cooled, nucleons and antinucleons annihi-lated one another. Fortunately, not all of them were annihilated. We were left with a tiny nucleon excess from which matter has formed. Was the universe created lopsided, or was the small matter-antimatter imbalance forced upon it? Explaining such a development is more satisfying than merely accepting it. Soon after the violation of time-reversal symmetry was detected, Andrei Sakharov (who was both the father of the Soviet hydrogen bomb and a champion of peace) showed how this tiny effect, long ago, may have made possible the existence of matter in today's universe.

Nucleon Number and Lepton Number

The Feynman diagrams describing the weak interactions shown in Figures 14.9 and 14.10 display an important property: Nucleon lines and lepton lines enter and leave the weak interaction blob, but they never appear or disappear. This property can be formulated more generally in terms of quantum numbers and conservation laws. Electric charge Q is an example of a conserved quantum number. Quantum electrodynamics demands, and experiments confirm, that in all reactions among particles, the sum of the charges of the initial particles is equal to the sum of the charges of the final particles.

Let's introduce two more quantum numbers: *nucleon number* $\mathcal{N}$ and *lepton number* $\mathcal{L}$. Each nucleon carries nucleon number $\mathcal{N} = 1$ and each antinucleon carries $\mathcal{N} = -1$. Similarly, each lepton (electron or neutrino) carries $\mathcal{L} = +1$ and each antilepton (positron or antineutrino) carries $\mathcal{L} = -1$. Because β^- decay produces an electron and an antineutrino, and β^+ decay produces a positron and a neutrino, these processes conserve lepton number $\mathcal{L}$. Table 14.1 shows the quantum number assignments of particles. Their antiparticles have equal and opposite quantum number assignments.

TABLE 14.1   Quantum Number Assignments

Particle	Lepton Number $\mathcal{L}$	Nucleon Number $\mathcal{N}$	Electric Charge Q
Proton	0	1	+1
Neutron	0	1	0
Electron	1	0	−1
Neutrino	1	0	0

The downfall of parity conservation and time-reversal symmetry reminded physicists that things are not always as they seem. Until they are put to the most rigorous experimental tests, nucleon number conservation and lepton number conservation cannot simply be accepted as inviolable laws of nature. The complete list of conserved quantities applicable to reactions involving nucleons, electrons, neutrinos, and their antiparticles includes energy, momentum, and angular momentum along with Q, and possibly $\mathcal{L}$ and $\mathcal{N}$. Any process consistent with all these conservation laws can and does occur in nature, although some processes are more likely to happen than others. Conversely, any process conflicting with a valid conservation law cannot take place.

EXAMPLE 14.3
Conserving Quantum Numbers

Test the following hypothetical decay modes for conservation of Q, $\mathcal{L}$, and $\mathcal{N}$:

$$^{14}C_6 \longrightarrow \begin{cases} ^{15}N_7 + e^- + \bar{\nu} \\ ^{14}N_7 + e^+ + \nu \\ ^{14}N_7 + e^- + \nu \end{cases}$$

Solution

The first process conserves Q because the parent nucleus has charge 6 (in units of e), which is the sum of the final particle charges. It conserves $\mathcal{L}$, which is 0 on the left and $0 + 1 - 1$ on the right. However, it does not conserve $\mathcal{N}$ because there are 14 nucleons on the left but 15 on the right. This process is forbidden. The second reaction conserves $\mathcal{L}$ and $\mathcal{N}$, but not $\mathcal{Q}$. The third conserves $\mathcal{N}$ and $\mathcal{Q}$, but not $\mathcal{L}$. All three processes are forbidden. The process that correctly describes the β decay of radioactive carbon is:

$$^{14}C_6 \longrightarrow {}^{14}N_7 + e^- + \overline{\nu}$$ ∎

Testing the Law of Conservation of Lepton Number

Scientific laws are put forward as hypotheses to be tested by experiments. One of the most sensitive tests of lepton number conservation involves the process of *double β decay*. Suppose that a nuclear species (A, Z) can neither α-decay to $(A-4, Z-2)$, nor β-decay to $(A, Z+1)$, nor electron-capture to $(A, Z-1)$. It can still be radioactive. If the nuclear masses satisfy the relation

$$M(A, Z) > M(A, Z+2)$$

the nucleus (A, Z) can decay by emitting two electrons and increasing its atomic number by two. The pattern of energy levels required for double β decay is shown in Figure 14.13.

Here are two nuclear processes by which Z may increase by two steps:

$$n + n \longrightarrow p + p + e^- + e^- + \overline{\nu} + \overline{\nu}$$

$$n + n \longrightarrow p + p + e^- + e^-$$

Both of these reactions conserve $\mathcal{N}$ and $\mathcal{Q}$. The first version of double β decay also conserves $\mathcal{L}$ because two leptons and two antileptons are produced. The second reaction, which is called no-neutrino double β decay, does not conserve $\mathcal{L}$ but causes it to increase by two units.

FIGURE 14.13
The atomic masses of ^{76}Ge and its neighbors. About 8 percent of natural germanium consists of the isotope ^{76}Ge. It cannot β-decay to either of its immediate neighbors. In fact, both ^{76}Ga and ^{76}As β-decay into germanium. However, the Ge isotope may jump two units of Z by the process of double β decay to become ^{76}Se. The half-life of this process is more than 10^{20} years. Isotopic masses are shown relative to the mass of the selenium isotope.

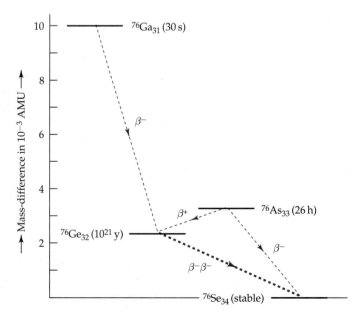

The law of conservation of lepton number permits the first reaction but forbids the second. Thus, the search for no-neutrino double-beta-decay can test the law. Dozens of nuclei are energetically permitted to double β decay, and some are known to do so. For example, $^{82}\text{Se}_{34}$ transforms itself, by the double β process, into $^{82}\text{Kr}_{36}$. Experimental data confirm the hypothesis of lepton number conservation. The allowed two-neutrino double β decay of selenium has been observed and measured in the laboratory. The mean life of ^{82}Se is an extraordinary 10^{20} years! (Thus, no more than one atom per mole decays in an hour!) Many groups of physicists are searching for the forbidden no-neutrino mode of double β decay, but no one has seen it so far. All evidence indicates that the conservation of lepton number is an exact symmetry of nature, but we don't yet know for sure.

Testing the Law of Conservation of Nucleon Number

Nobody has ever observed a process in which nucleon number changes. Perhaps they haven't looked hard enough. Perhaps nucleons decay but with an exceedingly long lifetime. Let's consider the following hypothetical decay scheme for protons:

$$p \longrightarrow e^+ + \gamma \tag{9}$$

Many other decay modes are possible, but this one (involving familiar particles) serves as an illustration. One thing is clear—the lifetime of the proton is surely very long because there are lots of protons still around in our 10^{10}-year-old universe.

In 1973, Howard Georgi and Sheldon L. Glashow put forward the first theory to unify all the elementary particle forces. Theories of this kind abolish nucleon number conservation. They predict that all matter is radioactive and that the mean life of a nucleon is about 10^{30} y. Could it be that diamonds are not forever? Two teams of scientists set up experiments to see for themselves, one deep within an Ohio salt mine and the other under a mountain facing the Japan Sea. "If protons must die," said Maurice Goldhaber, an old hand in the proton decay story, "let them die in my arms."

Large tanks of very pure water were built deep underground where they were shielded from cosmic rays. If a nucleon decayed in the water, a small amount of the energy of its decay products would be converted into a characteristic pattern of light that would be detected by many sensitive photodetectors surrounding the tanks (Figure 14.14, on page 596). If the nucleon lifetime were 10^{30} years, about one would decay each day per thousand tons of water. The tank had to be enormous. The American tank was a 20-m cube—the size of an apartment house. It contained 8000 tons of water. The Japanese tank was about a third the size but better instrumented. After years of searching, however, neither group detected any proton decays. The experiments proved that the mean life of the nucleon exceeds 10^{31} years.

Experiments with greater sensitivity are being carried out to test the conservation of lepton number and nucleon number. These conservation

FIGURE 14.14
The most sensitive search for proton decay was carried out by scientists from the University of California at Irvine, Brookhaven National Laboratory, Boston University and the University of Michigan. If a nucleon anywhere in a large underground tank of water had decayed, its decay products would have produced a tiny flash of light. The light would have been detected, and a computer could have constructed a picture of the decay process. The experiment was carried out for a decade, but no sign of proton decay was seen—one of my favorite theories was ruled out.

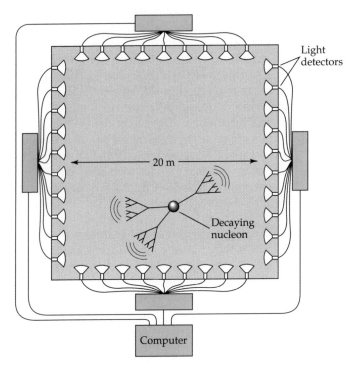

Light detectors

20 m

Decaying nucleon

Computer

laws are certainly approximately valid. Whether or not they are exact laws of nature remains to be determined. Theoretical arguments suggest that the proton must decay and that neither nucleon nor lepton numbers are always conserved.

Pions and Muons   In the 1930s and 1940s, cosmic ray physicists discovered many new particles. Cosmic rays were known to be of two sorts: an easily absorbed "soft" component and a more penetrating "hard" component that can be detected hundreds of meters below the ground. The former consists of electrons, positrons, and photons. The latter was mystifying. In 1937, Anderson and Neddermeyer at Cal Tech published a paper claiming the discovery of new particles of intermediate mass:

> Interpretations of the penetrating [tracks] encounter very great difficulties [unless they are produced by] particles of unit charge, but with a mass larger than that of an electron and much smaller than that of a proton.

Confirmations immediately appeared from groups at Tokyo and Harvard University. The discoverers of the new particles named them *mesotrons* from the Greek *mesos* meaning middle because they have middle-sized masses. Many physicists believed that the particles predicted by Yukawa had been found and sometimes referred to them as *yukons*. After all, the new particles had just the mass that Yukawa had foreseen. But nature had a big surprise in store for us. So did history: The denouement of the meson (truncated version of *mesotron*) story was deferred for the duration of the Second World War.

"BUT DON'T YOU SEE, GERSHON — IF THE PARTICLE IS TOO SMALL AND TOO SHORT-LIVED TO DETECT, WE CAN'T JUST TAKE IT ON FAITH THAT YOU'VE DISCOVERED IT."

© 1993 by Sidney Harris.

DO YOU KNOW

Why Proton Decay Cannot Power the Sun?

If a proton in the Sun decays by reaction 9 or in other ways, all of its rest energy is soon converted into heat. (Positrons produced by proton decay annihilate with electrons to become γ rays, whose energies would be transferred to the atoms in the Sun.) Imagine that proton decay is the source of the Sun's power. What is the lifetime τ of the proton if this be so? There are $N \simeq 10^{57}$ protons in the Sun. The number of protons decaying per second is N/τ. Each decay releases an energy $m_p c^2$, so the power resulting from proton decay is $P = N m_p c^2 / \tau$. The rest energy of a proton is $m_p c^2 \simeq 1.5 \times 10^{-10}$ J. We set P equal to the Sun's total power output of 3.8×10^{26} W to find that τ must be about 10^{14} years. Can astrophysicists be mistaken and proton decay, not nuclear fusion, be the power behind the stars?

If protons in the Sun decay, then so must those of Earth. Geologists know how much heat is welling up from Earth's interior. Some geothermal energy is used to generate electricity. Most of Earth's heat comes from radioactivity of the known quantities of uranium and thorium in its crust. Some additional heat trickles up from the still molten interior at a known rate. Because most geothermal power is accounted for, there cannot be another significant source of heat inside Earth. If proton decay is heating Earth, it cannot contribute more than an extra 10^{13} W. A simple calculation, as we did for the Sun, shows that the half-life of the proton must therefore exceed 10^{21} years. Simple geological considerations tell us that proton decay cannot power the Sun and the proton lifetime is at least a hundred-billion times longer than the age of the universe!

The particles making up the penetrating component of cosmic rays were found to be longer lived and more penetrating than expected. In 1946, an Italian collaboration proved that the particles discovered by Anderson and Neddermeyer were not the particles predicted by Yukawa. The observed particles could not mediate the nuclear force because they did not even respond to it.

Let me explain the trap nature prepared. Collisions of primary cosmic rays in the upper atmosphere produce Yukawa's particles in abundance. They are no longer called mesotrons or yukons, but rather *pions*, and they come in three varieties with different charges: positive pions π^+, negative pions π^-, and neutral pions π^0. Charged pions are each other's antiparticles. The neutral pion is its own antiparticle and decays very rapidly into two photons.* All three pions have masses of about 140 MeV/c^2. They comprise an isotopic triplet of similar particles just as neutrons and protons comprise an isotopic doublet.

Charged pions are unstable and short-lived. They decay via the weak force by a process akin to nuclear β decay. Their mean life is about 10^{-8} s and their decay products are almost always muons and neutrinos:

$$\pi^- \longrightarrow \mu^- + \overline{\nu} \qquad \pi^+ \longrightarrow \mu^+ + \nu$$

The penetrating component of cosmic radiation—the particles discovered by Anderson and Neddermeyer—are not pions but muons resulting from the decays of pions in the stratosphere! Pions are the strongly interacting particles predicted by Yukawa. The μ^- has no strong interactions. It is a lepton, like the electron and the neutrino.

Cecil Powell and his group of cosmic ray physicists discovered charged pions in 1947. "In recent experiments," they wrote, "we showed that charged mesons sometimes lead to the production of secondary mesons. We have now extended these observations by examining plates exposed in the Bolivian Andes at a height of 5500 m, and have found, in all, forty examples of the process leading to the production of secondary mesons."

Powell's group trekked to the high Andes because that's where the pions were. Photography is a powerful tool for both astronomy and physics. A charged particle passing through photographic emulsion produces a visible track when the plate is developed. Cloud chambers and photography were the principal tools of cosmic ray physicists. When accelerators were developed, these devices were supplanted by "bubble chambers" and sophisticated electronic techniques for detecting, identifying, and tracking energetic particles. Powell's paper continues:

> Our observations, therefore, prove that the production of a secondary meson is a common mode of decay of [primary] mesons. We represent the primary meson by the symbol π and the secondary by μ. It can thus be shown that the ratio m_π/m_μ is less than 1.45.

---

*Neutral pions were the first particles to be discovered at accelerators. They were observed in 1950 by the American physicist Jack Steinberger and his collaborators at a cyclotron in Berkeley, California.

The discovery of muons was a surprise. Today, we know what muons are, but we don't know what they are for in the grand scheme of things.* Muons are like fat electrons. That is, they are pointlike, seemingly elementary, electrically charged leptons. Like electrons and neutrinos, they are spin 1/2 particles that are not subject to the strong nuclear force. Their mass is about $200\,m_e$, or $105\ \mathrm{MeV}/c^2$, and they decay with a mean life of 2 microseconds according to the scheme

$$\mu^\pm \longrightarrow e^\pm + \nu + \overline{\nu}$$

For decades, physicists believed that nucleons and pions were as elementary as particles can be. The exchange of pions contributes to the force between nucleons, and for a time they were believed to be the ultimate nuclear glue. The list of particles and antiparticles encountered so far is not overwhelming:

- Nucleons (n and p) and antinucleons ($\overline{n}$ and $\overline{p}$)
- Leptons (e^-, μ^-, and ν) and antileptons (e^+, μ^+, and $\overline{\nu}$)
- Pions (π^+, π^0, and π^-)
- Photons (γ)

Strange Particles

It didn't take long, however, before particles were found that didn't fit into this simple picture. The trouble began two months after Powell's triumphant discovery of the pion. Once again, cosmic rays presented us with a puzzle and a challenge. G. D. Rochester and C. C. Butler, in Manchester, England, published a paper in 1947 entitled *Evidence for*

The first strange particles were detected at sea level in a cloud chamber. Another technique involves bringing photographic plates to a mountaintop. Cosmic rays passing through the plate produce tracks when the plate is developed: They photograph their own trips! C. F. Powell and his group of cosmic ray physicists at the University of Bristol published this photograph of a charged kaon (a strange particle) entering from the left and decaying into three charged pions. (One of the pions subsequently scattered from a nucleus.) *Source:* Department of Physics, University of Bristol/Courtesy, Meyers Photo-Art.

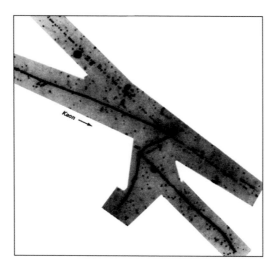

*The American physicist Luis Alvarez found a practical use for muons. In the late 1960s, he led an American-Egyptian project to search for hidden galleries within the Pyramid of Chephren. The team of scientists used cosmic ray muons to produce "X-ray" images of the pyramid. They proved there are no undiscovered treasures.

the Existence of New Unstable Elementary Particles. They had discovered charged and neutral *kaons*: spinless particles with masses between pions and nucleons and with mystifying properties. Soon afterward, other particles were discovered that decayed into nucleons and pions. None of these had been anticipated, and all of them displayed weird properties. They became known as *strange particles* and would fascinate and bewilder particle physicists for decades.

Strange particles were the tip of an enormous iceberg. The development of the cyclotron made available an intense and reliable source of energetic particles. By 1940, it was used to synthesize neptunium and plutonium, the first of the transuranic elements. As larger and larger accelerators were built, more and more particles were discovered: hundreds of them! Things would seem hopelessly complicated before nature's simplicity would emerge. In 1979, Philip Handler, the late president of the National Academy of Sciences, wrote about the scientific progress he had witnessed in his long career:

> Man learned for the first time the nature of life, the structure of the cosmos, and the forces that shape the planet, although the interior of the nucleus became if anything, even more puzzling.

Dr. Handler was not well informed about particle physics. Just a few years before, many of the wildest speculations of theoretical physicists had been assembled into what now seems to be a correct, complete, and coherent theory of the behavior of elementary particles. This *standard model* of elementary particle physics—the triumphant conclusion (for the moment) of the search for the basic building blocks of matter—is presented in Chapter 15.

Exercises

7. Extend Table 14.1 to include the quantum numbers of antileptons and antinucleons.

8. The following reactions are forbidden because they do not conserve one or more of lepton number, nucleon number, or electric charge. Which conservation laws are violated by each process? The symbols $\bar{p}$ and $\bar{n}$ denote the antiproton and antineutron.

   a. $\nu + p \longrightarrow e^- + n$
   b. $p \longrightarrow e^+ + \gamma$
   c. $\gamma + p \longrightarrow e^+ + n$
   d. $n + p \longrightarrow \bar{p} + \bar{n} + e^+ + e^+$
   e. $e^- + p \longrightarrow \bar{\nu} + n$

9. How do you suppose that no-neutrino double β decay can be distinguished experimentally from the two-neutrino process?

10. Give an example of a mode of proton decay other than reaction 9 that is compatible with electric charge conservation.

11. What is the mass in MeV/c^2 of a particle whose Compton wavelength is 1.2×10^{-15} m?

12. The decay $\mu \rightarrow e + \gamma$ has never been observed. Can you invent a new quantum number whose conservation explains why this process is forbidden?

13. Suppose that a pion at rest decays into a muon and a massless neutrino. What is the energy of the neutrino?

14. Consider the reaction $p + p \rightarrow p + p + \pi^0$. If the target proton is at rest, what is the minimum kinetic energy of the incident proton for which this reaction is energetically allowed?

14.3 Neutrino Astronomy (Optional)

The wondrous spectacle of the night sky gave birth to astronomy and its flawed sibling astrology. At first, astronomers studied the patterns and motions of heavenly bodies. Simple telescopic measurements led to Newton's grand synthesis. Centuries later, spectroscopy and photography enabled astronomers to determine the composition of stars and the velocities of galaxies. But modern astronomy is not restricted to the tiny slice of the electromagnetic spectrum eyes can see. Ultraviolet and infrared radiations have their secrets to tell. So do X rays and γ rays, radio waves, and microwaves. Modern astronomers study photons in a wavelength range spanning 24 powers of 10!

There are other ways to learn about the heavens. Meteors streak through the skies, and those falling to Earth as meteorites bring vital information about complex molecules made in space and complex nuclei made by ancient supernova. Cosmic rays introduced us to elementary particles and their interactions. The most recent addition to the astronomical arsenal is neutrinos. The science of neutrino astronomy began with the search for neutrinos produced by nuclear reactions in the solar core.

A New Science Is Born   Two American scientists, Raymond Davis, an experimenter, and John Bahcall, a theorist, were the founders of solar neutrino astronomy. "Theory and experiment depend on each other for their significance in solar neutrino research," they wrote. It is not enough to detect the neutrinos coming from the Sun—the result must be compared to theoretical expectations. They continued:

The early literature on nuclear fusion as the basis to solar energy production did not mention the possibility of testing the ideas by observing neutrinos. In the great papers by Bethe, neutrinos were not included specifically in the nuclear reactions... The principle of lepton number conservation was not clearly articulated and one was not required to balance leptons as well as nucleons.

By the early 1950s, Davis set about to search for solar neutrinos. He proposed to detect them through the reaction

$$\nu \text{ (from the Sun) } + {}^{37}\text{Cl} \longrightarrow e^- + {}^{37}\text{A} \qquad (10)$$

by which a solar neutrino strikes a chlorine atom converting it into a radioactive argon atom. The argon atoms would be collected and their individual decays would be detected and counted. His prototype experiment, completed in 1955, used a few tons of chlorine-rich cleaning fluid as a target. The reviewer of Davis's paper was amused:

Any experiment such as this, which does not have the requisite sensitivity, really has no bearing on the question of the existence of neutrinos. One would not write a scientific paper describing an experiment in which an experimenter stood on a mountain and reached for the moon, and concluded that the moon was more than eight feet from the top of the mountain.

Davis learned his lesson. His new experiment would be 10,000 times more sensitive. Before we describe his startling results, we return to the theory of stellar energy generation.

The Solar Furnace

In Chapter 13 , we explained how the Sun derives much of its power from the fusion reaction

$$^3\text{He} +^3 \text{He} \longrightarrow {}^4\text{He} + p + p + 12.9 \text{ MeV} \qquad (11)$$

The fuel to maintain this reaction is fueled by the two-step process:

$$p + p \longrightarrow {}^2\text{H} + e^+ + \nu \qquad (12a)$$

$$^2\text{H} + p \longrightarrow {}^3\text{He} + \gamma \qquad (12b)$$

To make a ^{4}He nucleus from four protons, two of them must be converted into neutrons by means of the weak interaction process (12a), releasing two positrons and two neutrinos. The synthesis of one ^{4}He nucleus produces a total of 28 MeV. The positrons annihilate with electrons in the Sun and contribute to its radiant energy production. The neutrinos, which carry about 2 percent of the Sun's radiant energy, are so weakly interacting that they easily pass through the Sun. About 6×10^{14} neutrinos from the pp reaction (12a) reach Earth per square meter per second. Day and night, enormous numbers of neutrinos stream harmlessly through our planet, our bodies, and Davis's huge tank of cleaning fluid. However, their energies (typically 0.3 MeV) are too small to induce reaction 10.

Many other thermonuclear reactions take place in the Sun. Among them is the side chain:

$$^3\text{He} + {}^4\text{He} \longrightarrow {}^7\text{Be} + \gamma \tag{13a}$$

$$^7\text{Be} + p \longrightarrow {}^8\text{B} + \gamma \tag{13b}$$

$$^8\text{B} \longrightarrow {}^8\text{Be} + e^+ + \nu \tag{13c}$$

$$^8\text{Be} \longrightarrow \alpha + \alpha \tag{13d}$$

These reactions produce a tiny fraction of the Sun's power. They are important because the neutrinos resulting from reaction 13c are more energetic than those produced by the dominant *pp* reaction and are therefore easier to detect. The mean energy of the ^8B neutrinos is about 6 MeV, but only 1 in 10,000 solar neutrinos is of this kind.

The end product of the Sun's thermonuclear reactions is ^4He. Because no $A = 8$ nucleus is stable, the Sun cannot pursue nuclear fusion beyond this point. All the larger elements that exist on Earth and in the Sun are remnants of the fiery deaths of earlier generations of stars.

Astrophysicists are confident of their knowledge of the processes that take place within the Sun and other stars, and they can explain stellar history and evolution. The Sun, when it runs out of hydrogen, will begin to collapse and heat up. When the core reaches a critical temperature, its helium will start to burn by the *triple α* process:

$$\alpha + \alpha + \alpha \longrightarrow {}^{12}\text{C} + \gamma$$

This process will proceed more like a bomb than a reactor and will some day cause the Sun to explode and engulf the planets. Stars larger than the Sun may withstand the triple-α process, but they face an even more calamitous death later on. They may become supernova.

Are the stars as well understood as we think? The light from a star comes from its surface, but the neutrinos produced by a star originate at its core. By detecting solar neutrinos on Earth and measuring their energies, scientists can confirm their theory of stellar energy generation. The chlorine experiment, to everyone's surprise, taught us that something is amiss in our understanding of either particles or stars.

The Chlorine Experiment

Since 1970, Davis has deployed a tank containing 600 tons of C_2Cl_4 deep underground at the Homestake Gold Mine in South Dakota. Reaction 10 can only take place if the neutrino's energy exceeds 1 MeV. Every few days, an energetic solar neutrino coming from ^8B decay succeeds in transmuting one chlorine atom into argon. Every month or so, for the last 20 years, Davis and his collaborators have been extracting and counting (and are still extracting and counting!) the argon atoms produced by solar neutrinos.

The good news is that Davis has observed neutrinos coming from the Sun. The experiment has detected decaying ^{37}A atoms at an average rate of about 12 per month. Thus, the Sun is most certainly a fusion reactor. The bad news is that Davis has seen too few neutrinos. Detailed astrophysical calculations by Bahcall and others imply that Davis should see about 30 events per month. Why has this not happened? The solar neutrino problem has three possible resolutions:

1. Davis erred and the experimental data are wrong. This is unlikely. The results of the chlorine experiment have been confirmed by other experiments.
2. Bahcall erred and the astrophysical calculation of the expected number of solar neutrinos is wrong. This is also unlikely. Other calculations confirm Bahcall's result.
3. The currently favored explanation for the observed discrepancy between experiment and theory is that something happened to the neutrinos on their way out of the Sun. There are three different kinds of neutrinos in nature, of which only one is involved in the process of β decay. Many physicists believe that solar neutrinos change their identities on their journey through the Sun and to Earth.

The truth is that scientists are not really sure about what is going on, although they hope to find out in the near future.

Other Experiments

I described the negative results of the search for nucleon decay in Section 14.2. The Japanese nucleon-decay detector (called *Kamiokande* because it is sited near the town of Kamioka) has been recycled—today, it searches for neutrinos from space. The most energetic neutrinos from ^{8}B decay produce the following reaction within this giant underground water tank:

$$\nu + e^- \longrightarrow \nu + e^-$$

Much of the neutrino's kinetic energy is transferred to the struck electron, which, in turn, produces a characteristic light signal in the tank. Kamiokande has detected solar neutrinos and confirmed the Davis result—only about half the predicted number of neutrinos were seen.

Two other large experiments sensitive to the copious, but lower energy, pp neutrinos are now under way. Both use gallium and the neutrino-induced reaction

$$\nu + {}^{71}\text{Ga} \longrightarrow e^- + {}^{71}\text{Ge}$$

In these experiments, radioactive germanium atoms produced by solar neutrinos are collected and counted, as in the chlorine experiment. Most of the Sun's neutrinos are energetic enough to react with gallium nuclei. One of these experiments, the GALLEX experiment, is primarily a European collaboration. It involves 30 tons of gallium (about the annual world production) placed in an underground laboratory under Gran Sasso Mountain near Rome. A second experiment, still known by the acronym SAGE, for Soviet-American Gallium Experiment, is being done in Russia. Published reports from both gallium experiments confirm that fewer solar neutrinos are seen than are expected.

Unless there is something wrong with the solar model, we are forced to believe that electron neutrinos are turned into other neutrino species on their way out of the Sun. This means that neutrinos must have nonzero masses and must be endowed with properties that are not yet fully understood. Within a few years, scientists at two new, large solar neutrino laboratories—one in Canada and one in Japan—will detect and measure thousands of solar neutrinos. Solar neutrino physics, which began as an attempt to confirm the theory of solar structure and evolution, has turned about completely. It is telling us things about neutrinos (or perhaps, something about the Sun) we could never have discovered without neutrino astronomy. Serendipity is rampant in science—no one can tell what surprises await the intrepid experimenter.

Supernova
Neutrinos

All stars produce neutrinos, but only the Sun is near enough to let us see its neutrinos—unless it's a star that suddenly becomes supernova. A supernova explosion releases 2×10^{47} J of energy in a few seconds—hundreds of times more than the Sun produces in its entire lifetime! Only a tiny fraction of this energy appears as starlight. Even so, the star becomes billions of times brighter than it was. Almost all the energy of a supernova is emitted in a few seconds in the form of neutrinos! Supernovae are rare occurrences; the average galaxy produces supernovae at a rate of a few per century. The last one known to have exploded in the Milky Way (our galaxy) was seen in 1604.

Thus, scientists were delighted on February 23, 1987, when a supernova appeared. The exploding star is in the Larger Magellanic Cloud, an appendix to the Milky Way, 160,000 light-years away. The new supernova was clearly visible to observers in the southern hemisphere—and to particle physicists in Japan and Ohio! The two great experiments searching for proton decay did not see what they set out to see; they saw supernova neutrinos instead. Within a 12-second interval, Kamiokande detected 12 neutrino events, the American group another 8. The results agreed with the predictions of astrophysicists, whose theory of stellar collapse was confirmed by the newborn science of neutrino astronomy. The next time a star in our galaxy explodes as a supernova—perhaps in a century, or maybe next year—neutrino astronomers will observe thousands of its neutrinos. The tiniest elementary particles will tell us what we cannot otherwise know about the death throes of a giant star.

Exercises

15. Explain why a fission reactor is a source of antineutrinos while the Sun is a source of neutrinos.

16. What is the mass of hydrogen in kilograms consumed by the Sun in each second? How much of the solar mass is converted into energy per second?

17. The reaction $p + p \rightarrow {}^2H_1 + e^+ + \nu$ is exothermic. In principle, the two protons in a hydrogen molecule could fuse, releasing 1.5 MeV of energy. Why is it that this process is never observed on Earth?

18. The flux of solar neutrinos on Earth is about 6×10^{14} per square meter per second. Their mean energy is 0.3 MeV.
    a. Estimate the number of neutrinos radiated by the Sun in one second.
    b. Estimate the power radiated by the Sun in the form of neutrinos.
    c. What fraction of the solar luminosity is the result of part b?

19. The radius of a nucleus containing A nucleons is about $1.2 \times 10^{-15} A^{1/3}$ m. What would the radius of the Sun become if it were compressed to nuclear density? (*Hint:* the Sun is made up of about 10^{57} nucleons.)

20. An exploding supernova releases 2×10^{47} J of energy. Calculate the mass in kilograms that has been converted into energy. Compare your result to the mass of the Sun.

Where We Are and Where Are We Going

Quantum electrodynamics is a precise and predictive theory describing the interactions of photons, electrons, and positrons. However, other forms of matter exist in the universe as well. In particular, there are neutrons and protons that somehow stick together to form nuclei. Are these elementary, or are they made from simpler things? What is the origin of the force that holds nucleons together? Radioactivity was understood as a nuclear transformation. The β process allows a proton to become a neutron, and vice versa. At the same time, a pair of leptons is produced. What is the fundamental mechanism underlying this process?

In Chapter 15, we shall answer all these questions. The 1970s was a time of explosive development of our knowledge of the microworld. The behavior of elementary particles results from the interplay of three fundamental forces of nature: electromagnetism and the strong and weak nuclear forces. Using quantum electrodynamics as a model, scientists developed a theory, called the standard model of elementary particle physics, that describes these forces. Experiment after experiment has confirmed the predictions of the standard model, which offers a consistent picture of all elementary particle phenomena. Although the model appears to be correct, it is also manifestly incomplete and some day will be replaced by a more powerful deluxe model. As questions about nature are posed and then solved, new and deeper questions arise. The more we learn about the universe, the more we want to learn.

15 The Standard Model—Where We Are Today

W e want to describe how we stand today in the age-old attempt to explain all of nature in terms of the workings of a few elements in infinite variety of combinations; in particular, what are the elements?

Richard P. Feynman, American physicist

Theorists and experimenter could lean back in their armchairs and gloat. The electromagnetic and weak theories had been unified, and a theory of the strong interactions had been devised. Taken together, these two theories are usually known as the "Standard Model." Within this model were two families of particles—the quarks and the leptons.

Barry Parker, American physicist

So, naturalists observe, a flea

Hath smaller fleas that on him prey;

And these have smaller still to bite 'em;

And so proceed ad infinitum.

Jonathan Swift, Irish author

The Tevatron collider at Fermilab, situated near Chicago, Illinois, is the most powerful accelerator in the world. Protons and antiprotons, whirling in opposite directions along 3-mile long circular rings, are made to collide within enormous detectors, thereby making possible the study of particle collisions at an invariant energy of 2000 GeV. *Source:* Fermi National Accelerator Laboratory/Photo Researchers.

Jean Baptiste de Lamarck, an eighteenth-century French naturalist, observed that "the most important discoveries of the laws, methods and progress of Nature have nearly always sprung from the observation of the smallest objects which she contains." Lamarck was speaking about biology, but his remark applies even more to modern physics. Particle physicists have developed a simple and comprehensive theory of the building blocks of matter and the rules by which they combine. Although hundreds of subatomic particles have been found, they are now known to be built from 12 basic constituents: the "matter particles." Section 15.1 is a brief introduction to the discipline of high-energy physics and its role in the discovery of the particles and forces of nature. Section 15.2 describes the classification of subatomic particles and some of their properties. The final two sections focus on the "force particles" that mediate the interactions of matter particles. Section 15.3 introduces quarks and the gluons that hold them together according to today's theory of the strong force, quantum chromodynamics. Section 15.4 explains how the weak nuclear force and electromagnetism were recognized to be parts of a unified electroweak force mediated by photons and their heavy cousins. These two theories comprise today's standard model of elementary particle physics and bring us to the current installment of the millenia-long search for the basic stuff of which matter is made. Many questions have been answered but even more remain for future scientists to tackle.

15.1 High-Energy Physics

The structure of the microworld was revealed by studying particle interactions at ever higher energies, as indicated in Figure 15.1. Scientists were led from atoms to nuclei to the constituents of nuclei. Particles once thought to be elementary—such as protons and neutrons—are now known to be composite systems. To appreciate this discovery and to see how it forms part of the standard model, we must understand the reasons physicists were driven to high energy. One has to do with quantum mechanics, the other with relativity.

FIGURE 15.1
The race to high energy—a century of exploration. As physicists explore the structure of matter, they are driven to higher energies. Optical spectroscopy (the study of photons of a few eV) led to quantum theory. X-ray spectroscopy (the study of photons with a few KeV) revealed the nuclear charge. Experiments at a few MeV exposed the structure of the nucleus. The standard model evolved from the study of particle collisions at several GeV and was confirmed by experiments at energies as high as 1 TeV. The big question—how particles got their masses—will be answered at the next generation of particle accelerators.

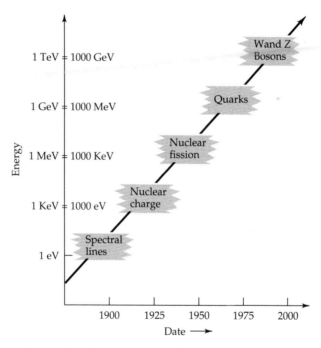

- To study the structure of the atom or its nucleus, probes must be used that are smaller than the systems being studied. We cannot "see" an atom because the wavelength of visible light is thousands of times larger than an atom. Beams of tiny particles must be used instead. However, quantum mechanics says that particles have wave properties. The de Broglie wavelength of a particle with momentum p is $\lambda = h/p$, where h is Planck's constant. The smaller the structure being observed, the higher must be the energy of the particles in the beam.

- Only a few of the particles of nature are among the constituents of ordinary matter. Many particles, like pions and muons, are unstable, with lifetimes much less than 1 s. Others, like positrons and antiprotons, are stable but are annihilated by encounters with ordinary matter.

Almost all so-called elementary particles must be made if they are to be known. According to the special theory of relativity, particles of every sort may be created by the collisions of ordinary particles. The heavier the particle to be created, the larger must be the energy of the colliding particles. In particular, the invariant mass of the colliding particles must be larger than the mass of the particle to be made.

The particles used by Rutherford and his contemporaries were not energetic enough to touch the nucleus and thereby measure its size. Cosmic rays are more energetic, and many new particles were discovered among them. However, they are too few and sporadic for use in controlled experiments. Further progress depended on the development of "atom-smashers" (or, more precisely, particle accelerators*) and particle detectors, sophisticated instruments with which the products of collisions can be detected and the data recorded. Before we proceed further along in energy, let's describe some of the tools of the high-energy physicist.

Particle Accelerators and Particle Detectors

The first experiments in particle physics used natural radioactivity or cosmic rays as the source of high-energy particles. Individual particles were detected by eye as they made tiny flashes of light upon striking a chemically coated screen.

By 1930, physicists used newly developing electronic technologies to build high-voltage devices that accelerated charged particles to energies as high as 1 MeV. These devices were difficult to build, dangerous to operate, and limited in the energies and beam intensities they could provide. Could it be possible, the American physicist Ernest O. Lawrence wrote, "to develop methods for the acceleration of charged particles that do not require the use of high voltages?" He answered his own question in 1932 by inventing and developing the *cyclotron:*

> The study of the nucleus would be greatly facilitated by the development of sources of high speed ions having kinetic energies in excess of 1 MeV, for it appears that such swiftly moving particles are best suited to the task of nuclear excitation. The straightforward method of accelerating ions through the requisite differences in potential presents great experimental difficulties associated with the high electric fields necessarily involved. The present paper reports the development of a method that avoids these difficulties by the means of the multiple acceleration of ions to high speeds without the use of high voltages.

Lawrence's trick was to confine particles with a magnetic field, and to accelerate them—bit by bit—by modest electric fields. Cyclotrons can accelerate particles to energies of 25 MeV and no further because relativistic

---

*The words *atom-smasher* and *atomic bomb* are misnomers. Atoms are easily disrupted by X rays or other radiations. Nuclear bombs and reactors work by smashing atomic nuclei. Accelerators make it possible for physicists to study the collisions of individual particles with enormous energies.

Early particle accelerators used very high voltages to accelerate charged particles. This accelerator was constructed in the 1930s to accelerate protons to energies of millions of electron volts. *Source:* Carnegie Institution of Washington, Department of Terrestrial Magnetism.

effects become important. Other kinds of accelerators were invented to get to larger energies. The maximum energy attained by accelerators has increased by a factor of about 25 per decade since 1930, with a consequent explosion of scientific discovery. For example, the 30-year-old proton accelerator at Brookhaven National Laboratory, New York, generates a pulse of 10^{12} protons every few seconds. Each proton has an energy of 30 GeV. Experiments performed at this facility have generated six Nobel Prizes.

Accelerator scientists have always aimed at the highest possible energies for fundamental research into the basic constituents of matter, but the technology involved has other important large-scale applications. Proton, pion, and electron beams are used by physicians to cure or ameliorate cancer and other diseases. In fact, 1 of every 8 Americans will receive medical treatment by an accelerator beam at some time in his or her life! Accelerators can determine the isotopic composition of materials to help identify their source and their date of origin. Industry uses accelerators to find flaws in manufactured goods and may someday use them to produce fuel for fusion reactors. Electron accelerators produce intense beams of monochromatic ultraviolet light that can be used to study the structure of materials. The demand for accelerators for applications to material science, nuclear science, medicine, chemistry, astrophysics, and industry has created a billion dollar industry.

Of course, accelerators would be useless without detectors to observe the results of particle collisions. When a charged particle passes through matter, it nudges atoms in its path, resulting in the production of ions or the emission of light. These effects can be amplified to produce an electronic signal or a visual image of a high-energy collision from which the energies, momenta, and identities of the produced particles can be determined.

Detector technology is also rich in spin-offs. The 1979 Nobel Prize in medicine was awarded to two physicists for their invention of computer-assisted medical imaging techniques. MRI scanners use nuclear magnetism to produce images, CAT scanners measure X rays, and PEP scanners measure γ rays from positron annihilation. All these procedures use detection and analysis techniques that have sprung from research in basic physics, not medical science. The 1992 Nobel Prize in physics was awarded to the French physicist George Charpak for his development of many of today's particle detectors. For the past decade, however, Charpak has applied his particle skills to the development of superior

The brain of a living person is revealed by Magnetic Resonance Imaging (MRI). Non–invasive scanning procedures have much in common with the techniques of high–energy physics. MRI works by tickling the nuclei of the atoms in the patient with electromagnetic fields and recording their response. A computer creates an image from this data. *Source:* © 1987 Ralph C. Eagle, Jr. MD/Photo Researchers.

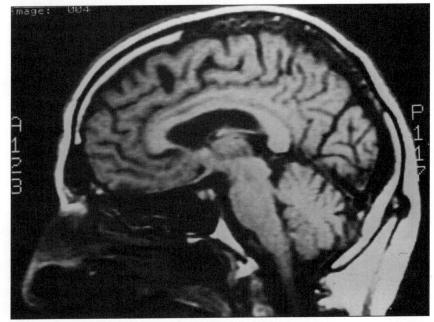

medical instrumentation. He invented a device that does radiochemical assays in hours rather than days. The journal *Physics in Medicine and Biology*, among others, is devoted almost exclusively to the medical applications of modern physics.

The Road to High Energy

Particle beams produced by early accelerators were used to explore the structure of the nucleus and to produce elements not found in nature. By the 1950s, accelerators and detectors had become sophisticated enough for a new breed of "high-energy physicists" to attack the problems of particle physics. We can catch the flavor of this new and exciting era by examining a few of its high points.

The American physicist Robert Hofstadter, working at Stanford University in the 1950s, studied the collisions of electrons with atoms. By measuring how many electrons were scattered at different angles, he determined the size of atomic nuclei and the distribution of positive charge within them. These experiments were performed at energies of 188 MeV and revealed the structure of atomic nuclei. For example, the nucleus of the hydrogen atom—the proton—was found to have a radius of about 10^{-15} m. Having shown that an individual nucleon is not a pointlike particle, Hofstadter (who won the Nobel Prize for his work) remarked that "it is fascinating to think of what the proton itself is built." Experimenters at Stanford would answer this question a decade later when a more powerful accelerator was built.

Although positrons were discovered in cosmic rays, many other forms of antimatter are produced and studied at powerful accelerators. Relativity and quantum mechanics demand that every particle has an antiparticle. For example, the antiparticle of the proton—the antiproton—has the same mass as the proton but a negative electric charge. Antiprotons could be produced by the reaction

$$p + p \longrightarrow p + p + p + \overline{p} \tag{1}$$

That is, a proton-proton collision could result in the creation of an additional proton-antiproton pair. This reaction would conserve charge and nucleon number but require a very energetic beam of protons that could be provided only by an accelerator.

EXAMPLE 15.1
Making Antiprotons

A beam of protons is directed at a target of liquid hydrogen. What minimum energy E must the protons have if they are to create antiprotons via reaction 1?

Solution

Reaction 1 can take place if the invariant mass M of the colliding protons is at least the sum of the masses of the final particles: $M \geq 4m_p$, where m_p is the proton mass. We may use equation 12.18 to determine M in terms of the initial energies and momenta. The projectile proton has energy E. Its momentum is $p = \sqrt{E^2/c^2 - m_p^2 c^2}$. The target proton is at rest. Its energy is $m_p c^2$ and its momentum is 0. Thus, the total energy of the initial system is $E + m_p c^2$ and the total momentum is p. M is given by

$$M = \sqrt{(E + m_p c^2)^2/c^4 - p^2/c^2} = \sqrt{2(1 + \gamma)m_p^2} \qquad (2)$$

To get the final result, we used the relation $E = \gamma m_p c^2$. If M is to be at least $4m_p$, then γ must be at least 7. To make antiprotons by reaction 1, the proton energy must be at least $7m_p c^2$, or 6.6 GeV.   ∎

Protons of the requisite energy did not become available until a proton accelerator called the *Bevatron* was completed in 1955 at Berkeley, California. One of the first experiments carried out at the bevatron led to the creation and observation of antiprotons, and to the award of a Nobel Prize to the leaders of the experimental team, Owen Chamberlain and Emilio Segré. The discovery of the antiproton marked the beginning of "big science," about which the French physicist Louis Leprince-Ringuet commented:

> What enormous complications! To construct an accelerator weighing thousands of tons and costing thousands [now billions!] of dollars and to assemble around this apparatus the finest technicians, the cleverest and best-trained physicists—all this demands an immense scientific and industrial potential. How far away we are from Henri Becquerel! Will it ever again be possible to discover simple phenomena with one man, a technical assistant, and a small apparatus? Or are we to become more and more involved in these cumbersome and complex organizations? For the present the truth comes from Berkeley. It is the only center equipped to produce antiprotons.

Even larger energies were required to reveal the existence of point-like constituents called *quarks* inside protons. In the late 1960s, Stanford's two-mile long linear electron accelerator produced electrons with energies of 17,000 MeV (or 17 GeV). These electrons were 100 times as energetic as those available to Hofstadter, and their collisions with protons produced many secondary particles. The experimenters focused on the scattering of the incident electrons and made an amazing discovery reminiscent of Rutherford's discovery of the atomic nucleus. The number of electrons deflected by large angles was far greater than it should have been if the proton were a structureless ball of positive charge. The data could be explained only if there were pointlike charged particles within the proton. These constituents of nucleons were later interpreted as quarks. For their discoveries, Jerome Friedman, Henry Kendall, and Richard Taylor were awarded the 1990 Nobel Prize.

The Stanford linear accelerator and other accelerators of its time were "fixed-target machines," in which energetic particles were made to collide with nucleons at rest. Much higher invariant energies are obtained by directing one beam of energetic particles against a second beam of energetic particles moving in the opposite direction. Using this strategy, physicists can observe tiny subnuclear structures and produce previously unknown particles. Three different kinds of particle colliders are now in use:

1. *Electron–positron colliding beams* These machines produce collisions between beams of electrons and positrons. The largest e^+–e^- colliders were deployed in 1989 at CERN, the European Center for Particle Physics in Geneva, Switzerland, and at the Stanford Linear Accelerator Center (SLAC). The European machine (known as LEP, for large electron–positron ring) is circular in shape and 27 km in circumference. At present, it directs a beam of 50 GeV electrons against a second beam of 50 GeV positrons, but a planned upgrade will double the beam energies. The American machine (known as SLC, standing for Stanford linear collider) produces fewer collisions, but makes up for that by having the colliding particles spin in the same direction. Many experiments confirming the standard model are being carried out at these facilities.

2. *Electron–proton colliding beams* Friedman, Kendall, and Taylor found evidence for quarks by studying the collisions of energetic electrons with a stationary proton target. Larger invariant masses are obtained if a beam of electrons or positrons can be made to collide with another circulating beam of protons. Just such a machine (nicknamed HERA) was completed in 1992 at Hamburg, Germany. HERA is the first and only machine of its kind. This two-ring device arranges for 30 GeV electrons or positrons to collide with 820 GeV protons moving in the opposite direction. The study of very high energy e^+–p and e^-–p collisions now being carried out at HERA will reveal many detailed properties of protons, such as the wave functions of its constituent quarks, and may lead to the discovery of new and unanticipated effects.

3. *Proton–proton or proton–antiproton colliding beams* The world's first proton–proton collider was built at CERN in the 1970s. It consisted of two rings, each containing a beam of 30 GeV protons. Collisions between these particles produced invariant masses of 60 GeV. A decade later, CERN scientists built a one-ring device in which oppositely moving beams of 300 GeV protons and antiprotons could be made to collide. Experimenters used this facility to discover W and Z bosons, the intermediaries of the weak force, discussed in Section 15.4. Both of these particles have masses of almost 100 m_p. Their discovery earned Nobel Prizes for the Italian physicist Carlo Rubbia and the Dutch engineer Simon Van der Meer. The *Tevatron Collider* was completed in the late 1980s at Fermilab, a national laboratory situated near Chicago. A beam of 1000 GeV (or 1 TeV) protons is made to collide with an oppositely moving beam of 1000 GeV antiprotons, which makes it the most powerful accelerator in the world by far. The invariant mass of the colliding particles is 2000 GeV/c^2. At present, the Tevatron Collider is hot on the trail of the last of the quarks—the yet-undiscovered *top quark* discussed in Section 15.3.

For reasons to be explained later, the Tevatron is not powerful enough to answer the question that now bedevils particle physicists: What is the origin of particle masses? American physicists had hoped to find the answer with the Superconducting Supercollider, but its construction was abruptly terminated by Congress in 1993. A smaller machine, Europe's Large Hadron Collider, may be big enough to unlock the secret of mass. However, this project is neither approved for construction nor funded. It will be many years, if ever, before an accelerator will be built that is powerful enough to provide the missing pieces for physicists to complete their picture of the microworld.

EXAMPLE 15.2
The Importance of
Colliding Beams

The Tevatron Collider generates collisions of 1000 GeV protons with 1000 GeV antiprotons and enables physicists to find out what happens at an invariant energy of 2000 GeV. What would the energy of a proton have to be to reach the same invariant energy in a collision with a proton at rest?

Solution

The invariant mass in a "fixed-target experiment" involving the collision of a projectile proton with energy $E = \gamma m_p c^2$ is given by equation 2. In this case, we set $M = 2000$ GeV/c^2 to obtain

$$2000 \text{ GeV}/c^2 \simeq \sqrt{2(1 + \gamma)m_p^2}$$

where $m_p \simeq 0.94$ GeV/c^2. Thus, $\gamma \simeq 2,300,000$. The required energy is over 2×10^6 GeV, far exceeding anything except the highest energy (and rarest) cosmic rays. The invention of colliders has allowed experimenters to leapfrog to otherwise inaccessible invariant energies. ∎

Exercises

1. With existing technology, an accelerator can be designed to create and store 10^{12} antiprotons per second. How long would it take such a machine, operating continuously, to produce 1 kg of antiprotons?

2. The LEP collider stores counterrotating beams of 50 GeV electrons and positrons. The circumference of the machine is 17 km. How many circuits of the accelerator does each particle make in the 12-hour period over which the beams may be maintained?

3. The HERA machine in Germany arranges for 30 GeV electrons to collide with an oppositely directed beam of 800 GeV protons.
   a. What is the invariant energy of the proton–electron collisions in GeV?
   b. If electrons are directed at a stationary hydrogen target, what must their energy be to achieve the same invariant energy as the HERA collider?

4. The Z^0 boson (whose mass is about 90 GeV/c^2) is produced at LEP by the reaction $e^+ + e^- \to Z^0$ when each of the oppositely moving particles has an energy of 45 GeV and the invariant energy coincides with the Z^0 rest energy. What energy would a positron need if it were to produce a Z^0 boson in a collision with an electron at rest?

15.2 Particle Names and Properties

In the early 1950s, the number of known elementary particles was relatively small, but it was growing. Aside from the constituents of atoms—electrons and nucleons—there were neutrinos, muons, and three kinds of pions (π^+, π^-, and π^0). Several additional particles (in the so-called strange particles) had been spied among cosmic rays. Things would get more complicated, however, before they became simple again.

Beginning in the 1960s, an astonishing variety of short-lived particles were produced and detected at accelerators. Nucleons and pions were joined by innumerable cousins and nephews with seemingly equal claims to elementarity. Dozens, then hundreds, of new particles were found. Annual lists of elementary particles became thicker and thicker and began to resemble the catalog of the Bronx Zoo. Two points of view competed with one another. Some physicists believed that all these particles were equally elementary, while others, appalled by the teeming nuclear democracy, insisted on the existence of a small number of basic building blocks from which all matter is built and on a simple law underlying the strong nuclear force. The truth lay somewhat in between. Most of the particles once regarded as elementary are made up of a few kinds of quarks, which cannot be isolated from the particles they form.

Fermions and Bosons

Physicists classify particles in much the way biologists classify life forms. Instead of plants and animals, physicists speak of *fermions* and *bosons*. Every particle, or system of particles, is either a fermion or a boson. Relativistic quantum mechanics relates the notion of spin to the Pauli exclusion principle: Fermions satisfy the exclusion principle, bosons do not. Particles with half-integer spin (1/2, 3/2, and so on) are called fermions in honor of the Italian-American physicist Enrico Fermi. Particles with integer spin (0, 1, 2, and so on) called bosons in honor of the Indian physicist Satendra Bose. Nucleons and electrons are spin 1/2 fermions, as are their antiparticles. Spin 1 photons and spin 0 pions are bosons, as are *deuterons* (nuclei made of one proton and one neutron) and hydrogen atoms (made of one proton and one electron). Any composite system containing an odd number of fermions is itself a fermion; any composite system containing an even number of fermions is a boson.

No more than one fermion can occupy the same quantum state. Atoms are mostly empty space: The rigidity of solids and the relative incompressibility of liquids result from the exclusion principle and the fact that electrons are fermions. The shell structure of atomic electrons is a consequence of the exclusion principle, as is the shell structure of nucleons in a nucleus. Many bosons can crowd into the same quantum state, producing bizarre effects. When helium (whose atoms and nuclei are bosons) is cooled below 2 K, it becomes a *superfluid* and creeps out of any containing vessel. Lasers are devices in which many photons in the same state encourage the emission of others, thereby producing an intense, narrow, and very useful beam of monochromatic light.

Hadrons and
Leptons

Electrons and protons are both spin 1/2 fermions. In other respects, though they are very different. Protons are much heavier than electrons. Further, protons interact with one another (and with neutrons) through the strong nuclear force, while electrons have weak and electromagnetic interactions but do not participate in the strong nuclear force. Spin 1/2 fermions without strong interactions are known collectively as *leptons*. This small class of particles includes electrons and neutrinos. There are six known lepton species in nature, each carrying one unit of lepton number $\mathcal{L}$. Their antiparticles (*antileptons*) carry lepton number $\mathcal{L} = -1$. Experimental evidence suggests that lepton number is exactly conserved.

The word *hadron* refers to any strongly interacting and seemingly elementary particle, whether it be a fermion or a boson. Protons are hadrons, as are antiprotons and pions. In fact, there are hundreds of different kinds of hadrons. All known hadrons carry nucleon number $\mathcal{N} = +1$, -1, or 0. For this reason, it is convenient to define three subclasses of hadrons: baryons, antibaryons, and mesons:

Baryons [from Greek *barys* (meaning heavy) + *-on*] are any strongly interacting particles (or hadrons) with $\mathcal{N} = 1$, such as neutrons or protons. *Antibaryons* are the antiparticles of baryons. They are hadrons with $\mathcal{N} = -1$. The properties of antiprotons and antineutrons ($\overline{p}$ and $\overline{n}$) are routinely studied at high-energy physics laboratories. Baryons and antibaryons are fermions.

Mesons [from Greek *meso* (meaning middle) + *-on*] are strongly interacting particles with $\mathcal{N} = 0$. The term was coined in 1939 by H. J. Bhaba, who wrote: "The name 'mesotron' has been suggested for the new particle found in cosmic radiation with a mass intermediate between that of the electron and proton. The 'tr' in the word is redundant since it does not belong to the Greek root. The 'tr' in neutron and electron belong, of course, to the roots 'neutr' and 'electra'. It would therefore be more logical and shorter to call the new particle a meson." Pions are mesons and every meson is a boson.

Table 15.1 displays the relation among these concepts. Every observed particle belongs in one of the six categories. Every particle is either a fermion or a boson. Fermions include baryons, leptons, and their antiparticles. Baryons, antibaryons, and mesons are strongly interacting hadrons. A particle of each category is given in parentheses in the table. The photon is the mediating particle of the electromagnetic force. How nuclear forces are mediated is discussed in Section 15.4. We shall learn in

TABLE 15.1   Particle Taxonomy

Fermions	*Bosons*	*Fermions*
$\mathcal{N} = +1$ Baryons (Protons)	$\mathcal{N} = 0$ Mesons (Pions)	$\mathcal{N} = -1$ Antibaryons (Antiprotons)
$\mathcal{L} = +1$ Leptons (Electrons)	Force Particles (Photons)	$\mathcal{L} = -1$ Antileptons (Positrons)

Section 15.3 that hadrons (baryons, mesons, and antibaryons) are not elementary. Their constituents—quarks and antiquarks—cannot be isolated from the hadrons they form. Before we proceed, let's get better acquainted with the most commonplace and best-studied hadrons, as well as the six leptons.

Some of the Hadrons

Because the gram is a huge unit on a subatomic scale, high-energy physicists specify particle masses in MeV/c^2. In terms of these units, the proton and neutron masses are:

$$m_p = 938.28 \text{ MeV}/c^2 \qquad m_n = 939.57 \text{ MeV}/c^2 \qquad (3)$$

The proton is either absolutely stable or very long-lived. Its lifetime is known to exceed 10^{31} years. An isolated neutron β^- decays with a lifetime of nearly 15 min. Protons and neutrons form an isotopic doublet of similar baryons.

Pions are much less massive than nucleons. The charged pions π^+ and π^- (which are each other's antiparticles) have masses of 139.58 MeV/c^2. The neutral pion π^0 has a somewhat smaller mass of 134.97 MeV/c^2. The three pions form an isotopic triplet of similar mesons. Many different hadrons can be produced by the action of the strong force in the collisions of energetic nucleons. In particular, pions may be produced through reactions such as:

$$p + p \longrightarrow p + n + \pi^+$$
$$p + p \longrightarrow p + p + \pi^0$$
$$p + n \longrightarrow p + p + \pi^-$$

Notice that electric charge Q and nucleon number $\mathcal{N}$ are conserved by these reactions. Many other collisions of energetic particles can produce pions, for example:

$$e^+ + e^- \longrightarrow \pi^+ + \pi^-$$
$$\gamma + p \longrightarrow n + \pi^+$$
$$\overline{p} + n \longrightarrow \pi^+ + \pi^- + \pi^- + \pi^0$$
$$e^- + p \longrightarrow e^- + p + \pi^0$$

Pion production in particle collisions allows physicists to produce beams of charged pions with which to study their interactions with matter. When energetic pions strike nucleons, additional pions and other particles are created. However, you cannot buy pions at your corner drugstore because they are unstable particles. Charged pions decay by means of the weak interactions and have lifetimes of about 26 ns (nanoseconds). Neutral pions decay even more rapidly by means of electromagnetic interactions. The π^0 lifetime is only 10^{-16} s. The principal decay modes of pions are

$$\pi^+ \rightarrow \mu^+ + \nu, \qquad \pi^0 \rightarrow \gamma + \gamma, \qquad \pi^- \rightarrow \mu^- + \overline{\nu}$$

The negative muon μ^- is a lepton whose mass is 105.66 MeV/c^2—basically, it's just an obese electron.

Most hadrons have lifetimes much shorter than pions because they decay via the strong force. For example, an isotopic quartet of particles designated by Δ have masses of about 1230 MeV/c^2 and spins of $\frac{3}{2}$. These particles were first produced in the 1950s through the reactions:

$$\pi^+ + p \rightarrow \Delta^{++}, \qquad \pi^+ + n \rightarrow \Delta^+, \qquad \pi^- + p \rightarrow \Delta^0, \qquad \pi^- + n \rightarrow \Delta^-$$

All four of these particles have lifetimes of about 10^{-23} s, characteristic of strong decay processes.

Some hadrons are produced via the strong force but decay via the weak force with lifetimes of at least 10^{-10} s. These hadrons are called *strange particles*. Among them are an isotopic doublet of strange mesons (K^+ and K^0, with masses about 500 MeV/c^2) and a singlet strange baryon (Λ^0, with mass about 1115 MeV/c^2). These particles are produced by reactions such as

$$\pi^- + p \longrightarrow K^0 + \Lambda^0$$

Strange particles are copiously produced in high-energy collisions and yet they are relatively long-lived.

All of the Leptons

In 1962, an experimental team from Columbia University (in New York City) and Brookhaven National Laboratory (on Long Island) showed that neutrinos created in β decay are different from those produced by pion decay. The three senior investigators, Leon Lederman, Mel Schwartz, and Jack Steinberger, shared the 1989 Nobel Prize in physics for this discovery. Their paper began:

> In the course of an experiment at the Brookhaven [proton accelerator], we have observed the interaction of high-energy neutrinos with matter. These neutrinos were produced primarily as the result of the decay of the pion:
>
> $$\pi^\pm \longrightarrow \mu^\pm + (\nu/\overline{\nu}) \tag{4}$$
>
> It is the purpose of this Letter to report some of the results of this experiment including [our] demonstration that the neutrinos we have used produce [muons] but do not produce electrons, and are hence very likely different from the neutrinos involved in β decay .... The essential scheme of the experiment is as follows: A neutrino beam is generated by decay in flight of pions. The pions are produced by 15 Gev protons [from the accelerator] striking a beryllium target. The resulting particles moving in the general direction of the detector strike a 13.5 m thick iron shield wall at a distance of 21 from the target. Neutrino interactions are observed in a 10-ton aluminum spark chamber located behind this shield.

The experimental layout is shown in Figure 15.2. Pions produced in the target decay en route to the iron shield. The only particles passing through the shield are the weakly interacting neutrinos and antineutrinos arising from reaction 4. Suppose that β decay involves ν_e and pion decay

FIGURE 15.2
The discovery of the muon neutrino. A beam of 30 GeV protons extracted from the synchrotron strikes a beryllium target producing many charged pions. Many of the pions decay into muons and neutrinos as they travel 21 m from the target to an iron shield. All of the particles except for neutrinos and antineutrinos are absorbed within a 13.5-m thick iron shield. Neutrino events are observed and measured by a 10-ton neutrino detector. The experiment showed that the collisions of muon neutrinos produce muons but not electrons.

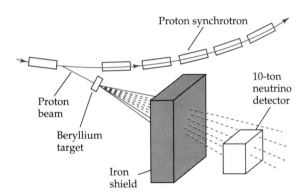

involves v_μ. Energetic neutrinos and antineutrinos would be expected to produce reactions by which they could be observed in the detector:

$$v_\mu + n \longrightarrow p + \mu^-, \qquad \bar{v}_\mu + p \longrightarrow n + \mu^+ \tag{5a}$$

$$v_e + n \longrightarrow p + e^-, \qquad \bar{v}_e + p \longrightarrow n + e^+ \tag{5b}$$

The discovery paper continued:

> If there is only one type of neutrino [and $v_\mu = v_e$], then neutrino interactions should produce muons and electrons in equal abundance. In the event that there are two neutrinos [and $v_\mu \neq v_e$], there is no reason to expect any electrons at all.

A total of 29 neutrino-induced events were observed in which a muon was produced via reactions 5a. No events were seen in which an electron was produced via reactions 5b. The experimenters concluded that there are at least two kinds of neutrinos: the electron neutrino v_e and the muon neutrino v_μ. In terms of these neutrino states, β decay and pion decay are written

β^- decay $\quad n \longrightarrow p + e^- + \bar{v}_e \qquad \pi^-$ decay $\quad \pi^- \longrightarrow \mu^- + \bar{v}_\mu$
β^+ decay $\quad p \longrightarrow n + e^+ + v_e \qquad \pi^+$ decay $\quad \pi^+ \longrightarrow \mu^+ + v_\mu$

The existence of two neutrinos suggests that there are two different leptonic quantum numbers, *electron number* $\mathcal{L}_e$ and *muon number* $\mathcal{L}_\mu$. Electrons and electron neutrinos are assigned $\mathcal{L}_e = 1$ and $\mathcal{L}_\mu = 0$. Muons and muon neutrinos are assigned $\mathcal{L}_\mu = 1$ and $\mathcal{L}_e = 0$. Electrons are produced by the interactions of electron neutrinos; muons by the interactions of muon neutrinos. In all particle processes that have been studied, electron number and muon number are independently conserved. For example, neither of the decay processes, $\mu^- \rightarrow e^- + \gamma$ or $\mu^- + p \rightarrow e^- + p$, have been seen. They conserve total lepton number, but they would change electron and muon numbers.

Nature had another surprise in store for us. A third charged lepton was discovered in 1975, the *tau lepton*. Its mass is about 1.8 GeV (about 17 times the muon mass). Because electrons and muons have their own neutrinos, it stands to reason that the tau lepton does as well. There must be a third neutrino species: the tau neutrino v_τ. Experiments are now under way to observe the effects of v_τ directly and prove that they really exist.

Experimental results from LEP (the electron–positron collider in Geneva) and convincing arguments based on cosmology show that the number of different neutrino species is three—there is no room for a fourth neutrino. It seems there are a total of six different leptons in nature forming three pairs. Each of the pairs carries its own lepton number and the three lepton numbers appear to be conserved:

$$\begin{pmatrix} e^- \\ v_e \end{pmatrix} \quad \begin{pmatrix} \mu^- \\ v_\mu \end{pmatrix} \quad \begin{pmatrix} \tau^- \\ v_\tau \end{pmatrix}$$

EXAMPLE 15.3
Leptonic
Weak Decays

The weak force responsible for nuclear β decay is not just a nuclear force: It acts on leptons and hadrons. Weak interaction processes can be purely hadronic (such as the decay of a strange meson into two pions, $K^0 \to \pi^+ + \pi^-$), or semileptonic (such as β decay, which involves both leptons and hadrons), or even purely leptonic (such as the decay of the muon into an electron, a neutrino, and an antineutrino).

a. Specify the neutrino identities in muon decay.

b. Give an example of a purely leptonic decay of τ^-.

c. Give an example of a semileptonic decay of τ^-.

Solution

a. The negative muon decay scheme is $\mu^- \to v_\mu + e^- + \bar{v}_e$. The initial state has $\mathcal{L}_\mu = 1$ and $\mathcal{L}_e = 0$. So does the final state.

b. The decay mode $\tau^- \to v_\tau + \mu^- + \bar{v}_\mu$ conserves all three lepton numbers.

c. The decay mode $\tau^- \to v_\tau + \pi^-$ is both allowed and observed. It accounts for about 11 percent of all τ decays.   ∎

Exercises

5. A charged pion decays into a muon and a massless neutrino. If the pion decays from rest, what is the energy of each of the final particles? (Use $m_\pi = 140$ MeV/c^2 and $m_\mu = 105$ MeV/c^2.)

6. An energetic positron strikes an electron at rest and produces a pair of muons: $e^+ + e^- \to \mu^+ + \mu^-$. What is the minimum energy of the positron?

7. Which of the following are bosons and which are fermions?
    a. An α particle
    b. A hydrogen atom
    c. A nucleus containing Z protons and N neutrons
    d. A nitrogen atom whose nucleus contains seven protons and seven neutrons
    e. An atom of ^{2}H

8. Which of the following six diatomic molecules are bosons and which are fermions: HH, HD, HT, DD, DT, DT? (D stands for ^{2}H and T stands for ^{3}H)

9. What are the particles produced by the decay of a positive muon?

10. The τ^- can decay into leptons or into leptons and hadrons. Give examples of each type of decay other than those in Example 15.3. Explain why the muon cannot decay into leptons and hadrons.

11. Explain why nobody has ever observed the decay of a negative muon into two electrons and a positron.

15.3   The Story of the Strong Force

Most of the particles we've met so far are important to life on Earth. Molecules, mice, and men are made of nucleons and electrons. Photons make light and hold atoms together. Pion exchange between nucleons contributes to the force holding nuclei together. Neutrinos were essential to the synthesis of other elements long ago and are essential to the operation of the solar furnace. If there were no neutrinos, Earth—were it to exist at all—would be a lifeless ball of frozen hydrogen. Every particle had a role to play—until muons showed up. When the American physicist Isadore Rabi (one of the founders of CERN) first learned about muons at a Chinese lunch, he retorted: *"Who ordered that?"* The muon is not the only unexpected and apparently purposeless particle that has shown up, and we still can't answer Rabi's question. However, these "odd" particles have guided physicists toward the next layer of the subatomic onion.

Strange Particles and Strangeness

In the late 1940s, cosmic ray physicists observed peculiar events in photographic emulsions and cloud chambers that were attributed to *strange particles*, particles whose properties were unusual and hard to explain. Some of the events showed the production and decay of strange mesons (charged and neutral kaons) with about half the proton mass. Others revealed six different strange baryons with masses ranging from 1.1 to 1.3 GeV/c^2. Why are strange particles strange? First, strange particles are often produced by the collisions of high-energy cosmic ray or accelerator hadrons. It follows that they partake in the nuclear force—they are strongly interacting hadrons. However, strange particles are never produced one at a time. They are always produced in pairs by a process dubbed *associated production*. Second, all strange particles are unstable. For example, the positive kaon K^+ decays into many different channels including $\pi^+ + \pi^0$ and $\mu^+ + \nu_\mu$. The lifetimes of strange particles are typically 10^{-10} s. They are produced copiously and in pairs through the strong force, but they decay in isolation through the weak force.

The discoveries of strange particles and muons changed the course of particle physics. These particles are not atomic constituents; they must be produced by cosmic rays or large accelerators. High-energy physicists were diverted from the study of matter-as-found to that of matter-as-made.

Francis Bacon, writing long before any elementary particles were known, concluded that "there is no excellent beauty that hath not some strangeness in proportion." The American physicist Murray Gell-Mann introduced *strangeness* into physics as a new quantum number S akin to charge and nucleon number. He assigned a value of S to every hadron. Pions and nucleons, which are not strange particles, were assigned $S = 0$. The positive kaon K^+ (a strange meson) was assigned $S = +1$, and its antiparticle, the negative kaon K^-, received $S = -1$. The associated production of strange particles was explained by means of an approximate law of strangeness conservation. Gell-Mann assumed that strangeness is conserved in all processes caused by the strong force.

However, Gell-Mann realized that strangeness, unlike nucleon number and electric charge, is not conserved in all particle processes. S changes by ± 1 as strange particles decay. The strong force conserves strangeness, but the weak force can change strangeness by one unit (as in kaon decay) or not at all (as in pion decay). No decay process has ever been seen in which strangeness changes by more than one unit. The rules were arbitrary but their consequences were powerful, predictive, and confirmed by observations.

TABLE 15.2   The Spin $\frac{1}{2}$ Baryon Octet

Particle	Symbol	Charge	Strangeness	Mass in GeV/c^2
Nucleon	(p, n)	$0, +1$	0	0.938
Lambda	(Λ^0)	0	-1	1.115
Sigma	$(\Sigma^+, \Sigma^0, \Sigma^-)$	$-1, 0, +1$	-1	1.190
Xi	(Ξ^0, Ξ^-)	$-1, 0$	-2	1.320

Table 15.2 shows the quantum number assignments of the nucleons and their closest relatives, the six short-lived spin 1/2 strange baryons. These eight particles are grouped into isospin multiplets containing several particles with almost the same mass: n and p form an isotopic doublet; Σ^+, Σ^0, Σ^- form an isotopic triplet; and Ξ^0, Ξ^- form an isotopic doublet. The lone Λ^0 is a singlet. The mass splittings within isotopic multiplets (such as the mass difference between Σ^+ and Σ^-) range from 1 to 10 MeV/c^2. However, the splittings between different isotopic multiplets are 100 MeV/c^2 and more. The same pattern applies to mesons: K^+ and K^0 have almost the same masses, but they are about 350 MeV/c^2 heavier than pions. The conservation of strangeness by collision processes explains which reactions take place and which do not.

EXAMPLE 15.4
Conserving
Strangeness

Strange particles can be created by the strong force in accordance with the law of strangeness conservation. Consider the five conceivable results to pion-nucleon collisions:

$$\pi^+ + p \longrightarrow K^+ + p \qquad\qquad (6a)$$

$$\pi^+ + p \longrightarrow \pi^+ + \Sigma^+ \qquad\qquad (6b)$$

$$\pi^+ + p \longrightarrow K^+ + \Sigma^+ \qquad\qquad (6c)$$

$$\pi^- + p \longrightarrow K^+ + \Sigma^- \qquad\qquad (6d)$$

$$\pi^- + p \longrightarrow K^- + \Sigma^+ \qquad\qquad (6e)$$

Each of these reactions conserves charge and nucleon number. Because the weak force is too feeble to affect the collision process, strangeness must be conserved as well. Which of these processes can take place?

Solution   The K^+ has $S = 1$ and the K^-, Σ^+, and Σ^- have $S = -1$. Because the colliding particles have no net strangeness, the net strangeness of the products must vanish as well. This is not true for reactions 6a and 6b in which only one strange particle is produced. These are forbidden reactions. In reactions 6c and 6d, two particles of opposite strangeness are made: These are allowed reactions that can and do take place. They are examples of the associated production of two strange particles. Reaction 6e yields two particles of the same strangeness and is forbidden.   ■

The Eightfold Way

In 1961, Gell-Mann* devised a mathematical theory (which he playfully called the *eightfold way*) governing the organization of hadrons and explaining many of their properties. A consequence of this theory is that particles with the same spin must appear in groups with specific quantum number assignments. The particles in each group may be plotted as points on a graph whose axes are charge and strangeness. The eightfold way determines the possible shapes of the patterns. In Figure 15.3, we see that particles with the same spin do form simple geometrical figures. Mesons form hexagons, as do nucleons and their strange spin 1/2 relatives. The spin 3/2 baryons form a triangular pattern. According to the eightfold way, all hadrons had to belong to groups of particles forming simple geometrical patterns in charge and strangeness. Furthermore, relations were predicted among particle masses, magnetic properties, decay schemes, and interactions with one another. That is, the theory made quantitative predictions that could be tested by experiments.

When the eightfold way was proposed, the predicted patterns were incomplete. There were gaps in Gell-Mann's figures, just as there were gaps in Mendeleev's table a century earlier! Seven of the eight spinless mesons were known when Gell-Mann wrote: "The most clear-cut prediction is the existence of [an eighth spinless meson] which should decay into two photons like the π^0." The predicted η particle was found a few months later. Three members of the spin 3/2 baryon decimet were also still at large. The discovery of two of them was reported at a major conference in 1962. "We should look for the last particle, called, say

*The Israeli physicist Yuval Ne'eman, then a research student of the Pakistani Nobelist Abdus Salam, published the same idea at about the same time.

FIGURE 15.3
a. The spin $\frac{1}{2}$ baryon octet. Each of the eight particles is plotted against charge (the horizontal axis) and strangeness (the vertical axis). A hexagonal pattern with two particles in the center is formed. *b.* The spin 0 meson octet. The hexagonal pattern is the same as that of the baryon octet, but it is centered about the point $S = Q = 0$. *c.* The spin $\frac{3}{2}$ baryon decimet. In this case, the ten particles form a triangular pattern. (The asterisks distinguish the spin $-\frac{3}{2}$ particles from their $-\frac{1}{2}$ namesakes.)

	$Q = -1$	$Q = 0$	$Q = +1$
$S = 0$		n	p
$S = -1$	Σ^-	Λ, Σ^0	Σ^+
$S = -2$	Ξ^-	Ξ^0	

a. Spin $\frac{1}{2}$ Baryon Octet

	$Q = -1$	$Q = 0$	$Q = +1$
$S = +1$		K^0	K^+
$S = 0$	π^-	η, π^0	π^+
$S = -1$	K^-	K^0	

b. Spin 0 Meson Octet

	$Q = -1$	$Q = 0$	$Q = +1$	$Q = +2$
$S = 0$	Δ^-	Δ^0	Δ^+	Δ^{++}
$S = -1$	Σ^{*-}	Σ^{*0}	Σ^{*+}	
$S = -2$	Ξ^{*-}	Ξ^{*0}		
$S = -3$	Ω^-			

c. Spin $\frac{3}{2}$ Baryon Decimet

DO YOU KNOW

How the Omega-Minus Particle Was Found?

Scientists accepted the eightfold way when the omega-minus particle (Ω^-) was discovered in 1964 at Brookhaven National Laoratory in New York. The team of 33 researchers was led by the American physicist Nicholas Samios. They wrote:

> Among the multitude of [hadrons] which have been discovered recently, the $\Delta(1238)$, $\Sigma^*(1385)$ and $\Xi^*(1530)$ can be arranged as a decuplet with one member still missing. [Figure 15.4] illustrates the positions of the 9 known resonant states and the postulated 10th particle plotted against mass and charge. [The missing] particle (which we call Ω^-, following Gell-Mann) is predicted to be a negatively–charged singlet with strangeness –3."

Protons from the Brookhaven 30 GeV synchro-

tron were directed at an external metal target so as to produce a wide variety of particles

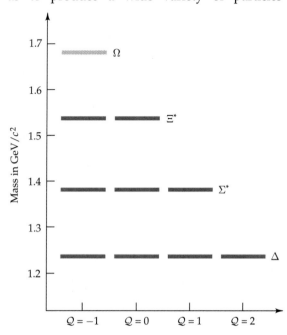

The discovery of the Ω^- baryon. *a.* The actual bubble chamber photograph. *b.* The interpretation of the event. Solid lines are observed tracks. Dotted lines indicate unseen neutral particles whose trajectories have been deduced. An incoming K^- (1) strikes a proton producing a K^+ (2), a K^0 (not seen), and a short track (3) corresponding to the discovered Ω^-.   The Ω^- promptly decays into a Ξ^0 and a π^- (4). The unseen Ξ^0 decays into a π^0 and a Λ^0. The decay of the Λ^0 produces a pion (5) and a proton (6). The two gammas

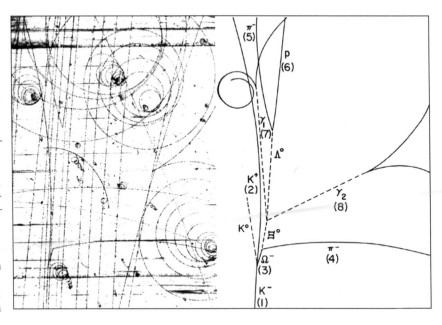

from the decay of the π^0 each produce electron-positron pairs, shown at the termini of the unseen tracks (7) and (8). *Source:* Courtesy N. P. Samios, director, Brookhaven National Laboratory.

with different energies and identities. A system of magnets extracted a beam of 5 GeV negative kaons from among the produced particles.   The K^- beam entered Brookhaven's liquid hydrogen–filled 80-inch bubble chamber.   Kaons striking protons in the chamber produced more particles.   A bubble chamber is a device in which charged particles leave tracks of tiny bubbles when they traverse the chamber.   About 100,000 photographs were taken of individual events.   The researchers searched the pictures for events corresponding to the production and decay of an Ω^- particle.   They found only one event meeting their criteria, but one was enough to prove that the predicted particle exists. The solitary Ω^- event is shown in the diagram, and its interpretation is shown in the accompanying diagram. There was enough information in the tracks for the experimenters to deduce exactly what had happened:

> The mass of particle 3 is computed to be 1686± 12 MeV/c^2. [Its] mean lifetime was computed to be 10^{-10} s; consequently we may assume that it decayed by a weak interaction with $\Delta S = 1$ into a system with strangeness minus 2. In view of the properties of charge, strangeness and mass established for particle 3, we feel justified in identifying it with the sought–for Ω^-.

Thousands of Ω^- particles have been seen since. The Ω^- mass is 1672 MeV/c^2, its lifetime is about 10^{-10} s, and its decay into $\Xi^0 + \pi^-$ is one of several observed decay modes. The properties of this particle confirm its interpretation as the lightest baryon made of three strange quarks.

Figure 15.4 (at left)

The members of the spin $\frac{3}{2}$ baryon decimet are plotted versus charge and mass. The last of the 10 particles, the Ω^- baryon, was discovered in 1964. Notice that the four adjacent isotopic multiplets in the baryon decimet are approximately equally spaced (about 150 MeV apart) in mass. The greater the magnitude of the strangeness of a particle, the heavier it is.

Omega-minus (Ω^-), with strangeness minus three," said Gell-Mann at the conference. "At 1685 MeV/c^2, it would be metastable and should decay by weak interactions into $\pi^- + \Xi^0$." Experimental physicists did look and within a year found the Ω^- particle. It had just the properties Gell-Mann foresaw.

Many more hadrons were discovered, such as octets and singlets of mesons with spins 1 and 2. They fit snugly into the patterns of the eightfold way. Mesons belonged to families of eight forming hexagons, or were singlets with $S = Q = 0$. Baryons fit into singlets, octets, or triangular decimets: They formed patterns of one, eight, and ten particles. The reason for this soon became clear: The notions of isospin and strangeness, as well as the successes of the eightfold way, follow from the quark substructure of hadrons.

The Quark Hypothesis

The eightfold way not only gathered particles into multiplets but predicted many of their properties. The success of the theory hinted that baryons and mesons are made of even more elemental constituents. Gell–Mann proposed a specific model of hadron substructure in 1963. He worked out the properties of hypothetical hadron constituents and proposed the name *quarks* for them.* If hadrons were to be built from quarks, the following simple rules had to be obeyed:

- Quarks are spin 1/2 fermions carrying nucleon number $\mathcal{N} = 1/3$.
- Each baryon is made of three quarks.
- Each meson is made of one quark and one antiquark.
- Each antibaryon is made of three antiquarks.
- No other combination of quarks forms an observable particle. In particular, individual quarks may not be isolated.
- Quarks come in several varieties, called quark *flavors*.

Many of the properties of quarks follow from the observed properties of hadrons. Experimenters noticed a remarkable fact about baryons. Among the hundreds of known baryons, none have charges greater than +2 or less than −1. That is, the charge of every baryon is one of four values ranging from −1 to +2 in unit steps. Interpreted in terms of quarks, this fact means that every quark either has charge 2/3 (in which case, three of them have $Q = +2$) or has charge −1/3 (in which case, three of them have $Q = -1$). The intermediate values of baryon charges correspond to mixed quark compositions:

$$2/3 + 2/3 - 1/3 = 1 \quad \text{and} \quad 2/3 - 1/3 - 1/3 = 0$$

*George Zweig, who has since become a biologist, made a very similar proposal at the same time. Incidentally, Gell–Mann insists that *quarks* rhyme with *forks*, not with *larks*.

Two different quark flavors are required for the construction of pions, nucleons, and other nonstrange particles. They are the *up quark u*, with charge $+2/3$, and the *down quark d*, with charge $-1/3$. For example, the proton is made of two up quarks and one down quark, while the neutron is made of one up quark and two down quarks:

$$p \equiv uud \qquad n \equiv ddu$$

A third quark flavor was needed to describe strange particles. Baryons with strangeness $S = 0, -1, -2$, and -3 had been seen. The third *strange quark s* was assigned strangeness $S = -1$. Just as the negative muon is a heavier version of the electron, the strange quark is a heavier version of the down quark. Strange particles are simply particles containing one or more strange quarks.* The quantum number assignments of Gell-Mann's three quarks are given in Table 15.3.

TABLE 15.3 Quark Quantum Number Assignments

Flavor	Charge	Nucleon No.	Strangeness
Up	2/3	1/3	0
Down	−1/3	1/3	0
Strange	−1/3	1/3	−1

Building Hadrons from Quarks

Hadrons are built from quark combinations in accord with Gell–Mann's rules. Mesons contain a quark and an antiquark. The positive pion contains an up quark u and the antiparticle a down quark $\bar{d}$. The negative pion contains a down quark and an anti-up quark. The quark composition of kaons is similar, with a strange quark replacing the down quark:

$$\pi^+ \equiv u\bar{d} \qquad \pi^- \equiv d\bar{u}$$

$$K^+ \equiv u\bar{s} \qquad K^- \equiv s\bar{u}$$

The neutral pion is a bit more complicated: the π^0 corresponds to the superposition state $u\bar{u} - d\bar{d}$.

Baryons are made of three quarks much like a three–scoop ice cream cone may be made of chocolate, vanilla, and strawberry ice cream. There are precisely ten possible combinations of quarks or ice cream flavors. The quark constitution of the possible baryon states are shown in Table 15.4.

---

*When the notion of strangeness was proposed, the K^+ was arbitrarily assigned $S = +1$. That's why the strange quark carries negative strangeness. A similar accident centuries earlier led to a convention for electrical currents opposite to the flow of negatively charged electrons.

TABLE 15.4   Baryon Quark Configurations

	$Q = -1$	$Q = 0$	$Q = +1$	$Q = +2$
$S = 0$	ddd	ddu	duu	uuu
$S = -1$	sdd	sdu	suu	
$S = -2$	ssd	ssu		
$S = -3$	sss			

Every observed baryon must correspond to one of the above combinations, and every combination must correspond to a baryon. That was almost precisely the situation in 1963: Every known baryon fit into one of the ten pigeonholes. The only missing particle was the triply-strange Ω^-, or sss baryon, whose existence Gell-Mann had predicted before he invented quarks.

All of the successes of the eightfold way—and more—were explained by the quark hypothesis. The quantum numbers of a hadron are determined by its quark substructure. Furthermore, the pattern of hadron masses is understood in terms of constituent quark masses. Up and down quarks have approximately the same mass, but the strange quark is heavier. That's why hadrons appear in isospin multiplets whose members have nearly the same mass. Neutrons and protons are similar in mass because they are made of light u and d quarks. The Λ and Σ baryons are heavier because they contain one s quark. The Ξ baryons, which contain two s quarks, are heavier yet. The greater the strangeness of a particle, the larger is its mass.

When hadrons collide with other particles, they can scatter, change their identities, or produce new particles. Reactions among hadrons can be described with Feynman diagrams showing their constituent quarks. Figure 15.5 illustrates three processes: (a) the elastic scattering of a neutron by a proton, (b) the production of a pion by the collision of a photon with a proton, and (c) the associated production of two strange parti-

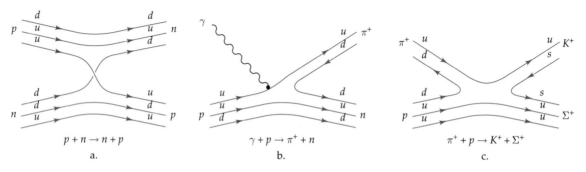

FIGURE 15.5 *a.* A proton and neutron scatter from one another by exchanging quarks. In this diagram, the proton loses an up quark and gains a down quark: It is transformed into a neutron. The quark–antiquark pair exchanged between the nucleons has the quantum numbers of a charged pion. *b.* A photon strikes an up quark within a proton. The creation of a quark–antiquark pair leads to the production of a positive pion and the transformation of the struck proton to a neutron. *c.* The antiquark in an incident π^+ annihilates a quark in the struck proton. The creation of a strange quark and its antiquark leads to the associated production of two strange particles: K^+ and Σ^+.

cles in a pion–proton collision. In each case, a baryon is indicated by three quark lines and a meson by a quark line and an antiquark line. In any process involving strong and electromagnetic forces, quarks may be created or annihilated in pairs or they may be interchanged between hadrons. However, quarks cannot change their identity and quark lines cannot terminate or appear in the diagram.

At first, few physicists accepted Gell-Mann's quark hypothesis—not even Gell-Mann himself. In his paper proposing quarks, he wrote: "It is fun to speculate about the way quarks would behave if they were physical particles instead of purely mathematical entities." A decade later, the experimental evidence for them became compelling. Aside from their utility for explaining hadron masses and quantum numbers, here are three good reasons to believe in quarks:

1. *Scattering experiments*   The proton cannot be taken apart into its constituent quarks because individual quarks cannot be isolated. However, experimenters have done the next best thing. Hard and point-like constituents of nucleons are "seen" by measuring the scattering of high-energy electrons, muons, neutrinos, and antineutrinos from them.

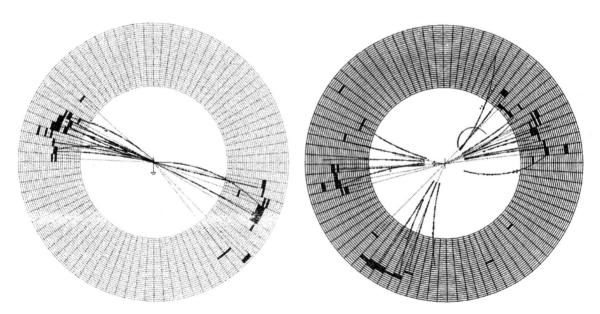

A 15 GeV positron collides with a 15 GeV electron moving in the opposite direction. Solid lines show charged particles leaving visible tracks in the (unshown) inner part of JADE detector. Dotted lines are the trajectories of neutral particles that reveal themselves in the outer portion of the detector. *a*. A two-jet event: The collision creates a quark and an antiquark, each of which becomes a narrow jet of hadrons. *b*. A three-jet event: The collision creates a quark, an antiquark, and a gluon, and results in the production of three narrow jets of hadrons. Gluons are the intermediaries of the strong force. They cannot be seen as isolated particles, but they can be seen as jets. *Source:* DESY, Hamburg, Germany.

2. *Jets*   When a high-energy electron and positron collide, they can annihilate into a quark–antiquark pair: $e^+ + e^- \rightarrow q + \bar{q}$. Because individual quarks cannot be isolated, many hadrons are produced at the collision point. However, the hadrons retain a "memory" of the original quark and antiquark—the emerging particles form a pattern of narrow jets, as shown in the accompanying photos.

3. *Excited states*   A hadron made up of quarks is a quantum-mechanical system with many energy levels. Consequently, hadrons must have excited states. The proton, for example, should be regarded as the ground state of a system of two up quarks and one down quark. It must—and indeed it does—have many excited states, some of which are shown in Figure 15.6.

FIGURE 15.6
Every composite system must have excited states. Those of atoms are never more than a few eV apart. Those of nuclei are a few MeV apart. The energy levels of hadrons can be hundreds of MeV apart. Some of energy levels of the nucleon and the Λ baryon are shown. In both cases, energies are shown relative to the ground state of the system (938 MeV for the nucleon and 1115 MeV for the Λ). Each energy level was once regarded as an elementary particle.

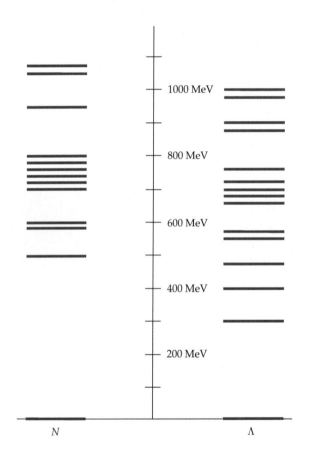

Quarks in "Colors"

Quarks are fermions satisfying the Pauli exclusion principle. No two quarks can occupy the same quantum state. However, the structure of baryons seemed to violate this doctrine. The Ω^- particle, for example, is formed of three s quarks with their spins pointing in the same direction—three quarks all happily ensconced in the same quantum state. Impossible!

The solution is simple but radical: Quarks carry yet one more quantum attribute called *quark color.** A quark of a given flavor (say, up) comes in three different colors (say, red, green and blue). The Ω^- baryon is made of three strange quarks: one red, one green, and one blue. There are not just three kinds of quarks but nine, corresponding to three flavors and three colors. The quark commandments had to be supplemented by an additional principle. Observed particles are "colorless" in the following sense: Baryons contain one quark of each color. (More technically, the wave function of the quarks in a hadron must be antisymmetric when quarks are interchanged. This property is relegated to the arena of color.) Mesons are composed of an equal, and therefore colorless, superposition of the three colors:

$$\text{Meson} \equiv q_{\text{red}}\,\overline{q}_{\text{red}} + q_{\text{blue}}\,\overline{q}_{\text{blue}} + q_{\text{green}}\,\overline{q}_{\text{green}}$$

Quark color seemed crazy at first but has become the keystone of today's successful theory of the strong force: *quantum chromodynamics* or QCD. As its name suggests, the theory is founded on an analogy with quantum electrodynamics, or QED, where the role of electric charge is played by quark color. The force between quarks is mediated by the exchange of massless spin 1 *gluons* between colored quarks, just as the electromagnetic force is mediated by the exchange of massless spin 1 photons between charged particles. However, gluons act upon, and can change, the quark's color. A red quark can absorb a gluon and become a blue or green quark. There are eight varieties (or colors) of gluons, which couple to different combinations of quark color. A second profound difference between QCD and QED is that gluons are colored while photons are electrically neutral. That is, gluons (but not photons) are coupled to themselves. The fundamental acts of becoming (shown in Figure 15.7) describe the couplings of gluons to quarks and to other gluons.

FIGURE 15.7
The fundamental acts of QCD quarks are denoted by colored arrows, gluons by wavy colored lines. *a.* The emission or absorption of a gluon by a quark. The color of the quark may change, but not its flavor. *b.* Gluons can interact with each other, unlike photons. For this reason, the force between quarks does not satisfy an inverse–square law: It does not decrease as quarks are separated from each other.

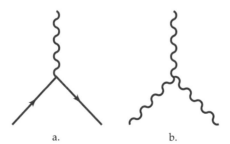

a.   b.

Because gluons exert forces on each other, the force between quarks has a different character from the force between charged bodies: It does not fall off at large distances, but remains constant. The attractive force between two tiny quarks in a hadron is about the same as the weight of a 30-ton boulder!

*Once again, physicists have preempted an everyday word (like *up* or *charm*) for their own contrary purposes. Quark colors have no relation to the colors of light.

No one has seen an isolated quark, and QCD explains why no one will. Quarks are confined to hadrons because the color force does not fall off at large distances. Imagine trying to remove a quark from the hadron of which it is a part. Suppose the quark is struck by an energetic electron. As the struck quark moves away, it is linked to the remainder of the hadron by a "spring" consisting of the constant color force. When enough potential energy accumulates in the spring to create a new quark and a new antiquark, the spring snaps (Figure 15.8c). All we have done is create a new hadron. Individual quarks and the gluons that hold them together cannot be seen as isolated particles in their own right. They exist as parts of the particles they make up. Hadrons are made of quarks held together by gluons, but neither of these fundamental constituents can be observed in isolation. Several analogies may make this notion more palatable. Every magnet has two opposite poles. Break the magnet in two and each fragment will have two poles. Any attempt to isolate a pole of a magnet is doomed. Similarly, a piece of string has two ends, never just one. So it is that observable particles may contain three quarks, but never one or two.

Physicists are currently using supercomputers to try to calculate the properties and behavior of hadrons from scratch. However, the calculations are simpler and more reliable for high-energy processes, where the effects of quarks flitting about within hadrons are seen most clearly. The success of these calculations confirms QCD as the origin of the strong nuclear force. The color force imprisons quarks to form the multitude of apparently elementary hadrons we see. Furthermore, it generates the nuclear force acting between color-neutral hadrons. Thus, a further analogy

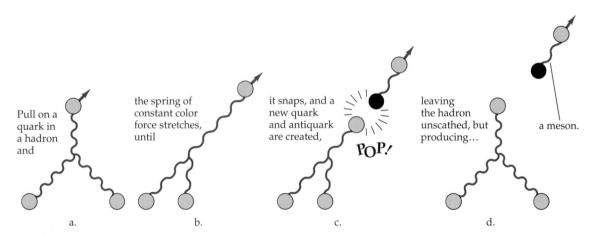

Pull on a quark in a hadron and — the spring of constant color force stretches, until — it snaps, and a new quark and antiquark are created, POP! — leaving the hadron unscathed, but producing... — a meson.

a.   b.   c.   d.

FIGURE 15.8 *a.* An attempt is made to extract a quark from a baryon. (Because quarkscrews are not readily available, we imagine that one of the quarks is struck by another particle.) *b.* As the struck quark is displaced, the spring of constant color force is stretched. *c.* When the potential energy of the spring is large enough, the spring snaps and a quark–antiquark pair is produced. *d.* The newly created quark restores the integrity of the struck baryon. The newly created antiquark combines with the struck quark to form a meson.

may be drawn between quantum chromodynamics and quantum electrodynamics:

- The primary role of QED is to bind electrons to nuclei to form atoms. A small residue of the electric force acts among electrically neutral atoms and lets them combine and react to form molecules. That is, QED describes atoms, molecules, and chemical phenomena.
- The primary role of QCD is to bind quarks together to form nucleons (and other hadrons). A small residue of the color force acts among colorless nucleons and lets them combine and react to form nuclei. That is, QCD describes hadrons, nuclei, and nuclear phenomena.

The Charm Quark and the Bottom Quark

As it turned out, three quark flavors were not enough to construct all the hadrons. Two more have been discovered: the *charm quark c* and the *bottom quark b*. This story began on November 11, 1974, when two laboratories reported the discovery of a new particle.

At Brookhaven National laboratory, a team of experimenters led by Samuel C. C. Ting directed a beam of 30 GeV protons against a beryllium target. Many particles were produced by these collisions, but the experimenters focused on the production of high-energy electron–positron pairs. That is, they studied the reaction

$$p + \text{Be} \longrightarrow e^+ + e^- + \text{other particles}$$

They measured the momenta of the electrons and positrons and, for each event, determined the invariant mass of the electron–positron pair. A graph of the number of events versus invariant mass revealed a large and narrow peak at an invariant energy of 3.1 GeV (Figure 15.9, on page 636). The data were interpreted as follows: Some of the collisions created an unstable J particle:

$$p + \text{Be} \rightarrow J + \cdots$$

Subsequently, the J particle decayed into an electron–positron pair with invariant energy Mc^2, where M is the mass of the J particle: $M = 3.1 \text{ GeV}/c^2$.

Meanwhile, electron–positron collisions were being studied at the Stanford Linear Accelerator Center by an experimental group directed by Burton Richter. The collisions between oppositely directed beams of electrons and positrons sometimes produce hadrons. The experimenters were measuring the rate at which such events were seen versus the energy of the colliding particles. At an invariant energy of 3.1 GeV, the event rate leaped up by a factor of 100. The data were interpreted in terms of a new and unstable particle Ψ with a mass of $M = 3.1 \text{ GeV}/c^2$. When the invariant energy of the colliding particles was equal to Mc^2, the reaction $e^+ + e^- \rightarrow \Psi$ took place. Subsequently, the Ψ particle decayed into hadrons, which were detected.

The discoveries were announced on the same day on the East and West Coasts. The two experiments were very different, but the J and Ψ

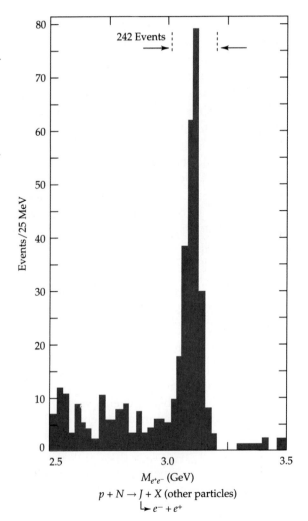

particles are the same. The leaders of the experimental teams, Ting and
Richter, shared the 1976 Nobel Prize in physics for their discovery of
what is now called the J/Ψ particle. Its double name is unique among
particles and honors both experimental groups.

There were over a hundred known hadrons in 1974, so what was
the big deal? The most surprising property of the J/Ψ particle was
its narrow width in energy, which meant its lifetime was 1000 times
longer than those of run-of-the-mill hadrons. The anomalous lifetime
was explained by quantum chromodynamics under the hypothesis that
the J/Ψ particle is made of a quark and an antiquark of a new flavor:

$$J/\Psi \equiv c\bar{c}$$

where c is the heavy $Q = 2/3$ *charm quark*. My colleagues and I had predicted and named the fourth quark years before, but not everyone believed our arguments—until the J/Ψ particle was discovered. The charges of c quarks and u quarks are the same, $Q = 2/3$. However, the mass of the c quark is 1.5 GeV/c^2, much larger than the masses of the u, d, and s quarks.

Subsequent experiments found additional particles containing just one charmed quark, such as the mesons:

$$D^+ \equiv c\bar{d} \quad \text{and} \quad F^+ \equiv c\bar{s}$$

and the baryon $\Lambda_c^+ \equiv udc$. The study of charmed particles proved that there was a fourth quark flavor and led to widespread acceptance of the quark hypothesis and quantum chromodynamics. For this reason, the discovery of the J/Ψ particle is sometimes referred to as the *November Revolution* of particle physics.

In 1977, history repeated itself. Leon Lederman and his team at Fermilab discovered the fifth quark flavor. They observed a particle with a mass of about 10 GeV/c^2 consisting of a bottom quark b and its antiquark, $b\bar{b}$. The charge of the bottom quark is $-1/3$ and its mass is about 5 GeV/c^2. Particles containing just one bottom quark have also been observed, such as the meson $B^- \equiv b\bar{u}$ and the baryon $\Lambda_b^0 \equiv bud$.

The standard model demands the existence of a sixth quark flavor—the *top quark* t. Particles containing top quarks have not yet been observed, but the mass of the top quark is known to exceed 100 GeV/c^2. However, physicists are confident that the t quark exists and will soon be found.

Thus, there are at least six quark flavors: up, charm, and top with $Q = 2/3$, and down, strange, and bottom with $Q = -1/3$. In Section 15.4, we shall see how they are linked together in pairs by the weak force:

$$\begin{pmatrix} u \\ d \end{pmatrix} \quad \begin{pmatrix} c \\ s \end{pmatrix} \quad \begin{pmatrix} t \\ b \end{pmatrix}$$

Because they are made of a quark and an antiquark, the J/Ψ particle and other particles with the quark composition $c\bar{c}$ are known as states of *charmonium*, in analogy with positronium. Particles with quark composition $b\bar{b}$ are sometimes called states of *bottomonium*. These particles provided new and effective proving grounds for the standard model. Today, ten energy levels of the $c\bar{c}$ system and a dozen of the $b\bar{b}$ system have been found and studied. The energy-level diagrams for these states are shown in Figure 15.10 (on page 638). Their masses (or invariant energies) are relatively easy to calculate because heavy quarks move slowly within hadrons. The agreement between theory and experiment is spectacular and provides the most convincing proof that quarks, gluons, and quantum chromodynamics are the correct explanation of hadrons and their interactions.

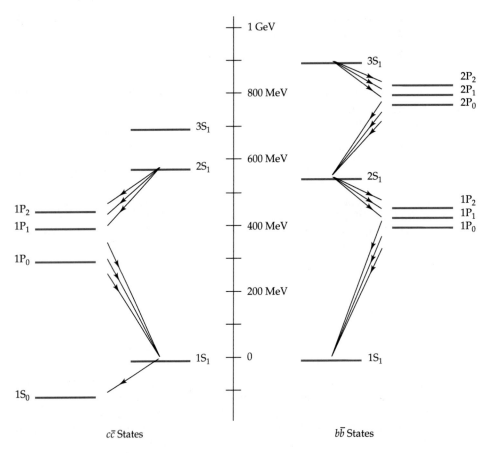

FIGURE 15.10 The energy levels of charmonium and bottomonium. Mesons made of a heavy quark and its antiquark are described by the Schrödinger equation. The observed states of the $c\bar{c}$ and $b\bar{b}$ system are shown. They have neither charge, nor strangeness, nor nucleon number. The remaining quantum numbers of the states are indicated: A prefixed integer is the principal quantum number, subscripts denote total angular momenta, and S (for 0) and P (for 1) denote the orbital angular momenta of the quark–antiquark system.

Exercises

12. Explain why the following decay modes do not take place:
    a. $\Xi^0 \longrightarrow \pi^- + p$
    b. $\Xi^0 \longrightarrow K^- + p$
    c. $\Xi^0 \longrightarrow \pi^- + \Sigma^+$

13. Samios observed the decay of the Ω^- into $\Xi^- + \pi^0$. What are two other allowed decay modes of the Ω^- particle?

14. Explain why the decay $\Sigma^+ \rightarrow K^+ + n$ does not take place.

15. Explain why the reaction $\pi^- + p \rightarrow K^+ + K^-$ does not take place.

16. The quark composition of $\overline{K}$ is $s\overline{d}$ and that of K^- is $s\overline{u}$. Draw quark diagrams for the following strong-interaction reactions:
    a. $\overline{K}^0 + p \longrightarrow \Lambda^0 + \pi^+$
    b. $K^- + p \longrightarrow \Xi^- + K^+$

17. How does the quark model explain the facts that no mesons carry nucleon number but all baryons carry nucleon number 1?

18. The charmed quark c carries charm $\mathcal{C} = 1$ and charge $\mathcal{Q} = 2/3$. What are the possible electric charges of $\mathcal{C} = 1$ baryons?

19. Describe a hypothetical experiment to search for fractionally charged particles within ordinary matter.

20. In a fictitious "dwark" theory, up dwarks have charge 1 while down and strange dwarks have charge 0. The proton is made of 1 up dwark and 2 down dwarks. Where does dwark theory go wrong?

15.4 The Story of the Weak Force

Strong and electromagnetic forces result from the exchange of spin 1 bosons—gluons and photons, respectively. Could it be that a similar mechanism underlies the weak force? Yukawa suspected as much; his pion was intended to generate both the strong and weak interactions. So did Fermi, Schwinger, Lee, and Yang. However, a genuine theory of the weak force did not emerge until the 1970s—at roughly the same time that QCD, the theory of the strong force, was formulated.

In the processes of nuclear β decay, one of the nucleons in a nucleus changes its identity as a lepton pair is produced:

$$n \longrightarrow p + e^- + \overline{\nu}_e \quad \text{or} \quad p \longrightarrow n + e^+ + \nu_e$$

Having discovered that nucleons have a quark substructure, we may interpret the β process at a more fundamental level as a transformation of a quark within the decaying nucleon:

$$d \longrightarrow u + e^- + \overline{\nu}_e \quad \text{or} \quad u \longrightarrow d + e^+ + \nu_e \tag{7}$$

Changing a down quark to an up quark converts a neutron to a proton, and vice versa. These transformations result from the weak force, which is now known to be mediated by a heavy, charged boson W^+ (and its antiparticle W^-). The fundamental acts underlying β decay and other weak interactions are the emission or absorption of a W boson by a quark or a lepton (Figure 15.11, on page 640).

Beta decay proceeds through a two-step process involving the production and decay of a W boson:

$$d \longrightarrow u + W^-, \quad W^- \longrightarrow e^- + \overline{\nu}_e,$$

or

$$u \longrightarrow d + W^+, \quad W^+ \longrightarrow e^+ + \nu_e$$

FIGURE 15.11
Fundamental acts of the weak force. The wavy line denotes a W^+ boson proceeding in the direction of the arrow. In *a.*, a W^+ is emitted by an up quark, which becomes a down quark of the same color. In *b.*, a W^+ boson is emitted by a down quark, which becomes an up quark of the same color. The weak interaction can change quark flavors but not quark colors. Parts *c.* and *d.* show how an electron neutrino may be transformed into an electron and vice versa. The absorption (or emission) of a W^- boson is described by the same diagram as the emission (or absorption) of a W^+ boson.

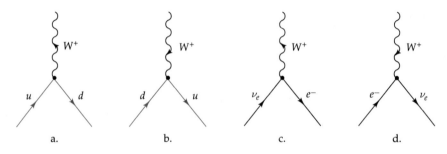

In these reactions, the W boson is not present as an observable particle. It is a *virtual particle*, like the photons exchanged by charged particles or the gluons exchanged between quarks. The energies involved in particle or nuclear decays are far too small to permit the actual emission of a W boson. The processes of $\beta^\pm$ decay are represented by the Feynman diagram shown in Figure 15.12.

Let's see how the muon fits into the picture. Recall that the muon is a charged lepton, like the electron. It was discovered in cosmic rays as the decay product of copiously produced and strongly interacting pions:

$$\pi^- \longrightarrow \mu^- + \bar{\nu}_\mu \qquad \pi^+ \longrightarrow \mu^+ + \nu_\mu$$

FIGURE 15.12
The Feynman diagrams describing β decay. *a.* A down quark in a neutron becomes an up quark as it emits a W^- boson. Notice that the absorption of a W^+ boson is equivalent to the emission of a W^- boson. The virtual W^- boson is transformed into an electron and its antineutrino. *b.* An up quark in a proton becomes a down quark as it emits a W^+ boson. The virtual W^+ boson is transformed into a positron and an electron neutrino.

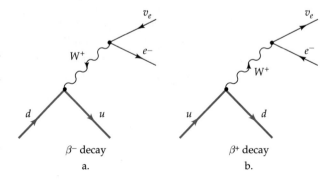

An additional fundamental act, shown in Figure 15.13, is needed to describe the decays of pions and muons. Figure 15.14a combines two fundamental acts into a Feynman diagram describing the decay of the muon: $\mu^- \rightarrow e^- + \bar{\nu}_e + \nu_\mu$. Alternatively, the muonic act is combined with the quark act in Figure 15.14b to obtain a Feynman diagram describing the process $u + \bar{d} \rightarrow \mu^+ + \nu_\mu$ corresponding to the decay of a charged pion made of a $u\bar{d}$ pair.

Another important and experimentally observed process describes the fate of a negative muon when it comes to rest in matter. The μ^-

FIGURE 15.13
A muon can exchange
its identity with a muon
neutrino as it interacts
with a *W* boson.

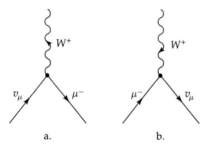

forms an electrically bound state with an atomic nucleus: a *muonic atom.*
The nucleus can capture the muon by a process analogous to electron
capture:

$$\mu^- + (A, Z) \longrightarrow \nu_\mu + (A, Z-1)$$

or

$$\mu^- + u \longrightarrow \nu_\mu + d$$

The Feynman diagram corresponding to μ capture is shown in Figure
15.14c.

The diagrams shown in Figures 15.12 and 15.14 are each composed
of two of the three fundamental acts. A number associated with each
act measures its strength. These numbers (or "coupling strengths") are
analogous to the electric charges governing the couplings of photons to
charged particles and can be determined from the measured rates of β
decay, muon decay, and muon capture. Physicists were astonished to
find that the coupling strengths of *W*s to electrons, muons, and quarks
are nearly equal to one another. The weak force was discovered to be
a *universal* property of quarks, muon-type leptons, and electron-type
leptons.

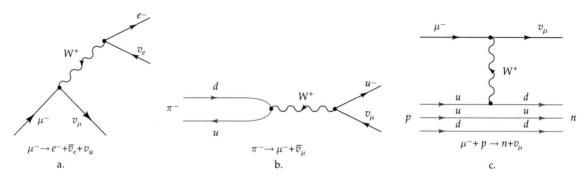

FIGURE 15.14 *a.* Feynman diagram showing muon decay $\mu^- \longrightarrow e^- + \bar{v}_e + v_\mu$. *b.* Feynman diagram showing
pion decay: $\pi^- \longrightarrow \mu^- + \bar{v}_\mu$ *c.* Feynman diagram showing muon capture: $\mu^- + p \longrightarrow v_\mu + n$

The Electroweak Synthesis

In the 1950s, several suggestive similarities—and as many profound differences—characterized the intermediate boson model of the weak force and the theory of quantum electrodynamics:

- The photon and the hypothetical W boson were both spin 1 particles. However, the photon was massless while the W boson (since it had not been seen) had to be heavy. In particular, if the weak coupling strength were set equal to the electric charge, the W boson had to have a mass about 100 times greater than the proton mass.

- The couplings of photons and W bosons were universal. The electric charges of the proton, electron, and muon were equal in magnitude and so were the couplings of the W boson to these particles. However, the couplings of the photon were mirror symmetric and parity-conserving, while those of the W led to parity-violating effects in weak interactions.

- There was a consistent theory of the interactions of photons with charged particles, but there was no such theory of the weak force. Nonetheless, Schwinger (my thesis advisor at the time) suggested the possibility of a unified theory of the weak and electromagnetic forces. Just such a theory emerged over the next decade and through the work of many physicists.

The parity-violation puzzle was solved in 1961. The $W^\pm$ bosons had to be supplemented with an additional heavy and uncharged companion, the Z^0 boson. This particle mediates a different kind of weak interaction by which particles can interact without changing their charges. In particular, neutrinos should interact with hadrons without being converted to charged leptons:

$$\nu + p \longrightarrow \nu + p, \qquad \nu + p \longrightarrow \nu + n + \pi^+, \quad \text{etc.} \qquad (8)$$

In these "neutral-current processes," neither quarks nor leptons change their identities as momentum is exchanged between them. New fundamental acts describing the emission and absorption of Z^0 bosons had to be added to the rules of the game.

In 1967, the Pakistani physicist Abdus Salam and my high school chum Steven Weinberg proposed a simple scheme by which W^+, W^-, and Z^0 bosons could acquire large masses while the photon stayed massless. According to their view, an intrinsic and perfect symmetry among the weak intermediaries and the photon was manifest when the universe was young and very hot. As the universe cooled, a phenomenon similar to crystallization took place. When a crystal emerges from a salt solution, it chooses certain planes and axes in space, even though the laws of physics are the same in all directions. In the hot young universe, the four intermediate vector bosons were massless. As the universe cooled, the symmetry among the four bosons was destroyed: Ws and Zs developed masses while photons did not. This phenomenon is known as "spontaneous symmetry breaking" and is a central feature of today's electroweak synthesis. In 1971, the Dutch physicists Gerhard 'tHooft and

Four large detectors have been installed at the large electron-positron collider LEP at CERN. They are called ALEPH, DELPHI, OPAL and L3. Each detector has different capabilities and has recorded about a million Z decays. This photograph shows the 8000 ton magnet, which is part of the L3 detector, and some of the 479 physicists making up the L3 collaboration. *Source:* Courtesy Samuel C. C. Ting/CERN.

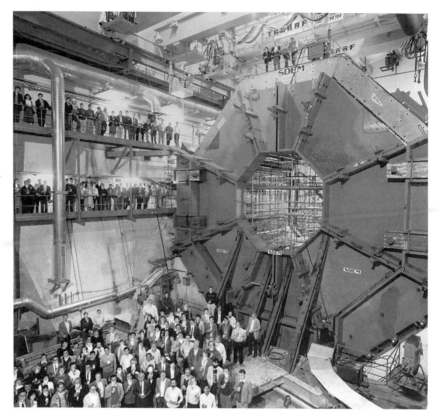

Tini Veltman proved that the mass-generating mechanism of Weinberg and Salam made sense. They produced a set of rules for manipulating the Feynman diagrams of the electroweak model and showed that the theory is mathematically consistent.

At this point, experimenters took up the challenge. Teams of scientists at the two largest accelerators began to search in earnest for neutral-current phenomena, such as reaction 8, mediated by the then–hypothetical Z^0 boson. The race was won by scientists working at the European Particle Physics Laboratory (CERN) in 1973, and their results were soon confirmed by scientists at Fermilab. A decade later, decisive proof of the validity of the electroweak theory was obtained. W and Z bosons were produced and detected at the CERN proton–antiproton collider. These particles have since been studied at the American proton–antiproton collider at Fermilab, at the Stanford linear collider (SLC), and at the CERN electron–positron collider called LEP. The mass of the $W^\pm$ boson is 80.6 GeV/c^2 and that of the Z^0 is 91.2 GeV/c^2. Millions of Z^0s have been produced and studied by four teams, each consisting of hundreds of scientists, working at LEP. About 100 different decay schemes of this particle have been identified. All of its properties agree with the predictions of the standard model. The photo, showing part of one of the four large particle detectors at LEP, indicates the magnitude of the experimental endeavor.

The Standard Model

There are three ingredients to today's standard model of elementary particle physics:

1. *Quantum chromodynamics* is the mechanism of the strong force. It acts via the exchange of massless spin 1 gluons between quarks. QCD lets quarks change their colors but not their flavors.

2. *The electroweak theory* is the mechanism of the weak and electromagnetic forces. It acts via the exchange of massless spin 1 photons (electromagnetism) and massive spin 1 Ws and Zs (the weak force) between two quarks, two leptons, or a quark and a lepton. The electroweak force lets quarks change their flavors but not their colors.

3. *The periodic table of fundamental fermions* describes the particles that participate in the preceding forces. They are subdivided into two categories: quarks (which may emit and absorb gluons) and leptons (which may not). Three quarks have charge 2/3 (in units of e) and three others have charge $-1/3$. Three leptons have charge -1 and three others (neutrinos) have no charge at all. Quarks (each in three colors) and leptons appear in families, as shown in Table 15.5.

TABLE 15.5   Periodic Table of Quarks and Leptons

	$Q = -1$	$Q = -1/3$	$Q = 0$	$Q = 2/3$
First Family	e^-	down	ν_e	up
Second Family	μ^-	strange	ν_μ	charm
Third Family	τ^-	bottom	ν_τ	top

The particles of the first family are sufficient to explain the structure and behavior of ordinary matter such as is found on Earth and in the stars: Two quarks explain the makeup of atomic nuclei, electrons complete the atom's structure, and neutrinos enable the solar furnace to operate. These are the *relevant* particles.

The particles of the second family explain the exotic particles seen in cosmic rays and at accelerators: muons and their neutrinos, as well as hadrons with or without strangeness and charm. None of these particles have any practical applications as yet. The relevant and the irrelevant are in balance.

Then there is the third family of particles, none of them with any obvious purpose.*

---

*Relevance is in the eye of the beholder. The particles of the second and third families are, of course, relevant to and demanded by the one and true theory of nature that may one day be revealed to us.

Each family of fundamental fermions consists of a quark with charge 2/3, a quark with charge $-1/3$, a charged lepton, and a neutrino. The neutrinos are known to have small masses. They may even be massless. Aside from the neutrinos, corresponding members of the three families are sequentially heavier:

$$m(\tau) \gg m(\mu) \gg m(e)$$

$$m(b) \gg m(s) \gg m(d)$$

$$m(t) \gg m(c) \gg m(u)$$

We have no idea why this is, nor can we calculate or relate the quark and lepton masses from first principles.

The electroweak theory allows leptons in the same family to change into one another by emitting or absorbing a W boson:

$$e^- \Longleftrightarrow \nu_e, \quad \mu^- \Longleftrightarrow \nu_\mu, \quad \tau^- \Longleftrightarrow \nu_\tau$$

Quarks in the same family are also linked together by W exchange:

$$d \Longleftrightarrow u, \quad s \Longleftrightarrow c, \quad b \Longleftrightarrow t$$

If this were the whole story, strange particles would have no way to decay: They would be stable. In fact, though, the weak force also links together quarks of different charge belonging to different families, but with reduced strengths.

Are there more families of quarks and leptons? The answer seems to be no. Each family includes a light or massless neutrino. However, two independent arguments lead us to believe that there are only three neutrino species, and hence only three families of fundamental fermions. One of these arguments is a direct experimental measurement of the number of neutrinos that was performed at CERN and at SLAC in 1989. Another is more indirect and pertains to the Big Bang origin of the universe. The number of neutrino states affects the rate of expansion of the early universe, which in turn affects the amount of helium in the universe today. The amount of helium observed in the universe implies that the number of different neutrino flavors cannot exceed three. Cosmology—the theory of the structure and evolution of the universe—is telling us something profound about nature of elementary particle physics.

According to the standard model, all the forces of nature except gravity arise from the exchange of spin 1 bosons associated with the mathematical symmetries of the theory. These particles—photons, $W^\pm$s, Zs, and gluons—are collectively known as *gauge bosons*. In addition to these force particles, the theory includes 12 different quarks and leptons.* It is called a model and not a theory because many mysteries remain to be solved:

• Why do the fundamental fermions come in families, and why are there precisely three of them? Why do quarks come in 3 colors, not 2 or 17? Do neutrinos have mass?

*If quark color is taken into account, the number of fundamental fermions is doubled.

- The standard model involves 18 adjustable parameters. I mean by this the values of particle masses and the strengths of their couplings. The theory would seem to make perfect sense whatever the values of these quantities. The big question is why they have the values they do.

- These questions cannot be approached until the nature of spontaneous symmetry breaking is understood. When the universe was young, all particles were massless and there was a perfect symmetry between weak and electromagnetic forces. What broke the symmetry? Fortunately, we know the energies at which the answer lies. The next great accelerator has been designed specifically to find the origin of particle masses. All we know for sure is that there must be at least one more species of particle in our bestiary. It may be the elusive spin 0 *Higgs boson* implied by Salam and Weinberg's picture of symmetry breaking, or perhaps it is something else that we are not smart enough to imagine.

- The standard model works too well! Large accelerators were built to test the model and search for new and inexplicable phenomena—and experiments confirm the predictions of the model in exquisite detail. There have been no anomalous results or surprising revelations. *W*s and *Z*s, charmed particles, and particles containing bottom quarks are produced in enormous numbers at accelerator laboratories and they behave just as the theory says they should. To answer the vexing questions that remain, we must build a better model. But to do so, we must find the flaws and weaknesses of the standard model. Because none have appeared so far, perhaps we have reached as far as we can to unravel the nature of matter. Such sentiments have often been expressed in the history of physics, but they have always turned out to be wrong. I am confident that nooks and crevices will be found, and that physicists will climb to a higher base station on their quest for a Theory of Everything.

- Finally, we come to the greatest mystery of all—the nature of the gravitational force. Newton's theory is good, Einstein's is better, yet neither of them is a quantum theory. Until very recently, all attempts to create a quantum theory of gravity have failed. Many theoretical physicists today are working on *string theory*, an ambitious attempt to incorporate what is already known with a quantum theory of gravity. String theorists view particles as tiny loops of vibrating string and space–time as a portion of space with many more dimensions. At this time, however, as string theorists struggle with its mathematical intricacies, string theory is in its infancy.

Exercises

21. About 1 in 1000 muons bound to hydrogen nuclei suffers capture rather than decay: $\mu^- + p \rightarrow \nu_\mu + n$. What is the recoil energy of the neutron in this reaction in the frame in which the initial particles are at rest? (Neglect the small mass difference between neutrons and protons, and assume that the neutrino is massless.)

22. Give an example of an allowed decay mode of the charged pion aside from $\pi^+ \rightarrow \nu_\mu + \mu^+$.

23. A muon at rest decays into three light particles. Show that it is possible for the electron to carry off half of the invariant energy $m_\mu c^2$ of the muon. (*Hint:* In this calculation, the effect of electron or neutrino masses may be neglected.)

24. To explain the fact that strange particles are unstable, it must be possible for a strange quark to emit a W^- (or absorb a W^+) and become an up quark. Consider the decay schemes: $\Sigma^+ \rightarrow n + e^+ + \nu_e$ and $\Sigma^- \rightarrow n + e^- + \bar{\nu}_e$. Explain why the first decay mode is forbidden while the second is allowed.

25. Suppose that pions of energy E are produced by cosmic ray collisions at an altitude of 30 km. The pion mass is 140 MeV and its lifetime is 2.6×10^{-8} s. What must E be if half of the pions are to reach the surface of Earth. (Ignore the possibility that the pions collide with air nuclei on their downward voyage.)

Where Are We Going?

This text has come to an end but the search for the basic building blocks of matter and the rules they obey continues. The goal of physical science is a unified description of matter and energy in all its forms—from the tiniest particles to the universe itself. Let us recall some of the many links that have been forged between the large and the small.

- In the seventeenth century, Newton showed that heavenly bodies obey the same laws of motion as objects on Earth.
- In the eighteenth century, Lavoisier and his contemporaries explained fire, water, earth, and air in terms of unseen atoms of chemical elements.
- In the nineteenth century, astronomers used the principles of atomic physics to show that stars are formed of the same elements as we are.
- In the twentieth century, astrophysicists used the principles of nuclear physics to explain how stars shine and why they sometimes explode.

In 1929, the American astronomer Edwin Hubble observed spectral lines of distant galaxies to be Doppler-shifted to the red. He realized from this that galaxies are hurtling away at velocities proportional to their distances from us. This means that the universe is expanding, somewhat like the fragments of an exploded bomb. Hubble's discovery poses two questions, one concerning the birth of the universe, the other its fate. The universe is now cold and sparse, but closer to the moment of its birth it was inconceivably hot and dense. At that time, the seeds that grew into galaxies must have been present. What were they, and how did they generate the structures astronomers see? Will the universe continue to expand forever in the "big chill," or will the pieces someday fall together again in the "big crunch"? These questions are related to one another and to a third question: How much matter is in the universe?

The search for answers to these questions has led to a remarkable convergence between particle physics (the study of the smallest features of matter) and cosmology (the study of the birth and evolution of the universe). A particle accelerator is like a microscope because it reveals the smallest structures in matter. Let me explain why a particle accelerator is also like a time machine that can help us understand what the universe was like when it was young. When the universe was less than a million years old, it was too hot for atoms to exist. When it was a few minutes old, the first atomic nuclei came into being. A fraction of a second after the Big Bang, the universe was teeming with all the particles and antiparticles in our bestiary. As more is learned about elementary particles, scientists are better able to trace the history of the universe back to the moment of its creation and to understand the origin of stars and galaxies.

As cosmologists pursue their dream, they are led to a surprising result. Today's picture of the birth of the universe requires 100 times more matter than is seen in the form of stars. The amount of matter in the universe must be just enough to stave off the "big crunch". If their theories are correct, most of the matter of the universe is dark and invisible. Meanwhile, astronomers, from an entirely different tack, confirmed the existence of dark matter. Studies of galactic motions show that most galaxies are much heavier than the total mass of the luminous stars within them. Their huge halos of dark matter cannot hide their gravitational effects.

What is the dark matter of the universe? It cannot be gas or dust and it seems unlikely to consist of burnt-out stars or jupiter-sized planets. In fact, it doesn't seem to be ordinary matter at all. Some scientists believe that dark matter consists of massive neutrinos, while others opt for black holes. However, it is possible that the dark matter of the universe is in the form of particles lying beyond the standard model. If this is true, the search for the basic building blocks of matter has hardly begun.

A The Particle Physicist's Guide to the Greek Alphabet

α: **Alpha** rays are helium nuclei emitted by nuclei in the process of α decay, one of the three forms of natural radioactivity.

β: **Beta** rays are energetic electrons emitted by nuclei in the process of β decay, another form of natural radioactivity.

γ: **Gamma** rays are energetic photons emitted by nuclei in the process of γ decay. Each form of radioactivity exemplifies one of the the three kingdoms of matter: α particles are *hadrons*, β particles *leptons*, and γ particles *gauge bosons*.

δ: **Delta** something (whether in upper or lower case) often stands for a small change in that something, as in δx or Δy. Moreover, the delta function $\delta(x)$, which is zero everywhere except at $x = 0$, where it is infinite, is a must if you are a theoretical physicist, and δ rays are important to experimenters.

ϵ: **Epsilon** signifies a numerical quantity that is, or may be made, arbitrarily small. To the great Hungarian bachelor mathematician Paul Erdös, children are epsilons.

ζ: **Zeta** was the name suggested by its "discoverers" for a particle that turned out not to exist after all. It has not been popular since.

η: **Eta** denotes the eighth member of the spinless meson octet whose existence was predicted by Murray Gell-Mann in 1961. Capitalized, η becomes H to honor Johns Hopkins University where the η particle was discovered.

θ: **Theta** should remind you of trigonometry, because it is a very popular choice for the measure of an angle.

ι: **Iota** looks very much like an i that someone forgot to dot.

κ: **Kappa,** like its predecessor, is too much like its Latin equivalent to be of great use, although I once invented the κ particle. It doesn't exist.

λ: **Lambda** is for the length of a wave. Big Λ is the lightest of the strange baryons.

μ: **Mu** stands for the muon, of which Rabi asked "Who ordered that?" half a century ago. We still don't know the answer.

ν: **Nu** can stand for any of the three known species of neutrinos: ν_e, ν_μ, and ν_τ. Moreover, μ's and ν's often appear as numerical subscripts, as in the basic equation of general relativity: $R_{\mu\nu} - \frac{1}{2}g_{\mu\nu} R = -\kappa\, T_{\mu\nu}$.

ξ: **Xi** is easy to say ("zy") but much too hard to pen in lower case. Capital Ξ represents a baryon with two units of strangeness.

o: **Omicron,** for all the world, looks like an o.

π: **Pi** denotes the ratio of the circumference of a circle to its diameter and it is a most important meson to boot. The pion usually decays into a muon and a neutrino: $\pi \rightarrow \mu + \nu$.

ρ: **Rho** can be a density. The ρ-meson is made of the same quarks as the pion but with their spins lined up.

σ: **Sigma** signifies the spin of a particle. Big Σ is a mathematical sum or one of three strange baryons: Σ^+, Σ^0 or Σ^-.

τ: **Tau** is the heaviest of three known charged leptons, except when it's a fixed time interval (like a half-life).

υ: **Upsilon** denotes a particle made of a b quark and its antiquark discovered by Leon Lederman in 1977 after a false start. The original was known as the 'Oops, Leon!'

ϕ: **Phi** is an angle when you've got more than one and have already used θ—for example, $\sin(\theta + \phi) = \sin\theta\cos\phi + \sin\phi\cos\theta$, a memorable trigonometric identity.

χ: **Chi** is a conveniently uncommitted symbol often standing for what you just thought of.

ψ: **Psi** is preëmpted for quantum–mechanical wave functions, although it could be an angle when you've got more than two. The J-particle (first seen in New York) is the same as the Ψ-particle (spied at the same time in California). Today it's known as the J/Ψ or "gypsy" particle.

ω: **Omega** is the overworked last letter of the Greek alphabet: ω is both a meson and a favorite for frequencies expressed in radians per second. Capitalized, Ω can be the ratio of the mean mass density of the universe to its critical value, or the triply-strange baryon discovered by the Greek–American physicist Nicolas Samios. In 14th-century England, omega was denoted by oo, as in

"I am alpha and oo, the bigynnyng and the endyng…"

(John Wyclif, 1382).

B Physical Constants

FUNDAMENTAL CONSTANTS

Avogadro's number: $N_A = 6.022 \times 10^{23}\, mol^{-1}$

Boltzmann constant: $k = 1.381 \times 10^{-23}\, J\,K^{-1}$

Constant in Coulomb's Law: $k_c = 9 \times 10^9\, N\,m^2\,C^{-2}$

Speed of light (exact): $c = 299{,}792{,}458\, m\,s^{-1}$

Planck's constant: $h = 6.626 \times 10^{-34}\, J\,s$

Planck's constant (reduced): $\hbar = h/2\pi = 1.055 \times 10^{-34}\, J\,s$

Electron charge (magnitude): $e = 1.602 \times 10^{-19}\, C$

Electron mass: $m_e = 9.109 \times 10^{-31}\, kg = 511\, keV\,c^{-2}$

Proton mass: $m_p = 1.673 \times 10^{-27}\, kg$
$$= 938.3\, MeV\,c^{-2}$$

Atomic mass unit: $amu = 1.661 \times 10^{-27}\, kg$
$$= 931.5\, MeV\,c^{-2}$$

Newton's constant: $G_N = 6.673 \times 10^{-11}\, m^3\,kg^{-1}\,s^{-1}$

DERIVED CONSTANTS AND CONVERSIONS

Fermi: $f = 10^{-15}\, m$

Light year: $ly = 9.46 \times 10^{15}\, m$

Absolute zero: $0\,K = -273.15°C$

Gas constant: $R = k\,N_A = 8.315\, J\,mol^{-1}\,K^{-1}$

Molar volume (at STP): $N_A\,k\,(273.15\,K)/1\,atm = 22.414 \times 10^{-3}\, m^3$

Bohr radius: $r_B = \hbar^2/(m_e\,k_c\,e^2) = 5.292 \times 10^{-11}\, m$

Electron Compton radius: $\bar{\lambda} = \hbar/m_e\,c = 3.862 \times 10^{-13}\, m$

Stefan–Boltzmann constant: $\sigma = \pi^2 k^4/60\,\hbar^3\,c^2$
$$= 5.671 \times 10^{-8}\, W\,m^{-2}\,K^{-4}$$

Rydberg energy: $Ryd = m_e\,k_c^2\,e^4/2\hbar^2 = 13.606\, eV$

Electron volt: $1\,eV = 1.6022 \times 10^{-19}\, J$

ASTRONOMICAL and GEOLOGICAL CONSTANTS

Astronomical unit:	1 A.U. $= 1.496 \times 10^{11}\, m$
Solar mass:	$M_\odot = 1.989 \times 10^{30}\, kg$
Solar radius:	$R_\odot = 696,000\ km$
Solar luminosity:	$L_\odot = 3.827 \times 10^{26}\, W$
Solar constant:	$S = 1368\ W\, m^{-2}$
Earth mass:	$M_\oplus = 5.974 \times 10^{24}\, kg$
Earth equatorial radius:	$R_\oplus = 6378\ km$
Age of universe:	$T_U = 1.5 \pm .5 \times 10^{10}\ yrs$
Standard air pressure:	$P = 1.0133 \times 10^5\ N\, m^{-2}$
Gravitational constant at sea level:	$g = 9,807\ m\, s^{-2}$
The year of time:	Tropical yr = 31, 556, 926 s
	Sidereal yr = 31, 558, 150 s
	Calendar yr = 31, 536, 000 s

C Periodic Table of the Elements

Group

Noble Gases (18)

Atomic number ⟶ 11
Symbol ⟶ Na
Atomic mass ⟶ 22.99

Atomic masses are based on carbon-12. Numbers in parentheses are mass numbers of most stable or best known isotopes of radioactive elements.

Transition Elements

Period	IA(1)	IIA(2)	IIIB(3)	IVB(4)	VB(5)	VIB(6)	VIIB(7)	(8)	(9)	(10)	IB(11)	IIB(12)	IIIA(13)	IVA(14)	VA(15)	VIA(16)	VIIA(17)	(18)
1	1 H 1.008																	2 He 4.003
2	3 Li 6.941	4 Be 9.012											5 B 10.81	6 C 12.01	7 N 14.01	8 O 16.00	9 F 19.00	10 Ne 20.18
3	11 Na 22.99	12 Mg 24.31						VIII					13 Al 26.98	14 Si 28.09	15 P 30.97	16 S 32.06	17 Cl 35.45	18 Ar 39.95
4	19 K 39.10	20 Ca 40.08	21 Sc 44.96	22 Ti 47.90	23 V 50.94	24 Cr 52.00	25 Mn 54.94	26 Fe 55.85	27 Co 58.93	28 Ni 58.7	29 Cu 63.55	30 Zn 65.38	31 Ga 69.72	32 Ge 72.59	33 As 74.92	34 Se 78.96	35 Br 79.90	36 Kr 83.80
5	37 Rb 85.47	38 Sr 87.62	39 Y 88.91	40 Zr 91.22	41 Nb 92.91	42 Mo 95.94	43 Tc 98.91	44 Ru 101.1	45 Rh 102.9	46 Pd 106.4	47 Ag 107.9	48 Cd 112.4	49 In 114.8	50 Sn 118.7	51 Sb 121.8	52 Te 127.6	53 I 126.9	54 Xe 131.3
6	55 Cs 132.9	56 Ba 137.3	57* La 138.9	72 Hf 178.5	73 Ta 180.9	74 W 183.9	75 Re 186.2	76 Os 190.2	77 Ir 192.2	78 Pt 195.1	79 Au 197.0	80 Hg 200.6	81 Tl 204.4	82 Pb 207.2	83 Bi 209.0	84 Po (210)	85 At (210)	86 Rn (222)
7	87 Fr (223)	88 Ra 226.0	89** Ac (227)	104 Unq (261)	105 Unp (262)	106 Unh (263)	107 Uns (262)	108 Uno (265)	109 Une (266)									

Inner Transition Elements

	58	59	60	61	62	63	64	65	66	67	68	69	70	71
Lanthanide Series 6 *	Ce 140.1	Pr 140.9	Nd 144.2	Pm (145)	Sm 150.4	Eu 152.0	Gd 157.3	Tb 158.9	Dy 162.5	Ho 164.9	Er 167.3	Tm 168.9	Yb 173.0	Lu 175.0
	90	91	92	93	94	95	96	97	98	99	100	101	102	103
Actinide Series 7 **	Th 232.0	Pa 231.0	U 238.0	Np 237.0	Pu (244)	Am (243)	Cm (247)	Bk (247)	Cf (251)	Es (252)	Fm (257)	Md (258)	No (259)	Lr (260)

Atomic number (top left) is the number of protons in the nucleus. Atomic masses (bottom) are relative to the mass of the carbon 12 atom, which is assigned a mass of exactly 12 atomic mass units. In most cases, atomic masses are weighted by isotopic abundance in the Earth's crust. For very radioactive elements, the numbers shown in parentheses are the atomic masses of their longest-lived isotopes. Because of priority disputes or lack of confirmation, the names for elements beyond $Z = 103$ are not yet set in stone. Among several possibilities put forward are:

104 = Rutherfordium	105 = Hahnium	106 = Alvarezium
107 = Nielsbohrium	108 = Hassium	109 = Meitnerium

Glossary of Selected Terms

Every discipline has its own jargon and physics is no exception. Some technical terms honor great scientists of the past, like Coulomb's law *and the* curie. *Others are ordinary words that have been given special meanings, like* charm *or* spin. *In many cases, entirely new words were coined to describe natural phenomena, like* quark *and* entropy. *Some of the more important words and phrases of physical science are defined below.*

Aberration of starlight
The shift of a star's position in the sky due to Earth's motion and light's finite speed. James Bradley discovered the effect in 1725 and correctly explained it two years later. He showed how the measurement of stellar aberration provides a determination of the speed of light.

Ampère's law
Expresses the mutual force between two long, straight, parallel wires carrying currents I and I' that are a distance d apart: $F = 2 \times 10^{-7} I\, I'\, L/d$, where L is the length of the wires. Parallel currents attract one another; anti-parallel currents repel.

Angular momentum
A conserved quantity like linear momentum. For a body of mass M moving at a speed v in a circular orbit of radius R, its magnitude is $m\,v\,R$, and it points in the direction of your right thumb when your fingers describe the orbit. Although a complete discussion lies beyond the scope of this book, angular momentum plays a central rôle in the development of quantum mechanics.

Antiparticles
To every particle seen in nature there corresponds an antiparticle with the same mass and opposite electrical properties. For reasons that are beginning to be understood, the early universe contained a bit more matter than antimatter, and what is left today is almost exclusively matter made up of protons, neutrons, and electrons. Each of their antiparticles has been made and carefully studied at physics laboratories.

Alpha particle
The nucleus of the helium atom, consisting of two protons and two neutrons. It is emitted from certain radioactive nuclei in the process of alpha decay.

Atomic number Can be given several entirely equivalent definitions that vary only in level of reductionism: (1) the place of an atom in the periodic table; (2) the number of electrons in it; (3) the nuclear charge; or (4) the number of protons in the nucleus. (At an even more profound level, it is $\frac{2}{3}$ the number of up quarks in the atom less $\frac{1}{3}$ the number of down quarks.)

Atomic weight The mass of an atom expressed in units wherein that of the common isotope of carbon is 12. The atomic weight of any isotopically pure chemical element is nearly equal to the number of nucleons in its nucleus.

Avogadro's hypothesis Equal volumes of any two gases or mixtures of gases, at the same temperature and pressure, contain equal numbers of molecules. What was originally an ad hoc hypothesis has been recognized for a century to be a derived consequence of the kinetic theory of gases. Avogadro's *number*, a closely related notion, is the number of molecules in a mole (or gram molecular weight) of an element or a compound.

Baryon Any hadron that is also a fermion. Equivalently, a particle made from three quarks.

Beta decay A form of radioactivity in which a neutron in the nucleus becomes a proton. At the same time, an energetic electron (or beta particle) is emitted along with an antineutrino.

Big bang cosmology Most physicists and astronomers believe, for three good reasons, that our universe began in a hot big bang about ten billion years ago. (1) Because of the observed rate of expansion of the universe according to the Hubble law; (2) because of the observed flux of background microwave radiation—the last gasp of the big bang; and (3) because of agreement between theoretically computed and observed abundances of the light chemical elements that were formed when the universe was only a few minutes old.

Binding energy If two or more particles are bound together, the composite system necessarily has a mass smaller than the sum of the masses of the constituents. The product of this mass defect with c^2 is called the binding energy of the system. The binding energy of the hydrogen atom (a composite of an electron and a proton) is $\sim 13.6\,eV$. On the other hand, the difference in mass between a uranium atom and its fission products is several hundred million eV.

Black holes Although they almost certainly exist, they have never been decisively detected. According to the general theory of relativity, a massive body curves space around it. If it is sufficiently massive but small, the curvature becomes so great that nothing, not even light, can escape from it. Black holes may result from the implosive catastrophe of a supernova. They are believed to lie at the heart of long-extinct quasars and today's "active galactic nuclei." There may be a relatively small black hole (with the mass of a million suns) deep within our own galaxy.

Boltzmann constant This fundamental constant k expresses the proportionality between the Kelvin temperature T of a gas and the mean kinetic energy E of the gas molecules: $kT = \frac{2}{3}E$. The product of k with Avogadro's number is the universal gas constant R.

Boson   A particle (either composite or elementary) with integer spin, such as the photon, the deuteron, and the hydrogen atom.

Boyle's law   "There is a spring to the air," Boyle remarked. He found that a cylinder of air confined by a piston behaves like a spring—that the air within exerts a restoring force when the piston is displaced from its equilibrium position. The pressure of a gas (when it is kept at a fixed temperature) is inversely proprtional to the volume it occupies.

Caloric   The word coined by Lavoisier to denote a hypothetical (and nonexistent!) heat fluid that he included in his list of chemical elements. Particles of caloric were believed to be attracted by ordinary atoms but repelled by one another. The caloric hypothesis was abandoned in 1789 when Benjamin Thompson announced, on the basis of exhaustive and compelling experiments, that "all attempts to discover any effect of heat upon the apparent weight of bodies will be fruitless."

Cathode rays   Stream through an evacuated tube from cathode to anode when a large voltage is applied to the device. Thomson showed that cathode rays consist of charged particles (electrons) and measured their mass and charge.

Charles' law   If a sample of gas is kept at constant pressure, its volume V increases by about $1/300$ for every Celsius degree it is heated. This result is independent of the chemical composition of the gas! The law is even simpler when it is expressed in terms of the Kelvin temperature T: at fixed pressure, $V \propto T$.

Charm   Gell-Mann, when he invented quarks, proposed that they come in three different "flavors." In 1970, indirect theoretical arguments suggested that there must exist in nature a fourth quark flavor, which was called *charm*. Hadrons formed from charmed quarks were discovered in 1974. Since then, decisive experimental evidence has appeared for a fifth quark flavor, and there are compelling theoretical arguments for yet one more.

Chemical element   To the ancient Greeks, fire, water, earth, and air were the substances from which all bodies were made. Alchemists added mercury (the metallic essence), sulfur (the essence of fire), and others. The modern notion of chemical elements dates to Boyle, and their interpretation as aggregates of identical atoms to Dalton. Lavoisier's list of 30 chemical elements included light and caloric. Many of today's list of 110 elements are synthetic.

Circular acceleration   A body moving at speed v in a circle of radius R experiences a centripedal (*i.e.*, toward the center) acceleration of magnitude v^2/R. A charged particle moving in the xy plane under the influence of a constant magnetic field in the z direction experiences a constant acceleration lying in the xy plane but perpendicular to its velocity. Thus, the particle describes a circular orbit.

Collisions, elastic and otherwise The molecules of air often collide with one another elastically, that is, with no conversion of kinetic energy into other forms of energy. For these collisions, both momentum and kinetic energy are conserved. When ordinary objects collide, some of the kinetic energy becomes heat. If two real billiard balls strike one another on an ideal frictionless table, momentum is conserved, but the final kinetic energy is less than the initial kinetic energy.

Combining volumes, law of In 1808, Gay-Lussac showed that when hydrogen and oxygen react to form water, their combining volumes are in the ratio 1:2. Further experiments led him to a remarkably simple statement: "It appears to me that gases always combine in the simplest proportions when they [react], and we have seen in reality in all the preceding examples that the ratio of combinations 1 to 1, 1 to 2, or 1 to 3." The universal validity of this law led Avogadro to his famous hypothesis.

Conservation laws An attribute of matter is said to be conserved if the quantity of it within any closed surface S changes only by virtue of the flow of the attribute through the surface. Thus, a conserved quantity in an isolated system cannot change at all. Linear momentum, angular momentum, and mass energy satisfy absolute conservation laws as a result of the homogeneity and isotropy of space–time. There are only a very few other exact conservation laws. Electric charge is almost certainly conserved; baryon number and lepton number may be.

Copernicus Put forward a heliocentric model of the solar system in 1543. By permitting the Earth to move, he obtained a far simpler and more elegant model of planetary motions than that of Ptolemy. Yet Martin Luther found Copernicus a fool and a heretic, some Jewish communities forbade the teaching of the Copernican system, and the Roman Catholic church put his writings on the index of forbidden books.

Cosmic background radiation A detectable relic of the big bang. When the early universe cooled sufficiently for atoms to form, it became transparent and was filled with black body radiation at a temperature of $\sim 3000\,K$. At present, this radiation has been red shifted from visible wavelengths to the microwave domain, which is to say, to $\sim 3\,K$, the present mean temperature of the universe.

Coulomb's law Expresses the mutual force acting between point charges of magnitudes Q and Q' at a distance R from one another: $F = k_c\,Q\,Q'/R^2$. Like charges repel; unlike charges attract.

Curie A unit of radioactivity, the amount of any material that undergoes 3.7×10^{10} radioactive disintegrations per second.

Cyclotron A circular accelerator generating particle energies from 10^6 to $10^8\,eV$, in which charged particles injected at the center are accelerated spirally outward by an alternating electric field. They move in a plane normal to a fixed magnetic field that causes them to spiral.

Dark mass problem Observations of the mutual motions and the rotational speeds of galaxies reveal the existence of far more matter in the universe that can be accounted for by stars, dust, and gas. Possibly, this dark matter is of a kind that has not yet been seen on Earth. Possibly, there is enough dark matter in the universe to terminate its expansion and lead to an eventual "big crunch."

De Broglie wavelength   Just as light has both particle and wave properties, so do electrons, protons, and any other particles or systems of particles. De Broglie showed that the "matter waves" associated with a body of mass m and speed v have wavelength given by $\lambda = h/mv$, where h is Planck's constant. For macroscopic bodies, λ is exceeding tiny and quantum effects are irrelevant. On the other hand, the λ of electrons in an atom is comparable to the size of the atom and quantum theory is paramount.

Diffraction   When a beam of light is passed through a pair of narrow slits, a complex pattern of bands appears on a distant screen. Where the light waves coming from the two slits add constructively, a bright band occurs. Where they add destructively, a dark band is seen. In contrast to the photoelectric effect, this behavior is most easily understood if light is regarded as a wave.

Dirac equation   The Schroedinger equation offered a correct quantum–mechanical description of slowly moving (that is, non-relativistic) electrons. Dirac devised a relativistic equation for the electron's wave function that incorporated the notion of electron spin and predicted the existence of the electron's antiparticle.

Doppler shift   A source emits waves (either sound or light) with frequency f in its rest system. An observer moving toward the source with speed v observes waves of higher frequency. If v is much smaller than s (the velocity of the waves), the shifted frequency is $f' \simeq (1 + v/s)f$. Conversely, if the observer and source are receding from one another, the perceived frequency is shifted downward.

Eightfold Way   A classification of hadrons introduced by Gell-Mann in 1961 in which all baryons belong to super-multiplets with 1, 8, or ten members, and all mesons to super-multiplets of 1 or 8. The successes of this model are understood today in terms of the quarks.

Electric battery   In 1791, Galvani described his observations of the muscular contractions produced in the bodies of recently killed frogs by static electricity and by contact with a scalpel. These results led Volta to the construction in 1799 of his "pile," consisting of disks of copper and zinc with wet cloth between each pair. This was the word's first electric battery, and it soon became a powerful tool in the hands of the British chemist Humphry Davy.

Electrolysis   Involves the passage of a current through an *electrolyte* an aqueous solution of a salt or a molten salt. Positive ions accumulate at the cathode where they acquire one or more electrons to become neutral atoms. Negative ions are deposited as neutral atoms at the anode. One faraday of electric charge (a mole of electrons, consisting of $\sim 96{,}500$ coulombs) must pass through the circuit to liberate one mole of a monovalent element.

Electroweak theory   A mathematically consistent and experimentally verified description of both the electromagnetic and weak nuclear forces. Electroweak theory and quantum chromodynamics are the ingredients of today's very successful "standard theory" of all elementary particle phenomena.

Energy levels   Any quantum mechanical system (like an atom, a molecule, a nucleus, or a particle built from quarks) has a variety of quantum states, each with a definite energy. The state of lowest energy is the ground state; those of higher energy are excited states. Bohr first deduced a formula for the energy levels of an electron bound to a proton (a hydrogen atom).

Entropy   A quantitative measure of the degree of disorder of a system. According to the second law of thermodynamics, the entropy of any isolated system cannot decrease.

Falling bodies   Galileo first realized that a freely falling body on Earth experiences a downward acceleration g that is independent of its mass, shape, composition, or horizontal motion. Its height is thereby given by the formula $h(t) = h_0 + v_0 t - \frac{1}{2} g t^2$, where h_0 and v_0 are its height and vertical speed at $t = 0$.

Faraday's law of electromagnetic induction   Describes how a changing magnetic field generates electrical effects. If a magnet is poked into a loop of wire, a current flows through the loop. This law underlies the design and construction of dynamos (to convert mechanical energy into electric energy) and the motor (to convert electric energy into mechanical energy).

Fermion   A particle (either composite or elementary) with spin of half an odd integer. Protons, neutrons, and electrons are spin $\frac{1}{2}$ fermions. So is a singly ionized helium atom.

Feynman diagrams   Pictorial representations of fundamental processes involving elementary particles in the context of quantum field theory.

Fission and fusion   The most stable nuclei (those with the most binding energy per nucleon) are the midsized ones around iron in the periodic table. Therefore, energy is released when a large nucleus (like that of uranium) fissions into two smaller fragments. This process also releases neutrons that may be absorbed by other uranium nuclei, causing them to fission in a chain reaction. Controlled nuclear fission is used to generate power. Uncontrolled, it is the basis of a nuclear weapon. Energy is also released when small nuclei fuse to produce larger and better bound nuclei. Nuclear fusion is the mechanism by which stars shine. Controlled nuclear fusion has not yet been achieved, but it may someday become a safe and inexhaustible source of electric power.

Galilean principle of relativity This principle, first stated by Galileo, denies the existence of a preferred inertial frame of reference. The laws of physics must be exactly the same to any uniformly moving observer.

Gamma decay A radioactive process by which an excited nucleus emits an energetic photon (or gamma ray) and falls to a state of lower energy.

Gauge theory The formal mathematical structure underlying our understanding of electromagnetism and both the strong and weak nuclear forces. All of the elementary particle forces are dynamical consequences of underlying symmetries of nature and are mediated by bosons of unit spin: the so–called gauge bosons (photons, gluons, $W^{\pm}$'s and Z^0's).

Gluons Massless bosons of unit spin that are exchanged between quarks to generate the strong nuclear force. They are the inisolable gauge particles of quantum chromodynamics.

Ground state See Energy levels.

Hadrons Seemingly elementary particles that are made up of quarks. Three quarks form a *baryon*; for example, two up quarks and a down quark form a proton. Three antiquarks form an antibaryon. A quark and an antiquark may combine together to form a *meson*. Because quarks are fermions, so are baryons and antibaryons. Mesons, on the other hand, are fermions.

Half life See Radioactive decay law.

Heat, latent Usually, the temperature of a system rises when heat is added to it. This in not the case, however, when the system changes its state. (A glass of ice water at $0° C$, for example, remains at that temperature until all the ice melts, just as boiling water remains at $100° C$ until it all boils away.) The heat of fusion of a solid is the heat per unit mass needed to melt it. The heat of vaporization of a liquid is the heat per unit mass needed to boil it.

Heat, specific When heat is added to a system and no change of state occurs, its temperature rises. The specific heat of a material is the quantity of heat needed per unit mass to raise its temperature by one Celsius degree. The specific heat of water is about one calorie per gram degree, and that of most other materials is much smaller.

Hooke's law An empirical law that the force exerted by a spring is proportional to the distance it is stretched or compressed. When a fish is hung from a spring scale, it comes to rest when the gravitational force (its weight) equals the force exerted by the spring. Thus, the displacement of the spring is roughly proportional to the weight of the fish. Most states outlaw spring balances in fish stores.

Hubble's law States that distant galaxies are receding from us at speeds that are roughly proportional to their distances from us. It was proposed by Edwin Hubble in 1929 to explain the red shifts he observed of galactic spectra. If the universe is expanding, it must have had a beginning. The Hubble constant, expressing the relation between recessional speed and distance, provides us with an estimate of the age of the universe: the time since the big bang.

Ideal gas law This law incorporates the results of Boyle, Charles, and Avogadro into a single formula, $PV = nRT$, where P is the pressure, V the volume and T the Kelvin temperature of the gas. R is a universal constant (equal to the product of Avogadro's number and Boltzmann's constant) and n is the number of moles of the gas in the sample. Although no gas is truly ideal, the ideal gas law offers a more than adequate description of most gases under ordinary conditions.

Ions An atom, normally electrically neutral, may become *ionized*. If it acquires one or more extra electrons, it becomes a negative ion or *cation*. If it loses one or more electrons it becomes a positive ion or *anion*. The atoms of a salt in solution are ionized, and ions are copiously produced by electric discharges and by energetic charged particles traversing matter.

Isospin Elementary particles occur in multiplets whose members have different electric charges but nearly the same mass. The neutron and the proton, for example, comprise an isospin doublet, and the three mesons π^+, π^- and π^0 belong to an isospin triplet. The appearance of isospin multiplets is a consequence of the fact that the up and down quarks have small masses compared to the intrinsic mass scale of quantum chromodynamics.

Isotopes Nuclear species are classified by the number of nucleons (neutrons and protons) A of which they are made, and by the nuclear charge Z (the number of protons). Atoms made of nuclei with the same Z but different A are chemically similar isotopes. For example, ^{235}U is a rare but readily fissionable isotope of uranium ($Z = 92$) whose most abundant isotope is ^{238}U.

Kepler's three laws Deduced from Brahe's painstaking (pre-telescopic) planetary observations: (1) planets travel in ellipses with the Sun at one focus; (2) the line connecting the Sun to a planet sweeps out equal areas in equal times; and (3) the square of the periods of the various planetary orbits varies with the cube of the orbital radii. Their derivation from first principles awaited Newton's hypothesis of universal gravitation.

Lens, gravitational The general theory of relativity was confirmed in 1919 with the observation that starlight is bent by the Sun. Distant galaxies bend the light coming from even more distant quasars and act as lenses to produce several distinct quasar images.

Leptons Seemingly elementary spin $\frac{1}{2}$ fermions that do not partake in the strong nuclear force. Six varieties of leptons are known: three of them are charged (including the electron) and three neutral (the neutrinos).

Lodestones "Lode" is old English for way or journey, hence "lodestar" for a star that shows the way. Ancient mariners fashioned compasses from naturally occuring magnets, which is why they are called lodestones. Some lodestones are magnetized along the Earth's magnetic field, but some lie in the opposite direction. That is how we know that Earth's magnetic field has often reversed itself.

Lorentz contraction   According to the special theory of relativity, a body moving at velocity v is foreshortened by the factor $\sqrt{1 - v^2/c^2}$.

Lorentz transformations   Allow us to relate measurements of space and time in one inertial coordinate system to those in another.

Magnetic monopoles   The hypothetical magnetic analogs to electric charges. If such things existed, we could construct a magnet with only one pole. All experimental searches for magnetic monopoles have failed.

Mass, conservation of   Lavoisier concluded that the mass of the reactants in a chemical process equals the mass of the products, and that mass may neither be created nor destroyed. This principle is not exactly true unless we take into account the (limited) interconvertability off mass and energy.

Maxwell's equations   Put together in a consistent fashion all of the empirical laws governing electrical and magnetic phenomena, including those of Coulomb, Ampère, and Faraday, and the conservation of electric charge. Furthermore, they demand the existence of electromagnetic waves, of which X rays and light and radio waves are particular examples.

Mechanical equivalent of heat   Joule wrote in 1843: "We shall be obliged to admit that Count Rumford is right in attributing the heat evolved in boring cannon to friction ... I [am] satisfied that the grand agents of nature are, by the Creator's fiat, *indestructible*; and that whenever mechanical [energy] is expended, an exact equivalent of heat is *always* obtained." Today, the basic unit of energy is called the joule, and the mechanical equivalent of one calorie is 4.184 *J*.

Meson   Any hadron that is also a boson. Equivalently, a particle made of a quark and an antiquark.

Michelson–Morley experiment   In 1887, Michelson and Morley demonstrated that the speed of light, as measured on Earth, is unaffected by Earth's motion through the heavens.

Mole   Has two equivalent definitions. It is the amount of a compound equal in grams to its molecular weight, or it is Avogadro's number of molecules of the compound.

Momentum   The linear momentum of a particle, classically, is the product of its mass and its velocity. (If the particle is rapidly moving, this approximate expression must be replaced by an exact relativistic formula.) If a system of colliding bodies is isolated from external forces, the vector sum of all the momenta cannot change: momentum is conserved.

Multiple proportions, the law of   Suppose that one substance can combine with another to form more than one compound. Let the ratio of combining masses be p in one case and

q in the other. Although neither p nor q is in general a ratio of simple integers, their ratio p/q must be. This law guided Dalton to his precise version of the atomic hypothesis.

Muons   Charged leptons with masses ~ 200 times that of electrons. They are unstable particles with a half–life of ~ 2 microseconds.

Neutral currents   A form of the weak force whose existence was demanded by the electroweak theory. These currents were first detected at CERN in 1973.

Neutron   An electrically neutral particle with mass nearly equal to that of the proton. Neutrons and protons ("nucleons") are the constituents of atomic nuclei. They are not elementary particles, but composite systems made of quarks.

Neutrinos   Massless (or nearly massless) neutral fermions coming in three known species: electron neutrinos ν_e, muon neutrinos ν_μ, and tau neutrinos ν_τ. In the process of nuclear β^- decay, an electron and an electron antineutrino are created. In β^+ decay, a positron and an electron neutrino are created. Neutrinos have been detected at nuclear reactors, at accelerator laboratories, in cosmic rays, from the Sun, and from a supernova.

Newton's three laws of motion   Form the logical basis of classical mechanics: (1) a freely moving body continues to move in a straight line at constant speed; (2) the net force acting on a body equals the product of its mass and its acceleration; and (3) to every action there is an equal and opposite reaction.

Newton's universal law of gravitation   Two spherical bodies of mass M and M' that are separated by a distance R mutually attract one another with a force $F = G_N M M'/R^2$ where G_N is a universal constant. With this one hypothesis and his laws of motion, Newton explained the motions of moons, comets, and planets.

Nuclear force   The force that holds the nucleus together was once thought to be "fundamental" and due to the exchange of "elementary" mesons between "elementary" nucleons. Today these particles are known to be made of quarks bound by the color force resulting from the exchange of gluons according to quantum chromodynamics. Its pale residue binds colorless nucleons into nuclei, just as a residue of the electromagnetic force binds uncharged atoms into molecules.

Nuclear reactions   In chemical reactions, such as $2\,HCl + Zn \rightarrow H_2 + ZnCl_2$, forces among valence electrons lead to recombinations of the participating atoms. The nucleus plays an entirely passive role. In nuclear reactions, nucleons are exchanged and the identity of the participating nuclei change. In 1919, Rutherford turned nitrogen nuclei into oxygen nuclei by bombarding them with α particles. Nuclear fission and the synthesis of trans-uranic elements are other examples of contrived nuclear reactions. Radioactive elements, produced by nuclear reactions, are often used for medical diagnoses and treatment, like the radioactive iodine "cocktail" that cured President Bush's goiter. And, of course, the Sun shines on us by virtue of the nuclear transformations taking place in its core.

Nucleons A catch-all for neutrons and protons, the particles from which atomic nuclei are built. Alternatively, nucleons are the lowest quantum states of non-strange baryons.

Oersted Discovered the first empirical connection between electricity and magnetism. By observing the deflection of a compass needle by a nearby current-carrying wire, he showed that electricity produces magnetic effects.

Ohm's law An empirical and approximate relation between the electric current I flowing though a device and the voltage drop V across it: $V = IR$, where the constant of proportionality R is a characteristic of the device. The unit of resistance with I measured in amperes and V in volts is called the ohm. The power P dissipated by the device is the product of current flow and voltage drop. Using Ohm's law, we obtain $P = V^2/R$. Because the voltage at an outlet is fixed by the utility company, the power consumed by a lamp or heater is *inversely* proportional to the resistance of the device.

Oxidation Lavoisier showed that the most familiar chemical reaction, combustion, is the combination of the fuel with that part of the air he named *oxygen.*

Pauli exclusion principle Not more than one electron can occupy the same quantum state at the same time. For example, in the helium atom, two electrons (one spinning up, the other down) are in the same orbital state. The valence shell is filled, and helium is consequently chemically inert. The apparent solidity of ordinary matter (although it is mostly empty space) results from the Pauli principle, because all of the lowest energy states of the constituent electrons are full up.

Periodic table The properties of the chemical elements show curious regularities with increasing atomic weights. For example, that of bromine is nearly the average of those of its chemically similar cousins, chlorine and iodine. On the basis of many such hints, Mendeleev created a systematic table of all the known elements in 1869. In it, chemical valencies are seen to vary periodically with atomic weight (or, as we now know, atomic number). He left places for unknown elements whose properties he correctly predicted in advance of their discoveries. One of the greatest triumphs of quantum chemistry is the explanation of the table's validity from the Schroedinger equation and the known structure of atoms.

Phlogiston Prior to (and even for a time after) Lavoisier's explanation of combustion as oxidation, scientists believed that combustion involved the release of a hypothetical (and nonexistent!) fluid called phlogiston.

Photoelectric effect Einstein earned a Nobel prize for his theoretical analysis of the photoelectric effect. When light of sufficiently high frequency impinges on a metal, electrons are driven off. The photoelectric current varies directly with the intensity of the radiation. However, the kinetic energies of the emitted electrons are determined by the frequency according to:

$E = h f - W$, where W is an energy (the work function) of the metal. Experimental confirmation of this prediction proved that light can display particle-like behavior.

Photon   Light, although neither particle nor wave, can display the properties of each. In interference and diffraction phenomena, it is characterized by a frequency f and a wavelength $\lambda = c/f$. In the photoelectric effect and in scattering phenomena, it behaves like a beam of particles (called photons) of energy $E = h f$ and momentum $p = h/\lambda$.

Pion   The lightest meson, or particle made up of one quark and one antiquark. Pions come in three charge states: positive, negative, and neutral.

Positron   The electron's antiparticle has the mass of the electron but the opposite electric charge. It was discovered in cosmic rays in 1932 and was the first observed form of antimatter.

Probability interpretation of wave function   The solutions of the Schroedinger equation determine the possible energy levels of the system, but what is the physical interpretation of the wave function $\Psi(x)$? Born argued that the positive quantity $P(x) = |\Psi(x)|^2$ should be regarded as the relative probability that the particle whose wave function is Ψ is at the point x. In other words, the particle is likely to be found where P is large, and is unlikely to be where P is small.

Proton   The nucleus of ordinary hydrogen and a constituent of all other nuclei. See Neutron.

Quantization of angular momentum   Bohr argued that only certain orbits of an electron in a hydrogen atom are allowed: those in which its angular momentum J is an integer multiple of $\hbar$. It is a general feature of quantum mechanics that J for any particle or system of particles must be an integer (or half integer) multiple of this quantity. Not only is the magnitude of J quantized, but so are the directions in which it may point: a system of angular momentum J can be in any one of $2J + 1$ quantum states. All this is in stark contrast to classical thinking: surely a baseball can have any spin at all which can point in any direction at all. (In fact, J is so enormous for a macroscopic body that its quantization is undetectable.)

Quantum chromodynamics (QCD)   Our current mathematical framework for the description of the strong nuclear force. Colored quarks are held together to form colorless hadrons by the exchange of gluons. A small residue of the color force acts among nucleons to hold them together as nuclei. A curious (and unsettling) feature of the theory is that neither quarks nor gluons can be seen as particles in their own right, but only as parts of hadrons.

Quantum field theory   The mathematical framework on which our understanding of elementary particles is founded. It represents a synthesis of quantum mechanics and Einstein's special theory of relativity.

Quarks   Among modern candidates for the most basic building blocks of all matter. Up and down quarks are the constituents of nucleons, out of which all of the atomic nuclei are built. Four other "flavors" of quarks, all of them heavier than ordinary quarks, are known to exist. Hadrons containing one or more of these quarks (that is, *strange, charmed, top, or bottom*) are unstable particles. Particles containing each of these quark flavors, except for the top quark, have been seen and studied in the laboratory.

Radioactive decay law   Rutherford and Soddy showed that every radioactive species decays exponentially; that is, the probability remains constant that a given nucleus decays in the next second. Consequently, every radioactive sustance is characterized by a *half life* the period of time during which half of the nuclei decay. The half lives of naturally occuring nuclei vary from billions of years to fractions of a second. The exponential decay law applies not just to radioactive nuclei, but to the excited states of any quantum mechanical system.

Radioactivity, varieties of   There are three forms of natural radioactivity. A radioactive nucleus (A, Z) may emit an α particle to become the nucleus $(A-4, Z-2)$. In beta radioactivity, A stays fixed but Z increases by one unit (with the simultaneous creation of an electron and antineutrino) or decreases by one unit (with the creation of a positron and neutrino). Often, an α or β decay leaves the daughter nucleus in an excited state. The subsequent emission of a γ ray (with no change in A or Z) is the third form of radioactivity.

Relativity, the special theory   Maxwell's description of electromagnetism conflicted with Galileo's principle that the laws of physics must be the same to all inertial observers. Einstein proposed this new view of space and time to restore consistency. He was forced to the view that the speed of light in vacuum is the same to all uniformly moving observers, but that measurements of spatial distances and time intervals are not.

Relativity, the general theory   Special relativity demands that the laws of physics are the same to all *inertial* observers. Yet, "if a person falls freely he will not feel his own weight," Einstein observed in 1907. Is this not "a powerful argument for the fact that the relativity postulate must be extended to coordinate systems which, relative to each other, are in non-uniform motion?" In 1915, Einstein invented a field theory of gravity that is entirely coordinate system independent: the general theory of relativity. In it, the "force of gravity" merely reflects the curvature of space–time caused by a nearby mass. The new theory explained a small anomaly in Mercury's orbit and has passed many precise observational tests since.

Schroedinger equation   This is the equation satisfied by a quantum mechanical wave function. Its solutions, in the case of the hydrogen atom, include the "stationary states" that Bohr conjectured. The discovery of this equation provided a firm mathematical framework to quantum mechanics.

Spin and statistics   Every simple system (molecule, atom, nucleus or particle) in a definite quantum state has an angular momentum J that is either an integer or half–integer multiple of $\hbar$. If $J/\hbar$ is an integer, the system is a "boson" and satisfies Bose–Einstein statistics. If it is half an integer, the system is a "fermion" and satisfies Fermi–Dirac statistics. Fermions (like electrons and nucleons) obey the Pauli exclusion principle.

Stefan–Boltzmann law The power radiated by an ideal black body increases with the fourth power of its Kelvin temperature; that radiated per unit area is given by $\sigma T^4 \simeq 5.67 \times 10^{-8} T^4 W$, where σ is the Stefan–Boltzmann contant. The Sun is well described as a black body at a temperature of $5800 T$, and the relation tells us that each square centimeter of its surface is as bright as six 1000-watt lamps.

Strangeness Certain unstable elementary particles seemed to behave inexplicably. Today we know that a strange particle is simply one containing a strange quark-bearing -1 unit of strangeness. The weak nuclear force allows a strange quark to become an ordinary down quark or, equivalently, a strange particle to decay.

Superconductivity A phenomenon by which certain materials, at low temperature, offer no resistance to the passage of an electric current.

Thermodynamic equilibrium A system is said to be in thermodynamic equilibrium when no heat flows from one part to another (it is at a uniform temperature) and no changes are taking place. A glass of iced tea on a summer's day is not at equilibrium until the ice melts and the tea is at room temperature.

Thermodynamic laws There are three of them. The first law states that energy may neither be created nor destroyed. The second law states that an isolated system must tend towards greater disorder or, equivalently, that we cannot extract the thermal energy resident in matter without paying a cost in energy. The third law says that we may never attain a temperature of absolute zero. Together they are the basis to the science of thermodynamics.

Time dilation A consequence of the special theory of relativity. Suppose that one of two identical clocks is at rest relative to an observer and the other is moving at velocity v. The moving clock will appear to be slow. A time interval Δt on the stationary clock corresponds to the shorter interval $\sqrt{1 - v^2/c^2}$ on the moving clock.

Torricellian column In 1643, Torricelli discovered that a column of mercury in an inverted tube that is sealed at the top falls to a height (indicating the ambient air pressure) leaving a vacuum above it. The device could be used as a barometer (to measure time variations of the air pressure), or as an altimeter (because air pressure falls with altitude). This discovery led scientists to investigate the properties of air as a material substance and, in particular, led Boyle to discover the first quantitative law describing the physical properties of gases.

Uncertainty relations In 1927, Heisenberg pointed out that certain quantities, like a body's position x and momentum p, cannot both be known with arbitrary precision.

If Δx is the position uncertainty and Δp is the momentum uncertainty, then their product must be greater that a certain value: $\Delta x \, \Delta p > \frac{1}{2}\hbar$.

W and Z bosons   The existence of these particles, which mediate weak nuclear forces, was predicted by the electroweak theory. They were first produced and seen at the CERN proton–antiproton collider in 1983. The electrically–charged $W^{\pm}$ have masses of $80.6 \, GeV/c^2$ while the Z^0 mass is $91.2 \, GeV/c^2$. They are the most massive of known "elementary" particles, and their masses accord with theoretical expectations.

Wave properties   A general wave form may be expressed as a superposition of waves with different wavelengths. The frequency of such a sinusoidal wave is its velocity divided by its wavelength. For either sound or light waves moving through air, the velocity is independent of wavelength. Ocean waves behave differently: those of greater wavelength travel noticeably faster.

X rays   This form of radiation was discovered by Roentgen in 1896. Electromagnetic waves with wavelengths comparable to atomic sizes are produced when outer atomic electrons fall into quantum states left vacant by the ejection of inner electrons. The X portion of the electromagnetic spectrum lies between ultraviolet radiation and gamma rays.

Zeeman effect   Faraday's last experimental endeavor was to search for an effect of a magnetic field upon the spectral lines of a flame. The final sentence of his seven-volume diary concludes, "Not the slightest effect ... was observed." The challenge was met by Zeeman in 1896. He discovered that a single spectral line splits into several nearby components when the source is placed in an intense magnetic field. To explain the Zeeman effect in detail, Pauli introduced "a peculiar, not classically describable, two–valuedness" of the properties of the electron, which Goudsmit and Uhlenbeck would later identify as electron spin.

Index